Springer Series in Materials Science

Volume 233

The Springer Series in Materials Science covers the complete spectrum of materials physics, including fundamental principles, physical properties, materials theory and design. Recognizing the increasing importance of materials science in future device technologies, the book titles in this series reflect the state-of-the-art in understanding and controlling the structure and properties of all important classes of materials.

More information about this series at http://www.springer.com/series/856

Anthony E. Hughes · Johannes M.C. Mol
Mikhail L. Zheludkevich
Rudolph G. Buchheit
Editors

Active Protective Coatings

New-Generation Coatings for Metals

Springer

Editors

Anthony E. Hughes
Commonwealth Scientific and Industrial Research Organisation
Clayton
Australia

Johannes M.C. Mol
Department of Materials Science and Engineering
Delft University of Technology
Delft
The Netherlands

Mikhail L. Zheludkevich
Department of Ceramics and Glass Engineering
University of Aveiro
Aveiro
Portugal

Rudolph G. Buchheit
Fontana Corrosion Center
The Ohio State University
Columbus
USA

ISSN 0933-033X ISSN 2196-2812 (electronic)
Springer Series in Materials Science
ISBN 978-94-017-7538-0 ISBN 978-94-017-7540-3 (eBook)
DOI 10.1007/978-94-017-7540-3

Library of Congress Control Number: 2015959943

Printed on acid-free paper

This Springer imprint is published by SpringerNature
The registered company is Springer Science+Business Media B.V. Dordrecht

Contents

Contributors

F. Andreatta Dipartimento Politecnico di Ingegneria e Architettura, Università di Udine, Udine, Italy

R.G. Buchheit Materials Science Engineering, The Ohio State University, Columbus, OH, USA

Joseph Byrom Department of Coatings and Polymeric Materials, North Dakota State University, Fargo, ND, USA

Philip Casey Commonwealth Scientific and Industrial Research Organisation, Manufacturing, Clayton, VIC, Australia

I.S. Cole Manufacturing Flagship, Commonwealth Scientific and Industrial Research Organisation, Melbourne, Australia

L. Fedrizzi Dipartimento Politecnico di Ingegneria e Architettura, Università di Udine, Udine, Italy

H.R. Fischer TNO Technical Sciences, Materials for Integrated Products, AP Eindhoven, The Netherlands

G.S. Frankel Fontana Corrosion Center, The Ohio State University, Columbus, OH, USA

S.J. García Novel Aerospace Materials, Delft University of Technology, Delft, The Netherlands

Victoria Gelling Department of Coatings and Polymeric Materials, North Dakota State University, Fargo, ND, USA

Y. Gonzalez-Garcia Materials Science and Engineering Department, Delft University of Technology, Delft, The Netherlands

Guido Grundmeier Technical and Macromolecular Chemistry, University of Paderborn, Paderborn, Germany

Dominic Richard Harris Commonwealth Scientific and Industrial Research Organisation, Manufacturing, Clayton, VIC, Australia

A.E. Hughes Institute for Frontier Materials, Deakin University, Waurn Ponds, Geelong, Australia; Mineral Resources, Commonwealth Scientific and Industrial Research Organisation, Clayton, VIC, Australia

Niteen Jadhav Department of Coatings and Polymeric Materials, North Dakota State University, Fargo, ND, USA

K.H. Leong PETRONAS Research, Kajang, Selangor DE, Malaysia

J.M.C. Mol Department of Materials Science and Engineering, Delft University of Technology, Delft, The Netherlands

M.F. Montemor CQE, DEQ—Instituto Superior Técnico, Universidade de Lisboa, Lisboa, Portugal

Mark Nichols Ford Motor Company, Dearborn, MI, USA

Berkem Oezkaya Technical and Macromolecular Chemistry, University of Paderborn, Paderborn, Germany

Ozlem Ozcan Technical and Macromolecular Chemistry, University of Paderborn, Paderborn, Germany

Abhijit Suryawanshi Department of Coatings and Polymeric Materials, North Dakota State University, Fargo, ND, USA

Janice Tardiff Ford Motor Company, Dearborn, MI, USA

Herman Terryn Delft University of Technology, Delft, The Netherlands; Vrije Universiteit Brussel, Brussel, Belgium

Russell J. Varley CSIRO Manufacturing, Clayton South MDC, VICTORIA, Australia

Peter Visser AkzoNobel Performance Coatings, Sassenheim, The Netherlands

Sam Yang CSIRO, Clayton, VIC, Australia

M.L. Zheludkevich Institute of Materials Research, Helmholtz-Zentrum Geesthacht, Zentrum für Material- und Küstenforschung GmbH, Geesthacht, Germany; Department of Materials and Ceramic Engineering, CICECO-Aveiro Institute of Materials, University of Aveiro, Aveiro, Portugal

Chapter 1
Introduction

A.E. Hughes, J.M.C. Mol, M.L. Zheludkevich and R.G. Buchheit

1.1 Introduction

This book is about reviewing and exploring the opportunities for introducing increased functionality into coating systems. The intention is to not only provide an account of recent developments, such as other books of this genre, but to also review, in detail, the nature of the problems that we are trying to address in advancing the materials science of coating systems. To this end we examine the fundamentals of corrosion science, the materials science of coatings, and developments in characterisation methods that all help to give an overview of the field. Importantly, we also look at the needs of industry with respect to developments in the coatings field.

Coating systems are used to protect, decorate, or add other functionalities, such as wear and scratch resistance, to a substrate surface. For some applications only cosmetic functions such as colour and sheen (reflectivity) are required and these can

A.E. Hughes (✉)
Mineral Resources, Commonwealth Scientific and Industrial Research Organisation, Private Bag 10, Clayton Sth, Melbourne 3169, VIC, Australia
e-mail: Tony.hughes@csiro.au

A.E. Hughes
Institute for Frontier Materials, Deakin University, Geelong Waurn Ponds Campus, Waurn Ponds, VIC, Australia

J.M.C. Mol
Mechanical, Maritime and Materials Engineering, Delft University of Technology, Delft, The Netherlands

M.L. Zheludkevich
Helmholtz Zentrum Geesthacht, Magnesium Innovation Centre, Institute of Materials Research, Max-Planck-Str. 1, Geesthacht, Germany

R.G. Buchheit
Materials Science Engineering, The Ohio State University, Columbus, OH, USA

A.E. Hughes et al. (eds.), *Active Protective Coatings*,
Springer Series in Materials Science 233, DOI 10.1007/978-94-017-7540-3_1

be achieved in a single coating layer. However, in most cases, a coating system should provide a greater level of functionality and contains specific additives that all play a role in its performance including final appearance and durability. In these instances the durability and appearance requirements are separated into different coating layers: a multi-layer coating system. There are many application areas where two functionalities, such as corrosion protection and cosmetic appearance, are incorporated into a two layer system consisting of a primer and topcoat respectively. These two layer systems are used in a wide variety of applications from domestic housing to aircraft. In other instances the two functionalities are spread among even more layers, with the auto industry being a good example, where they use up to five layers to achieve the final auto finishes (see Chap. 13).

The materials science of coatings has seen an explosion in activity in recent decades due to a complicated mixture of drivers. One of the major, and earliest, drivers was the environmental imperative to reduce the amount of volatile organic chemicals as well as dangerous inorganic inhibitors used in the coatings industry [1]. In this latter category chromate-based chemicals have been a target for replacement for many years now. An emerging, additional driver is to reduce corrosion costs since they constitute a significant proportion of GDP of many industrialised countries [2]. This is particularly true for military airfleet operators who are looking to significantly reduce maintenance costs [3, 4]. In terms of technological changes to coating systems, the need to replace chromate in coatings has resulted in the search for alternative inhibitors and delivery systems to ensure an optimised and targeted response to corrosion and coating damage. The reduction in maintenance costs has driven research into the incorporation of signalling or sensor systems into coatings [5, 6]. These should, at a minimum, indicate where either corrosion or a defect in the coating has occurred and call for adequate and timely maintenance interventions. Another driver is the extensive introduction of hybrid structures to different industries especially in vehicle design (aircraft and auto) where light weighting means an improvement in energy efficiency and a reduction in CO_2 emissions. These hybrid structures include mixed metal components and composite materials (metal and polymers). Such mixed structures require new coating systems applicable for a mixture of substrates and capable of actively preventing, physical, chemical and galvanic driven degradation.

The breadth of the research into coatings may not have been so wide if not for substantial advances in a range of fields including, physics, chemistry, materials science and characterisation techniques which have underpinned much of the activity in the coatings area. Arguably this rapid expansion in materials science has been facilitated in two stages. First, the robotics used in high throughput experimentation, which was developed for drug [7, 8] and then polymer discovery and optimization [9, 10], has led to the development of many new materials as well as approaches to experimentation in materials science. This has been applied to polymer coatings [11, 12] and inhibitors [13–20]. Second, rapid development in high speed electronics through miniaturisation have revolutionised data collection and processing meaning that large volumes of data can be collected with relative ease. Even techniques for high throughput screening of results have been developed, meaning scientists need only focus on interpretation of relevant results rather than all the data. High speed computational modelling approaches are also developing to the point where they are nearly routine techniques [21].

1.2 The Problem

To begin this exciting journey into the materials science of coatings, it is important to define and understand the target of our research. At a glance this may seem obvious; the objective is to build smart coatings that have the ability to self repair and increase the longevity of the substrate that the coating is protecting. However, once we ask the question "OK, where do we start?", then we discover the very wide range of applications to which current coating technologies are applied and the equally wide range of solutions that have been developed. These range from coatings for simple cosmetic appearance to thick protective coatings loaded with other fillers and inhibitors designed to protect in extreme environments such as on oil rigs or in industrial applications. An additional problem that is often overlooked is the testing requirements and industry standards that have been developed for different sectors of the coatings industry. The effect of standards testing should not be underestimated in introducing new technologies and this will be addressed in the next section.

To define the problem that we're attempting to address by building multifunctionality into coatings, let's begin by breaking it down into (i) externally introduced damage from the in-service environment and (ii) protection of the substrate. In the former case in-service environmental damage can be caused by aerosol deposition, UV exposure and mechanical damage to name a few sources. In the latter case, protection can be provided at many levels as outlined below.

To provide a context for this discussion let's examine a typical coating system used in much of the aerospace industry as depicted in Fig. 1.1. This type of system (primer and topcoat) is common to many other application areas. In detail, the coating consists of a topcoat which serves as a barrier, a carrier of pigment and the polymer matrix which has some wear resistance. The primer is epoxy-based and is loaded with inhibitor (sparingly soluble chromates), fillers and pigments. (Water uptake in the primer is rapid and is followed by chromate release [22–25]) The surface of the metal has usually undergone treatment which, in the case of aluminium, could be either anodising or chromate conversion coating (as depicted in Fig. 1.1). This surface treatment is largely to achieve the desired level of paint adhesion but also to provide additional protection.

There are different aspects to the level of repair that this coating system provides which are depicted in Fig. 1.2. If we begin at the metal/conversion coating interface, there is a level of "pre-emptive healing" (purple double arrow in Fig. 1.2). This means that when water has permeated the primer, inhibitor can diffuse to the interface and provide additional protection to the surface. This cannot be observed in the chromate case because of the presence of either the anodised coating or the conversion coating. In this instance there is already chromate at the interface

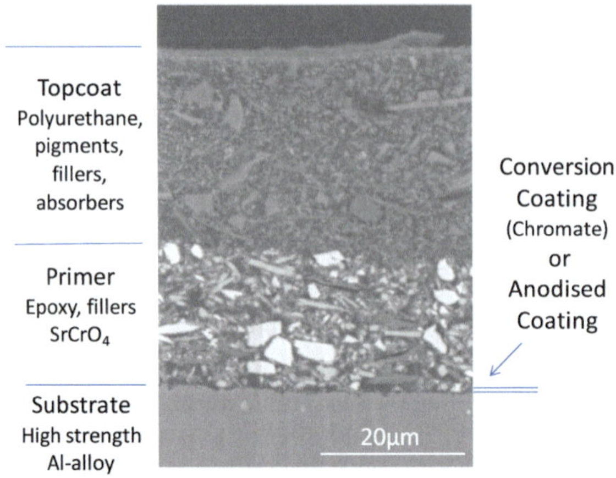

Fig. 1.1 Typical backscatter electron image of a cross section of an aerospace protective coating for the exterior of commercial aircraft and some military aircraft. The chromate conversion coating is typically 0.1–0.2 μm whereas the anodised layer is up to a few microns thick

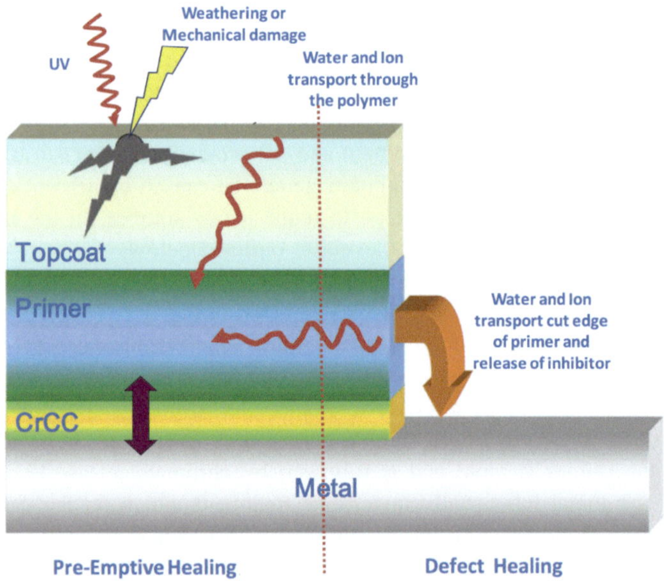

Fig. 1.2 Schematic of damage and healing mechanisms

because of these coatings (anodised coatings are often sealed in chromate [26]). It has, however, been observed with a Ce-dibutylphosphate (dbp)—inhibited epoxy painted onto aluminium 2024-T3 which had only been abraded [27] where Ce and P were observed at the interface after immersion of the primer into water. This appeared to improve the filiform corrosion resistance of the primer.

The other type of protection this system provides occurs when there is a defect in the paint system. If the defect penetrates down to the metal then the primer will also be ruptured and chromate will be released preventing or arresting corrosion in the defect [28–30]. This provides a functional repair since it prevents further corrosion.

So the problem that we're trying to address needs to include the current functionalities and desirably, include some additional ones. The first would be to protect the topcoat and the primer from degradation due to weathering. This repair is largely targeted at the polymer system for which there are a number of approaches spanning scales from the molecular level to the tens of microns level [31–33]. The other functionality that could be envisaged is "defect filling" where the void left by the mechanical damage is "backfilled" to a level where it fully regains its cosmetic appearance. This is not a trivial task for a coating where the volume of the defect is of similar size to the surrounding paint. The full cosmetic recovery would seem to suggest that a delivery system like a vascular network might be required. An overview of approaches to embodying these functionalities into coatings is provided in the next section. These are variously called smart, self healing, regenerative or active coating systems. For the purposes of simplicity we will use the term "smart" coatings in this chapter, although we'll talk about self healing in the following sections.

1.3 What Are Smart Coatings?

Smart coatings are emerging as a major activity in the coatings research area. As Challener [34] observes "There are as many definitions of smart coatings as there are scientists conducting research in this very broad field" and perhaps this is true given the broad ranges of technologies that are being utilised to increase the functionality of coatings. So to begin this chapter, let's examine the capability of current technology and look at additional functionality that might be applied to increase the performance of these coatings.

Within the scenario described above, Fig. 1.3 outlines one possible classification system for smart coatings. At the simplest level, coatings form a barrier between the external environment and the substrate material. This type of passive barrier depicted on the left of Fig. 1.3 is monofunctional and once breached has no response mechanism to provide further protection against corrosion. These monofunctional barrier coatings are only used in application areas where the probability of failure (during application and lifetime) is low and the subsequent consequences of failure are not significant. Such a paint system might be a simple house paint where a breach causes no immediate threat to the structural integrity, but has some loss of cosmetic appeal.

The next classification for coatings are those that provide a global but primitive response to damage. These types of coatings have the capability to provide responsive repair, but the response mechanism is neither mediated nor controlled. By way of demonstration let's examine the chromate inhibited primer as described in the previous section. Water permeates epoxy-based coatings used for primer applications relatively quickly [22, 27] beginning the process of solubilising the

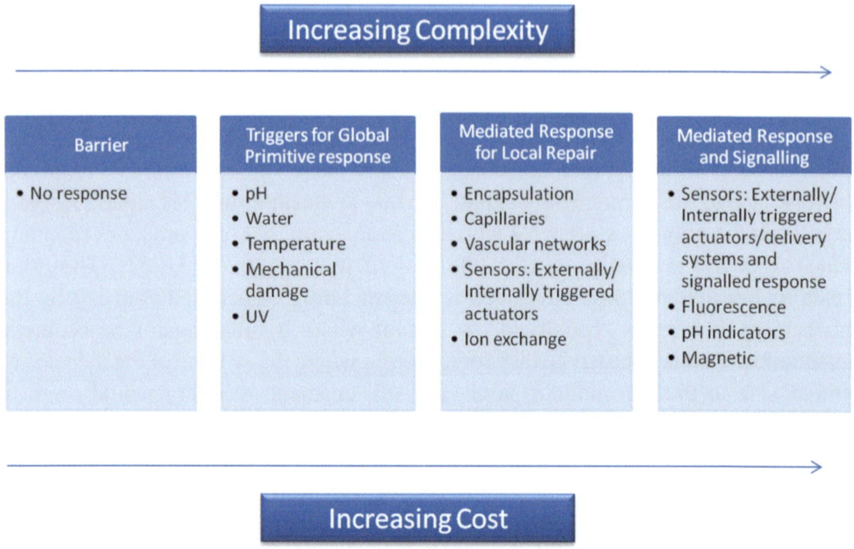

Fig. 1.3 Hierarchy of responses that could be adopted for smart coatings

strontium chromate inhibitor (the most common chromate inhibitor in paints). The strontium chromate can be solubilised anywhere in the primer (i.e. globally) where water is present rather than where required at an active corrosion site where the pH has decreased due to anodic dissolution; hence the label primitive. So in the case of strontium chromate some of it may be wasted because soluble chromate may be released at the defect long after the repair has been effected. In this instance this is because the restoration of a more neutral pH upon repair does not prevent the further release of chromate. Moreover, uncontrollable solubility can increase osmotic blistering effects leading to faster degradation of the coating system. However, the solubility of chromate decreases with increasing pH and speciation changes as well from dichromate to chromate so even this primitive response has nuances to it that allows for dosing when it is required [35]. Many inhibitors will have similar solubility characteristics and speciation changes which may be used to trigger a response more targeted to corrosion. Speciation is important in other transition metal oxyanion systems such as the vanadates system [36].

The next level of complexity (the third block in Fig. 1.3) is to add a mediated response for local repair, i.e. design a "smart" coating with a mechanism that only releases the amount of material required for repair or healing and then stops. Here the focus is on the development of smart delivery systems in order to regulate the location, the movement and the amount of repair material to match the level of damage. Encapsulation is an obvious method to form a barrier between the damage site and the active functional agent which responds to some aspect of the external damage. pH-sensitive capsules are an obvious candidate since a local acid environment develops during pitting corrosion. Furthermore, there is experience in the development of capsules that dissolve (or resist) acid environments from the pharmaceutical industry. Polyelectrolyte-based shells are among the most used candidates in this case. A second common example is the use of ion exchange materials such as clay materials or synthetic Layered Double Hydroxides (LDH) [37–39]. In this instance the LDH-structure is loaded with an anionic inhibitor which is triggered for release when chloride is present and exchanged into the LDH releasing the anionic inhibitor. Such a mechanism provides an active healing response to the aggressiveness of the environment and prevents uncontrollable leaching. Perhaps a third example is the use of internal sensors and actuators as proposed by Kendig et al. who suggested that "micro-machines" might be used to deliver inhibitor to corrosion sites [5]. These machines would use the potential of the Al (metal) to Al^{3+} reaction.

The right hand side of Fig. 1.3 shows a fourth level of smart coatings including a triggered response for release of the healing agent as well as a signalling response to indicate that an event has occurred that has required healing. This class of self repairing and indicating coatings is a very complex system from a materials development perspective. In a primitive sense many current systems already operate at this level to some extent since the development of corrosion through a defect results in inhibitor release and the optical signalling of the corrosion event is manifested through the presence of corrosion product which is often a different colour to the metal. However, the external signalling required for the right hand side of Fig. 1.3 should be concomitant with the healing event so that the remedial action can be monitored. The nature of the signalling is important. Visual signals are of limited use since many defects are either small or occur in structures that cannot be observed visually such as in aircraft wing structures or on reinforcement in concrete. In the future it may be possible to measure these responses internally in the same way mechanical damage is assessed in prototype space craft studies [23]. Visual signalling also degrades the cosmetic appearance of the coating system. Therefore signals may need to be obtained wirelessly or via a network such as an optical fibre network. For most structures any of these paths is possible, for aircraft both paths may be an issue since wireless signals may interfere with avionics and optical fibres cause extra complications because of layout, particularly if they are laid within a sealant in butt joints [40].

Another important aspect for smart coatings, especially when active protection from coatings is considered, is compatibility of their components. For example, nano-containers which can be used for encapsulation of active agents can negatively affect the barrier properties of the polymer layer as well as its adhesion to the metal surface if introduced in an improper way. Figure 1.4 schematically demonstrates three different situations. In the case of low compatibility the total performance of the system diminishes when nanocontainers, for example, are added to the polymer matrix in spite of the fact that an additional active protection component has been introduced. The decrease of adhesion and barrier properties is simply not compensated by the active protection effect. The situation of limited compatibility is observed when the protection system with an active component outperforms the passive coating in a certain range of concentrations. While high compatibility can

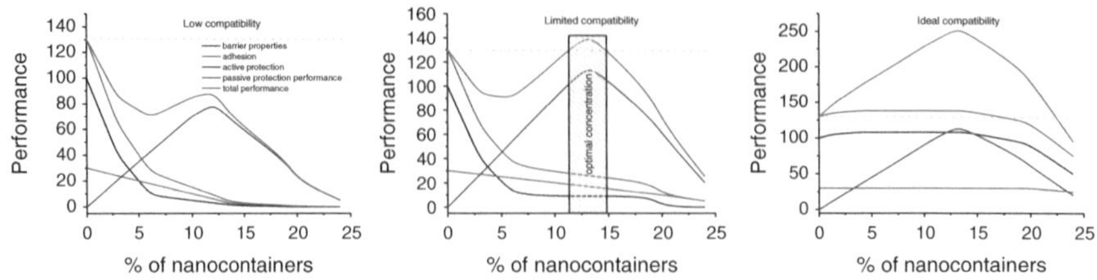

Fig. 1.4 Effect of compatibility of nano-containers with polymer matrix on performance of the protection coating system

ensure a synergistic effect of active nanocontainers with passive polymer coating leading to significantly improved performance and well defined self-healing ability.

Furthermore, going from left to right in the diagram represents a technological pathway of increasing complexity. The increasing technological complexity makes the engineering of the coating more complicated. It also raises issues about longevity of both the shelf life of the formulation as well as its durability as a coating. Increasing the complexity also suggests that the more complicated coating systems, may be considerably more expensive from an initial design and production investment point of view. The increased complexity and cost inevitably means that the most expensive coating systems are used in the highest value added applications where the cost of the coating is small compared to the overall cost of the structure and its maintenance operations, such as in the case of aircraft for example.

1.4 Biomimicry

The term *smart coating* is often used interchangeably with self healing coating which itself implies something of a biomimetic capability to restore the coating properties. To some extent primitive "healing" can be built into coating systems as discussed above and elaborated in the chapters below, but achieving a true biomimetic technology while a worthy aspiration, is difficult in practice and current developments fall well short of real biological systems. Part of the reason for this is that from a materials science perspective the human body, for example, is in a continuous state of flux since there is an ongoing exchange of atoms and molecules into the body from the environment, such as oxygen from the atmosphere as well as molecules from food, which then undergo chemical reaction to provide energy and materials for body maintenance. Reaction products are then released back into the environment where other organisms make use of many of them and so the cycle continues throughout the biosphere. This exchange provides energy and the basic chemical building blocks that help keep the human body in a dynamic equilibrium that epitomises the normal healthy state. Furthermore it's not clear what the biolomimetic equivalent of a coating is; for example, is it the skin, cell lipid bilayer or some other entity (bones being the most common).

By contrast, current materials design works on a principle of structured materials where atoms and molecules are retained as part of the material structure for the lifetime of the structure. Van der Zwaag highlights this point in comparing design strategies for self healing systems to traditional materials design [32]. So once the material begins to experience bond breakage through oxidation or other chemical or physical attack, and material loss due to ageing or mechanical damage, then the lifetime of the material will be reduced without intervention to remediate the damage. Industry performance standards also have a significant role to play on the acceptability of new materials. Self healing systems may be at a significant disadvantage when assessed against current industry standards since they are geared towards assessment of the virgin properties and do not take account of recovery properties that are part of the design philosophy for self healing materials.

Of course the materials science of coatings doesn't only draw on *analogies* with self healing in biological systems, but also borrows from the *actual* science. So in later chapters there will be reference to inhibitor delivery systems which draws on work from the pharmaceutical industry [41]. It is worth being aware of some fundamental differences between the ability to deliver drugs to a specific target in the human (or animal body) and being able to deliver targeted healing in coatings systems. This awareness helps to define the scope of the challenge of incorporating self healing mechanisms and technologies into coating systems. There are two key aspects of pharmaceutical delivery systems; one is the encapsulating technology for the drug and the other delivery mechanism within the body. Clearly, the capsule technology can be transferred from the pharmaceutical industry to coating applications assuming compatible chemistry. The delivery within the human body however, relies strongly on the physiology and biochemistry of the body. In terms of physiology, there are ready made systems within the body such as the digestive tract and cardiovascular system for transporting drugs to the desired target; these do not exist in coatings. The trigger for release of the drug is, in many instances, facilitated by very specific, surface biochemistry. For example, in gene therapy, non-viral vectors (containers carrying the drugs) can be coated with biomolecules such as galactopyranoside which recognises asialoglycoprotein receptors on liver cells and therefore provides a great method for targeting that organ [42, 43]. It remains to be seen whether analogous transport paths exist or can be built into coating systems and we are only at the beginning of devising very specific chemistry triggers for inhibitor release. More will be said on this in Chaps. 8 and 10. Of course, as seen above, from a practical perspective, very complicated systems may be too expensive or too incompatible to introduce into coatings.

1.5 The Structure of the Book

The intention of this book is to provide an overview of advances in multifunctional, self healing approaches as applied to coatings technology.

The first section of the book starts with an examination of the fundamentals of corrosion processes leading into atmospheric corrosion. These chapters define the problem that is being addressed in research focused on the development of new coating systems. The last two chapters in the first section look at the general mechanisms of inhibition and the science and technology of coatings into which inhibitors are incorporated to prevent corrosion.

The second section of the book explores advances in the development of self healing coatings in depth. This, of course, includes the developments in new inhibitors as well as in the polymeric resins in which they are incorporated. Inhibitor design itself is undergoing transformation with the development of multifunctional inhibitors that can perform as both cathodic and anodic inhibitors and perhaps even cause functional repair to the polymer resin. The search for new multifunctional inhibitors has been facilitated with the development of high throughput techniques for the evaluation of individual and combinations of inhibitors [13–15, 17–20, 44–47]. Key aspects of new coatings include both the ability of the polymer to repair itself and the controlled release of the inhibitor so that only the dose required to prevent corrosion is delivered to the attack site. Moreover, additional active mechanisms of corrosion prevention can also be incorporated. These mechanisms include preventive entrapment of aggressive corrosive anions and the water repelling from defects. While self healing of polymeric coatings draws strongly on self healing strategies for bulk polymers [12, 32, 33], there are also active protection strategies that are peculiar to coating applications. One of the most extensively studied systems in this latter category are conducting polymers and perhaps the use of silanes that form thin, but protective coatings. For inhibitor release, there has been broad consideration of a range of delivery systems. Much of this work draws from both the pharmaceutical industry and drug delivery systems as well as the food industry [48] where encapsulation is used extensively for a range of reasons: the encapsulated materials can be protected from moisture, heat or other extreme conditions, thus enhancing their stability and maintaining viability [48]. Encapsulation in foods is also utilized to mask odours or preservation of tastes. The use of encapsulation for sweeteners such as aspartame and flavours in chewing gum is well known [48]. With respect to the pharmaceutical industry, drug delivery is aimed at the digestive system and particularly the stomach either to bypass it or deliver in it for orally taken drugs. Fortuitously, the stomach is essentially an aqueous, hydrochloric acid environment, not unlike active corrosion sites, so developments in the pharmaceutical industry can be very easily applied to delivery systems for inhibitors with the right types of triggers, in principle.

The third section of the book deals with how to measure self healing in coating systems. Some of the most important developments in this area are based on electrochemical techniques particularly scanning localized techniques. Techniques such as the scanning vibrating electrode and the scanning electrochemical microsope provide new insight into the level of functional (performance) recovery when a defect occurs through a coating and the self repairing mechanism is activated. Physico-chemical techniques are often used in conjunction with electrochemical techniques to provide an overall characterisation of materials. Vibrational and electron spectroscopies, for example, can provide useful information on the nature of chemical changes that occur during the healing process. Developments in phase and adsorption contrast X-ray absorption and computed tomography (CT) from these experiments are emerging as powerful tools not only for materials characterisation but also to observe and measure healing effects. The combination of CT and electrochemical techniques is an emerging area but should provide a complete repertoire of tools to explore the ability of new self healing systems.

The fourth and final section of the book looks at the industry requirements from new mutlifuntional coatings. As pointed out above compatibility of self healing mechanisms and components needs to enhance the overall properties and performance of the system rather than undermine it. In this section there are contributions from industries that have a range of performance requirements including, automotive, aerospace and heavy industry. While these industries can have very different coating formulations, there are many common needs across them.

References

1. P.T. LaPuma, J.M. Fox, E.C. Kimmel, Chromate concentration bias in primer paint particles. Regul. Toxicol. Pharmacol. **33**(3), 343–349 (2001)
2. G.H. Koch, M.P.H. Brongers, N.G. Thompson, Y.P. Virmani, J.H. Payer, *Corrosion Costs and Preventative Strategies in the United States*, F.H.A. Department of Transport, Editor. 2002, (NACE International, Houston)
3. Anon., Success stories: air force: material substitution and new sealing technologies keep airframes flying longer. AMPTIAC Quarterley. **7**(4), 57–61
4. R.A.C Laudise, *Aging of U.S. Air Force Aircraft*. (National Research Council (United States), Washington D.C, 1997)
5. M. Kendig, M. Hon, L. Warren, 'Smart' corrosion inhibiting coatings. Prog. Org. Coat. **47**(3–4), 183–189 (2003)
6. A. Trinchi et al. Distributed quantum dot sensors for monitoring the integrity of protective aerospace coatings, in *2012 IEEE Aerospace Conference* (2012)
7. J.R. Broach, J. Thorner, High-throughput screening for drug discovery. Nature **384**(6604), 14–16 (1996)
8. O.G. Schramm et al., Polymeric nanocontainers with high loading capacity of hydrophobic drugs. Soft Matter **5**(8), 1662–1667 (2009)
9. H.Q. Zhang et al., Combinatorial and high-throughput approaches in polymer science. Meas. Sci. Technol. **16**(1), 203–211 (2005)
10. R. Hoogenboom, M.A.R. Meier, U.S. Schubert, Combinatorial methods, automated synthesis and high-throughput screening in polymer research: Past and present. Macromol. Rapid Commun. **24**(1), 16–32 (2003)
11. S. Bode et al., Self-healing metallopolymers based on cadmium bis(terpyridine) complex containing polymer networks. Polym. Chem. **4**(18), 4966–4973 (2013)
12. S.J. Garcia, H.R. Fischer, S. van der Zwaag, A critical appraisal of the potential of self healing polymeric coatings. Prog. Org. Coat. **72**(3), 211–221 (2011)
13. B.D. Chambers, S.R. Taylor, The high throughput assessment of aluminium alloy corrosion using fluorometric methods. Part II—A combinatorial study of corrosion inhibitors and synergistic combinations. Corros. Sci. **49**(3), 1597–1609 (2007)

14. B.D. Chambers, S.R. Taylor, High-throughput assessment of inhibitor synergies on aluminum alloy 2024-T3 through measurement of surface copper enrichment. Corrosion **63**(3), 268–276 (2007)
15. B.D. Chambers, S.R. Taylor, The high throughput assessment of aluminium alloy corrosion using fluorometric methods. Part I—Development of a fluorometric method to quantify aluminium ion concentration. Corros. Sci. **49**(3), 1584–1596 (2007)
16. Chambers, B.D., *The Discovery and Investigation of Synergistic Combinations of Corrosion Inhibitors for AA2024-T3*. (University of Virginia, Charlottesville, VA, 2007)
17. P.A. White et al., A new high-throughput method for corrosion testing. Corros. Sci. **58**, 327–331 (2012)
18. S.J. Garcia et al., Validation of a fast scanning technique for corrosion inhibitor selection: influence of cross-contamination on AA2024-T3. Surf. Interface Anal. **42**(4), 205–210 (2010)
19. P.A. White et al., High-throughput channel arrays for inhibitor testing: Proof of concept for AA2024-T3. Corros. Sci. **51**(10), 2279–2290 (2009)
20. T.H. Muster et al., A rapid screening multi-electrode method for the evaluation of corrosion inhibitors. Electrochim. Acta **54**(12), 3402–3411 (2009)
21. X.Z. Wang et al., Robust QSAR model development in high-throughput catalyst discovery based on genetic parameter optimisation. Chem. Eng. Res. Des. **87**(10A), 1420–1429 (2009)
22. F.H. Scholes et al., Chromate leaching from inhibited primers—Part I. Characterisation of leaching. Prog. Org. Coat. **56**(1), 23–32 (2006)
23. Hoschke, N., et al., Structural Health Monitoring of Space Vehicle Thermal Protection Systems, in *Structural Health Monitoring: Research and Applications*, ed. by W.K. Chiu and S.C. Galea (Trans Tech Publications Ltd, Stafa-Zurich, 2013), pp. 268–280
24. L. Fedrizzi et al., EIS study of environmentally friendly coil coating performances. Prog. Org. Coat. **29**(1–4), 89–96 (1996)
25. P. Carbonini et al., Electrochemical characterisation of multilayer organic coatings. Prog. Org. Coat. **29**(1–4), 13–20 (1996)
26. S. Wernick, R. Pinner, P.G. Sheasby, *The Surface Treatment and Finishing of Aluminium and its Alloys*, 5th edn. (Finishing Publications and ASM International, Teddington, 1987)
27. J. Mardel et al., The characterisation and performance of Ce(dbp)3-inhibited epoxy coatings. Prog. Org. Coat. **70**(2–3), 91–101 (2011)
28. S.A. Furman et al., Corrosion in artificial defects II. Chromate reactions. Corros. Sci. **48**(7), 1827–1847 (2006)
29. R.L. Howard et al., Inhibition of cut edge corrosion of coil-coated architectural cladding. Prog. Org. Coat. **37**(1–2), 83–90 (1999)
30. I.M. Zin et al., The mode of action of chromate inhibitor in epoxy primer on galvanized steel. Prog. Org. Coat. **33**(3–4), 203–210 (1998)
31. S.J. Garcia, Effect of polymer architecture on the intrinsic self-healing character of polymers. Eur. Polymer J. **53**(1), 118–125 (2014)
32. Fischer, H.R., Nat. Sci. **2**, 873–901 (2010)
33. D.Y. Wu, S. Meure, D. Solomon, Self-healing polymeric materials: A review of recent developments. Prog. Polym. Sci. **33**(5), 479–522 (2008)
34. C. Challener, The intelligence behind smart coatings. J. Coat. Technol. 2006(January), 50–55
35. M.W. Kendig, R.G. Buchheit, Corrosion inhibition of aluminum and aluminum alloys by soluble chromates, chromate coatings, and chromate-free coatings. Corrosion **59**(5), 379–400 (2003)
36. K.D. Ralston et al., Corrosion inhibition of aluminum alloy 2024-T3 by aqueous vanadium species. J. Electrochem. Soc. **155**(7), C350–C359 (2008)
37. A.N. Salak et al., Anion exchange in Zn-Al layered double hydroxides: In situ X-ray diffraction study. Chem. Phys. Lett. **495**(1–3), 73–76 (2010)
38. M.L. Zheludkevich et al., Active protection coatings with layered double hydroxide nanocontainers of corrosion inhibitor. Corros. Sci. **52**(2), 602–611 (2010)

39. S.K. Poznyak et al., Novel Inorganic Host Layered Double Hydroxides Intercalated with Guest Organic Inhibitors for Anticorrosion Applications. ACS Appl. Mater. Interfaces **1**(10), 2353–2362 (2009)
40. G. McAdam et al., Fiber optic sensors for detection of corrosion within aircraft. Struct. Health Monit. **4**(1), 47–56 (2005)
41. S.K. Sahoo, V. Labhasetwar, Nanotech approaches to delivery and imaging drug. Drug Discov. Today **8**(24), 1112–1120 (2003)
42. M. Nishikawa et al., Hepatocyte-targeted in vivo gene expression by intravenous injection of plasmid DNA complexed with synthetic multi-functional gene delivery system. Gene Ther. **7** (7), 548–555 (2000)
43. I. Roy et al., Calcium phosphate nanoparticles as novel non-viral vectors for targeted gene delivery. Int. J. Pharm. **250**(1), 25–33 (2003)
44. S.R. Taylor, B.D. Chambers, Identification and characterization of nonchromate corrosion inhibitor synergies using high-throughput methods. Corrosion **64**(3), 255–270 (2008)
45. S.R. Taylor, B.D. Chambers, The discovery of non-chromate corrosion inhibitors for aerospace alloys using high-throughput screening methods. Corros. Rev. **25**(5–6), 571–590 (2007)
46. B.D. Chambers, S.R. Taylor, M.W. Kendig, Rapid discovery of corrosion inhibitors and synergistic combinations using high-throughput screening methods. Corrosion **61**(5), 480–489 (2005)
47. S.J. Garcia et al., The influence of pH on corrosion inhibitor selection for 2024-T3 aluminium alloy assessed by high-throughput multielectrode and potentiodynamic testing. Electrochim. Acta **55**(7), 2457–2465 (2010)
48. B.F. Gibbs et al., Encapsulation in the food industry: a review. Int. J. Food Sci. Nutr. **50**(3), 213–224 (1999)

Part I
Fundamentals

Chapter 2
Fundamentals of Corrosion Kinetics

G.S. Frankel

2.1 Introduction

Corrosion, the environmental degradation of materials, is a complex process that depends on details of the environment and material, and is controlled by underlying thermodynamic and kinetic factors. The ultimate goal of a corrosion engineer is to predict and control the rate of corrosion. To do so requires a thorough understanding of the thermodynamic and kinetic fundamentals. This chapter will review the electrochemical kinetics important to the overall corrosion process, including activation controlled kinetics, mixed potential theory, and mass transport effects.

The chapter will begin with a description of the typical anodic and cathodic half reactions that comprise the corrosion phenomenon. This is followed by an introduction to the most fundamental type of electrochemical kinetics, activation controlled kinetics, which are applicable in situations where the rate is controlled by the charge transfer reaction. The Tafel equation describes many of the reactions involved in corrosion. With the basis of simple electrochemical kinetics, it is then possible to present mixed potential theory, which predicts corrosion potential and rate on the basis of the kinetics and thermodynamics of all of the reactions occurring on an electrode surface. The Evans diagram, a graphical representation of this situation, is introduced to, clarify the mixed potential condition. The rate of corrosion is often controlled by mass transport of species to the surface from the bulk solution, and the kinetics of mass transport are the next topic of discussion. The inhibition of corrosion, in which the kinetics of one or more of the corrosion reactions are altered by the presence of species added to the environment, will then be covered. Passivity, or the spontaneous formation of a thin protective surface film, can be viewed as a form of anodic inhibition and is introduced at the end of this chapter.

G.S. Frankel (✉)
Fontana Corrosion Center, The Ohio State University, Columbus, OH 432310, USA
e-mail: frankel.10@osu.edu

A.E. Hughes et al. (eds.), *Active Protective Coatings*,
Springer Series in Materials Science 233, DOI 10.1007/978-94-017-7540-3_2

2.2 Corrosion Reactions

Corrosion often involves oxidation of metal atoms to form ionic species with higher oxidation state and the liberation of electrons. For a generic metal M:

$$M \rightarrow M^{z+} + ze^- \quad (2.1)$$

Examples of real metals:

$$Fe \rightarrow Fe^{2+} + 2e^- \quad (2.2)$$

$$Zn \rightarrow Zn^{2+} + 2e^- \quad (2.3)$$

$$Al \rightarrow Al^{3+} + 3e^- \quad (2.4)$$

These are called half-cell reactions because the electrons liberated by the oxidation reaction must be consumed by a reduction reaction occurring on the same electrode. A reduction reaction that is common in acids is hydrogen evolution:

$$2H^+ + 2e^- \rightarrow H_2 \quad (2.5)$$

The complete corrosion reaction for Zn in an acid would be the sum of the oxidation and reduction reactions:

$$Zn + 2H^+ \rightarrow Zn^{2+} + H_2 \quad (2.6)$$

The sites for the oxidation reactions are called anodes, and the sites for the reduction reactions are called cathodes. Anodes and cathodes can be spatially separated at fixed locations associated with heterogeneities on the electrode surface. Alternatively, the locations of the anodic and cathodic reactions can fluctuate randomly across the sample surface. The former case results in a localized form of corrosion, such as pitting, crevice corrosion, intergranular corrosion, or galvanic corrosion, and the latter case results in nominally uniform corrosion.

In most environments of interest to corrosion there is not a large concentration of metal ions acting as cathodic reactants (*i.e.* the reverse of reaction (1)). The important cathodic reactions in corrosion mechanisms involve water. There are two primary cathodic reactions, each of which takes a different form in acids or bases. The first is the hydrogen evolution reaction, in acids takes the form shown in Eq. 2.5. In neutral or basic solutions the hydrogen evolution reaction is:

$$2H_2O + 2e^- \rightarrow H_2(g) + 2OH^- \quad (2.7)$$

Each of the half reactions given in Eqs. 2.1–2.7 will be in equilibrium at a specific potential called the reversible potential, E^{rev}. When all reactants and products are in their standard states with activity of unity, the potential reaches a special value of reversible potential called the standard potential, E^0. The well-known electromotive force series is a list of half reactions ordered by their standard potentials. These standard potentials are derived from the free energies of formation of the species involved in the reactions.

For conditions under which the reactants and products are not in their standard states, which is the typical condition of corrosion, $E^{rev} \neq E^0$, and E^{rev} can be calculated from the Nernst equation. The Nernst equation for the hydrogen evolution reaction given in Eq. 2.5 at room temperature is:

$$\begin{aligned} E^{rev}(H^+/H_2) &= E^0(H^+/H_2) + (0.059/2)\log\left[H^+\right]^2 \\ &= 0.059 \log\left[H^+\right] \\ &= -0.059\ \mathrm{pH} \end{aligned} \tag{2.8}$$

It can be easily shown that the Nernst Equation for the base form of the hydrogen evolution reaction, Eq. 2.7, is also $E^{rev} = -0.059$ pH.

When dissolved oxygen gas is present in the aqueous solution, the following oxygen reduction reactions are possible, with Eq. 2.9 dominant in acids and Eq. 2.10 in neutral or basic solutions:

$$O_2 + 4H^+ + 4e^- \rightarrow 2H_2O \tag{2.9}$$

$$O_2 + 2H_2O + 4e^- \rightarrow 4OH^- \tag{2.10}$$

The Nernst equation for the acid form of the oxygen reduction reaction is:

$$\begin{aligned} E^{rev}(O_2/H_2O) &= E^0(O_2/H_2O) - (0.059/4)\log\left[H^+\right]^{-4} \\ &= 1.229 - 0.059\ \mathrm{pH} \end{aligned} \tag{2.11}$$

As for the case of the hydrogen evolution reaction, the Nernst Equation for the base form of the oxygen reduction reaction is the same as that for the acid form. The base form of either the hydrogen evolution reaction or the oxygen reduction reaction is obtained from the acid form by adding a multiple of the water dissociation reaction:

$$H_2O \rightarrow H^+ + OH^- \tag{2.12}$$

This chapter deals primarily with the corrosion of a fictitious metal M with no relevant microstructural heterogeneities. In reality, engineering alloys have complex microstructures and surface features that can have a large effect on the progression of the corrosion processes. It was described above, that metal corrosion takes place by a combination of an anodic half reaction such as given in Eqs. 2.1–2.4, and a cathodic half reaction such as in Eqs. 2.5, 2.7, 2.9, and 2.10. The locations of the local anodes and cathodes depend on the surface and microstructural details and determine the form of corrosion that results.

The corrosion reactions will tend to occur on the most reactive spots on a surface. An ordered list of increasing reactivity might be: crystal ledges, crystal kinks, emerging dislocations, high index grain faces, impurities, grain boundaries, inclusions and particles, second phases, crevices and cracks. For example, for a very pure metal that is single crystalline and well annealed with a well-prepared surface, the reactions might take place on crystalline ledges and kink sites. The form of corrosion might be etch pits at dislocations, grain boundary etching, or second phase attack, depending on what are the most susceptible sites present on the surface. Because most alloy surfaces have heterogeneities that provide reactive sites and surface films, it is rare for attack to be dominated at the less reactive sites like emergent dislocations.

The anodic and cathodic reactions will take place at separate sites, but those sites might be extremely close. For the current to pass from cathode to anode, there must be a potential difference. However, if the anodes and cathodes are extremely close, the potential difference will be extremely small. Nominally uniform corrosion will occur if the local anodic and cathodic sites continually move across the surface owing to small changes in surface reactivity as the corrosion process proceeds. If the location of the anodic reaction does not continually move, but stays at localized sites, the form of attack will be localized. For example, etch pits will form at dislocations if those sites remain the most active sites for dissolution. Extensive localized attack in the form of pits, crevices, intergranular corrosion or cracks can occur if the separation of the anodic and cathodic reactions is sustained.

The cathodic reactions involving water in Eqs. 2.5, 2.7, 2.9, and 2.10 all result in an increase in production of OH^- or the consumption of H^+. Therefore, the pH tends to increase in the region near a sustained cathode. In contrast, the pH tends to decrease in the region near a sustained anode, such as a pit or crevice owing to metal cation hydrolysis:

$$\begin{aligned} M^{n+} + nOH^- &\rightarrow MOH^{n-1} + (n-1)OH^- \rightarrow M(OH)_2^{n-2} + (n-2)OH^- \ldots \\ &\rightarrow M(OH)_n \end{aligned} \tag{2.13}$$

Hydrolysis will consume hydroxide ions, causing the pH to decrease. Complete hydrolysis of the cations to form hydroxide, the right most product in Eq. 2.13, does not generally occur. Each of the steps of the hydrolysis process will reach equilibrium, with different equilibrium constants for different cations. If cation hydrolysis were able to go to completion and complete mixing of the anode and cathode environments was possible, every hydroxide ion produced or H^+ ion consumed by the cathodic reaction would be balanced by an equal consumption of hydroxide ion by hydrolysis and there would be no pH change. However, because of the lack of complete cation hydrolysis, the pH of a closed corroding system always increases.

For the case of localized corrosion, in which there is sustained spatial separation of the anodic and cathodic reactions, a gradient in chemistry as well as potential will exist between the anode and cathodic sites. Migration of anions toward the anode will also occur owing to the potential gradient. Therefore, the local anode environment might be enriched with both chloride anions and H^+ cations, forming a very aggressive solution. This will tend to prevent repassivation and sustain the localized corrosion.

2.3 Activation Controlled Kinetics

Consider a metallic electrode immersed in a corrosive aqueous environment. Anodic and cathodic reactions will occur spontaneously on the electrode surface, possibly resulting in corrosion of the electrode. The resulting potential of the electrode will be different from the reversible or equilibrium potentials of each of the reactions occurring on the surface. If the concentrations of the reactants and products at the electrode surface are the same as in the bulk solution, the difference in potential from the reversible potential for a given reaction is called activation overvoltage or charge transfer overvoltage, η. It stems from the fact that the rate of charge transfer at the electrode-electrolyte interface is not infinitely fast. Note that the concentration at the electrode surface equals the bulk concentration when the rate of mass transport is fast compared to the rate of charge transfer. This aspect will be discussed in more detail later.

For a reaction in which the rate is limited by activation overvoltage, the relationship between the rate of reaction, which can be expressed by a current density i, and the driving force for the reaction, or potential E, is given by the Butler-Volmer equation:

$$\begin{aligned} i &= i_0 \exp\left[\frac{\alpha n F(E - E_{rev})}{RT}\right] - i_0 \exp\left[\frac{-(1-\alpha) n F(E - E_{rev})}{RT}\right] \\ i &= i_0 \exp\left[\frac{\alpha n F \eta}{RT}\right] - i_0 \exp\left[\frac{-(1-\alpha) n F \eta}{RT}\right] \end{aligned} \tag{2.14}$$

where F is the Faraday constant = 96,487 C/equivalent, n is the charge on the ion in equivalents/mol, R is the gas constant = 8.314 J/mol-K, and α is the unitless charge transfer coefficient. The value of α is usually close to 0.5, but must be between 0 and 1.

For a sufficiently large value of anodic polarization from the reversible potential (overpotential $\eta_a > \sim 50$ mV), the first term on the right side of Eq. 2.14 dominates the second term. Therefore, at large overpotentials, the Butler-Volmer equation simplifies to:

$$i_{net} = i_0 \exp\left[\frac{\alpha n F \eta_a}{RT}\right] \tag{2.15}$$

Rearranging, one gets the Tafel equation:

$$\eta_a = b_a \log(i/i_0) \qquad (2.16)$$

where $b_a = 2.3\ RT/\alpha nF$ is the anodic Tafel slope. For $\alpha = 0.5$ and $n = 1$, $b_a = 0.12$ V (or V/decade). A similar equation is found for cathodic activation polarization:

$$\eta_c = -b_c \log(|i|/i_0) \qquad (2.17)$$

Corrosion conditions typically are far removed from the reversible potentials for any of the reactions. Therefore, Tafel kinetics are usually an accurate description of corrosion kinetics for conditions under which mass transport limitations are not important.

2.4 Mixed Potential Theory

Under conditions of importance in corrosion, the predominant cathodic reaction is typically one of the hydrogen or oxygen reactions described above. Furthermore, the electrode potential is usually far from the reversible potentials for the any of the reactions occurring on the surface.

The principle of charge conservation dictates that, to avoid the accumulation of charge on a freely-immersed electrode, the sum of all of the oxidation currents must equal the sum of all of the reduction currents. Any electrode immersed in an environment will naturally have a potential, called the corrosion potential, that fulfills this requirement. Therefore, at the corrosion potential:

$$\Sigma I_a + \Sigma I_c = 0 \qquad (2.18)$$

Note that reduction currents (I_c) are negative. The corrosion potential is also called the open-circuit potential, free potential, or rest potential. The corrosion potential is a mixed potential indicating that its value depends on the rate of the anodic as well as the cathodic reactions. Furthermore, if the corrosion cell involves one dissolution reaction and one cathodic reaction, the corrosion potential will be between the reversible potentials of the two half reactions.

The situation can best be visualized using an Evans diagram, Fig. 2.1. In this figure, the Tafel lines for the anodic and cathodic branches of the hydrogen reaction are shown together with the Tafel lines for the dissolution and plating of a metal with reversible potential below that for the hydrogen reaction. These are the possible electrochemical reactions for an active metal in acid containing little dissolved metal ion, where the primary cathodic reaction is hydrogen evolution. In this figure, $E^{rev}_{M/M+}$ and $E^{rev}_{H/H+}$ represent the reversible potentials for the metal M dissolution and hydrogen evolution reactions, respectively; $i_{0,\ M/M+}$ and $i_{0,H_2/H+}$ represent the exchange current densities for metal dissolution and hydrogen evolution on M,

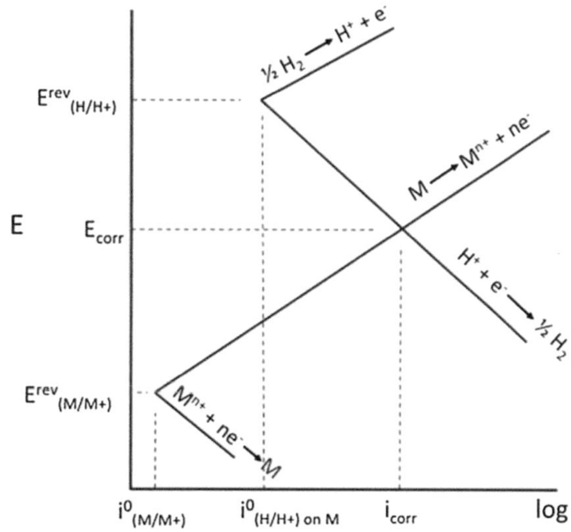

Fig. 2.1 Schematic Evans diagram for the corrosion of metal M by an acid showing the application of mixed potential theory

respectively. The principle of charge conservation, Eq. 2.18 can be applied. Since the reactions are all occurring on a single electrode of given area, Eq. 2.18 can be written in terms of current density, i = I/A:

$$\Sigma i_a + \Sigma i_c = 0 \tag{2.19}$$

The total oxidation and reduction current densities will be equal at the point where the anodic line for the metal dissolution reaction intersects the cathodic line for hydrogen evolution. The potential at which these lines intersect is the corrosion potential. The rate of the anodic reaction at the corrosion potential is the corrosion rate (corrosion current density). It can be seen that the corrosion potential always takes a value between the reversible potentials for the two partial reactions.

The corrosion potential and current density can shift with time if the surface or solution changes and the Tafel lines representing the reactions change. Note that the electrode is not in equilibrium at the corrosion potential because net changes occur: metal is oxidized and water or other oxidants are reduced. The corrosion potential and current density are influenced by both *thermodynamics*, through the reversible potentials of the various reactions, and by *kinetics*, through the exchange current densities and the Tafel slopes of the reactions.

Polarization curves of metals in solution can be determined by potentiodynamic polarization. Using a potentiostat, a counter electrode, and a reference electrode, the potential of a sample is scanned or stepped in small increments over a range from about 250 mV below the corrosion potential to well above the corrosion potential. The potentiostat determines the current passed at the electrode surface as a function of the potential. The current density is easily calculated and is often plotted as a function of potential on a semi-log plot like that shown in Fig. 2.2. Such a plot is called a polarization curve. Note that the absolute value of the current is plotted, as

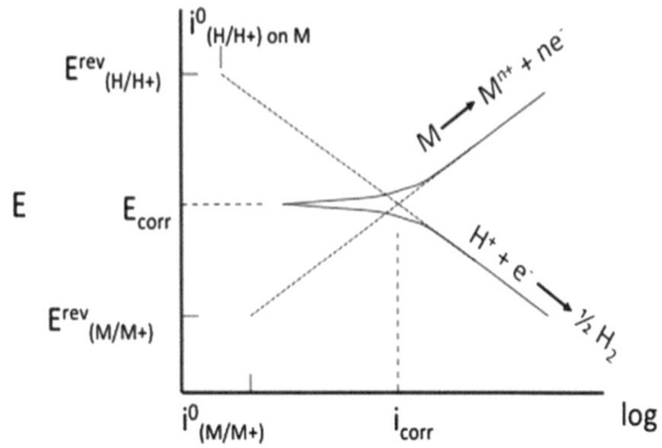

Fig. 2.2 Schematic measured polarization curve for metal M in an acid showing E_{corr} and Tafel extrapolation for determination of i_{corr}

the log of a negative number is undefined. The net measured current for the case of a metal corroding at a mixed potential in a solution containing an oxidizing agent will look like the curve presented in Fig. 2.2. Far from the corrosion potential, one of the reactions, either the cathodic or anodic reaction, occurs at a much greater rate, and so the relationship tends to follow the straight line representative of Tafel kinetics for the anodic reaction at high potentials and for the cathodic reaction at low potentials. At the corrosion potential, $I_a = I_c$, so the measured or net current equals zero, the log of which is $-\infty$. Therefore the potential at which the curve points to the left denotes the corrosion potential. Extrapolation to the corrosion potential of the linear regions of the semilogarithmic plot, or Tafel regions, gives the corrosion rate, Fig. 2.2. In the region close to the corrosion potential, the anodic and cathodic reactions occur at similar rates so the measured current, which is a sum of the two, deviates from the Tafel relationship.

The schematic polarization curve shown in Fig. 2.2 plots potential and current on the ordinate and abscissa axes, respectively. This is a common plotting format but is not intuitive because nowadays the measurements are typically performed by controlling the potential and measuring the current. This format has historical precedence because, before the advent of potentiostats, polarization measurements were made by controlling current and measuring potential. It is equally common to plot polarization curves with the independent variable, current, on the ordinate and the potential on the abscissa, as they should be based on the measurement. Experienced practitioners can transform such plots readily.

For a corroding system in which a single anodic reaction and a single cathodic reaction occur, such as illustrated in Fig. 2.2, the relationship of the net current as a function of potential will be:

$$i_{net} = i_{corr} \exp\left[\frac{2.3(E - E_{corr})}{b_a}\right] - i_0 \exp\left[\frac{2.3(E - E_{corr})}{b_c}\right] \tag{2.20}$$

where b_a and b_c are the anodic and cathodic Tafel slopes, respectively. The form of this relationship is similar to the Butler-Volmer equation, Eq. 2.14, which describes the kinetics of a single half reaction at potentials above and below the reversible potential for that reaction. In contrast, this equation, which has no commonly-referred to name, describes the net current when two different half reactions occur on a single electrode surface. Non-linear least squares fitting of the $i_{net}(E)$ data set to Eq. 2.20 by the analysis software provided by the potentiostat manufacturer is the most common means to evaluate corrosion data as it provides values for the constants in the equation, i_{corr}, E_{corr}, b_a, and b_c, in an automated fashion.

2.5 Mass Transport Controlled Kinetics

For conditions in which the surface concentration of a reactant is not the same as the bulk concentration, the equations describing the reaction kinetics must be altered. The oxygen reduction reaction, Eqs. 2.9 and 2.10, is an important reaction in corrosion that often meets this condition because the reactant, molecular oxygen, is consumed at the surface, which reduces its surface concentration. This then requires the transport of oxygen from the bulk through a layer adjacent to the surface. For this condition, the cathodic Tafel equation, which is similar to Eq. 2.15, must be altered to include the concentration of oxygen in the bulk and at the surface, c_b and c_s, respectively:

$$i = -i_0\left(\frac{c_s}{c_b}\right)\exp\left[-\left(\frac{(1-\alpha)nF}{RT}\right)(E - E^{rev})\right] \tag{2.21}$$

The equation shows that the rate of the charge transfer reaction at the interface depends not only on the applied potential but also on the concentration of reacting species prevailing at the electrode surface.

Consider now a metal sample being corroded by a generic oxidizing agent, B, which is being transported from the bulk and consumed at the surface. For neutral species such as oxygen, or charged species present in small amounts in the presence of a supporting electrolyte, the contribution of migration to transport is small. For the subsequent discussion it will be assumed that near the electrode surface there is a stagnant layer of electrolyte of thickness δ, which is called the Nernst Diffusion layer or, somewhat loosely, the diffusion layer. Within the stagnant diffusion layer only diffusion contributes to the flux of the reacting species, whereas outside the diffusion layer no concentration gradients exist and convection is the only transport mechanism for the reacting species.

The reaction of species B at the surface will reduce its concentration and the transport of B from the bulk to the surface through the diffusion layer can be described by the electrochemical equivalent of Fick's first law:

$$i = \frac{-nFD_B(c_{B,b} - c_{B,s})}{\delta} \tag{2.22}$$

where i is the current density associated with the reduction of B, F is the Faraday constant, and D_B is the diffusion coefficient of B. The effect of convection on the electrode reaction can be understood from this equation by considering that increased convection decreases δ and thus increases the flux of B and its reduction rate i.

Solving Eq. 2.22 for surface concentration of B gives

$$c_{B,s} = c_{B,b} + \frac{i\delta}{nFD_B} \tag{2.23}$$

Note that in this expression i is a negative quantity as it describes a reduction reaction so it is equivalent to write it as:

$$c_{B,s} = c_{B,b} - \frac{|i|\delta}{nFD_B} \tag{2.24}$$

The surface concentration of the reacting species therefore decreases with increasing current density.

As the cathodic current density increases, the surface concentration of the reacting species, $c_{B,s}$, decreases until reaching a value of $c_{B,s} = 0$. The current associated with this condition is the maximum possible reduction rate of B and is called the limiting current density i_{lim}.

$$i_{lim} = \frac{-nFD_B c_{B,b}}{\delta} \tag{2.25}$$

Dividing Eq. 2.22 by Eq. 2.25 yields:

$$\frac{i}{i_{lim}} = 1 - \frac{c_{B,s}}{c_{B,b}} \tag{2.26}$$

which can be rearranged to give:

$$\frac{c_{B,s}}{c_{B,b}} = 1 - \frac{i}{i_{lim}} \tag{2.27}$$

Substituting into Eq. 2.20 gives:

$$i = i_0\left(1 - \frac{i}{i_{lim}}\right)\exp\left[\frac{-(1-\alpha)nF}{RT}(E - E^{rev})\right] \tag{2.28}$$

This can be rearranged to yield Eq. 2.29, which describes the current-potential relationship for a cathodic reaction controlled by combined charge transfer kinetics and mass transport.

$$i = \frac{i_0 \exp\left[\frac{-(1-\alpha)nF}{RT}\left(E - E^{rev}\right)\right]}{1 - \frac{i_0}{i_{lim}} \exp\left[\frac{-(1-\alpha)nF}{RT}\left(E - E^{rev}\right)\right]} \tag{2.29}$$

If $i_{lim} >> i_0 \exp\{[-(1-\alpha)nF/RT](E - E^{rev})\}$, then the denominator of Eq. 2.29 goes to 1 and this equation becomes the Tafel equation. However, if $i_{lim} << i_0 \exp\{[-(1-\alpha)nF/RT](E - E^{rev})\}$, this equation becomes $i = i_{lim}$ and the reaction rate is independent of potential.

Another way to look at the situation of combined influences of charge transfer overvoltage and mass transport is to rearrange Eq. 2.29 to get:

$$E - E^{rev} = \eta = -\frac{RT}{(1-\alpha)nF} \ln\frac{i}{i_0} - \frac{RT}{(1-\alpha)nF} \ln\left(1 - \frac{i}{i_{lim}}\right) \tag{2.30}$$

The first term on the right hand side is the charge transfer or activation overvoltage and the second term is concentration overvoltage. The total overvoltage is seen to be a sum of the two. The concentration overvoltage goes to negative infinity as the current approaches the limiting current, Fig. 2.3. According to Eq. 2.25, the i_{lim} depends on diffusivity, diffusion layer thickness, and bulk concentration of the reacting species. Diffusion is a thermally activated process and the diffusion layer thickness depends on the solution velocity, so environment temperature and agitation will affect i_{lim}. The charge transfer (activation) and concentration overvoltages combine as shown in Fig. 2.3. This is the typical polarization behavior observed for a system under mixed charge transfer and transport control.

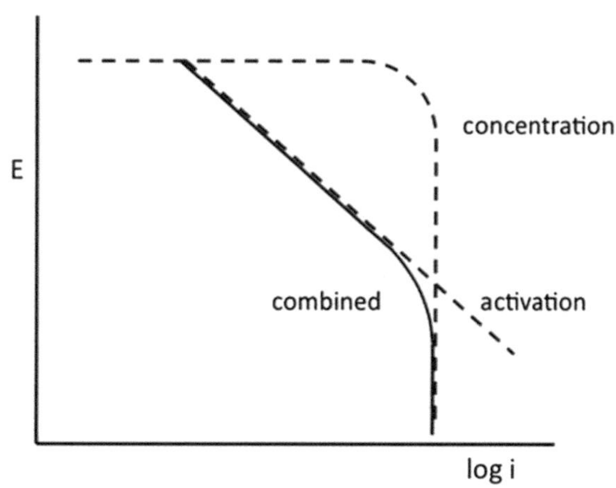

Fig. 2.3 Representation of the combine effects of activation- and concentration-controlled kinetics

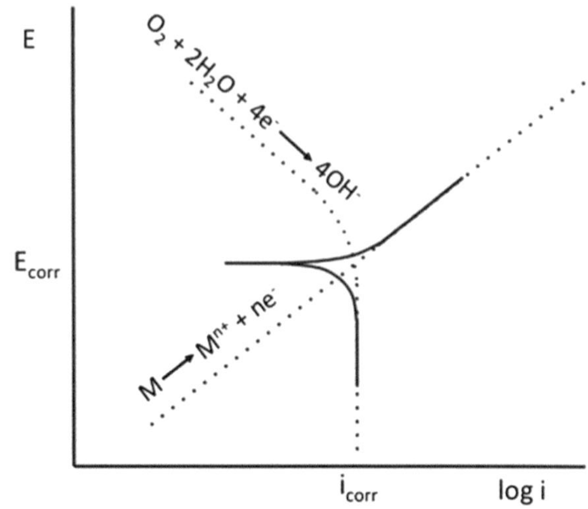

Fig. 2.4 Schematic measured polarization curve for the corrosion of metal M in an environment where the rate of reaction is controlled by the diffusion limitation of O_2 to the surface

Mass transport is of primary importance for the rate of corrosion in environments with limited cathodic reactant, for example in a neutral solution containing dissolved oxygen as reacting species. The maximum corrosion rate in such a situation is given by the limiting current density of the cathodic reactant to the surface,

$$i_{corr} = i_{lim} \tag{2.31}$$

An example is corrosion of steel in an aerated neutral solution where the rate of oxygen reduction reaction is largely controlled by mass transport. An Evans diagram for this situation is given in Fig. 2.4. Similarly, the rate of corrosion of iron in dilute hydrochloric acid is limited by the rate of the mass transport of the proton to the iron surface.

2.6 Inhibition

Corrosion inhibitors are chemicals added to an environment to decrease the rate of corrosion of metals exposed to that environment. The most common applications of inhibitors are limited-size or closed systems such as acid pickling baths or closed-loop water cooling systems. Acidic pickling baths are used to chemically dissolve the iron oxide scale on steel, and organic inhibitors are added to adsorb onto the metal surface after dissolution of the scale and thereby block the electrochemical attack of the steel. Closed-loop water cooling systems used, for instance, to extract heat from engines are typically neutral or alkaline solutions to which inorganic molecules are added to promote oxide formation on contacting metal surfaces or to buffer the pH.

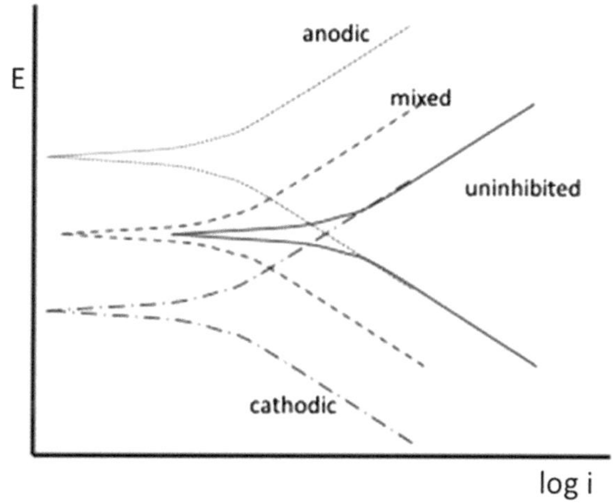

Fig. 2.5 Schematic representation of the effects of *anodic*, *cathodic* and *mixed inhibition*

Inhibitors are also added as soluble pigments in protective organic coatings, or paint. The poly-electrolyte formed when the paint is saturated with water, protects the metal and can release inhibitive species into local environments such as delaminations or through-scratches so as to protect the exposed metal.

The effect of the inhibitor on the corrosion process can be understood and classified by comparing polarization curves in environments with and without the inhibitor added. Figure 2.5 shows schematic examples of the types of inhibition and their effects on the polarization curves. Inhibitors can be classified as an anodic, cathodic, or mixed inhibitor as shown in this figure. As the names imply, an anodic inhibitor reduces the rate of the anodic reaction while having less of an effect on the cathodic reaction, a cathodic inhibitor reduces the rate of the cathodic reaction while having less of an effect on the anodic reaction, and a mixed inhibitor reduces the rates of both reactions equally. The corrosion potential will often be different in a solution containing inhibitor compared to the value in an uninhibited solution, and the nature of the change will be different for the different classes of inhibitor. E_{corr} will increase for an anodic inhibitor, decrease for a cathodic inhibitor, and not change for a mixed inhibitor. As long as the corrosion rate is indeed reduced, the shift in E_{corr} can be used as a diagnostic for the type of inhibitor.

2.7 Passivity

In the field of corrosion passivity describes a state in which the corrosion rate of a metal substrate is very low because of the presence of a thin oxide or oxy-hydroxide layer on the surface that deactivates or passivates the metal. The best passive films, such as those that form on Al and other so-called valve metals like Ti, Zr, Nb, etc. are very thin, on the order of a few nm. Stainless steels have high corrosion

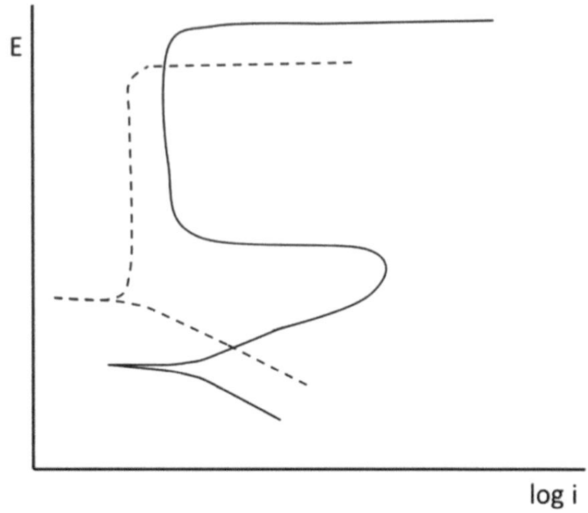

Fig. 2.6 Schematic polarization curves for passive samples. *Solid line* is for a metal showing an active/passive transition and *dashed line* is for a spontaneously passive metal

resistance owing to the presence of a Cr-rich passive film on their surfaces. The highly corrosion resistant engineering alloys achieve that property by passive films.

Passivity can be considered a form of anodic inhibition because the rate of metal dissolution is greatly reduced by the passive film. As mentioned above, the inorganic corrosion inhibitors added to neutral environments such as cooling water systems act by improving the passivity of the metals in contact with them. Schematic polarization curves for passive materials are shown in Fig. 2.6. The solid line represents a metal exhibiting active-passive behaviour. As the potential increases above the corrosion potential, the current increases exponentially, as would be expected for an actively dissolving metal. However, at a potential called the Flade potential, the current trend reverses and can decrease by many orders of magnitude because of the formation of a passive film. In the passive region, the current density tends to have little dependence on potential. In comparison with an extrapolation of the current behaviour from the actively dissolving region, the current density in the passive region is decreased enormously. In certain conditions, current can increase again at a higher potential owing to the breakdown of the passive film by localized corrosion or by transpassive reactions such as oxygen evolution or Cr dissolution as chromate. For a case such as shown by this curve, passivity can only be achieved by applying an anodic potential, which is called anodic protection, or by the addition of oxidizing agents that would increase the corrosion potential to a value in the passive region as described next.

The dashed line in Fig. 2.6 shows the polarization curve for a spontaneously passive metal. For this material, the active-passive transition is at lower potentials and is not observd because it is dominated by the cathodic reaction in that potential region. The intersection of the curve for the cathodic reaction with the anodic polarization curve is in the passive region such that the material is spontaneously passive at the open circuit potential. In this case, i_{corr} is limited by the anodic reaction and is equal to the passive current density.

The curves shown in Fig. 2.6 were arbitrarily drawn such that the spontaneously passive metal has lower values of corrosion current density, passive current density and breakdown potential, which is given by the potential at which there is a sharp increase in current. These values will depend on the material properties and environmental severity.

As shown in Fig. 2.6, the dissolution current increases rapidly with increasing potential after breakdown of the passive film. An example of such a system is the corrosion of high strength Al alloys in chloride-containing environments, which takes the form of pitting or intergranular corrosion when passive film breakdown occurs. The rate of corrosion in such a case is often controlled by the cathodic reaction, which is oxygen reduction in many natural aerated environments. Therefore, inhibition of oxygen reduction is an important strategy for limiting the extent of corrosion in natural environments. Inhibiting species can be added to the environment or to protective films to be released into a local environment as needed. Figure 2.7 shows how inhibition of the oxygen reduction reaction can greatly reduce the rate of corrosion for a material like a high strength Al alloy. For an uninhibited case, the dashed curve representing the kinetics of the oxygen reduction reaction intersects the curve representing the anodic behaviour of the Al alloy at a potential above the breakdown potential. In this case, the alloy will undergo localized corrosion at the open circuit potential. In contrast, inhibition moves the oxygen reduction curve to the left so that the intersection with the anodic curve occurs below the breakdown potential. In the inhibited case, localized corrosion is not spontaneous at the corrosion potential, and the corrosion rate is given by the passive current density.

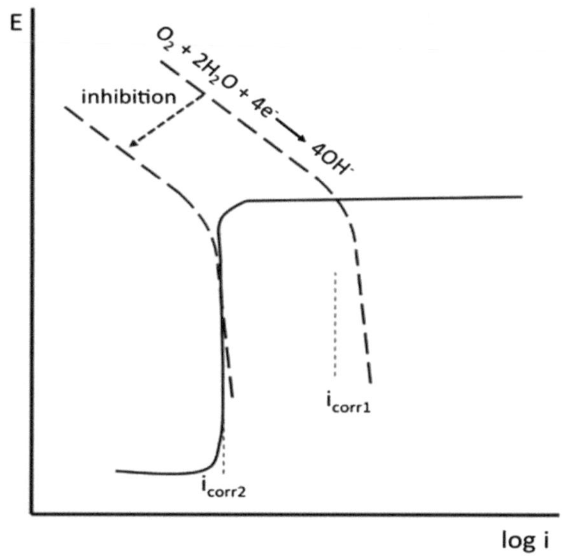

Fig. 2.7 Schematic representation of the effect of *inhibition* of the oxygen reduction reaction on the corrosion of a metal such as high strength Al alloy, which undergoes localized corrosion

Recommended Reading

1. D.A. Jones, *Principles and Prevention of Corrosion*, 2nd edn. (Prentice Hall, 1996)
2. D. Landolt, *Corrosion and Surface Chemistry of Metals* (CRC Press, 2007)
3. E. McCafferty, *Introduction to Corrosion Science* (Springer, 2010)
4. M. Fontana, N. Greene, *Corrosion Engineering* (McGraw-Hill, New York, 1978)
5. F. Mansfeld (ed.), *Corrosion Mechanisms* (Marcel Dekkar, New York, 1987)
6. P. Marcus (ed.), *Corrosion Mechanisms in Theory and Practice*, 2nd edn. (Marcel Dekkar, New York, 2002)
7. J.O'M. Bockris, A.K.N. Reddy, *Modern Electrochemistry*, vol. 2 (Plenum Press, 1970)

Chapter 3
The Atmosphere Conditions and Surface Interactions

I.S. Cole

3.1 Introduction

This chapter will address the aspects of atmospheric interaction with the surface and corrosion that need to be considered in developing new smart coatings. Recently [1], there have been a number of reviews on the atmosphere and corrosion but to the author's knowledge none that are particular to the development of smart coatings. Most of the atmospheric factors that promote corrosion of bare metals will also control the degradation of coatings and corrosion of coated metals, however some factors will affect coatings to a greater extent (notably UV). However, the conditions that develop at the coating/atmosphere interface may be substantially different to those that occur at the interface of uncoated metal and the atmosphere.

This chapter will thus look at:

(1) Factors controlling the deposition of pollutants on the surface of a coated metal.
(2) Factors that control the local microclimate at a coating surface and the formation of moisture layers
(3) Models of moisture and pollutant movement through coatings.

I.S. Cole (✉)
Manufacturing Flagship, Commonwealth Scientific and Industrial Research Organisation, Melbourne, Australia
e-mail: Ivan.cole@csiro.au

A.E. Hughes et al. (eds.), *Active Protective Coatings*,
Springer Series in Materials Science 233, DOI 10.1007/978-94-017-7540-3_3

3.2 Factors Controlling the Deposition of Pollutants on the Surface of a Coated Metal

In discussing pollutant deposition, it is necessary to consider pollutant deposition via aerosol deposition, or via rain deposition plus interaction of gaseous pollutants with a coated surface. In considering aerosol deposition it is important to consider the source of aerosols, their transport, chemical reactions during transport and their deposition. In the case of rain, the critical factors are changes in rain chemistry and rain deposition and retention on a surface. For gaseous interaction, two cases are relevant, deposition into a moisture film on a coating and deposition and reaction with a dry coating.

The major sources of aerosols [2] are marine aerosols and industrial aerosols and while aerosols may range in size from 0.1 to up to 100 μm, fine aerosol (<2.5 μm) will generally not deposit onto surfaces and so are not of interest in coating degradation.

Marine aerosols, which may originate from whitecaps [3] (breaking waves that generate white foam) as particles torn from the crests of such ocean whitecaps, or from breaking waves on the shore (hereafter referred to as surf produced aerosol) [3]. Fitzgerald [4] found that ocean-produced aerosols had size ranges from <1 to greater 50 μm with the modal frequency range being <1 μm and more than 50 % of aerosols have a size that is less than 3.73 μm, while McKay et al. [3] found that the size distribution for surf-produced aerosols was from 5 to >50 μm with the modal frequency being 25 μm and 50 % of aerosols having a size range that is less than 28 μm. Thus surf tends to produce large coarse aerosols, while ocean activity produces mostly medium sized aerosols. Of course, as the major constituents of sea salt are hygroscopic, marine aerosols will absorb moisture when the ambient RH exceeds the deliquescent RH (DRH) of the constituent salts (75 % at 20 °C for NaCl, and 35 % at 20 °C for $MgCl_2$). Thus marine aerosols increase their diameter as they wet by 20 % at RH of 60–140 % increase at a RH of 90 %.

Industrial aerosols form from a process of nucleation, condensation and accumulation (by impact with other other aerosols and absorption of gasses) with the aerosols in the "accumulation stage" having size ranges from 0.1 to 2.5 μm. Studies show that land based aerosols have low pH values, that depend on size and composition reflecting the absorption of gaseous species so that in Hong Kong [5], aerosols consisted of NH_4^+ and SO_4^{2-} at a ratio consistent with letovicite ($(NH_4)_3H(SO_4)_2$), and at pH values from −1 to 1.5 while Yokohama, Japan [6], the aerosol major constituents were NH_4^+, SO_4^{2-} and NO_3^- and the aerosol pH was between 2.0 and 2.2.

3.3 Aerosol Chemistry and Reactions in Moisture Layers

The chemistry of aerosols may be significantly altered while in the atmosphere, both from aerosol coagulation and from interactions (primarily absorption) with atmospheric gases. The concentration of an absorbed gas in a liquid is related to the gas partial pressure by Henry's Law:

$$A(g) \leftrightarrow A(aq) \tag{3.1}$$

$$[A(aq)] = H_A p_A \tag{3.2}$$

where $[A(aq)]$ is the aqueous phase concentration (in mol L^{-1}), p_A is the partial pressure in the gas phase (in atm), and H_A is the (temperature dependent) Henry's law coefficient for species A [7]. According to Seinfeld and Pandis [2] species with $H_A < 1000$ exist mainly in the gas phase, species with H_A coefficients between 1000 and 10,000 are moderately soluble in water, and those with $H_A > 10{,}000$ are considered very soluble. Henry's law coefficients for common gases at 298 K are given in Table 3.1, and while it is evident that many common gases (e.g. CO_2, SO_2) may be regarded as relatively insoluble, the absorption of these gases may be significantly increased by secondary reactions (e.g. dissociation and oxidation), which can decrease the concentration of dissolved species, allowing further absorption. The extent of these secondary reactions and thus the value of Henry's law is highly dependent on pH

A key case where secondary reactions dramatically increase the absorption of a gas is that of SO_2 absorption. The following reaction sequence occurs after SO_2 is absorbed [2]:

$$SO_2(g) + H_2O \leftrightarrow SO_2 \cdot H_2O \tag{3.3}$$

Table 3.1 Henry's law coefficients for common gases (in water at 298 K). Reprinted From Seinfeld et al. [2]. Reproduced with kind permission of John Wiley & Sons

Species	H_A (M atm^{-1})
O_2	1.3×10^{-3}
O_3	1.1×10^{-2}
NO_2	1.0×10^{-2}
CO_2	3.4×10^{-2}
SO_2	1.23
NH_3	62
OH	25
HCl	727
HCOOH	3.6×10^{3}
CH_3COOH	8.8×10^{3}
NO_3	2.1×10^{5}
HNO_3	2.1×10^{5}

$$SO_2 \cdot H_2O \leftrightarrow H^+ + HSO_3^- \quad (3.4)$$

while

$$HSO_3^- \leftrightarrow H^+ + SO_3^{2-} \quad (3.5)$$

The balance between HSO_3^- and SO_3^{2-} depends on pH, with the former dominating at pH values of 2–7, and the latter dominating above this pH range [2]. These dissociation reactions may be taken into account by defining an effective Henry's law coefficient that represents the equilibrium dissolved species when subsequent aqueous reactions are considered. The effective Henry's law coefficient for SO_2 ranges from 1.23 M atm^{-1} at a pH of 1 to >10^7 M atm^{-1} at a pH of 8. This has a dramatic effect on S(IV) concentrations in solution, which for a mixing ratio of 200 ppb (a typical value of a mildly industrial environment of SO_2 at 298 K) would increase from 3×10^{-7} to 3×10^{-3} M as pH is increased from 0 to 6.[1] In the case of NH_3, the Henry's law coefficient is 62 M atm^{-1}, and it is more readily absorbed than for SO_2. In a similar way NH_3 and CO_2 absorption is highly dependent of subsequent aqueous reaction and thus highly pH dependent.

Another group of gases whose absorption is important to atmospheric corrosion are those such as O_3, H_2O_2 or NO_x which may oxidise or catalyse the oxidation of dissolved species. Nitric acid is extremely water soluble and fully dissolves in all pH > 1 solutions. Ozone has a very low solubility, as indicated by its very low Henry's law coefficient [2]. In order to determine the balance between the gas phase and the absorbed phase for a number of compounds, oxidation reactions also need to be included (as the atmosphere is in general strongly oxidising). An important example is the absorption of SO_2, where the oxidation of SO_3^{2-} or other forms of S (IV) to SO_4^{2-} or other forms of S(VI) is critical, as it allows the continued absorption of SO_2, [14]. This oxidation may occur via a variety of mechanisms [8], including reactions with O_3, H_2O_2 and O_2 (catalysed by Mn(II), Fe(III) and NO_2).

In summary, the pH and species concentration in aerosols will depend on the effective Henry's law coefficient and thus on the extent of secondary reactions. Therefore, while the pure Henry law coefficient for SO_2 absorption is relatively low (1.23 M atm^{-1} compared with 727 M atm^{-1} for HCl), secondary reactions raise the effective Henry's law coefficient and the speciation between HSO_4^- and SO_4^{2-} leads to significant buffering at low pH. This combination makes SO_2 levels a critical factor in aerosol-induced corrosion. The high Henry's law coefficients of HCl, HNO_3 and NO_3 indicate that they are readily absorbed and thus can acidify aerosols. Oxidants such as O_3 (and NO_2) are not readily absorbed, however they remain important as their relatively low levels in solution can catalyse the oxidation of S(IV) to S(VI). In fact, the reported [9] synergistic effects between SO_2, NO_2 and O_3 in determining the extent of corrosion are likely to arise, at least in part, from the role of NO_2 and O_3 in promoting the oxidation of S(IV) and thus the continued absorption of SO_2 into moisture layers. The absorption of alkali precursor gases is also critical, and NH_3 is the most important of these due to its relatively high

Henry's law coefficient and its relative abundance. Thus, the balance between acid and alkali precursors needs to be considered in estimating the likely acidification of aerosol or rain water.

3.4 Key Aqueous/Gas Reactions

As indicated above, aqueous gas reactions are critical in controlling aerosol and rainwater chemistry. In this short review it is not possible to discuss in detail all the aqueous/gases systems of importance to atmospheric chemistry. Instead this chapter will highlight two of importance to the degradation of coatings. The sulfuric acid ammonia—water system which controls the chemistry of a major group of industrial aerosols and the absorption of gases into marine aerosols.

3.4.1 Sulfuric Acid–Ammonia–Water System

This system is of prime importance in determining the chemistry of aerosols in industrial locations. Due to their high solubility, both H_2SO_4 and NH_3 will be readily absorbed into aerosols and will react according to form a series of solid and dissolved species, including ($(NH_4)_3H(SO_4)_2$), $(NH_4)_2SO_4$, NH_4HSO_4, NH_4^+, SO_4^{2-}, HSO_4^- and NH_3). It is illuminating to consider the system at low relative humidity (RH) (30 %) and a high RH (75 %) at a H_2SO_4 concentration of 10 μg m^{-3}. At low RH and low ammonia levels, the aerosols consist primarily of H_2SO_4, with some NH_4HSO_4(s) and significant water (H_2SO_4 attracts a significant amount of water even at low RH) [10]. When the ammonia/sulfuric acid ratio increases above a molar ratio of 0.5, NH_4HSO_4(s) becomes the dominant species, and H_2SO_4 and associated water levels fall to zero [10]. As the molar ratio approaches 1, the salt letovicite $(NH_4)_3H(SO_4)_2$(s) forms and gradually replaces NH_4HSO_4(s). At a molar ratio of 1.25–1.5, $(NH_4)_3H(SO_4)_2$(s) goes from being the dominant to the sole species, and for molar ratio from of 1.5–2 ammonium sulfate salts form. At molar ratios of 2 or greater, ammonium sulfate salt is the sole species.

At 75 % RH the results are very different, the system is affected by the deliquescence of NH_4HSO_4 (which has a deliquescent relative humidity (DRH) of 40 %) and $(NH_4)_3H(SO_4)_2$ (DRH = 69 %), and so is in general a solution of NH_4^+ and a sulfate species (whose form depends on ammonia levels). Thus, the aerosol is a liquid solution of H_2SO_4 at low ammonia levels, HSO_4^- at moderate ammonia levels, SO_4^{2-} at ammonia/sulfate ratios above 1.5, while the solid $(NH_4)_2SO_4$ forms at ratios above 2 when there is sufficient ammonia to completely neutralize the sulfate [10].

It is evident from the above discussion that the composition and hydrogen ion concentration within a wet aerosol will depend critically on the relative balance of H_2SO_4 and NH_3 in the atmosphere. The hydrogen ion concentration will determine the extent of coating degradation promoted by the deposition of the aerosol, and thus the extent of corrosion. This result can be generalized to environmental systems containing a range of S-containing gases and ammonia, and emphasizes the need for an understanding of both acid and alkali precursors in understanding the corrosiveness of an environment.

3.5 Reactions in Marine Aerosols

Marine aerosols may be acidified by the absorption of acid precursor gases as outlined above. In particular, in wet marine aerosols, H_2SO_4 (or other strong acids such as HNO_3 that have lower vapour pressures relative to HCl) may be absorbed into the aerosol, where it will react with NaCl and acidify the solution, causing HCl to volatilize [11]:

$$H_2SO_4(g) + NaCl \leftrightarrow NaHSO_4 + HCl(g) \tag{3.6}$$

This produces the acid salt $NaHSO_4$ which may later dissociate. Thus, gaseous HCl may enter the atmosphere through the dechlorination of airborne sea salt particles. However, as it is relatively soluble ($H_A = 727$ M atm^{-1}), it is readily re-absorbed into atmospheric water droplets. In turn, these HCl-rich droplets may then react with and promote the absorption of other strong acids which may later dissociate:

$$HCl(aq) + HNO_3(g) \leftrightarrow HCl(g) + HNO_3(aq) \tag{3.7}$$

$$HCl(aq) + H_2SO_4(g) \leftrightarrow HCl(g) + H_2SO_4(aq) \tag{3.8}$$

In effect HCl will act as a buffer and prevent the acidity of marine derived aerosols falling significantly below a pH of 2. However the gaseous HCl may then be re-absorbed and so such cycling of HCl can thus lead to further acidification of both aerosols and rain or cloud droplets in marine locations [12], with significant contamination by industrial or urban pollutants.

In fact, marine aerosol are initially alkaline [13] immediately after breaking free from ocean waves (pH = 9.5. at 76.2 % RH) but this pH may subsequently reduced to below 5 within 15 min as a result of aqueous phase reactions (particularly the absorption of H_2SO_4) [14]. In fact, Keene et al. [15] recently measured the pH of the larger sea-salt fractions (geometric mean diameter (GMD) ≥ 2.9 μm) and found that the median pH ranged from 3.1 to 3.4; the median pH for submicron fractions was ≤ 1.6. Thus, the experimental data reinforces the observations made on the basis of the equations controlling aerosol chemistry. Both industrial and marine aerosols may be acidified primarily by the absorption of either H_2SO_4 or SO_2 (and subsequent liquid phase reactions), and the acidified aerosol obtained may be significantly buffered.

3.6 Aerosol Transportation

3.6.1 General Scale of Movement of Gases and Particles

In the atmosphere, a series of processes occur in parallel—gaseous species, aerosols and drops are constantly reacting, while the air is in constant motion on scales of a centimeter or less (tiny eddies) to continental dimensions. At the same time, species are entering the atmosphere from a variety of natural and man-made sources, or are leaving the atmosphere through scavenging by raindrops or by impact with the ground or ground-based objects. The period during which a species stays in the atmosphere is referred to as its residence time (which is closely related to the reactivity of the species), and residence time is closely related to the distance the species may travel from its source. In Fig. 3.1, the temporal scales (residence times) and spatial scales are plotted for a range of chemical species [2]. The hydroxyl radical (formed by photolysis of O_3 and H_2O_2) is highly reactive and thus has a residence time of less than one second and it will travel no more than meters from its source. In contrast, N_2O has lifetimes in tens of years and it will be mixed throughout the global atmosphere (the major pathway for N_2O loss is photo dissociation in the stratosphere).

The spatial impact of a species may be increased by its reaction with other entities. For example both gaseous SO_2 and H_2SO_4 are readily absorbed in aerosols that may have residence times of months and which can be dispersed distances from 10 to 100's of kilometers from their source, depending on size. Such aerosols can then either form cloud condensation nuclei (CCN) or be scavenged by either these nuclei or by raindrops, and thus acidification of cloud droplets or rain may occur a significant distance from the original source of DMS.

Figure 3.1 thus has significant implications for atmospheric corrosion, as it highlights that highly reactive species such as hydroxide radicals will travel extremely short distances, oxidants such as NO_x and H_2O_2 limited distance (although tropospheric O_3 may travel further), gaseous SO_2 short distances, and aerosols will travel moderate distances from their source. Thus, direct impacts of

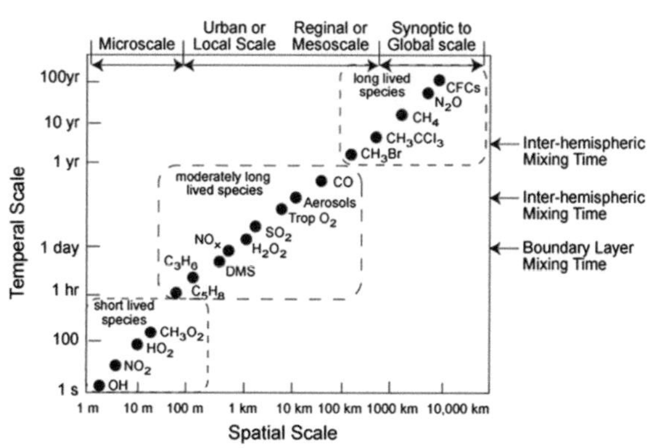

Fig. 3.1 Spatial and temporal scales of variability for atmospheric constituents (from Seinfeld and Pandis [2])

gaseous species on metallic corrosion will be limited to tens of kilometers from source, while indirect effects via aerosols or rain deposition may be at appreciable distances from source.

3.6.2 Analytical Approach to Gas Transport

The transport of gases in a turbulent fluid can be modeled [2] using a Gaussian distribution. The analysis indicates that dispersal of gases from their source would be highly dependent on source height and plume stability. For low-height sources (<10 m) such as short stacks, the maximum pollutant levels will occur at 100–500 m of the source, while for plumes emitted from sources at a height of 100 m the maxima may occur up to 10 km from the source. Thus, gaseous pollutants may directly impact on metal structures for distances up to tens of kilometers from their source. However, they may also be absorbed in aerosols and transported significantly greater distances.

3.6.3 Aerosol Transport

The corrosion science literature has concentrated on the transport of marine aerosols, and within corrosion science there has been little work on the deposition of industrial aerosols, which, as outlined previously, are significantly finer (generally <2.5 μm diameter). The factors influencing the transport of marine aerosols will also influence industrial aerosols, but industrial aerosols would be less affected by the various deposition processes and would thus have significantly larger residence times and be transported longer distances.

As indicated above aerosols may transport industrial acids much greater distances than the gas alone could travel, for example, work by Schwartz [16] indicates that industry-produced acids may be deposited up to 1000 km from their source. This is confirmed by Poste and Bridgman [17], who studied rainwater and fog acidity at two locations in northern New South Wales, Australia, approximately 200 and 500 km from any industrial location, and approximately 40 and 60 km from the coast, respectively. Both sites exhibited high levels of salts of marine origin and significant sulfates from anthropogenic sources. The most common explanation for this enhanced transportation is that SO_2/DiMethylSulphide or DMS (DMS is produced by biological activity in shallow seas) may be absorbed in cloud droplets or wet aerosols (including marine aerosols) and transported via the aerosols (subsequently the aerosols may be scavenged by fog or rain).

Analytical models of aerosol transport can provide a more accurate picture of aerosol dispersion. Cole et al. [18]. developed a computation fluid dynamics (CFD) model of marine aerosol transportation, whereby aerosols are lifted by diffusion, convected by wind and dragged down by gravity. The study indicates that coarse aerosols produced by the surf will generally be transported only 1–2 km before being scavenged or falling to the ground, and thus their residence time is only likely to be of the order of 1–20 min. In contrast, finer ocean-produced aerosols may travel hundreds of kilometres from the coast, and thus have residence times of hours to days. As Chameides and Stelson [19] indicate that it takes 5–15 min for the original alkaline pH of a marine aerosol to be reduced to below 5, there would not be any residence time limitation on absorption phenomena for ocean-produced aerosols, but such limitation may exist for surf-produced aerosols. In fact, work by Cole et al. [20] found that the chemistry of deposited material on salt candles and sample plates close to surf beaches closely reflected marine concentrations, and did not show any significant non-marine concentration effects while that at a distance where only ocean produced aerosols would be expected could show significant non-marine sulfate concentrations [21].

3.6.4 *Deposition of Pollutants*

Traditionally, deposition onto surfaces has been classified as wet or dry deposition, with rain and snow being defined as wet deposition, and deposition by particulates and aerosols (even when these are wet) being defined as dry. Deposition modes can also be defined as:

- Deposition of snow, rain or fog.
- Deposition of particulates or aerosols.
- Deposition of gases into wet surfaces.
- Deposition of gases onto dry surfaces.

3.6.4.1 Rain

Raindrops are generally assumed to have a 100 % efficiency of deposition onto a surface. However, once a raindrop falls onto a surface, it may splash off the surface and break into smaller drops, which could re-enter the airflow and not redeposit. Analyses by Cole and Paterson [22] indicated that although splash does occur, it does not lead to significant loss of deposition efficiency for rain falling onto stationary objects. However, it can be quite significant for fast-moving objects (e.g. airplanes in flight).

3.6.4.2 Gas and Particle Deposition

Similar principles can be applied to both the dry deposition of gases and to the deposition of particles. In a generic formulation, common in atmospheric physics, but pioneered in corrosion by Spence et al. [23], the deposition depends on the upstream concentration of the depositing species by the deposition velocity.

A variation to this approach is that of Cole et al. [24] in which the overall deposition velocity v can be calculated from the deposition velocities for the individual deposition mechanisms below:

- Gravitational settling.
- Turbulent diffusion.
- Laminar diffusion (also called Brownian deposition and diffusiophoresis).
- Thermophoresis (migration from high temperatures to low).
- Electrostatic attraction.
- Momentum-dominated impact.
- Vortex shedding (transport by transient laminar flows).
- Filtering (flow past or through raised fabrics).
- Photophoresis (motion generated by an intense beam of light).

However, whilst all these mechanisms can be important in interior spaces, in open spaces the prime mechanisms are gravity, momentum-dominated impact and turbulent diffusion. Figure 3.2 compares the efficiency of deposition of particles of increasing sizes onto a cylinder (typical of a salt candle, as described in ISO 9205) via momentum-dominated impact and turbulent diffusion. Efficiency is defined as the percentage of aerosol flux that would pass through the air space in the absence of an object. For very small particles, smaller than 0.1 or 0.01 μm in diameter, electrostatic attraction and laminar diffusion play a significant role in deposition. For small particles with a diameter of about 1 μm turbulent diffusion dominates, whilst for larger particles with a diameter of about 5 μm momentum-dominated impact dominates. For particles with a diameter above 20 μm gravitational settling becomes the prime deposition mechanism. As the particle diameter increases past 2.5 μm the deposition efficiency increases dramatically.

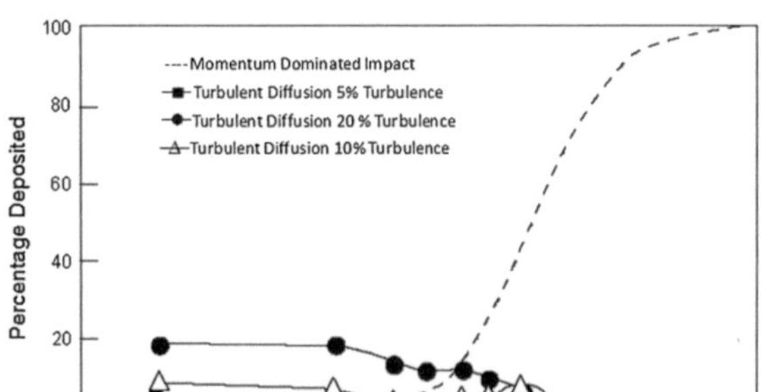

Fig. 3.2 Influences of momentum-dominated impact and turbulent diffusion on the deposition of aerosols (as a function of diameter in μm) onto a cylinder of arbitrary diameter at arbitrary wind speed, as a percentage of the aerosol flux that would pass through in the absence of the cylinder. The percent turbulence is the turbulence intensity (rms velocity/mean velocity) upstream

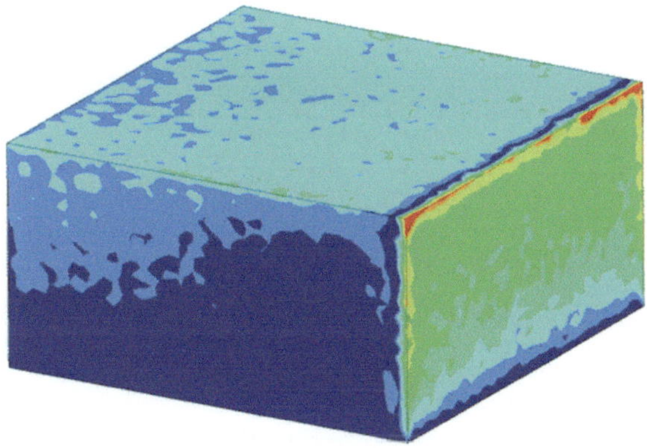

Fig. 3.3 Aerosol deposition on a building 10 m high and 20 × 20 m in plan. The flow is from *right* to *left*. *Blue* is low concentration and *red* is high concentration (from Cole et al. [25])

In fact, Cole et al. [25] the deposition onto common objects (plates, salt candles) in the open environment can be estimated by:

$$D = CVA\eta \tag{3.9}$$

where

$$\eta = 100I^{\beta}C_s \tag{3.10}$$

and I is turbulence, β is a constant (determined numerically to be 0.38), and C_s is the shape factor.

This model of deposition onto simple objects can be extended to more complex objects such as buildings or rows of buildings. The deposition onto an isolated building (10 m high and 20 × 20 m in plan) was computed using computational fluid dynamics and the result is given in Fig. 3.3 (from Cole et al. [25]). The randomness in Fig. 3.3 is caused by the stochastic nature of the simulation. The aerosol deposition rate is lowest near the ground, high near the upper edge of the front face and on the sides, and is highest on the upper corners and upper sides. The deposition rate is highly influenced by local wind turbulence, which is highest on corners and sides.

3.6.4.3 Deposition of Gases onto Wet and Dry Surfaces

The absorption of gases into moisture layers on metal surfaces is governed by the same factors that control absorption of gases into aerosols, and will in part be controlled by the Henry's law coefficient. In contrast, the surface absorption of gaseous species onto dry surfaces is controlled by normalized reactivity. Table 3.2 (from Seinfeld and Pandis [2]) presents the effective H_a and the normalized reactivity values for various gas species. The effective H_a in Table 3.2 takes into account reactions of the aqueous species. When H_a is high and the normalized

Table 3.2 Relevant properties of gases for dry deposition calculations (from Seinfeld and Pandis) [2]

Species	H_a^a (M atm^{-1}) at 298 K	Normalized reactivity
Nitric oxide	2×10^{-3}	0
Ozone	1×10^{-2}	1
Nitrogen dioxide	1×10^{-2}	0.1
Hydrogen sulfide	0.12	–
Ammonia	2×10^{4}	0
Nitrous acid	1×10^{5}	0.1
Sulfur dioxide	1×10^{5}	0
Hydrogen peroxide	1×10^{5}	1
Formic acid	4×10^{6}	0
Acetic acid	4×10^{6}	0
Hydrochloric acid	2.05×10^{6}	0
Nitric acid	1×10^{14}	0

[a]Effective H_a assuming a pH of 6.5

reactivity is low (e.g. sulfur dioxide), then deposition will be primarily through absorption into a moisture droplet (be it a wet aerosol or a surface moisture film). When H_a is low and the normalized reactivity is high (ozone), direct gaseous absorption onto dry surfaces will dominate. Table 3.2 indicates that those species of prime importance to atmospheric corrosion (ammonia, sulfur dioxide, formic acid and acetic acid) will all primarily be absorbed through the moisture layer. Of the nitrogen species, nitric acid will be readily absorbed into the moisture layer, while nitrogen dioxide will undergo limited interactions with both a dry and a wet surface. While the oxidizing species O_3 and hydrogen peroxide will be readily absorbed on a dry surface, their residence times in the atmosphere are extremely short, and thus the source would have to be very close to the metal surface for the concentrations of these gases to be significant. Thus for practical purposes, gaseous deposition onto a dry surface is of marginal importance to atmospheric corrosion.

3.6.4.4 Comparison of Deposition Modes

The relative significance of particulate deposition and gaseous deposition in an industrial environment clearly depends on the particulate and gas concentrations (aerosol deposition will dominate in marine environments). However, a comparison of deposition modes given typical compositions is useful. Consider the deposition onto a plate exposed at an angle of 45° in an exterior environment where the total concentration of particulates is 20 μg/m^3 (dry weight), and consists of ammonium sulfate, ammonium bisulfate or hydrogen sulfate, the wind speed is 3 m/s, and gaseous concentrations are CO_2 at 400 ppm, SO_2 at 75 ppb, NH_3 at 20 ppb, O_3 at 200 ppb and H_2O_2 at 10 ppb. Taking into account particle size and deposition efficiencies the deposition rate per hour would be 108 μg/dry weight per hour onto a

Table 3.3 Effect of particulates on moisture film composition

Particulate concentration ($\mu g/m^3$ dry weight)	Particulate type	Moisture film thickness (μm)	$[H^+]$ (M)	$[SO_4^{2+}]$ (M)
20	$(NH_4)_2SO_4$	100	–	7.5×10^{-4}
20	$(NH_4)_3H(SO_4)_2$	100	4×10^{-4}	8×10^{-4}
20	NH_4HSO_4	100	8.5×10^{-4}	8.5×10^{-4}
20	H_2SO_4	100	20×10^{-4}	10×10^{-4}
20	$(NH_4)_2SO_4$	500	–	1.5×10^{-4}
20	$(NH_4)_3H(SO_4)_2$	500	0.8×10^{-4}	1.6×10^{-4}
20	NH_4HSO_4	500	1.7×10^{-4}	1.7×10^{-4}
20	H_2SO_4	500	4.0×10^{-4}	2.0×10^{-4}
5	H_2SO_4	500	1×10^{-4}	0.4×10^{-4}

plate of size 100 × 150 mm at an angle of 45° to the vertical, while the change in chemistry of a pre-existing moisture film ($[H^+]$ and $[SO_4^{2-}]$ in M/L per hour) is given in Table 3.3.

Modeling by Cole et al. [26] estimated that for the gaseous compositions given above, the aqueous phase concentrations will be 1.7×10^{-6} M of $[H^+]$ and 1.2×10^{-4} M of $[SO_4^{2-}]$, or 0.63×10^{-4} M of $[H^+]$ and 0.34×10^{-4} M of $[SO_4^{2-}]$ in the absence of NH_3. Thus, the deposition of particulate onto a moisture film will substantially increase the ionic content and acidity of the film. If the moisture film persists for a number of hours, then particulate deposition is likely to dominate over gaseous absorption. If one considers the build-up of acidity (hydrogen ions) in a aqueous phase on a surface by acid rain or particulate deposition under the similar conditions as above then the fluxes are of similar orders of magnitude, but acidified rain is the major source of H^+ if the rain water pH is low and rainfall high, while particulate deposition is of prime importance if the rain fall is low and the concentration of acidified particulates is high. The analysis indicates that while dry deposition of gases onto metal surfaces is likely to be of marginal importance, rain deposition, particulate deposition and gaseous absorption into moisture films are all likely to be significant sources of both acidity and ion concentration in moisture on metal surfaces in industrial environments.

3.6.4.5 Cloud Nucleation and Rain Chemistry

Cloud nucleation occurs when aerosols undergo rapid growth in the presence of a supersaturation of water vapor. These aerosols are called cloud condensation nuclei (CCN). The supersaturation of water arises due to the steady decrease in temperature with distance from the ground, so that when a hot mass of air rises the RH of the air steadily increases. CCN will tend to be the larger sized aerosols (from coarse aerosols into the accumulation mode). Once cloud droplets have formed, they may

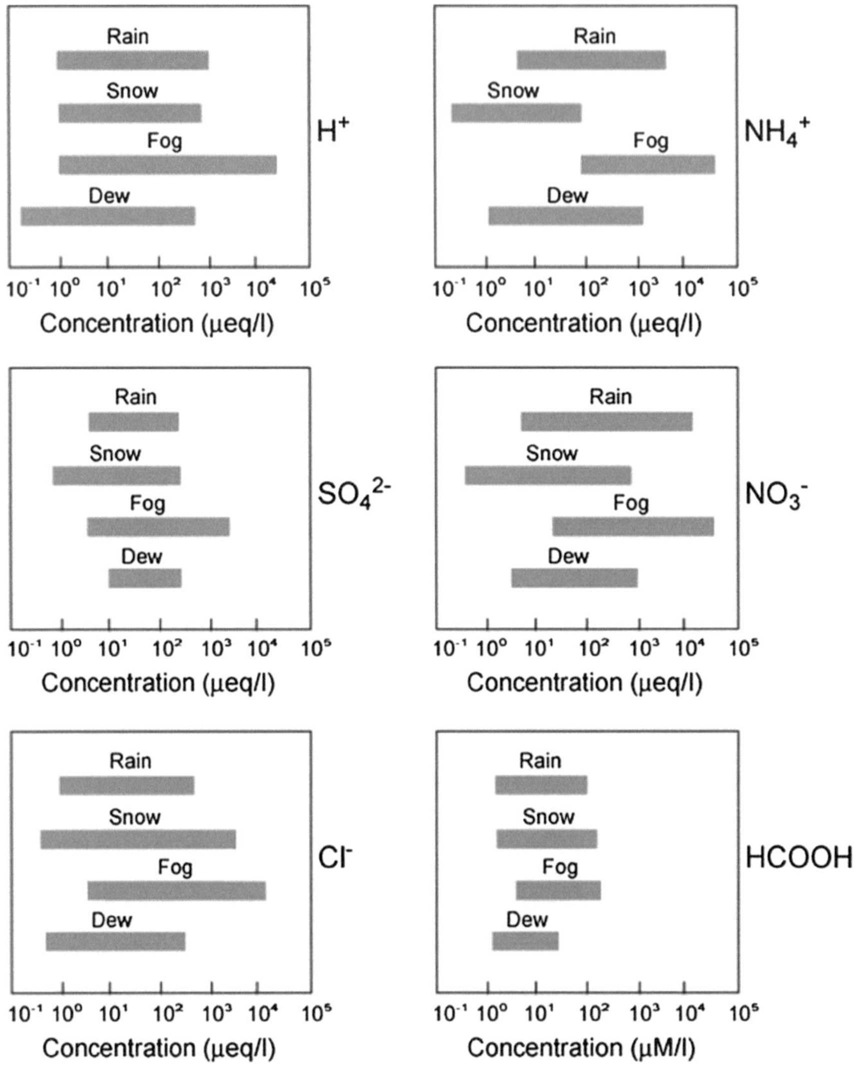

Fig. 3.4 Concentration ranges of ionic and molecular constituents of different types of atmospheric water (from Leygraf and Graedel [27])

continue to grow by scavenging the remaining aerosols in the clouds or by coalescence. Once a cloud droplet reaches a critical size and falls as rain, it may continue to scavenge aerosols or absorb gases until it reaches the ground.

Thus considerable concentrations of ionic species can build up in rainwater with typical concentrations being given in Fig. 3.4 [27]. However the formation and deposition of rain drops occurs over significant time and spatial scales so the composition of rainwater will reflect pollutant levels in a general area not only at the point of deposition. Fog arises when CCN form and grow close to the earth. A number of studies have identified the problem of acid fog, both within cities and in rural and forested areas in their vicinity. North American studies of fog water have indicated pH levels from 2.9 [28] to 3.6 [29] with high nitrate, ammonia and sulfate levels. The variation of chloride concentration in rainwater, shows a strong dependence on the distance from the location where the rainwater was sampled. Keywood et al. [30] have shown that chloride concentration falls in an approximately exponential fashion with peak values between 500 and 1000 μ eq/l at the coast to around 10 μ eq/l 500 km from the coast and less than one in the centre of the Australian continent.

3.6.4.6 Role of Rain and Wind in Surface Cleaning

Rain will clean a surface if the drop is of sufficient size to run off the surface. This size will be controlled by surface angle and the difference in interfacial contact angle between the front and back of the drop [31]. Studies show that drops grow by coalescence and absorbing raindrops impacting on them so that over time, the amount of water on the surface built up to a peak and then releases, giving periods of almost no runoff followed by periods of strong runoff [31]. Thus the rate of cleaning in a location depends on rain intensity and thus on climate zone. For example, compare Darwin (Australia) which has intense rain being in a tropical location, and Adelaide (Australia) which has rain of moderate intensity. It would take only a 7 min rain storm to clean a surface (<1 % of initial pollution) in Darwin but 43 min in Adelaide.

While in theory the removal of salts by wind could be significant, in practice it is only significant in climates where rain is very limited or when the wind carries abrasive material as experimental tests (wind tunnel) have demonstrated that very high winds are required to remove salt from a surface [22, 32] (from 9 m/s (dry salt) to 14 m/s (wet salt) and RH was ≤40 %).

3.6.5 *Forms of Moisture on Surfaces*

Moisture may form on a surface due to the local RH, via the deposition of rain, fog and wet aerosols, or due to wetting of hygroscopic salts on the surface.

3.6.5.1 Re-Wetting of Surfaces and Condensation

On a clean surface, condensation will occur when the air close to the surface is at 100 % RH. The temperature of the surface may vary from that of the ambient air, and thus the RH close to the surface may be different from the ambient air. The evaporation of aerosols or rain from surfaces may leave crystallized salts, many of

Table 3.4 DRH of common aerosols (at 20 °C) [33, 34]

Salt	DRH (%)
Na_2SO_4	84.2
NH_4Cl	80.0
$(NH_4)_2SO_4$	79.9
NaCl	75.3
$NaNO_3$	74.3
$(NH_4)_3H(SO_4)_2$	69.0
NH_4NO_3	61.8
$NaHSO_4$	52.0
$(NH_4)HSO_4$	40.0
$MgCl_2$	35.0

which may be hygrosopic. These salts will wet when the surface RH is greater than the DRH of the contaminant salts. The DRH of a number of common pollutants is given in Table 3.4. A study [21] demonstrated that the nature of the salts present (and thus the wetting point of the sensor) was highly dependent on location, with severe marine sites contaminated with $MgCl_2$ and NaCl, marine sites contaminated with NaCl, severe industrial sites contaminated with $(NH_4)HSO_4$, and moderate industrial sites with a marine influence contaminated with $(NH_4)_3H(SO_4)_2$, $(NH_4)_2SO_4$, $NaHSO_4$ and NH_4NO_3.

3.6.5.2 Surface Temperature of Plates

Metal and coated surfaces exposed to the environment will undergo a diurnal cycle, heating up during the day (due to solar heating) and cooling down at night (due to irradiation to the night sky), with both process modulated by convective interaction with air. The extent of both heating and cooling will depend on a range of environmental factors, including, level of global radiation (during the day) cloud cover (particularly at night), wind speed and the state (wet/dry, clean/contaminated) as the surface emissivity of the metal or coating [35].

Condensation and evaporation will tend to reduce this diurnal cycle; condensation will release latent heat restricting further undercooling, while evaporation of moisture films after sunrise will cool surfaces, reducing solar heating. As a result, it may take significant time for metal surfaces to dry after sunrise [35]. The rate of evaporation is clearly influenced by surface temperature, as discussed above. With clear skies, a modest wind (5 m/s) and not excessive humidity (75 %), a moisture film of 200 μm will have evaporated an hour after sunrise. Evaporation actually starts before sunrise as the minimum temperatures (under these conditions) generally occur an hour before sunrise. In contrast, on a cloudy, still day of high humidity (95 %), the same moisture film will still not have fully evaporated four hours after sunrise. The upper end of the distribution of wetted aerosols have dimensions greater than 100 μm and those will also take a number of hours to evaporate [35]. Rain drops evaporate in times a little longer than aerosols for modest winds but in still conditions rain drops may take 6–12 h to evaporate depending on the humidity [36].

Consistent with this mathematical analysis, Cole et al. [37] demonstrated from field studies that on wetness sensors can be divided in 3 modes: Mode A, Mode B and Mode C. Mode A are short daytime wetness events with an average duration of around 2 h (they result from either hygroscopic wetting of deposited salts or rainfall but are of short duration as daytime evaporation is rapid). Mode B are nighttime wetness events that occur when condensation occurs in the evening and evaporation the next morning. The duration of the wetness period varies depending on the severity of the location (and thus the nature of deposited salts) from an average of 10 h in benign (inland) sites to 20 h in severe marine locations. Mode C are wetness periods that last more than one day, they are generally associated with the deposition of salts with low DRH (e.g. $MgCl_2$) so that surface RH may not dry out the wetted salts on some days. Again the duration depends on the severity of the site raging from an average of 72 to 120 h in marine and severe marine sites.

3.6.5.3 Evaporation and Condensation on a Porous Surface

The above analysis refers to clean surfaces, however on a corroded surface or an aged paint the situation may be more complex. Firstly many oxides are themselves hygroscopic, and secondly many oxides and degraded films are porous and thus, when aerosols or raindrops are deposited onto such films, some of the solution will be absorbed into the pores and will subsequently evaporate, leaving contaminants within the pore structure. In these circumstances, capillary forces may alter the conditions for evaporation and condensation in the pores, making it possible for condensation to occur in an atmosphere of less than 100 % RH. These effects are best explored using the differential vapor pressure approach, where the relative lowering of the saturated vapor pressure in pores is given by Thomson's equation [38].

$$p = p_o e^{-2\sigma M/\rho RTr} \tag{3.11}$$

where p and p_o are the saturated vapor pressure above a concave meniscus of radius r and a plane surface respectively, σ is the surface tension of the liquid at absolute temperature T, ρ is its density and M is its molecular weight of the condensing gas, and R is the gas constant. It can be estimated that capillary radii of 360, 94 and 30 Å will reduce the relative humidity for condensation to 98, 90 and 70 % respectively. Correspondingly, the period that moisture remains in pores may be considerably greater than when it is free on the surface. For example, consider two plates exposed in Canberra, one with a perfectly flat surface and one with pores of 90 Å with both surfaces being free of contaminates. The percent time favourable for wetness a flat surface will be 4 % while on the surface with pore it will be 20 %.

3.6.6 Summary of Environmental Conditions

Pollutant deposition onto a surface may occur by gaseous absorption into a moisture film, deposition of aerosol or of rain. Gases, particularly reactive gases will not be transported far from their sources, aerosols may form relatively close to the pollutant source but may be transported significant distances, while the processes of rain drop formation are complex and will not bear a direct relation to pollutant sources. Thus, it is to be expected that the concentration of atmospheric gases will be highly influenced by local pollutant sources, aerosol concentration will also be influenced by local sources but concentrations will spread significantly from these sources while rain water chemistry will be affected more by the pollutants in a region rather than in one locality

In Table 3.5 some key parameters defining gaseous concentrations, aerosol and rain water chemistry are given for Marine, Industrial, Urban, Rural and Remote locations from data presented in this chapter. The table should only be taken as indicative and does not attempt to define the effect of particular local sources on pollutant levels.

Table 3.5 Summary of gaseous concentration, aerosol type and rain water composition (μ eq/l) for marine, industrial, urban, rural and remote locations

Location type	Marine	Industrial	Urban	Rural	Remote
Gaseous (mixing ratio)					
SO_2 (ppt)	260	1500	160–1500	160	20
H_2S (ppt)	65	365	365	35–60	3.6–7.5
NO_x (ppb)	0.2–1000	10–1000	10–1000	0.2–10	0.02–0.08
O_3 (ppb)	20–40	100–400	100–400	20–40	20–40
Aerosols					
pH	0–9.5	−1 to 2.4	1.9–3	–	–
Major species	NaCl, MgCl	H_2SO4, $NH_4)_3H$ $(SO_4)_2)$, $(NH4)_2SO_4$, NH_4HSO_4	NH_4NO_3, NH_4Cl $(NH_4)_2SO_4$	–	dust pollen plant waxes
Secondary species	Na_2SO_4, H_2SO_4, NH_4HSO_4, $(NH_4)_2SO_4$ and $NH_4NO_3.NaHSO_4$	–	$NaNO_3$	–	–
Rain water					
pH	4–5.6	4.2–7.3	4.4–6.1	3.6–5.8	5.6–6
Cl^-	100–1300	9–142	10–27	3–25	1–1300
NO_3^-	3–10	40–140	13–140	3–25	3–10
SO_4^{2-}	3–10	70–240	12–60	4–60	3–10
Na^+	100–1200	5–60	20–60	3–30	100–1200
NH_4^+	2–10	30–200	10–30	2–20	2–10
Ca^{2+}	–	20–300	2–35	2–20	80

The size and duration of moisture film formation will also depend on the source of the moisture. In the case of aerosol drops, these drops will only cover a small fraction of the exposed surface. For example, assuming that all drops have a uniform size consider two cases, a marine environment with the salinity of 100 mg/m^2 day and the aerosol drop diameter of 50 μm and a near coastal environment with a salinity of 10 mg/m^2 day and the aerosol drop diameter of 10 μm. After a night without raining aerosol droplets will occupy 2 and 1 % of the surface area of the exposed metal respectively. Thus aerosols can be treated as "isolated" on a surface. The surface coverage of rain drops however can be very high and depends of contact angles, surface angle etc., and there will be frequent coalescence between droplets so that droplet size will be limited by the maximum droplet size prior to drop run off (typically from 2–20 mm diameter depending on surface angle and difference between leading and following contact angle). Thus rain will deposit on a surface as a series of drops all below the critical size, in close proximity to each other. The critical drop size will increase as the surface ages and the contact angle is lowered.

The physical form and duration of the residency that deposited aerosols take on the surface will be very dependent on the type of deposition environment (i.e. whether it is a marine, industrial, urban, rural or remote location) as outlined in Table 3.6. For example in a marine environment a typical aerosol level will be 100 mg/m^2 day and the aerosols will range from 2–250 μm. If they deposit on a surface in day-light hours (Mode A) they will quickly evaporate (in a few minutes to a couple of hours). However if they deposit during the night they will remain during the night and then evaporate after sunrise (Mode B). For the larger aerosols on cool still days this evaporation could take some time while small aerosols will evaporate shortly after sunrise. However if the deposited aerosols are rich in $MgCl_2$ and it is a cloudy day so that high surface temperatures do not arise the aerosols may not evaporate during the day but may persist during the daylight hours, remain stable or grow at night and only evaporate on a clear day.

Table 3.6 Physical properties and typical duration of aerosols on surfaces

Location	Marine	Industrial	Urban	Rural	Remote
Typical aerosol level (mg/m^2 day)	100	15	10	5	3
Aerosol diameter (μm)	2–250	1–30	1–30	1–20	1–10
Time (hrs) to evaporate—Mode A	0.01–2	0.01–2	0.01–2	0.01–2	0.01–2
Mode B	10–20	10–14	10–14	10–14	10–14
Mode C	24–120	–	–	–	–
% coverage	1–8	4–18	2–12	2–6	2–6

3.7 Relevance of Environmental Conditions to Moisture Ingress and Inhibitor Stability

In discussing moisture ingress into films, two questions are important, what is the maximum possible volume within the paint film that could be occupied by the drop that was initially deposited on the surface and how long would it take for this state to arise.

To the author's knowledge, no analysis has been undertake of the residual drop volume that remains on a surface as a drop diffuses into the pores of a polymer film. It is likely that due to surface tension effects, there will be some residual surface moisture, however let it be assumed that all the droplet volume diffuses into the film. Then the wetted pore volume should equal the initial drop volume. Mardel et al. [39] found that pore volume in a two component epoxy (epoxy Epikote 828 with amine cross linker) was 2.2 wt% or given that the epoxy had a density of 1.4 then the pore volume was 3 % volume percent. This implies that the wetted area will occupy approximated 33 times the volume of the original drop. If it is assumed that the moisture diffuses into the film at a constant rate in all directions then the penetration in any direction will be approximately 2.9 that of the original dimension of the drop. Under these assumptions a Mode A aerosol that formed a hemispherical drop of radius 5 μm on the surface would form a hemispherical drop of radius of approximately 15 μm below the surface. Now this analysis needs to be undertaken in a more detailed way but it illustrates that smaller aerosols typical of non-marine locations may not lead to thorough wetting of a paint film while the absorption of larger aerosols typical of marine locations and rain droplets may.

The second issue is whether a droplet on a surface will reside for a sufficient time to diffuse into the surface before it is removed from that surface (by evaporation or roll off etc.). In terms of water diffusion and uptake through a polymeric film, Prosek and Thierry [40] suggested that the time for the solution to reach the metal surface through a coating and return to the external surface can be calculated using the equation [12]:

$$d = 2l = (Dt)^{0.5} \tag{3.12}$$

where d is twice the film thickness l, D is the diffusion coefficient, and t the time.

Diffusion coefficients of water in epoxy have been reported in the range of 1×10^{-9} to 4×10^{-8} cm^2/s by Philippe et al. [41], and 2.8×10^{-8} cm^2/s by Mardel [39] The diffusion of chloride is taken from Hu et al. [42] as 4.7×10^{-12} cm^2/s. In Table 3.7 the depth of penetration of moisture and chloride is calculated assuming Fick's law. If the coating is of the order of 80 μm, water will penetrate though the coating in around 60–120 min (assuming the two highest diffusion constants), although it will only penetrate around 20–30 μm in the same time assuming the lowest diffusion coefficient. Thus penetration of moisture through a paint film will only occur for droplets which persists for a number of hours. This implies that large aerosols deposited at night may penetrate through the coating but those deposited

Table 3.7 Possible diffusion depths for given droplet lifetimes and diffusion constants

		Water			Chloride
	Droplet lifetime	1.33×10^{-9}	9×10^{-9}	2.8×10^{-8}	4.68×10^{-12}
Mode A	1 min	2.8	7.3	13	0.2
	1 h	22	57	100	1.3
	2 h	31	81	142	1.8
Mode B	10 h	70	180	320	4.1
	14 h	82	213	375	4.9
	20 h	98	254	449	5.8
Mode C	24 h	107	278	491	6.4
	120 h	240	623	1100	14.2

during the day may not. Interestingly the diffusion of chloride will be only into the skin of the paint film even for the longest wetting periods reported in this paper.

Of course the data indicates that there are repeated cycles of wetting and drying which raises interesting questions on how moisture and salt migrate into paint films with repeated cycling. Conceptually there are three possibilities:

(1) The diffusing species enters in the wet cycle and completely exits in the dry cycle with the properties of the paint film remaining constant.
(2) The diffusing species enters in the wet cycle and completely exits in the dry cycle with the properties of the paint film change.
(3) The diffusing species enters in the wet cycle but does not completely exit in the dry cycle.

If possibility 1 were correct it would imply that chloride would not migrate through intact paint films and thus corrosion of intact paint films would be limited. However, studies have shown that the ingress of moisture into a polymer film may cause permanent changes to the film creating voids and additional moisture pathways so that subsequent moisture ingress can be more rapid [40]. Thus it is likely that with repeated cycling the properties of the polymer film may change. No studies have been undertaken (to the author's knowledge) that looked at moisture or salt retention with repeated cycling. This is certainly an area worthy of research and if it occurs would imply that chloride could incrementally migrate through a paint film and thus promote corrosion in an intact film.

As mentioned previously and outlined in Table 3.5, both industrial aerosols and marine aerosols may be acidified. There does tend to be a correlation between aerosol pH and size with the smaller aerosols having a lower pH. Rain drops will have a larger size and higher pH than aerosols. Thus at least three possible scenarios may arise

(1) Relatively small and acidic aerosols will be deposited onto a dry coating. The aerosol will only penetrate into the top of the coating where it may dissolve additives to the coating.

(2) Large aerosols and rain droplets may deposit onto the dry coating. Their size and residency periods will permit through penetration of the coating. The moderate acidity of the coating may assist the dissolution of additives. Through moisture penetration will permit electrochemical cells to be established that may also alter the pH of the pore water in the coating.
(3) If aerosols and rain droplets are deposited onto a wet coating then they may alter the pH of the pore water but the pH may have already have been affected by the establishment of electrochemical cells prior to the deposition.

3.8 Damage to Coatings

Environmental factors, both those mentioned in this paper (acid deposition, temperature, cyclic wetting) and those not discussed (UV), can cause significant degradation to coatings and clearly an understanding of their degradation is critical in coating design. This is a subject that has been treated by a number of substantive reviews and the reader is urged to consult [43–45].

3.9 Summary

This chapter has outlined atmospheric factors that may affect the function of smart coatings. The emphasis has been on the factors that affect the intensity, lifetime and chemistry of aerosols and rain drops that may deposit on coatings.

The paper outlined the source and chemistry of marine and industrial aerosols and how gaseous absorption can acidify both marine and industrial aerosols. Gaseous absorption is influenced not only by the rate that gases are absorbed into moisture films but by secondary reactions that transformed the absorbed species and allow absorption to continue. Thus aerosols and rain drops become enriched in sulfates via SO_2 absorption and transformation despite the fact that the Henry law coefficient that measure the rate of SO_2 absorption is relatively low. Aerosols may be transported significant distances from their sources and during this transport their chemistry may be altered and thus deposition of acidified aerosol may occur at a significant distance from the pollutant source. Similarly rain drops may be acidified through the acidification of cloud condensation nuclei and gaseous absorption during rainfall so that one again acidified rain may occur at some distance from pollutant sources. Both acid rain and deposition of acidified pH are major sources of the flux of acidic species onto a surface.

A surface may wet both from the deposition of rain and aerosols and from the rewetting of pollutants that had previously been deposited. Of particular importance is the deposition and retention of hygrosopic salts. Many aerosols and rainwater will leave behind such salts when they evaporate. Of particular importance are the ammonium sulfates, magnesium and sodium chlorides and sodium and ammonium nitrates. These salts will wet when the relative humidity at the coating surface exceeds that of the deliquescent RH of the salt (which may vary from 35 % for $MgCl_2$ to 84 % for Na_2SO_4). The surface RH depends both on the surface temperature of the coating and the ambient RH. Ambient RH and surface temperature will depend very strongly on the diurnal or daily climate cycle, with a typical Australian cycle being high RH and low surface temperatures at night and low RH and high surface temperatures during the day. Thus deposited rain drops and aerosols will tend to evaporate after sun-rise. Thus three modes of droplet duration on a surface associated with aerosol deposition.

(1) Mode A—Aerosols deposited during the day will generally evaporate quickly (minutes to a couple of hours)
(2) Mode B—Aerosols deposited or salts that rewet during the night will not evaporate to after sunrise (total life from 8–14 h)
(3) Mode C—Deposition of aerosols rich in $MgCl_2$ or rewetting of $MgCl_2$ salts may lead to drop lifetimes in excess of 24 h as daytime surface RH may not become lower than the DRH of $MgCl_2$

The different surface droplet modes and similar factors for rain water control the form that wetting of a coating may take and together with the chemistry of the aerosol or rain droplet this may affect the impact of the wetting event on the coating. Again three scenarios may occur:

(1) Relatively small and acidic aerosols will be deposited onto a dry coating. The aerosol will only penetrate into the top of the coating where it may dissolve additives to the coating.
(2) Large aerosols and rain droplets may deposit onto the dry coating. Their size and residency periods will permit through penetration of the coating. The moderate acidity of the coating may assist the dissolution of additives. Through moisture penetration will permit electrochemical cells to be established that may also alter the pH of the pore water in the coating.
(3) If aerosols and rain droplets are deposited onto a wet coating they may alter the pH of the pore water but the pH may have already have been affected by the establishment of electrochemical cells pore to the deposition.

References

1. I.S. Cole, N.S. Azmat, A. Kanta, M. Venkatraman, What really controls the atmospheric corrosion of zinc? Effect of marine aerosols on atmospheric corrosion of zinc. Int. Mater. Rev. **54**, 117–133 (2009)
2. J. Seinfeld, S. Pandis, *Atmospheric Chemistry and Physics: From Air Pollution to Climate Change* (Wiley Interscience, New York, 1997)

3. W.A. McKay, J.A. Garland, D. Livesley, C.M. Halliwell, M.I. Walker, The characteristics of the shore-line sea spray aerosol and the landward transfer of radionuclides discharged to coastal sea-water. Atmos. Environ. **28**, 3299–3309 (1994)
4. J.W. Fitzgerald, Marine aerosols - a review. Atmos. Environ. Part A-General Topics **25**, 533–545 (1991)
5. X. Yao, T.Y. Ling, M. Fang, C.K. Chan, Size dependence of in situ pH in submicron atmospheric particles in Hong Kong. Atmos. Environ. **41**, 382–393 (2007)
6. M. Takeuchi, H. Okochi, M. Igawa, Characteristics of water-soluble components of atmospheric aerosols in Yokohama and Mt. Oyama, Japan from 1990 to 2001, Atmos. Environ., **38,** 4701–4708 (2004)
7. S.N. Pandis, J.H. Seinfeld, Sensitivity analysis of a chemical mechanism for aqueous-phase atmospheric chemistry. J. Geophys. Res.-Atmos. **94**, 1105–1126 (1989)
8. R. Vogt, P.J. Crutzen, R. Sander, A mechanism for halogen release from sea-salt aerosol in the remote marine boundary layer. Nature **383**, 327–330 (1996)
9. M.R. Hoffmann, J.G. Calvert, in Chemical Transformation Modules for Eulerian Acid Deposition Models: Vol.2 - The Aqueous Phase Chemistry, (US Environmental Protection Agency Reserach Triangle Park NC, 1985)
10. J.E. Svensson, L.G. Johansson, A laboratory study of the effect of ozone nitrogen-dioxide, and sulfur-dioxide on the atmospheric corrosion of zinc. J. Electrochem. Soc. **140**, 2210–2216 (1993)
11. C. Pilinis, J.H. Seinfeld, Continued development of a general equilibrium-model for inorganic multicomponent atmospheric aerosols. Atmos. Environ. **21**, 2453–2466 (1987)
12. S.L. Clegg, P. Brimblecombe, A.S. Wexler, Thermodynamic model of the system H + -NH4 + -SO42–NO3–H2O at tropospheric temperatures. J. Phys. Chem. A **102**, 2137–2154 (1998)
13. W.C. Keene, R. Sander, A.A.P. Pszenny, R. Vogt, P.J. Crutzen, J.N. Galloway, Aerosol pH in the marine boundary layer: A review and model evaluation. J. Aerosol Sci. **29**, 339–356 (1998)
14. R. Sander, P.J. Crutzen, Model study indicating halogen activation and ozone destruction in polluted air masses transported to the sea. J. Geophys. Res.-Atmos. **101**, 9121–9138 (1996)
15. W.C. Keene, H. Maring, J.R. Maben, D.J. Kieber, A.A.P. Pszenny, E.E. Dahl, M.A. Izaguirre, A.J. Davis, M.S. Long, X.L. Zhou, L. Smoydzin, R. Sander, Chemical and physical characteristics of nascent aerosols produced by bursting bubbles at a model air-sea interface, J. Geophys. Res.-Atmos. 112 (2007)
16. S.E. Schwartz, Acid deposition - unraveling a regional phenomenon. Science **243**, 753–763 (1989)
17. D. Post, H.A. Bridgman, G.P. Ayers, Fog and rainwater composition in rural se australia. J. Atmos. Chem. **13**, 83–95 (1991)
18. I.S. Cole, D.A. Paterson, W.D. Ganther, Holistic model for atmospheric corrosion - Part 1-Theoretical framework for production, transportation and deposition of marine salts. Corros. Eng. Sci. Technol. **38**, 129–134 (2003)
19. W.L. Chameides, A.W. Stelson, Aqueous-phase chemical processes in deliquescent sea-salt aerosols–a mechanism that couples the atmospheric cycles of s and sea salt. J. Geophys. Res.-Atmos. **97**, 20565–20580 (1992)
20. I.S. Cole, W.D. Ganther, D. Lau, Field studies of surface cleaning and salt retention on openly exposed metal plates. Corros. Eng. Sci. Technol. **41**, 310–320 (2006)
21. I.S. Cole, W.D. Ganther, J.D. Sinclair, D. Lau, D.A. Paterson, A study of the wetting of metal surfaces in order to understand the processes controlling atmospheric corrosion. J. Electrochem. Soc. **151**, B627–B635 (2004)
22. I.S. Cole, D. Lau, F. Chan, D.A. Paterson, Experimental studies of salts removal from metal surfaces by wind and rain. Corros. Eng. Sci. Technol. **39**, 333–338 (2004)
23. J.W. Spence, F.H. Haynie, F.W. Lipfert, S.D. Cramer, L.G. McDonald, Atmospheric corrosion model for galvanized steel structures. Corrosion **48**, 1009–1019 (1992)

24. S. Cole I., D.A. Paterson, D. Lau, in *Holistic Modeling of Gas and Aerosol Deposition and the degradation of Cultural Objects*, Physical Techniques in the Study of Art, Archeology and Cultural Heritage (Elsevier, 2007)
25. I.S. Cole, D.A. Paterson, Holistic model for atmospheric corrosion - Part 5–Factors controlling deposition of salt aerosol on candles, plates and buildings. Corros. Eng. Sci. Technol. **39**, 125–130 (2004)
26. I.S. Cole, Mechanisms of atmospheric corrosion in tropical environments. Am. Soc. Test Mater. 33–47 (2000)
27. C. Leygraf, T.E. graedel, Atmospheric Corrosion (Wiley, New York, 2000)
28. D.J. Jacob, J.M. Waldman, J.W. Munger, M.R. Hoffmann, Chemical-composition of fogwater collected along the california coast. Environ. Sci. Technol. **19**, 730–736 (1985)
29. G.A. Isaac, P.H. Daum, A winter study of air, cloud and precipitation chemistry in Ontario, Canada. Atmos. Environ. **21**, 1587–1600 (1987)
30. M.D. Keywood, A.R. Chivas, L.K. Fifield, R.G. Cresswell, G.P. Ayers, The accession of chloride to the western half of the Australian continent. Aust. J. Soil Res. **35**, 1177–1189 (1997)
31. I.S. Cole, D.A. Paterson, Holistic model for atmospheric corrosion - Part 7 - Cleaning of salt from metal surfaces. Corros. Eng. Sci. Technol. **42**, 106–111 (2007)
32. T.H. Muster, I.S. Cole, Attachment efficiencies of salt aerosols onto infrastructure and implications for atmospheric corrosion. J. Electrochem. Soc. **152**, B125–B131 (2005)
45. I.N. Tang, On the equilibrium partial pressures of nitric-acid and ammonia in the atmosphere. Atmos. Environ. **14**, 819–828 (1980)
34. I.N. Tang, H.R. Munkelwitz, Composition and temperature-dependence of the deliquescence properties of hygroscopic aerosols. Atmos. Environ. Part A-General Topics **27**, 467–473 (1993)
35. I.S. Cole, D.A. Paterson, Mathematical models of dependence of surface temperatures of exposed metal plates on environmental parameters. Corros. Eng. Sci. Technol. **41**, 67–76 (2006)
36. I.S. Cole, R. Holgate, P. Kao, W. Ganther, The rate of drying of moisture from a metal-surface and its implication for time of wetness. Corros. Sci. **37**, 455–465 (1995)
37. I.S. Cole, W.D. Ganther, Experimental determination of time taken for openly exposed metal surfaces to dry. Corros. Eng. Sci. Technol. **41**, 161–167 (2006)
38. W. Thomson, Proc. Roy. Soc. Edinburgh (1928)
39. J. Mardel, S.J. Garcia, P.A. Corrigan, T. Markley, A.E. Hughes, T.H. Muster, D. Lau, T.G. Harvey, A.M. Glenn, P.A. White, S.G. Hardin, C. Luo, X. Zhou, G.E. Thompson, J.M.C. Mol, The characterisation and performance of $Ce(dbp)_3$-inhibited epoxy coatings. Prog. Org. Coat. **70**, 91–101 (2011)
40. T. Prosek, D. Thierry, A model for the release of chromate from organic coatings. Prog. Org. Coat. **49**, 209–217 (2004)
41. L. Philippe, C. Sammon, S.B. Lyon, J. Yarwood, An FTIR/ATR in situ study of sorption and transport in corrosion protective organic coatings–1. Water sorption and the role of inhibitor anions. Prog. Org. Coat. **49**, 302–314 (2004)
42. J.M. Hu, J.Q. Zhang, C.N. Cao, Determination of water uptake and diffusion of Cl- ion in epoxy primer on aluminum alloys in NaCl solution by electrochemical impedance spectroscopy. Prog. Org. Coat. **46**, 273–279 (2003)
43. G.P. Bierwagen, The science of durability of organic coatings - a foreword. Prog. Org. Coat. **15**, 179–195 (1987)
44. P.A. Sorensen, S. Kiil, K. Dam-Johansen, C.E. Weinell, Anticorrosive coatings: a review. J. Coat. Technol. Res. **6**, 135–176 (2009)
45. P.S. Sidky, M.G. Hocking, Review of inorganic coatings and coating processes for reducing wear and corrosion. Br. Corros. J. **34**, 171–183 (1999)

Chapter 4
Corrosion Inhibitors

F. Andreatta and L. Fedrizzi

4.1 Classification of Corrosion Inhibitors

4.1.1 Definition of Corrosion Inhibitor

A corrosion inhibitor is a chemical compound that is employed to reduce the corrosion rate of a metal exposed to an aggressive environment. The extent of the reduction of corrosion rate relative to the initial corrosion rate is indicated as inhibitor efficiency, which can be expressed as follows:

$$R_I = \frac{v_0 - v}{v_0}$$

where: R_I is the inhibitor efficiency, v_0 is the initial corrosion rate and v is the corrosion rate after the addition of the inhibitor.

The inhibitor efficiency depends on different parameters such as:

- Type of inhibitor and its concentration
- Aggressiveness of the environment
- Substrate

The classification of corrosion inhibitors can be done according to [1, 2] (Fig. 4.1):

- Effect of the inhibitor on partial electrochemical reactions
  - Anodic inhibitors
  - Cathodic inhibitors
  - Mixed inhibitors

F. Andreatta (✉) · L. Fedrizzi
Dipartimento Politecnico di Ingegneria e Architettura, Università di Udine,
Via del Cotonificio 108, Udine 33100, Italy
e-mail: francesco.andreatta@uniud.it

A.E. Hughes et al. (eds.), *Active Protective Coatings*,
Springer Series in Materials Science 233, DOI 10.1007/978-94-017-7540-3_4

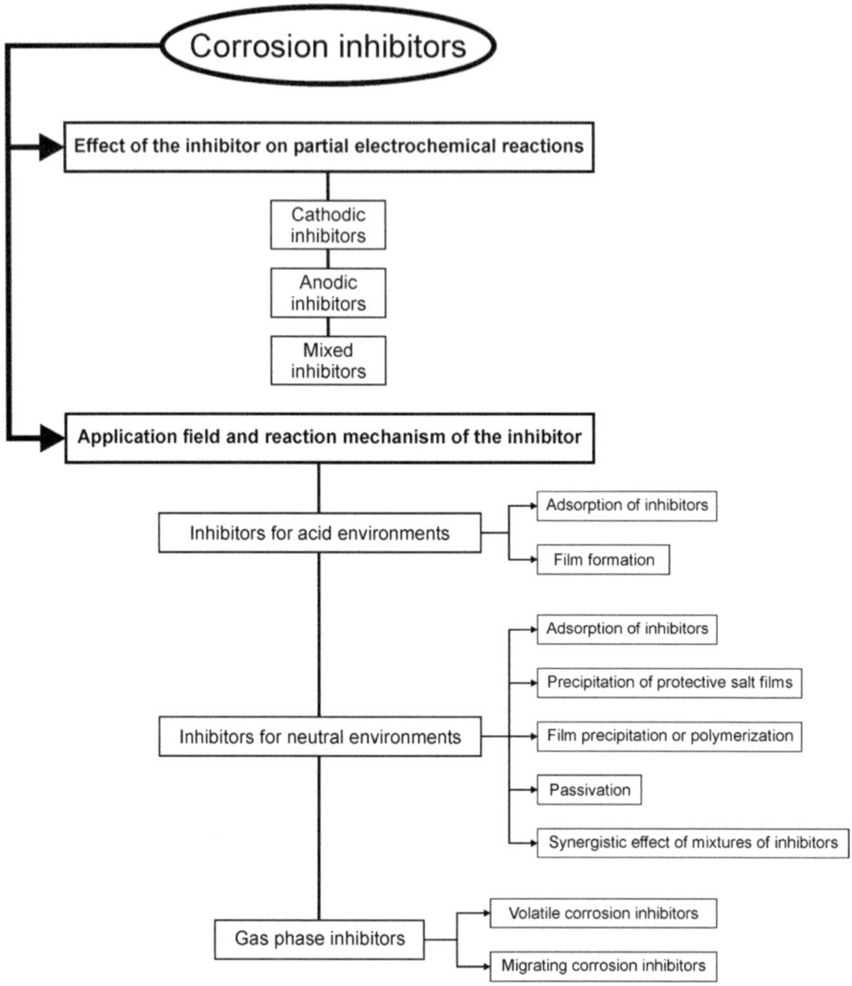

Fig. 4.1 Classification of corrosion inhibitors

- Application field
  - Inhibitors for acid environments
  - Inhibitors for neutral environments
  - Gas phase inhibitors
- Reaction mechanism of the inhibitor.

4.2 Effect of the Inhibitor on Partial Electrochemical Reactions

An inhibitor can be classified in terms of the mixed-potential theory considering its effect on partial oxidation and reduction reactions involved in the corrosion process. The mixed potential theory is based on the assumption that each electrochemical process can be divided in two or more partial oxidation and reduction reactions. Accordingly, the interaction of an inhibitor with a corroding system might affect only the cathodic or the anodic reactions. In addition, some inhibitors might have a combined effect on both anodic and cathodic reactions. Based on the effect on partial electrochemical reactions, corrosion inhibitors are classified as follows:

- Cathodic inhibitors
- Anodic inhibitors
- Mixed inhibitors.

4.2.1 Cathodic Inhibitors

In acid environments, the cathodic reaction is the reduction of hydrogen ions to hydrogen atoms, which combine forming molecular hydrogen (hydrogen evolution):

$$2H^{+} + 2e^{-} \leftrightarrow H_2$$

In neutral or alkaline environment, the cathodic reaction is oxygen reduction:

$$O_2 + 2H_2O + 4e^{-} \leftrightarrow 4OH^{-}$$

Cathodic inhibitors affect the partial reduction reactions reducing their current density with no or very limited effect on the anodic reaction. The decrease of the cathodic current density leads to a shift of the corrosion potential in the negative direction. The inhibition effect is associated with the decrease of the net corrosion current density, as can be seen in the Evans diagram shown in Fig. 4.2.

4.2.2 Anodic Inhibitors

In the case of a corroding metal, the oxidation reaction is given by:

$$M \leftrightarrow M^{n+} + ne^{-}$$

Anodic inhibitors reduce the current density of the partial oxidation reactions without affecting the partial reduction reactions. This is associated to the shift of the

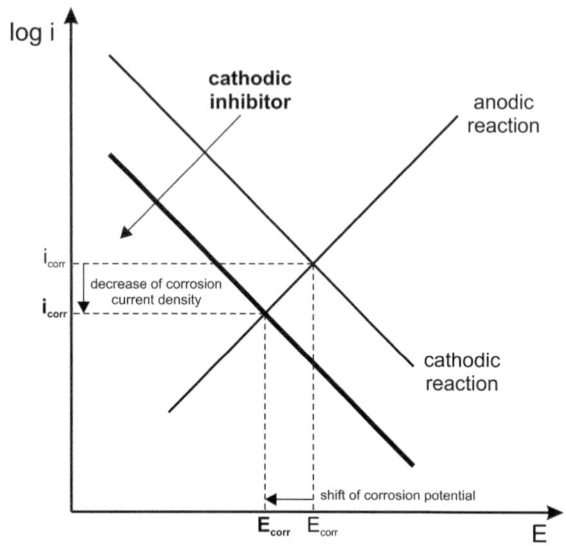

Fig. 4.2 Effect of a cathodic inhibitor on partial electrochemical reactions in the Evans diagram

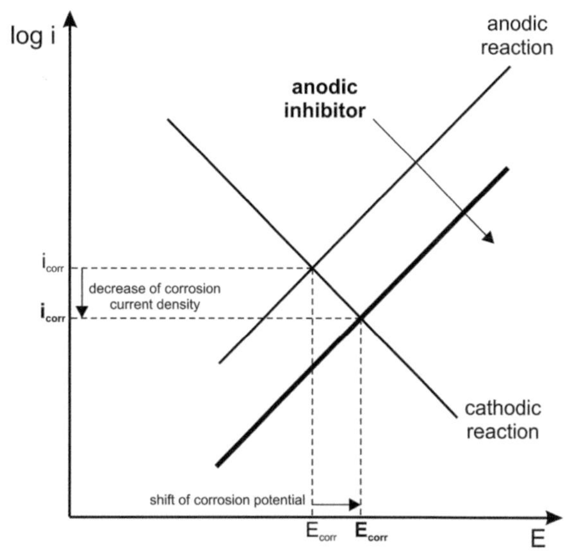

Fig. 4.3 Effect of an anodic inhibitor on partial electrochemical reactions in the Evans diagram

corrosion potential in the positive direction and to the reduction of the net corrosion current density (Fig. 4.3).

4.2.3 Mixed Inhibitors

Mixed inhibitors decrease the current density of both cathodic and anodic reactions (Fig. 4.4). In this case, the current density is decreased but the effect on the

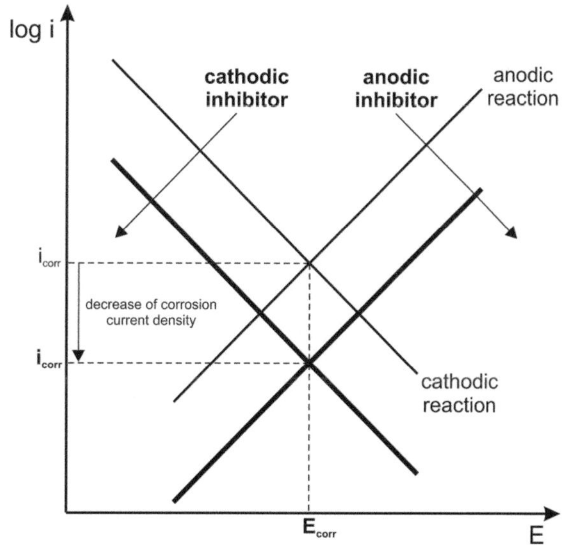

Fig. 4.4 Effect of a mixed inhibitor on partial electrochemical reactions in the Evans diagram

corrosion potential depends on the relative extent of the partial oxidation and reduction reactions. It can be expected that the effect on the corrosion potential is limited if the current reduction due to the inhibitor is similar for the anodic and cathodic reactions, as represented in Fig. 4.4. In contrast, the corrosion potential might shift in the negative direction if the effect of the inhibitor on the cathodic reaction is stronger than on the anodic direction (Fig. 4.5a). Similarly, the corrosion potential might shift in the positive direction if the effect on the anodic reaction is prevailing on that on the cathodic reaction (Fig. 4.5b).

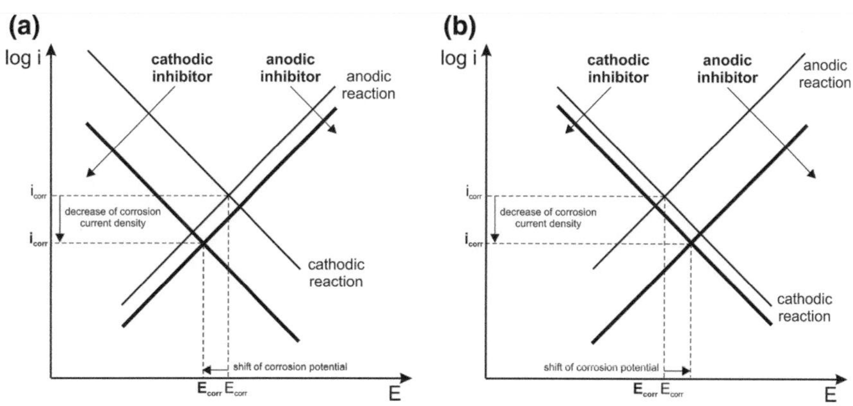

Fig. 4.5 Effect of a mixed inhibitor with prevailing effect on one of the partial electrochemical reactions in the Evans diagram

4.3 Inhibition Mechanisms

4.3.1 Inhibitors for Acid Environments

4.3.1.1 Adsorption of Inhibitors

Inhibitors employed in acid environments are usually organic compounds. These are often aromatic compounds or macromolecules with linear or branched structure. The inhibition effect is based on the ability to establish a link between the organic compound and the metal substrate, which is usually oxide-free in acid environments. The adsorption might be the result of the electrostatic interaction between ionic charges or dipoles and the electrically charged metal surface at the metal/electrolyte interface. Besides, the adsorption can be due to the formation of links between the inhibitor and the metal surface by electron transfer to the metal, usually in the case of organic molecules with lone pair electrons, multiple (double or triple) covalent bonds, aromatic rings. In the case of adsorption of organic molecules onto the metal, the working principle of inhibitors employed in acid environments is due to their structure, which contains a hydrophobic non polar group and a hydrophilic functional group. The functional group bonds to the metal substrate, while the hydrophobic group limits the access of water to the metal substrate. As a result, the electrochemical active area of the metal substrate is reduced resulting in the inhibition of corrosion processes. The adsorbed inhibitor might affect the cathodic reaction, the anodic one or present a synergetic effect on both reactions. Typical hydrophilic functions are amine, mercapto, hydroxyl, carboxylic and phosphate groups. These functional groups enable the inhibitor dissolution in solution along with the bonding ability to the substrate. The inhibitor efficiency is essentially affected by:

- strength of the bond between the functional group and the metal substrate
- steric behavior of the non polar hydrophobic group
- inhibitor concentration

Figure 4.6a shows experimental potentiodynamic polarization curves of carbon steel in 1 M HCl and with addition of furan derivatives: 2-methylfuran, furfuryl alcohol and furfurylamine) as reported in Ref. [3]. The furan derivatives can adsorb on the metal surface due to the lone electron pair available in the oxygen heteroatom of the furan group, as can be seen in the structure of the inhibitor molecules reported in Fig. 4.6a. In addition, the adsorption might be strengthened by the existence of additional donor atoms such as oxygen and nitrogen in the functional group. As a result, coordinate covalent bonds can be formed between the inhibitor molecules and the metal surface. The potentiodynamic polarization curves in 1 M HCl with addition of the three different furan derivates exhibit a significant decrease of the corrosion current density relative to the blank substrate (Fig. 4.6a). Moreover, this is associated to a marked reduction of the cathodic and anodic current density. The effect of the adsorbed inhibitor on the cathodic and anodic Tafel slopes is

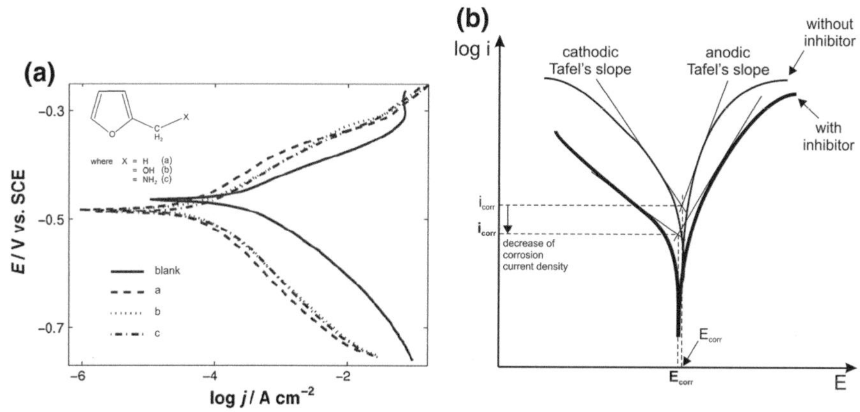

Fig. 4.6 Potentiodynamic polarization curves (potential vs. current density) of carbon steel (*blank*) in 1 M HCl solution and with addition of furan derivatives: 2-methylfuran (*curve a*), furfuryl alcohol (*curve b*) and furfurylamine (*curve c*) [3] (**a**); schematic representation of the effect of the furfuryl alcohol inhibitor on potentiodynamic polarization curves (current density vs. potential) (**b**)

schematically represented in Fig. 4.6b. It should be noted that Fig. 4.6a reports the potential versus the current density while Fig. 4.6b displays the current density versus the potential in order to follow the schematic representation of the effect of inhibitor in the Evans plots given above. The current density in the anodic and cathodic branch of the polarization curve is strongly decreased by the inhibitor. Moreover, cathodic and anodic Tafel's slopes significantly decrease due to the adsorption of the inhibitor. This indicates that furan derivatives behave as mixed inhibitors affecting both cathodic (hydrogen evolution) and anodic (Fe dissolution) reactions. The adsorption of the furan derivatives on the metal surface strongly decrease the corrosion current density indicating that the inhibitor efficiency is rather high. Moreover, it can be observed that inhibition adsorption has a limited effect on the corrosion potential in line with the behavior discussed above for mixed inhibitors (Figs. 4.4 and 4.5). The inhibitor efficiency calculated from potentiodynamic polarization curves in Fig. 4.6a is 84.7 % for 2-methylfuran (curve a), 92.55 for furfuryl alcohol (curve b) and 85.8 % for furfurylamine (curve c). The different inhibitor efficiency observed for the furan derivatives is related to the structural and electronic properties of the compounds. In particular, the functional group in the inhibitor molecule affects the adsorption on the substrate. Furfuryl alcohol and furfurylamine display a higher tendency to adsorb on the metal substrate than 2-methylfuran due to the presence of lone pairs of electrons in the oxygen atom and in the nitrogen atom of the functional group, respectively. In the case of furfuryl alcohol, the adsorption is further favored because this inhibitor tends to polymerize in acid solution improving the adsorption on the substrate [3].

4.3.1.2 Film Formation

Inhibition can be provided by precipitation of a protective film due to the reaction between the corrosion inhibitor and the dissolving metal. This leads to the formation of complexes on the metal substrate limiting further dissolution of the metal. The protective film formed on the metal surface acts as a physical barrier for the diffusion of aggressive species to the surface. Figures 4.7 and 4.8 display two examples of the effect of film formation induced on carbon steel by addition of decylamides of α-amino acid derivatives to an acid electrolyte, as reported in Ref.

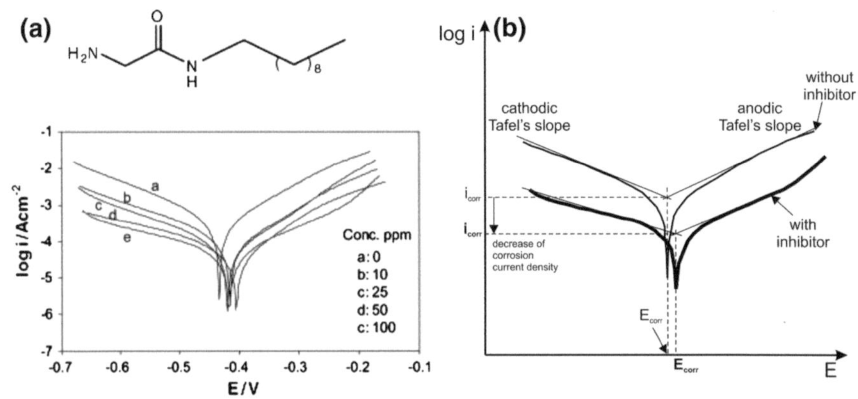

Fig. 4.7 Potentiodynamic polarization curves of carbon steel in 1 M HCl solution and with addition of different concentrations of decylamide of glycine (2-amino-N-decyl-acetamide) [4] (**a**); schematic representation of the effect of the inhibitor on potentiodynamic polarization curves (**b**)

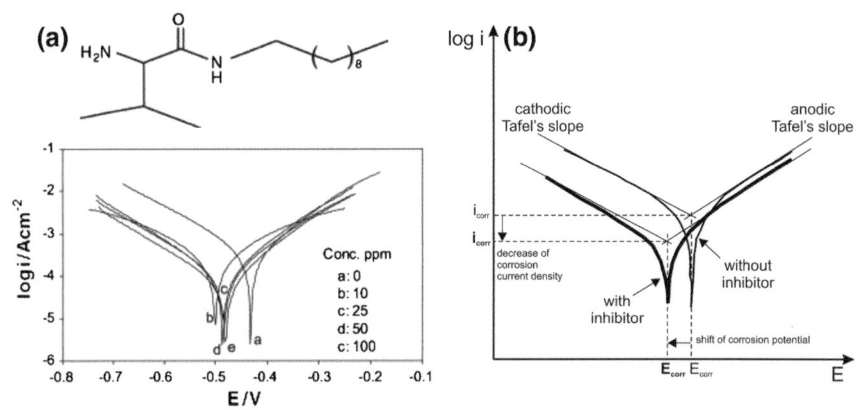

Fig. 4.8 Potentiodynamic polarization curves of carbon steel in 1 M HCl solution and with addition of different concentrations of decylamide of glycine (2-amino-N-decyl-3-methyl-butylamide) [4] (**a**); schematic representation of the effect of the inhibitor on potentiodynamic polarization curves (**b**)

[4]. Figure 4.7a shows the potentiodynamic polarization curves of carbon steel in 1 M HCl solution and in the same electrolyte with addition of different concentrations of decylamide of glycine (the structure of the molecule is reported in the figure). The addition of the inhibitor induces a decrease of anodic and cathodic current density associated to a strong decrease of the corrosion current density. This behaviour indicates that film formation on the metal substrate due to absorption phenomena tends to block the reaction sites strongly reducing cathodic and anodic processes. This is schematically represented in Fig. 4.7b. Moreover, it should be noted that the Tafels slopes of the polarization curves remain unchanged after the addition of the inhibitor suggesting that the reaction mechanism is not modified by the inhibitor. It can be considered that the decylamide of glycine behaves as a mixed inhibitor. However, it can be seen in Fig. 4.7b that there is a small shift of the corrosion potential in the positive direction after the addition of the inhibitor. This suggests that this inhibitor has a stronger effect on the anodic reactions. The filming behavior of the decylamide of glycine is due to the chemical bonding between the inhibitor and the metal substrate resulting in adsorption and film formation. Adsorption of the inhibitor takes place most likely through the ability of the nitrogen in the amino group of sharing lone pair electrons with the substrate. In addition, Fig. 4.7a clearly shows that the inhibitor efficiency increases with the concentration of decylamide of glycine in the acid solution. This indicates that the barrier effect observed for the film deposited on the substrate becomes stronger increasing the concentration of the inhibitor. The barrier effect of the film due to inhibitor adsorption is reported also for other inhibitors like triazole derivatives in acid environments [5].

Figure 4.8a shows the potentiodynamic polarization curves reported in Ref. [4] for carbon steel in 1 M HCl solution with addition of different concentrations of decylamide of valine (the structure of the molecule is reported in the figure). The decylamide of valine has an additional methyl group, as compared to the decylamide of glycine. The potentiodynamic polarization curves in Fig. 4.8a exhibit a marked reduction of the cathodic current density after the addition of the inhibitor. In contrast, the anodic current density is not significantly affected by the addition of the inhibitor. Moreover, the electrochemical behavior seems less affected by the inhibitor concentration for the decylamide of valine than for the decylamide of glycine. As can be seen in Fig. 4.8b, the cathodic and anodic Tafel slopes of the cathodic and anodic reactions do not change after the addition of the inhibitor, as already observed in Fig. 4.7b. The corrosion potential is shifted in the negative direction by addition of decylamide of valine to the 1 M HCl solution. The reduction of the cathodic current density and the shift of the corrosion potential in the negative direction is a clear indication that the decylamide of valine is a cathodic inhibitor, in accordance with the behavior discussed in Fig. 4.2. In this case, it can be expected that the barrier effect due to the film formation after the addition of the inhibitor is more efficient at the cathodic sites than at the anodic ones. The different electrochemical behavior observed for the decylamide of valine (Fig. 4.8a) than for the decylamide of glycine (Fig. 4.7a) is probably related to the different molecular structure of the inhibitors. The decylamide of glycine exhibits a linear structure (Fig. 4.7a), which favours adsorption phenomena on the metal substrate. In contrast, the decylamide of valine presents a methyl group in the α position hindering the formation of linear arrangements. This might reduce the adsorption of the inhibitor on the metal substrate with respect to the decylamide of glycine.

4.3.2 *Inhibitors for Neutral Environments*

4.3.2.1 Adsorption of Inhibitors

Inhibitors for neutral environments can provide corrosion protection with the adsorption mechanism discussed above for acid solutions. However, it should be taken into account that in neutral environments the cathodic reaction is usually oxygen reduction whereas in acid solutions the main cathodic reaction is hydrogen evolution. Moreover, in neutral environments the metals are usually covered by an oxide layer. This oxide layer is often not stable in acid environments in the case of many metals. Benzoates, tartrates and salicylate are typical inhibitors working with an adsorption mechanism in neutral environments. The adsorption of these inhibitors prevents the access of the electrolyte to parts of the metal surface blocking the corrosion processes. In the case of mild steel, it is reported that the stabilization effect of the native oxide layer is the main effect of adsorbed benzoates leading to the inhibition of the anodic reaction [6–10]. This is related to the formation of complexes between the benzoates and the products of the anodic reaction (Fe idroxides), which are adsorbed on the metal surface hindering further Fe dissolution

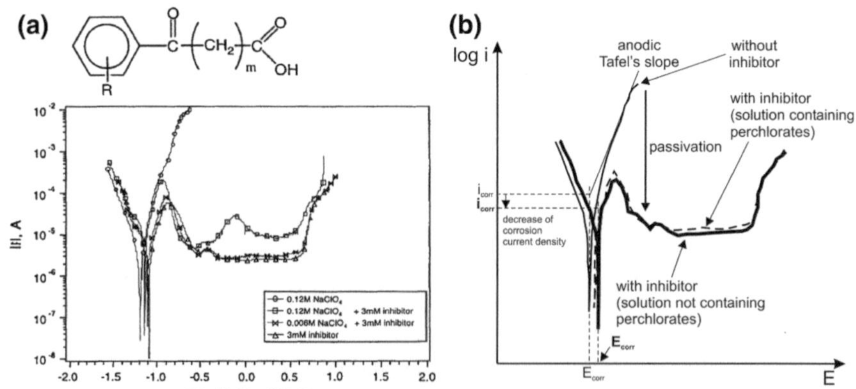

Fig. 4.9 Potentiodynamic polarization curves of low carbon steel (rotating disc electrode) acquired in 3 mM solution of an aromatic carboxylic acid inhibitor and in $NaClO_4$ solution with and without inhibitor [11] (**a**) and schematic representation of the effect of the inhibitor on potentiodynamic polarization curves (**b**)

[6]. As an example, Fig. 4.9a displays the effect of an aromatic carboxylic acid based inhibitor on potentiodynamic polarization curves of mild steel in neutral environment (pH = 7.5), as reported in Ref. [11]. The cathodic branch of the potentiodynamic polarization curve exhibits the same behavior in perchlorate solution, in the inhibitor solution and in the mixed solution containing the perchlorate and the inhibitor. This indicates that the carboxylic acid based inhibitor does not affect the cathodic reaction, which is oxygen reduction in the three electrolytes considered in the figure. The anodic branch of the potentiodynamic polarization of mild steel in perchlorate solution shows a typical active behavior with a marked increase of the current density for potentials more positive than the corrosion potential. In contrast, in the inhibitor solution not containing perchlorate, the anodic branch of the potentiodynamic polarization curve exhibits passive behavior. Moreover, this is accompanied by a shift of the corrosion potential in the positive direction and by a decrease of the corrosion current density. The same behavior is observed for the potentiodynamic polarization curve recorded in the mixed solution containing the 0.006 M perchlorate and the inhibitor. The passive current slightly increases when the chloride content is increased passing from 0.006 to 0.12 M. The behavior of mild steel in perchlorate solution and in the inhibitor solution with and without perchlorate is schematically represented in Fig. 4.9b. It can be observed that the main effect of the inhibitor is on the anodic branch of the polarization curve. Therefore, it can be stated that the aromatic carboxylic acid based inhibitor is an anodic inhibitor for mild steel in neutral environment. This is confirmed by the small shift of the corrosion potential in the positive direction. The mechanism proposed for the inhibition by the carboxylic acid based compound considers the adsorption of the inhibitor molecules on the dissolving Fe substrate, which involves the formation of intermediates between Fe and the inhibitor. These intermediates are adsorbed on the metal substrate favoring the passivation of the substrate. As a result, the intermediates adsorbed on the surface block further adsorption of aggressive ions and the iron dissolution. The potentiodynamic polarization curve recorded in the mixed solution containing the perchlorate and the carboxylic acid based inhibitor confirms the existence of competitive adsorption between the inhibitor and perchlorate ions, as can be seen in Fig. 4.9a. In the case of perchlorate, it can be stated that the inhibitor provides corrosion protection with an adsorption mechanism in the presence of aggressive species (perchlorate ions). In contrast, sulfate ions exhibit competitive adsorption with the inhibitor impairing the corrosion protection through the adsorption mechanism [1, 11]. Figure 4.10a shows the potentiodynamic polarization curves of mild steel in the presence of the same inhibitor considered in Fig. 4.9a, in sodium sulfate solution and in the mixed solution containing the sodium sulfate and the inhibitor. The anodic branch of the potentiodynamic polarization of mild steel in sodium sulfate solution shows active behavior. The anodic branch of the mild steel exhibits passive behavior in the inhibitor solution not containing sodium sulfate. This is in line with the behavior observed in perchlorate solution, where inhibitor adsorption leads to passivation of the metal substrate. In contrast, the potentiodynamic polarization curve recorded in the mixed solution containing the sodium sulfate and the carboxylic acid based

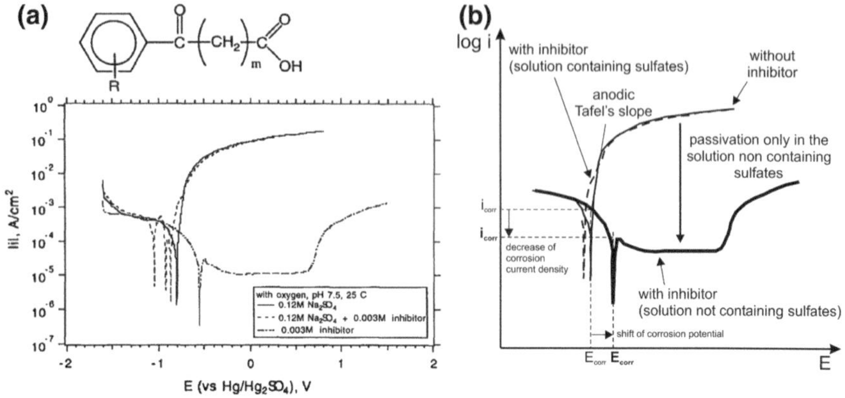

Fig. 4.10 Potentiodynamic polarization curves of low carbon steel (rotating disc electrode) acquired in 3 mM solution of an aromatic carboxylic acid inhibitor and in Na_2SO_4 solution with and without inhibitor [11] (**a**) and schematic representation of the effect of the inhibitor on potentiodynamic polarization curves (**b**)

inhibitor shows active behavior in the anodic branch, as can be seen in Fig. 4.10. This suggests that the inhibitor is not efficient in the presence of sodium sulfate. This is due to preferential adsorption of sulfate anions on the substrate leading to the dissolution of the iron (anodic reaction). This demonstrates that the type of anions in the electrolyte strongly affect the inhibitor efficiency.

4.3.2.2 Precipitation of Protective Salt Films

The inhibition in neutral environments can be due to precipitation of protective surface films. This can be obtained through the formation of an insoluble salt film on the metal surface. The salt films are usually thick as compared to adsorbed films discussed above. Inhibitors leading to precipitation of insoluble salt films are usually cathodic inhibitors because oxygen diffusion to the cathodic sites is limited or blocked by the film formed on the surface. However, it is also possible that the anodic processes are inhibited by the salt film. A typical example is the precipitation of insoluble hydroxides of zinc, magnesium, manganese and nickel. In this case, the hydroxide formation occurs at cathodic sites due to the reaction with hydroxyl ions produced by the cathodic reaction (oxygen reaction). The cathodic reaction leads to local alkalinization of the electrolyte promoting the formation of the insoluble hydroxides. Figure 4.11 shows an example of precipitation of a protective salt film on zinc, as reported in Ref. [12]. Two organic inhibitors, sodium benzoate (NaBz) and sodium N-dodecanoylsarcosinate (NaDS), were investigated in this work. Figure 4.11a shows the potentiodynamic polarization curves of zinc in 0.5 M NaCl solution with addition of different concentrations of sodium benzoate (NaBz) and sodium N-dodecanoylsarcosinate (NaDS), respectively. These inhibitors suppress the cathodic and the anodic reactions at 10^{-2} M concentration. The sodium

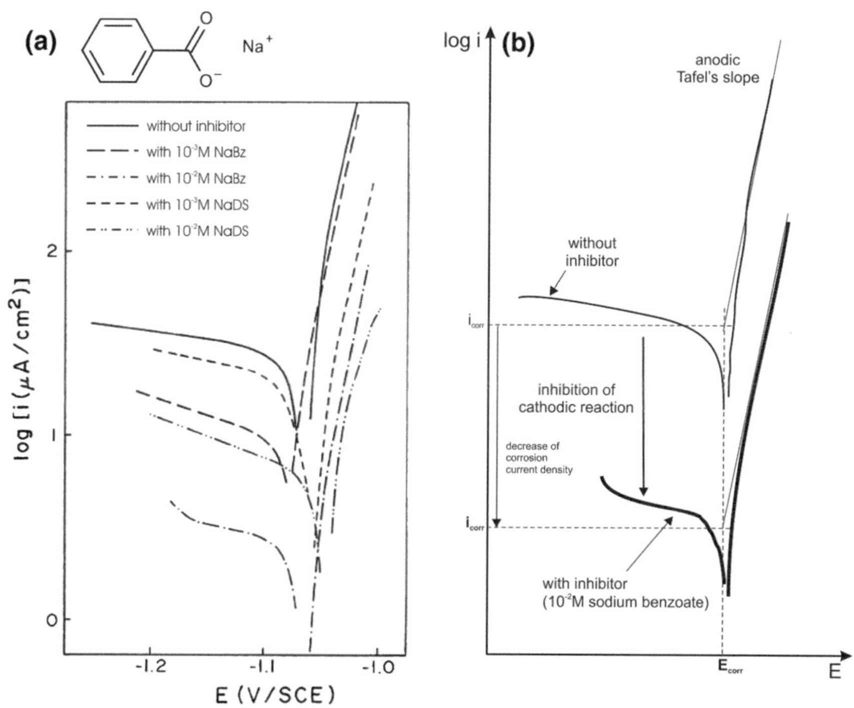

Fig. 4.11 Potentiodynamic polarization curves of zinc in aerated 0.5 M NaCl solution and in the same electrolyte with addition of 10^{-3} M and 10^{-2} M of sodium benzoate (NaBz) and sodium N-dodecanoylsarcosinate (NaDS) [12] (**a**); schematic representation of the effect of the sodium benzoate inhibitor on potentiodynamic polarization curves (**b**)

N-dodecanoylsarcosinate (NaDS) displays the same effect on potentiodynamic polarization curves at 10^{-3} M concentration, while the sodium benzoate (NaBz) inhibits mainly the cathodic process. The behavior observed in the potentiodynamic polarization curves of zinc in Fig. 4.11a is attributed to the precipitation of a protective film composed of zinc hydroxide, oxide and salt or complexes of the inhibitor ions. The effect of sodium benzoate at 10^{-2} M concentration is further considered in the scheme in Fig. 4.11b. It can be recognized that the precipitation of a film composed of zinc hydroxide and zinc salt or complex of benzoate strongly affects the oxygen reduction. In addition, there is a limited effect on the anodic reaction since the anodic current density is reduced with respect to the solution without inhibitor. As considered above, the precipitation of the salt film is induced by the cathodic reaction and therefore its formation tends to occur at cathodic sites suppressing the oxygen reduction. Nevertheless, the precipitation of a relatively thick film also might affect the anodic sites, mainly at relatively high inhibitor concentration. The behavior shown in Fig. 4.11b suggests that the sodium benzoate is mainly a cathodic inhibitor for zinc in NaCl solution although an effect on the anodic processes cannot be excluded.

4.3.2.3 Film Precipitation or Polymerization

The inhibition of corrosion processes in neutral environments can be obtained by film precipitation. This inhibition mechanism is based on the formation of surface layers by precipitation or polymerization. The layer formed on the metal substrate acts as a barrier for species involved in the cathodic reaction (oxygen) lowering the corrosion current density. Polyphosphates and organophosphates are typical inhibitors leading to film precipitation. These inhibitors form thin adherent films on cathodic sites preventing oxygen access at these locations. An example of this type of mechanism is the inhibition by Ce diphenylphosphate [13–15]. Figure 4.12a displays the potentiodynamic polarization curves in 1 M NaCl solution and in the presence of Ce diphenylphosphate inhibitor for AA2024-T3, as reported in Ref. [14]. The cathodic current density of AA2024-T3 in presence of the inhibitor is significantly reduced relative to the cathodic current density measured in NaCl solution. This is associated to a shift of the corrosion potential in the negative direction indicating a cathodic protection mechanism for Ce diphenylphosphate (Fig. 4.12b). The inhibition is confirmed by the observation of the sample morphology after potentiodynamic polarization (Fig. 4.12a). The inhibition is attributed to the formation of a protective film of Ce diphenylphosphate over the substrate. The inhibitor preferentially deposits at electrochemically active intermetallics with cathodic behavior. It is reported that the inhibitor undergoes hydrolysis at the cathodic sites leading to the formation of Al/Ce bimetallic bonds between the Al substrate and the inhibitor. This results in the formation of an insoluble protective film that induces the inhibition of the oxygen reduction reaction (cathodic reaction) [13]. It is also possible that the deposition involves the Al matrix for longer immersion times in the inhibitor solution leading also to an inhibition effect on anodic reactions [13].

The same inhibition mechanism of Fig. 4.12 is reported for triazole derivatives employed for corrosion protection of aluminium alloys [16]. Figure 4.13a shows the potentiodynamic polarization curves recorded in 3.5 % NaCl solution for AA2024-T3 with different additions of triazole. Figure 4.13b proves that the main effect of the triazole is the inhibition of the oxygen reduction reaction leading to the reduction of the corrosion current density and the shift of the corrosion potential in the negative direction. Triazole derivatives adsorb preferentially at cathodic sites (intermetallic compounds rich in Cu) in AA2024-T3 inhibiting the cathodic reaction [16]. In addition, a possible effect on the anodic reaction due to the deactivation of cathodic sites that reduces the susceptibility to pitting was reported [17]. This effect is not visible in the potentiodynamic polarization curves shown in Fig. 4.13.

4.3.2.4 Passivation

Passivation is a typical mechanism for corrosion inhibitors in neutral solutions. In this case, the passivation of the metal surface can be induced by oxidizing species like chromates, nitrites, nitrates and molybdates. In the case of a passivating metal

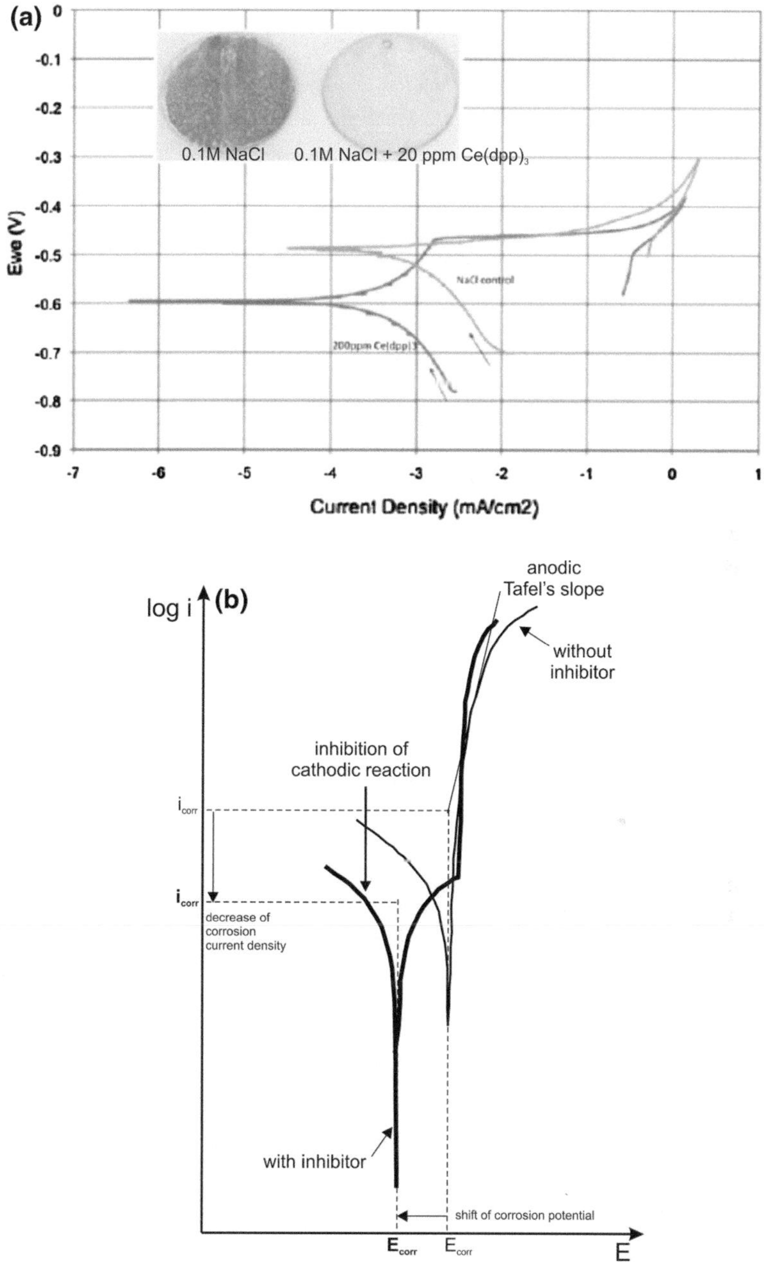

Fig. 4.12 Potentiodynamic polarization curves (potential vs. current density) and optical images of the samples after polarization (**a**) for AA2024-T3 in 0.1 M NaCl solution and in the presence of 200 ppm Ce diphenylphosphate inhibitor ($Ce(ddp)_3$) [14]; and schematic representation of the effect of Ce diphenylphosphate inhibitor on potentiodynamic polarization curves (current density vs. potential) (**b**)

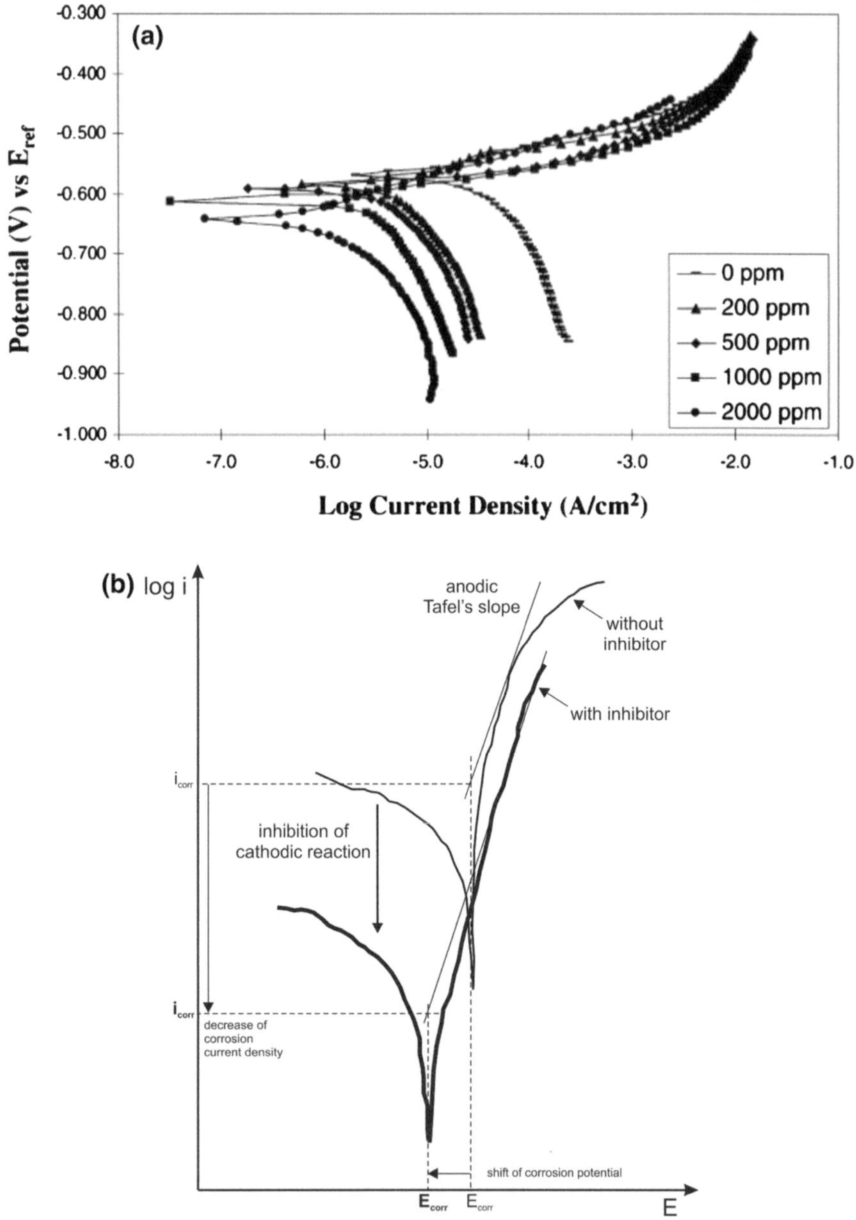

Fig. 4.13 Potentiodynamic polarization curves (potential vs. current density) for AA2024-T3 in 3.5 % NaCl solution polarization and in the presence of 200, 500, 1000 and 2000 ppm triazole inhibitor [16] (**a**); and schematic representation of the effect of triazole on potentiodynamic polarization curves (current density vs. potential) (**b**)

like stainless steel, the anodic branch of the potentiodynamic polarization curve can be divided in three regions: active, passive and transpassive (Fig. 4.14a). In the active region, the current density increases with increasing the potential. In the passive region, the current density remains low and does not significantly change the potential. The transition from the active to the passive region is usually characterized by 10^3 or higher reduction in current density. In the transpassive region, a marked increase of the current density is observed due to passivity breakdown. Figure 4.14 schematically shows the effect of the addition of oxidizing species on the electrochemical behavior of a passive metal. If the metal is in the active region, the addition of the oxidizer induces an increase of the cathodic partial current density leading to a shift of the corrosion potential from the active region to the passive region of the anodic branch of the polarization curve (Fig. 4.14a). The shift of the corrosion potential in the passive region of the anodic branch is associated with a strong decrease of the corrosion current density due to passivation of the metal surface, as shown in Fig. 4.14a. It should be noted that a further addition of oxidizing species might shift the corrosion potential in the transpassive region leading to a strong increase of the corrosion current density (Fig. 4.14b).

Chromates and nitrites are the most efficient oxidizing inhibitors for ferrous and nonferrous substrates. The mechanism of inhibition of these compounds is based on their ability to promote or strengthen passive behavior in the metal substrate. However, these compounds present severe environmental and health issues, which has driven research activities in the development of environmentally friendly alternatives. In the case steel, the chromates maintain iron in the passive state preventing the breakdown of the passive layer leading to localized attack. The chromates have been extensively described in literature for iron substrates [2, 18–20]. According to the work reported by Brusher et al. [18–20], the inhibition by chromates is based on the uptake of Cr on the iron surface leading to the deposition of a monolayer of Cr compounds. This was considered as an irreversibly adsorbed monolayer at the oxide/solution interface. Issacs et al. [21] proved that the Cr uptake on the metal substrate is the result of cathodic processes, which compete with oxygen reduction. Accordingly, the formation of a layer of Cr compounds is attributed to these cathodic processes rather than to the growth of a passive oxide film as initially proposed by Brusher et al. [18–20]. Therefore, it was concluded that the adsorption of chromates on the oxide is very weak. It is possible that the chromate ions are incorporated in the passive film, especially in solutions with high concentration of chromates. The structure of the passivating film determines a complex inhibition mechanism which is based on different contributions in the case of Fe substrates. The chromate ions might retard the dissolution of Fe preventing the breakdown of the passive film. An additional contribution to passivation is given by competitive adsorption of chromate ions with aggressive ions like chlorides strongly reducing the concentration of aggressive species at the metal/electrolyte interface. In addition, the existence of adsorbed chromate ions might promote the healing of the passive film at the sites where film undermining takes place. The repair of the weak spots can be promoted by the formation of a new passivating film of iron oxide at the location of the weak spots. In addition, the

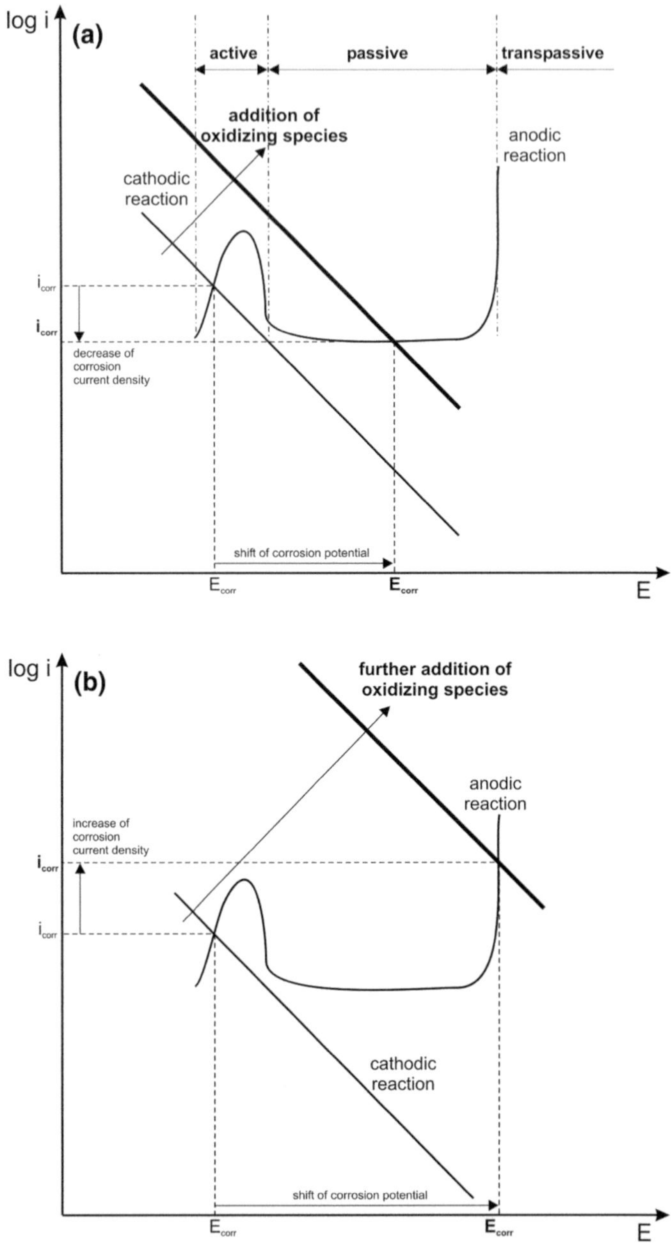

Fig. 4.14 Effect of an oxidizing inhibitor on the partial electrochemical reactions in the Evans diagram of a passivating metal: shift of the cathodic reaction in the passive region (**a**) and in the transpassive region (**b**)

healing might be due to the reaction of dissolving iron with the adsorbed chromate compounds leading to the formation of insoluble products with a plugging effect.

Chromates are employed as inhibitors also for nonferrous materials. As an example, chromates are extensively employed for corrosion inhibition of high strength aluminium alloys [22]. The mechanism of inhibition for aluminium alloys is based on the behavior of Cr^{6+} oxoanions that are reduced to Cr^{3+} anions at the sites of localized corrosion. The reduced species are adsorbed on the metal surface inhibiting the cathodic reduction of oxygen. In addition, chromates inhibit pit initiation at second phase particles (intermetallic compounds), which are initiation sites for localized corrosion (pitting). Moreover, the chromates modify the chemical composition of the oxides and intermetallic phases by adsorption and buffering. The adsorbed chromates limit the adsorption of chlorides, which promote the breakdown of the passive oxide layer. Based on the inhibition mechanism, it is generally recognized that inhibition of Al by chromates is due both to the inhibition of the oxygen reduction (cathodic inhibition) and inhibition of Al dissolution delaying the onset of pitting (anodic inhibition). Figure 4.15a displays cathodic potentiodynamic polarization curves of AA2024-T3 in 1 M NaCl solution and with different additions of sodium dichromate, as reported in Ref. [23]. The addition of chromate species to the neutral solution induces a marked reduction of the cathodic current density (Fig. 4.15b). The oxygen reduction reaction, taking place at potentials between −600 mV versus SCE and −1200 mV versus SCE, is strongly inhibited by the addition of chromates. In addition, it can be observed that the same effect is recorded for the hydrogen and water reduction reactions, which occur at potentials more negative than −1200 mV versus SCE. In AA2024-T3, the inhibition effect of chromates on the cathodic reaction is mainly related to the reduction of Cr^{6+} to Cr^{3+} at the sites of Cu-containing intermetallics, which support the cathodic reactions.

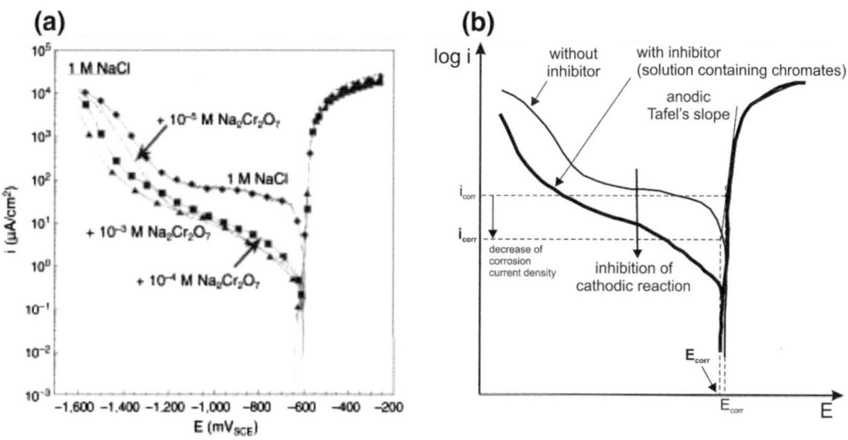

Fig. 4.15 Cathodic potentiodynamic polarization curves of AA2024-T3 in 1 M NaCl solution with addition of sodium dichromate [23] (**a**) and schematic representation of the effect of sodium dichromate on the cathodic potentiodynamic polarization curves (**b**)

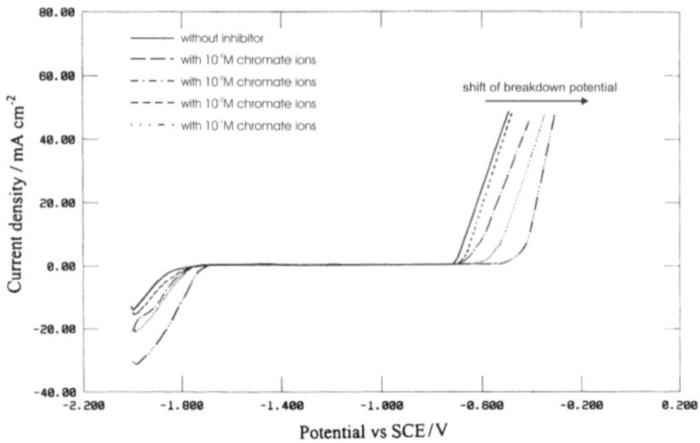

Fig. 4.16 Potentiodynamic polarization curves of pure aluminium in 0.5 M NaCl solution with different additions of chromate anions [24]

The reduction reaction involving the chromate species leads to the formation of adsorbed compounds that strongly inhibit the cathodic reaction [22]. The chromates might display a modest anodic inhibition in the case of aluminium alloys [22]. A shift of the breakdown potential in the positive direction is reported as the main effect of chromates on anodic polarization curves of pure aluminium in neutral 0.5 M NaCl solution [24] (Fig. 4.16).

4.3.2.5 Synergistic Effect of Mixtures of Inhibitors

Mixtures of corrosion inhibitors might provide a synergistic effect that can strengthen the protection of the substrate by combining different protection mechanisms. As an example, Fig. 4.17 shows the synergistic effect of sodium carboxylate and tolutriazol on zinc corrosion. Figure 4.17a displays the potentiodynamic polarization curves for Zn in an aggressive electrolyte (ASTM D 1384-87 solution) with addition of sodium heptanoate, tolutriazol, and mixtures of these inhibitors. Tolutriazol is a triazol derivative that forms a complex polymer film on zinc providing cathodic protection with the inhibition mechanism described in Fig. 4.13b. As it can be seen in Fig. 4.17b, tolutriazol strongly inhibits the cathodic reaction on zinc, while it has only a small effect on the anodic reaction. This results in a strong decrease of the corrosion current density and in the presence of a very small passive range. Sodium heptanoate is a carboxylate compound that provides anodic inhibition with the absorption mechanism discussed in Fig. 4.9b. The effect of this inhibitor is on the anodic reaction with no effect on the cathodic reaction, as it is shown in Fig. 4.17c. In particular, the potentiodynamic polarization curve recorded in the presence of sodium heptanoate exhibits an extended passive range. Figure 4.17d schematically represents the effect of the mixture of tolutriazol and

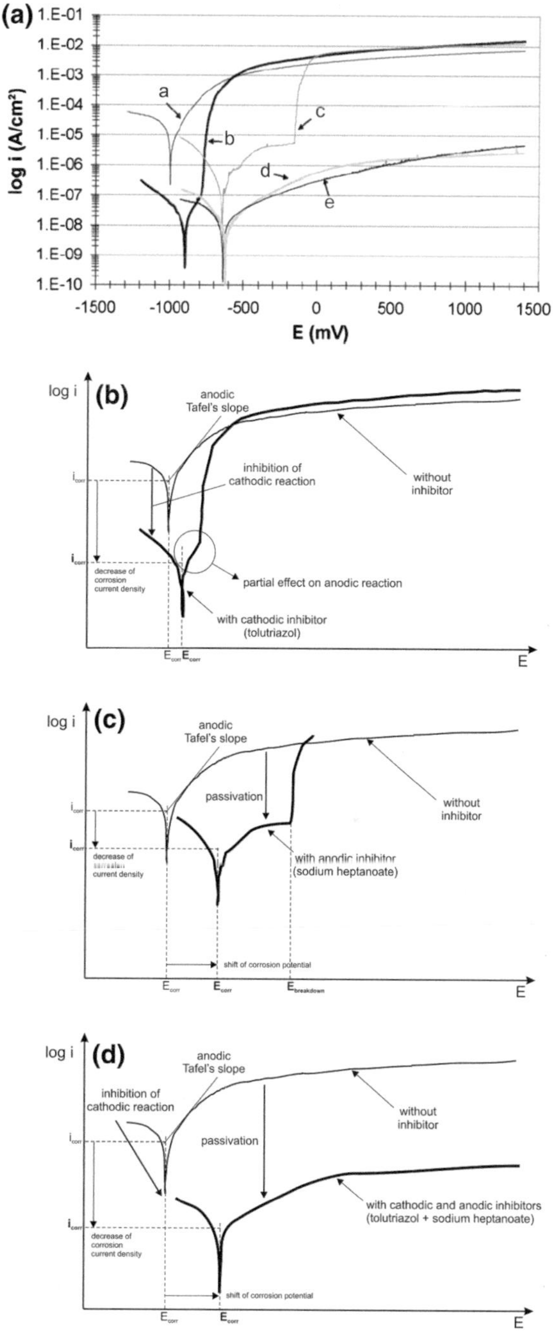

Fig. 4.17 Potentiodynamic polarization curves **a** of zinc in ASTM D 1384-87 solution (148 mg l^{-1} Na_2SO_4, 138 mg l^{-1} $NaHCO_3$ and 165 mg l^{-1} NaCl) (*curve a*) and in the same electrolyte with addition of 1 g l^{-1} of tolutriazol (*curve b*), of 0.05 mol l^{-1} of sodium heptanoate (*curve c*), of 0.05 mol l^{-1} of sodium heptanoate + 0.25 g l^{-1} of tolutriazol (*curve d*) and 0.05 mol l^{-1} of sodium heptanoate + 1 g l^{-1} of tolutriazol (*curve e*) [25]; schematic representation of the effect of sodium heptanoate (**b**), of tolutriazol (**c**) and of the mixture of these inhibitors (**d**) on potentiodynamic polarization curves

sodium heptanoate inhibitors. The potentiodynamic polarization curve recorded in the presence of the mixture of the inhibitors display a marked reduction of the cathodic current density and passive behavior in the entire anodic branch. This behavior is attributed to the synergistic effect of the inhibitors with tolutriazol mainly affecting the cathodic reaction and sodium heptanoate preventing Zn dissolution. It should be noted that the synergistic effect is evident by the strong reduction of corrosion current density in the presence of the mixture of inhibitors. In particular, this is lower than in the presence of the single inhibitors.

4.3.3 Gas Phase Inhibitors

4.3.3.1 Volatile Corrosion Inhibitors

Volatile corrosion inhibitors are mainly employed for temporary protection from atmospheric or gaseous corrosion [26, 27]. The packaging industry is an application field for this type of inhibitors. In addition, volatile corrosion inhibitors are often employed in closed environments like containers or boilers. Volatile inhibitors are typically organic compounds with low molecular weight and high vapor pressure. These include aliphatic, aromatic, cyclohexylamines, aminonitrobenzoates and heteroalkylated lower amines. The volatile inhibitors are transported by volatilization from a source to the metal substrate. The inhibitor can be adsorbed on the metal surface from the gas phase reducing the corrosion rate with an adsorption-based inhibition mechanism. The protection can be strengthened by the formation of an hydrophobic film, as in the case of amine-based inhibitors [28]. The protection by volatile corrosion inhibitors requires that the vapor pressure is sufficiently high to provide an adequate transfer of the inhibitor to the metal surface. This will guarantee that the concentration of the inhibitor at the site of corrosion is maintained high enough to ensure durable protection. Moreover, the efficiency of the inhibitor is related to the distance from the source and to the accessibility of the metal surface. The adsorption-based mechanism leads to inhibition of the anodic processes due to blocking of the anodic sites. This is similar to the protection mechanism discussed for the adsorption in neutral environments. In some case, it is possible to observe a synergistic effect of the inhibitor on both anodic and cathodic processes [26].

Figure 4.18a shows the potentiodynamic polarization curves of brass in simulated atmospheric water with and without 3-bis-diethylamino-propan-2-ol volatile corrosion inhibitor, as reported in Ref. [29]. The anodic branch of the potentiodynamic polarization curves is strongly modified by the addition of the volatile inhibitor, while the cathodic branch is not affected. Figure 4.18b shows that the inhibition of the anodic reaction is due to passivation, which is related to the adsorption of the inhibitor on the brass surface. The adsorbed inhibitor blocks the active sites reducing the anodic current density and shifting the breakdown potential in the positive direction.

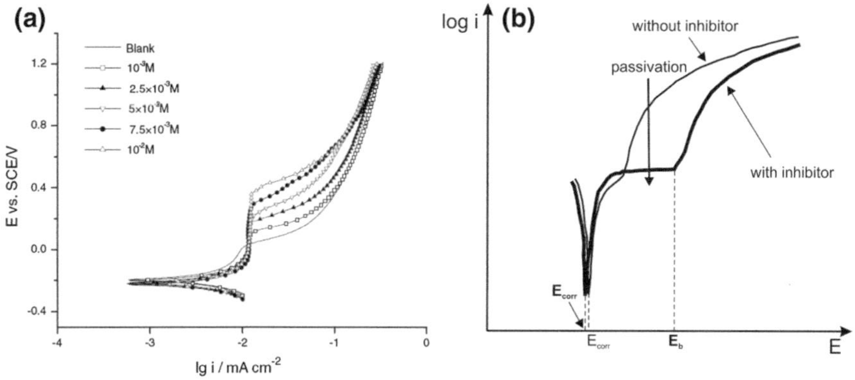

Fig. 4.18 Potentiodynamic polarization curves of brass in simulated atmospheric water without and with different concentrations of 1,3-bis-diethylamino-propan-2-ol volatile corrosion inhibitor [29] (**a**) and schematic representation of the effect of the inhibitor on potentiodynamic polarization curves (**b**)

4.3.3.2 Migrating Corrosion Inhibitors

Migrating corrosion inhibitors are employed for the protection of embedded steel rebar in concrete structures [30]. These migrating agents are able to penetrate into a concrete structure and protect the steel from corrosion due to chloride attack. The migration of the inhibitor takes place by diffusion in the liquid phase and in the gas phase. The liquid diffusion occurs via the moisture that is always present in concrete structures. The diffusion in the gaseous phase is due to the high vapor pressure of the inhibitor, as considered in the case of volatile corrosion inhibitors. Moreover, the diffusion is favored by the capillary structure of the concrete (porosity) and by the presence of microcracks. Migrating corrosion inhibitors are usually based on amino carboxylate chemistry. These can be incorporated as an admixture or can be surface impregnated on existing concrete structures. The migrating agents might affect both anodic and cathodic processes.

Figure 4.19 shows an example of the use of migrating corrosion inhibitors for the protection of steel rebar in concrete structures, as reported in Ref. [30]. The potentiodynamic polarization curves were recorded on bare rebar in a saturate Ca $(OH)_2$ solution containing chlorides (2000 ppm NaCl) and different concentrations of migrating agents (Fig. 4.19a). It should be noted that this experimental test enables only the study of the effect of the migrating corrosion inhibitor on the electrochemical behavior of the substrate, while it is not able to provide information on the migration of the inhibitor through the concrete structure. This was evaluated by other experimental methods, which evidenced the formation of a protective film on the steel rebar after impregnation of the surface of the concrete structure with the migrating corrosion inhibitor [30]. The bare steel rebar exhibits passive behavior in the alkaline $Ca(OH)_2$ solution. The potentiodynamic polarization curve recorded in the $Ca(OH)_2$ solution containing chlorides retains this passive behavior but

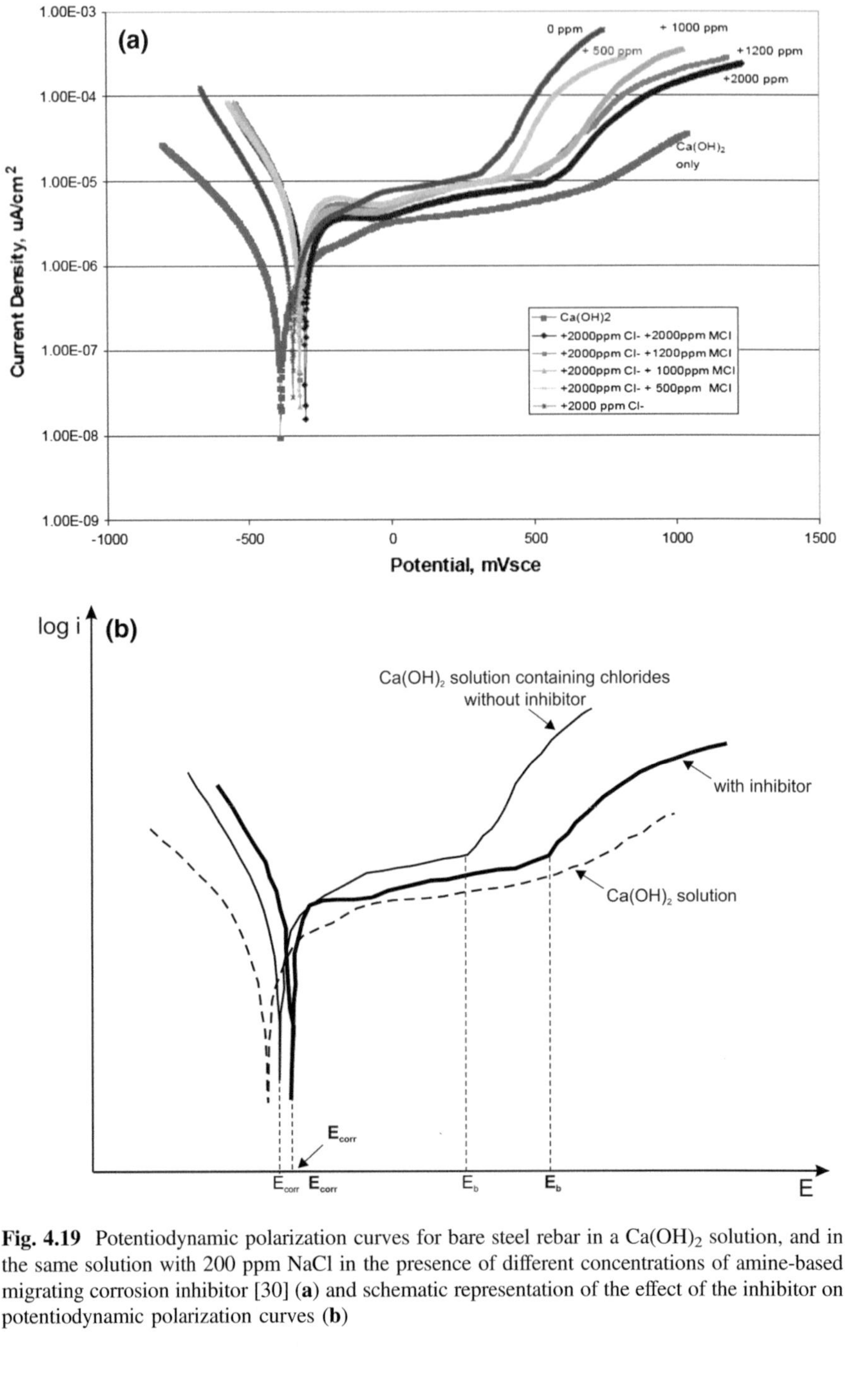

Fig. 4.19 Potentiodynamic polarization curves for bare steel rebar in a $Ca(OH)_2$ solution, and in the same solution with 200 ppm NaCl in the presence of different concentrations of amine-based migrating corrosion inhibitor [30] (**a**) and schematic representation of the effect of the inhibitor on potentiodynamic polarization curves (**b**)

passivity breakdown (E_b) is observed at 350 mV versus SCE due to the onset of pitting corrosion caused by chlorides. The addition of the migrating corrosion inhibitor progressively shifts the breakdown potential in the positive direction. The breakdown potential is 600 mV versus SCE with an addition of 2000 ppm migrating corrosion inhibitor. This indicates that the substrate retains its passive behavior in the presence of the inhibitor at higher potentials as compared with the system without inhibitor. This is schematically shown in Fig. 4.19b, where it can be seen that the inhibition is mainly due to a passivation effect related to the addition of the inhibitor, while the cathodic processes are not affected by the inhibitor.

References

1. D. Landolt, *Corrosion and Surface Chemistry of Metals* (EPFL press, Lausanne, 2007), p. 544
2. J.G.N. Thomas, in *Corrosion*, vol. 2 (Chapter 17), ed. by L.L. Shreir, R.A. Jarman, G.T. Burstein (Butterworh-Heinemann, Oxford, 1994)
3. E. Machnikova, K.H. Whitmire, N. Hackerman, Corrosion inhibition of carbon steel in hydrochloric acid by furan derivatives. Electrochim. Acta **53**, 6024–6032 (2008)
4. O. Olivares, N.V. Likhanova, B. Gomez, J. Navarrete, M.E. Llanos-Serrano, E. Arce, J.M. Hallen, Electrochemical and XPS studies of decylamides of a-amino acids adsorption on carbon steel in acidic environment. Appl. Surf. Sci. **252**, 2894–2909 (2006)
5. M. El Azhar, M. Traisnel, B. Mernari, L. Gengembre, F. Bentiss, M. Lagrenee, Electrochemical and XPS studies of 2,5-bis(n-pyridyl)-1,3,4-thiadiazoles adsorption on mild steel in perchloric acid solution. Appl. Surf. Sci. **185**, 197–205 (2002)
6. K. Takahashi, J.A. Bardwell, B. MacDougall, M.J. Graham, Mechanism of anodic dissolution of iron—II. Comparison of the behavior in neutral benzoate and acetate buffer solutions. Electrochim. Acta **37**, 489–494 (1992)
7. D. Eurof Davies, Q.J.M. Slaiman, Mechanism of the corrosion inhibition of Fe by sodium benzoate—I: The influence of concentration and pH in air-saturated solutions of sodium benzoate. Corros. Sci. **11**, 671–682 (1971)
8. Q.J.M. Slaiman, D. Eurof Davies, Mechanism of the corrosion inhibition of Fe by sodium benzoate—II: the inhibitive properties of sodium benzoate in de-aerated and air-saturated solutions. Corros. Sci. **11**, 683–692 (1971)
9. D. Eurof Davies, Q.J.M. Slaiman, Mechanism of the corrosion inhibition of iron by sodium benzoate—III: the role of oxygen. Corros. Sci. **13**, 891–905 (1973)
10. G. Reinhard, M. Radtke, U. Rammelt, On the role of the salts of weak acids in the chemical passivation of iron and steel in aqueous solutions. Corros. Sci. **33**, 307–313 (1992)
11. P. Agarwal, D. Landolt, Effect of anions on the efficiency of aromatic carboxylic acid corrosion inhibitors in near neutral media: experimental investigation and theoretical modeling. Corros. Sci. **40**, 673–691 (1998)
12. K. Aramaki, Effects of organic inhibitors on corrosion of zinc in an aerated 0.5 M NaCl solution. Corros. Sci. **43**, 1985–2000 (2001)
13. T.A. Markley, M. Forsyth, A.E. Hughes, Corrosion protection of AA2024-T3 using rare earth diphenyl phosphates. Electrochim. Acta **52**, 4024–4031 (2007)
14. T.A. Markley, J.I. Mardel, A.E. Hughes, B.R.W. Hinton, A.M. Glenn, M. Forsyth, Chromate replacement in coatings for corrosion protection of aerospace aluminium alloys. Mater. Corros. **62**, 836–840 (2011)
15. J.A. Hill, T. Markley, M. Forsyth, P.C. Howlett, B.R.W. Hinton, Corrosion inhibition of 7000 series aluminium alloys with cerium diphenyl phosphate. J. Alloy. Compd. **509**, 1683–1690 (2011)

16. H. Yang, W.J. van Ooij, Plasma-treated triazole as a novel organic slow-release paint pigment for corrosion control of AA2024-T3. Prog. Org. Coat. **50**, 149–161 (2004)
17. M.L. Zheludkevich, K.A. Yasakau, S.K. Poznyak, M.G.S. Ferreira, Triazole and thiazole derivatives as corrosion inhibitors for AA2024 aluminium alloy. Corros. Sci. **47**, 3368–3383 (2005)
18. D.M. Brasher, H.A. Kingsbury, D.A. Mercer, Passivation of iron by chromate solutions. Nature **180**, 27–28 (1957)
19. D.M. Brasher, H.A. Kingsbury, The study of the passivity of metals in inhibitor solutions, using radioactive tracers. Part 1—the action of neutral chromates on iron and steel. Trans. Faraday Soc. **54**, 1214–1222 (1958)
20. D.M. Brasher, D.A. Mercer, Radiotracer studies of the passivity of metals in inhibitor solutions. Part 3—influence of pH and oxygen content of chromate solutions on film growth on mild steel. Trans. Faraday Soc. **61**, 803–811 (1965)
21. H.S. Isaacs, S. Virtanen, M.P. Ryan, P. Schmuki, L.J. Oblonsky, Incorporation of Cr in the passive film on Fe from chromate solutions. Electrochim. Acta **47**, 3127–/3130 (2002)
22. M.W. Kendig, R.G. Buchheit, Corrosion inhibition of aluminum and aluminum alloys by soluble chromates, chromate coatings, and chromate-free coatings. Corrosion **59**, 379–400 (2003)
23. A. Sehgal, G.S. Frankel, B. Zoofan, S. Rokhlin, Pit growth study in Al alloys by the foil penetration technique. J. Electrochem. Soc. **147**, 140–148 (2000)
24. S. Zein El Abedin, Role of chromate, molybdate and tungstate anions on the inhibition of aluminium in chloride solutions. J. Appl. Electrochem. **31**, 711–718 (2001)
25. C. Georges, E. Rocca, P. Steinmetz, Synergistic effect of tolutriazol and sodium carboxylates on zinc corrosion in atmospheric conditions. Electrochim. Acta **53**, 4839–4885 (2008)
26. P.R. Roberge, *Handbook of Corrosion Engineering* (McGraw-Hill, New York, 2000), p. 838
27. B. Boyle, A look at developments in vapor phase corrosion inhibitors. Met. Finish. **102**, 37–41 (2004)
28. U. Rammelt, S. Koehler, G. Reinhard, Use of vapour phase corrosion inhibitors in packages for protecting mild steel against corrosion. Corros. Sci. **51**, 921–925 (2009)
29. G. Gao, C.H. Liang, 1,3-Bis-diethylamino-propan-2-ol as volatile corrosion inhibitor for brass. Corros. Sci. **49**, 3479–3493 (2007)
30. B. Bavarian, The efficacy of using migrating corrosion inhibitors (MCI 2020 & MCI 2020 M) for reinforced concrete, Report 1137, The Cortec Corporation (2004)

Chapter 5
Formulating Surface Coatings

Dominic Richard Harris and Philip Casey

5.1 Introduction

Self healing in inhibitive paint coatings tends to mean both a functional repair (e.g. smart ways of releasing inhibitors to stop corrosion) as well as cosmetic repair (e.g. void filling for scratches by reforming ruptured binds). One strategy that has been adopted to deliver these is the use of capsules that host or hold chemical moieties that on release contribute functionally to the self healing mechanism or another using nanoadditives for cosmetic protection or repair. Capsules can generally be categorized by their chemistry (e.g. inorganic or organic) and in formulating them into a surface coating they can generally be regarded in the same way an additive such as a pigment is regarded. Consequently, consideration of both material and production factors is important.

Inorganic "Capsules" covers materials that have been ion exchanged or have absorbed or adsorbed inhibitors (see Chap. 8). For layered compounds such as phyllosilicate and other clays, exfoliation and assembly into shell structures have also been tried. Sol-gel approaches also provide a very large number of possibilities.

Naturally occurring layered materials include zeolites, layered double hydroxides (LDH)(hydrotalcites) and various clays that can provide either a layered or tubular structure (e.g. Halloysite) (see Chap. 8). Manufactured materials include mesoporous materials (i.e. materials with pores between 2 and 50 nm) which are mostly oxides impregnated with inhibitors, inorganic oxyanions (vanadates, tungstates) and organic compounds (thiol and Azo compounds particularly heterocyclic). Exfoliated LDH may be used for both inhibitor as well as resin healing agents.

D.R. Harris · P. Casey (✉)
Commonwealth Scientific and Industrial Research Organisation, Manufacturing, Clayton, VIC, Australia
e-mail: Phil.casey@csiro.au

A.E. Hughes et al. (eds.), *Active Protective Coatings*,
Springer Series in Materials Science 233, DOI 10.1007/978-94-017-7540-3_5

Organic "capsules" tend to be used more for healing of the paint resin, although some mecaptans perform both functions. Polymer capsules include urea formaldehyde and melamine and polyelectrolytes including conductive polymers and polyanion/cations configurations (both resin and inhibitor) (see Chap. 8).

In this chapter, we consider coating formulations using *pigmented* coatings as an illustrative case as these are the most widely used commercially and provide a good basis on which to illustrate the guiding formulation principles pertinent to commercial production of self healing coatings. Formulation is especially important when taking an experimental outcome to commercial scale production and application. In a simple sense, a coating is a suite of ingredients (pigments, fillers, modifiers) held in a host matrix (polymer binder) which is applied to a substrate as a liquid (aqueous or non-aqueous) but which subsequently transforms into a solid.

In the surface coatings industry, formulation is particularly important as it determines not only final properties and attributes of the coating but also the manner and ease with which a coating may be manufactured, applied and transformed into its final form.

The most commercially relevant pigmented coatings are paints and printing inks: paints require durability and serviceability while inks require brightness of colour and suitability for their method of application. Architectural and decorative paints are predominately applied by brush, roller or airless spray while industrial automotive and aerospace coatings are applied by air-atomized spray, electrostatic spray, dip, or electro-deposition.

The key steps in formulation include:

(i) Component selection
(ii) Formulation and Production
(iii) Application
(iv) Property Testing

Some/all of these steps may be iterative but all implicitly require an understanding of the interactivity of both materials and processes.

This chapter focuses on the first two of these steps and aims to provide a consideration of

- Raw materials, composition and disperse phases.
- Methods of dispersing pigments
- The relationship between physico-chemistry and coating properties including
  - particle size, opacity and colour strength,
  - dispersion and colloidal stabilization,
  - dispersants and the effect that particle size and surface area has on the demand for dispersant,
  - hydrodynamic volume and mobility

Methods of application and property testing are not specifically considered here in any detail.

5.2 Component Selection: the Composition of Paints

Paints are blends of components and formulating and producing them requires the use of a large number of raw materials. In this regard, formulators live in a world of tens of thousands of raw materials, with each type usually available from several suppliers, in many grades with most, continually researched, developed and promoted to provide the opportunity for a new or improved coating. For example, there are over 400 varieties of titanium dioxide pigment alone, manufactured by about ten manufacturers using different processes, a diversity of TiO_2 composition, surface coating, surface treatment and particle size, each aimed at specific markets and applications. Similarly, there are over 350 varieties of acrylic resin dispersions with a plethora of secondary additives or modifiers such as detergents and emulsifiers [1]. The proliferation of raw materials is endemic to the industry and each new material that is developed (whether pigment, resin or other) is initially targeted to the application needs of the first customer. Subsequent development and refinement find additional customers and consequently the number of available raw materials is continually growing. The current catalogue is immense [2].

5.2.1 Resins

Polymer resins provide the essential properties and functionality for the utility of the coating and are the basis for the particular "type" or "line" of coating and the resin's chemistry will be determinant in this regard. Increasingly, formulations are based on aqueous rather than non-aqueous systems.

Generally, thermoplastic acrylic latexes predominate decorative coatings while thermosetting acrylics and urethanes are used in automotive and aerospace. Epoxies are used for infrastructure maintenance and nitrocellulose for wood finishing. The generic chemistry of some of these are presented in Fig. 5.1.

In organic compounds, covalent or coordinate bonds between two atoms have one or more *pairs* of shared electrons. During reaction, bonds are cleaved and undergo scission either by heterolysis (one atom takes both electrons to become negatively charged leaving the other positively charged) or homolysis (each atom takes an electron to become a free radical).

In heterolysis, polymerization propagates via 'step growth' mechanisms (condensation) with all molecules steadily reacting (to full conversion) gradually increasing molecular weight. Resins undergoing this process include polyesters, alkyds (fatty-acid modified polyesters), polyurethanes, epoxies, phenoxies and ketone/aldehyde (amino) resins. In coatings, such resins include heat cured thermosetting amino and acrylics and the two pack curing of polyurethanes and epoxies. The archetype alkyd resin dominating decorative paints in the 1950s and 1960s and still continuing to be widely used is based on a glycerol: phthalic anhydride: safflower oil fatty acid (1:1:1 molar ratio).

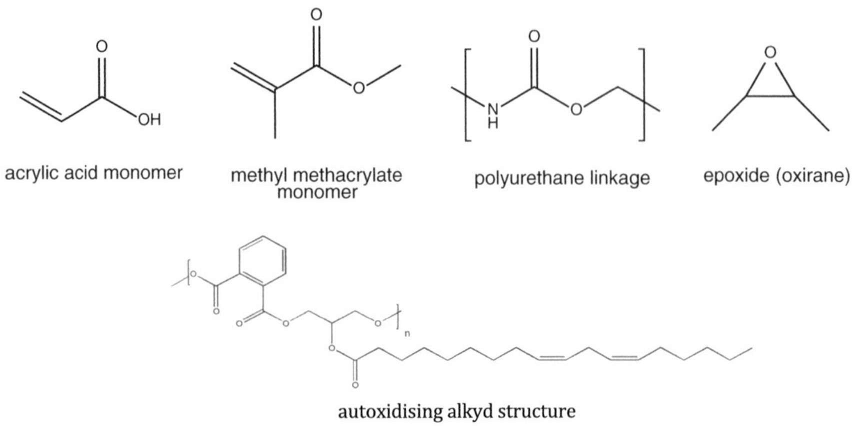

Fig. 5.1 Generic chemical structures

In homolysis, each atom accepts an electron to become a free radical. In polymerization of 'ethylenically unsaturated' monomers, these free radicals are generated by an initiator such as a peroxide or hydroperoxide. The rapid propagation of few (initial monomers is a 'chain reaction' mechanism (free radical or addition) up to termination and a *final molecular weight.* Such resins include hydrocarbons (polyethylene, polypropylene), vinyl—PVC—polyvinyl acetate—(including styrene), and acrylic resins polymethyl methacrylate, polyethyl acrylate, and in coatings; *autoxidation* of alkyds, and hydroperoxide initiated curing of 'polyester' resins.

Degradation is the reverse of these reactions with acrylics in particular 'unzipping' on the surface but not breaking down throughout, to shrink and crack, (e.g. old house oil paint, exacerbated by the presence of oxidation catalysts such as metals Co Mn) giving them advantageous durability.

Emulsion polymerized polymers, particularly acrylics, have dominated decorative paints since the 1970s. Emulsion polymerization starts in water with a solution of anionic surfactant (e.g. sodium lauryl sulphate which dissolves to form nano-sized micelles where the hydrophobes aggregate together) plus a free radical initiator such as persulphate. At elevated temperature, monomer (e.g. methyl methacrylate, ethyl acrylate, perhaps styrene) is slowly fed into the stirred reactor where it firstly forms droplets but then remarkably with very slight solubility in water migrates/diffuses to the hydrophobic centre of the micelles, where with the initiator begins free radical/addition/chain polymerization. With the incipient polymer, the hydrophobicity/lipophilicity of the micelle increases to attract more monomer. Additional surfactant (e.g. alkyl phenol 30 mol ethoxylate) aids the micellar solubilization and acts as a latex post stabilizer. Under exothermic control, polymerization continues to produce a solids loading of particles with 300 nm diameter of about 40 wt% and with a relatively monodisperse molecular weight (owing to confinement in the micelle).

New and improved resins are being continually developed particularly for service under extreme conditions and this is where self healing approaches complement and contribute to higher levels of performance.

5.2.2 *Pigments*

Pigments provide the "shades"/colours, or other variation, for each type/"line" of coating: opaque for paint, bright shade for inks, anticorrosive for primers, matting agents for low gloss, filling for surfaces.

Pigments, which are categorised by their Pigment Colour Number and Colour Index Number [e.g. Titanium Dioxide: Pigment White **6** (PW6), Colour Index (C I) 77891; Copper Phthalocyanine Blue–Beta form: Pigment Blue 15:3 (PB15:3), Colour Index (C I) 74,160], come in a broad range of colours with the final coating colour often utilising mixtures of tinting pigments added to a pigment base (typically white) [3]. Pigments may be supplied either as a powder or in a dispersed form.

5.2.2.1 White Pigments

Titanium dioxide (PW6) is the most widely used white pigment (typical particle size 0.18–0.25 μm). It is extracted from ilmenite and in its rutile form has a very high refractive index (RI = 2.6). At high solids loadings in a coating, light is scattered from, rather than transmitted through, the coating thus promoting its brightness and opacity. Zinc Oxide (PW4, particle size 0.2–0.3 μm) which has a much lower refractive index (RI = 2.00) may also be used in paints. Both TiO_2 and ZnO are good absorbers of ultra violet light, However, they are very photoactive and a coating will "chalk" on exterior exposure unless particles are coated with a barrier such as silica. Consequently, they provide a dual function of enhancing opacity while protecting the substrate from degradation by ultraviolet light.

Other "natural" white pigments used as functional fillers, "inert" or "extender" (of opacity) pigments with larger particle sizes (1–15 μm) and lower refractive indices (1.5–1.7) include dry ground sulphates (Barytes ($BaSO_4$), PW21B], layered silicates containing Na, Ca, Al, Mg [PW20 mica/feldspar, PW19 kaolin, PW26 talc], carbonates such as PW18 natural limestone/calcium carbonate/chalk and silica [PW27]. As indicated in the introduction some of these may serve as carriers for healing agents via absorption, surface adsorption or ionic exchange.

"Synthetic" white pigments with similar RIs to their natural counterparts include precipitated calcium carbonate [PW18], aluminium silicate [PW19] and synthetic "high brightness" kaolin (i.e. very white) [PW22], Blanc Fixe (precipitated $BaSO_4$); PW24 $Al(OH)_3$; hydrated magnesium silicate Talc [PW26] and synthetic silica gel [PW27].

Zinc Phosphate [PW32] plus a range of other phosphates are used as "non toxic anticorrosive pigment" in primers on metal surfaces coats.

5.2.2.2 Coloured Pigments

Inorganic coloured pigments range in colour across the spectrum. They are predominantly metal based compounds or complexes (predominantly transition metals such as Fe, Co, Cr, Ni, Cu, Zn, Ti, V) that include variant oxides, mixed oxides, titanates, chromates, sulphides, vanadates etc. Included in this category are black pigments based on carbon (e.g. carbon black) and special effects pigments such as pearlescent, fluorescent or heat reflective pigments.

The most common organic coloured pigments are the Azo pigments which are made by diazotization of an aromatic amine. Their characteristic colours are due to the azo (–N=N–) colour forming linkage and determined by the coupled organic moieties. Other common pigments are those based on phthalocyanine (complexed with metals). For a comprehensive list of pigments the reader is referred [2, 4–6].

5.2.3 Solvents

Solvents dissolve or carry dispersions of resins and are simultaneously used reduce viscosity for ease of application and assist in the control film formation. The range of chemicals that can act as solvents is broad. Water, of course, is the basis for aqueous dispersions while organic solvents are used in non aqueous dispersions. An example of the latter are the aliphatic hydrocarbons with aromatic content of 18 % of white spirit (mineral turpentine) that dissolve alkyd resins typically used in decorative coatings.

Historically, the glycol ethers such as Mono-Ethylene-Glycol-Ethyl-Ether (MEGEE, originally called "Cellosolve") and its variations of Methyl-, Ethyl-, Propyl-, Butyl-; and Mono-, Di- and Tri- Ethylene Glycol, have been important solvents, but are now found to be reprotoxic (hazardous to pregnancy) and are limited in use [7]. The propylene glycol ether analogues, however, are accepted as safe. Ketones and esters including acetate esters of the above glycol ethers are important for epoxies and polyurethanes, whereas active hydrogen (–OH) solvents: alcohols and glycol ethers are precluded by their reactivity. Aromatic toluene and xylene contribute to solvency as co-solvents when used in conjunction with oxygenated solvents.

The solvent plays a key role in application. Solvents for spray are carefully formulated with low boiling point ('fast') solvents for atomization while high boiling point ('slow') and strong solvents (for good dissolution) are used to promote flow and good film formation on the substrate surface. Fast aliphatic solvents contribute to atomization. Neither alcohol nor toluene or xylene dissolve the lacquer grades of nitrocellulose, but in combination they will dissolve them and thus alcohol is called a "latent" solvent. The term 'lacquer' refers to coatings drying solely by the release of solvent (called 'non-convertible' i.e. a thermoplastic coating) while the term 'enamel' means coatings which cure and cross link (by autoxidation, heat, or reaction, called 'convertible' i.e. a thermoset coating).

The coatings industry is continually seeking ways to reduce or eliminate the use of organic solvents and replacing them with aqueous based systems driven predominantly by environmental considerations.

5.2.4 *Additives*

Additives are ancillary materials that promote desired behaviour in formulations and commonly include *dispersants, surfactants and colloidal stabilizers*. For example, a 'simple' white aqueous acrylic paint requires a dispersant, a wetting agent (for pigment wetting and wetting greasy surfaces), an anti-settling agent, a defamer, a coalescent, a silicone additive for water resistance and organic liquids to provide lap time and freeze-thaw stability. Other optional additives may include a UV absorber, a cross-linker, additives for additional hiding power using vesiculated or hollow beads, a biocide for in-can protection or for protection of the coating, a perfume/reodorant, colour tinters to adjust colour and thickeners to provide control of flow.

5.2.5 *Encapsulated Liquids*

Microcapsules containing flowable polymerizable liquids, 'capsule-catalyst' or vascular; two pack/two component (known as 2 K) disperse phase e.g. epoxy and amino precursors, or two component polyurethane, may be used in self healing compositions. When the capsule is ruptured consequent of an external event (the damage), the embedded disperse phase (two component healing agent) is released such that they react and thus repair the damaged area [8].

5.3 Formulation and Production: Disperse Phases

Coating formulations are distinguished from other blended products such as pharmaceuticals or chemical specialities by the presence of a continuous phase (binders/resin/polymer dissolved in solvent) within which secondary phases are dispersed (disperse phase(s)). Dispersing these phases is typically not a facile process and careful consideration of both the processes and material requirements to effect good dispersion is necessary together with an understanding of how these will affect the final properties of the coating. It is not simply a matter of adding one component to the other. In paints, the most common approach to disperse pigments is by the process of grinding and for specialty coatings, such as those related to self-healing, this will also be important.

5.3.1 *Grinding and Dispersing Pigments*

Pigments and powders are incorporated into water, solvents and/or liquid resins by 'grinding and dispersion' processes which generally have two aspects: (i) mechanical separation and comminution (size reduction) of particles and (ii) colloidal stabilisation of the dispersion. Concurrent with the grinding process is the generation of new/fresh surfaces on the particles which are in turn amenable to chemical modification to assist in the dispersion process.

Grinding is controlled by both intrinsic and extrinsic factors. Intrinsic factors include such things as material chemistry, crystal structure, plasticity, ductility, particle size and morphology, while extrinsic factors include grinding media size, solids loading, the liquid medium, dispersing agents and the mill type, its operational parameters including temperature, grinding intensity and duration.

Most pigments are made by the manufacturer to a specific crystallite size (typically 0.1–0.3 μm) that is optimal for high gloss, good opacity and light and weather fastness. These properties are developed in dispersions by reducing/eliminating the propensity of the solids to aggregate and agglomerate which is a consequence of drying as the departing water/solvent leaves strong polar forces on the surfaces. Highly transparent pigments are of smaller size (0.05–0.08 μm) and many nanosized pigments, produced by proprietary methods, are commercially available.

There is a broad range of batch and continuous milling and mixing equipment that may be used to grind and/or disperse pigments in a host medium. The following section briefly overviews some of these and which are also illustrated in Fig. 5.2.

5.3.1.1 Batch Ball Mills

Batch ball mills are rotating horizontal cylinders (suspended on axles in plant, or sitting and running on a pair of rollers in the lab) that are ceramic or steel lined and contain ceramic pebbles or steel balls as grinding media. Here comminution relies on low frequency, high impact physical events and the predominating size reduction mechanism is material cleavage. These mills were the predominant grinding machinery in the 1950s and 1960s, for grinding solvent borne paints typically operating overnight (16 h).

Type of Mill	Mode of Operation	Milling Action	Operating Parameters
Batch Ball Mill	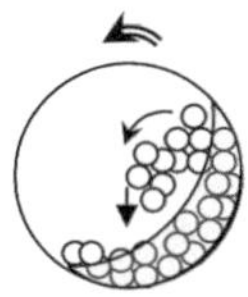	Low frequency; High Impact; Low shear; Particle Cleavage	Batch Mode, Low RPM (<500), medium to large Beads (cm)
Stirred Ball Mill	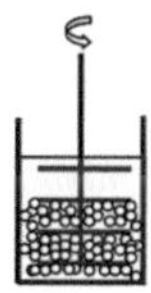	Moderate to High Frequency; Moderate impact; Medium Shear, Mixture of cleavage and attrition	Batch Mode, Moderate RPM (<1000), small to medium beads (3 10 mm)
Horizontal Bead Mill (annular gap)	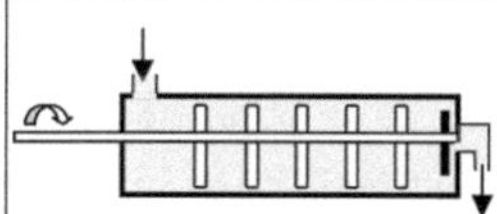	High Frequency; Low impact; High Shear, Predominantly attrition	Continuous Mode, Peripheral speed 10m/s., small to fine beads (<1 mm–10 mm)
Immersion Mill (Basket)	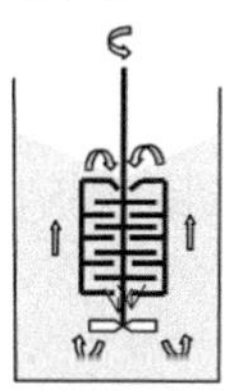	Very High Frequency; Low impact; High Shear, Attrition	Continuous Mode, small to fine beads (<1 μm–5 μm)

Fig. 5.2 Illustration of various types of ball mills

The "Attritor" is a stirred batch ball mill invented by Szegvari, now Union Process, of Akron Ohio. Here stirrer tips speeds, which rotate at moderately high speed, produce a regime of predominantly shear forces (rather than impact) and comminution is effected by "attrition" rather than cleavage as is the case in ball mills [9].

5.3.1.2 Continuous Mills

The first continuous mill was the "sand grinder" a *continuous high-speed* stirred media milling in vertical cylinders developed by DuPont in the 1950s based on the observation that grinding in ball mills was more effective when smaller ball sizes were used while accommodating the limitation of Stoke's law, where settling velocity is a function of the radius squared. Here, mill base was fed from a pump or a pressure pot into the bottom of the cylinder in which rotor bearing discs stirred small media (sand of about 0.8 mm diameter) with disengagement of beads from the mill base via a screen at the top [10].

5.3.1.3 Horizontal Bead Mills with 'Rotating Gap'

Later, machinery manufacturers built horizontal bead mills, with a rotor bearing various shapes of discs, stirring beads of 2–5 mm diameter, and with "rotating gap" bead disengagement. A disc on the end of the rotor shaft is adjacent to a similar sized annular aperture at the endplate of the shell with an adjustable ($\sim$0.25 mm) "rotating gap" to centrifuge and disengage the beads. The disadvantage of continuous mills (whether vertical or horizontal) is that the product can have a variable size distribution. This is because the residence time of the particles in the mill varies (some stay longer others exit more quickly) as a consequence of differences in fluid flow patterns in different parts of the mill chamber [11].

5.3.1.4 Immersion Mills

New mill designs have overcome the above limitation by adopting more sophisticated designs that simultaneously produce homogenous fluid flow regimes into and within the milling chamber as well as producing greater milling efficiencies by using a smaller milling chamber immersed in a larger chamber.

An example of this is a Basket Mill which combines expulsion and suction of the mill base through the milling "basket" which is immersed in the mill base ensuring that the entire base product passes through the grinding chamber many times during the operation and is progressively uniformly processed. An elegant variant is the immersion mill by Hockmeyer where the milling chamber (a basket) is immersed in the mill base contained in a larger vessel or chamber such that high throughput homogeneous fluid flows into and within the grinding chamber are promoted. These flow regimes produce an increased grinding efficiency and by using a combination of stationary and rotating pegs in the chamber aggressive shear zones are created that produce fine product with a narrow size distribution [12].

Pigments may be milled with fine (0.1–0.6 mm) beads in production scale bead mills made by high throughput, fine bead mills in which bead disengagement is effected by centrifugation. There are laboratory scale mills of these brands.

5.3.1.5 High Shear Rate Pre-milling

High speed mixing is performed in a regime that invokes high shear rates typically using various rotor configurations at high speed. It differs from milling in that no grinding media are used. High speed disc mixing began with the invention of the Cowles Dissolver: a disc with peripheral sectors alternately bent up and down, slightly inward from the tangent, rotating at the end of a shaft in the mill base at 25 m/s, inducing, in preferably dilatent millbases, high intensity mixing with cavitation. High Speed Mixing is now widely used for premilling mill base for other milling operations. Kinetic dispersers (Kady, Janke, Kunkel Ultra Turrax) have high speed nested slotted rotors and stators. The Silverson homogeniser has a high speed paddle rotor working inside a peripheral ring perforated with holes. There is a great diversity of high speed homogenisers, particularly at laboratory scale [13].

5.3.1.6 High Shear Stress Milling

Sigma bladed or Z-bladed mixers are batch mills, usually jacketed for heating or cooling, wherein two horizontal rotors, exchanging mill base between them, apply high shear forces to the mill base. They are commonly used for high viscosity slurries or dispersions. These mills have had the special application of "flushing" where organic pigment filter-press cake (wet) is mixed with gradual additions of resin into which the pigment transfers from the aqueous to the organic phase. Separated water is poured off by tilting the mill, and finally the mill is closed, heated via the jacket, and put under vacuum to remove all water. An advantage of flushing is that it avoids the aggregation and agglomeration caused by drying pigments to powder [14].

5.3.1.7 Testing the End Point

A simple but effective way of assessing whether a good dispersion has been achieved is to typically use a Hegman gauge or a variant of it. A Hegman gauge is a block of steel with one or two channels cut into it about 12 mm wide which taper from zero to 100 μm depth [15]. A sample of mill base is placed at the deep end and is scraped down the channels using a straight blade. A reading is taken where significant "nibs", "grits" or "seediness" appear. The fineness gauges are thus reading the proportion of a small but significant population of larger particles.

Paint pigments ground to a particle size required primarily for matte finish have a Hegman of ~25 μm and for gloss a Hegman ~12.7 μm. For viscous inks for example the limit has been 5 μm or "off the gauge" where the National Printing Ink Research Institute (NIPRI) production grindometer tapers from zero to 25 μm. More modern measures of particles size such as Transmission Electron Microscopy

(TEM), Scanning Electron Microscopy (SEM), X-Ray Diffraction (XRD) and light scattering or photon correlation spectroscopy (PCS) may reveal, when the dispersion has been approved by the Hegman gauge, undispersed primary crystallites and aggregates. Thus beyond the endpoint on a Herman gauge there is the possibility of more effective grinding towards the primary crystallite size as provided by the pigment supplier.

5.3.2 *Colloidal Stabilization*

Colloidal stabilization of the ground dispersion is necessary to prevent flocculation, i.e. reversion to a loose association of particles called floccules in the dispersion. A floccule behaves as a large but loosely held particle. Flocculation and undispersed agglomerates and aggregates are entirely counterproductive to all useful properties of coatings. It is typically observed as the contraction of the flocculated volume in a standing liquid leaving clear continuous phase, called *syneresis*, usually at the top but it can be on the bottom depending on the density of the dispersed phase. It is desirable to fully grind to provide for full deflocculation and colloidal stabilization in the liquid phase and subsequently in the solid coating. Some resins have adequate wetting and dispersion power, but improved dispersion and deflocculation are usually achieved by the addition of dispersants.

5.3.2.1 Settling/Sedimentation

A flocculated dispersion settles quickly and softly (can be easily stirred), owing to the large effective size of the floccules. A de-flocculated dispersion settles slowly and the sediment is hard, owing to finer particles becoming more compact. Formation of *sediment* may be ameliorated or prevented by using a higher viscosity medium or by rheological yield value imparted by ‘antisettling’ agents. Thickeners and rheology modifiers are added to achieve desired rheology typically using cellulosics, modified clays or more recently acrylics and urethanes in the form of associative thickeners which may be conveniently post-added and stirred into the liquid. Settling can manifest as a real problem during long distance transport.

5.3.2.2 Miscibility

Miscibility must be perfect, regardless of the effectiveness of dispersants, to avoid flocculation. One exception is that surfactants, which become concentrated at particle-liquid interfaces, need not be entirely miscible. Prediction of miscibility and compatibility was put on a quantitative basis by a classic article of Burrell

"Solubility Parameters for Film Formers" [16]. Hildebrand and Hansen predict solubility parameters using a 3D space based on the energy from dispersion, intermolecular and hydrogen bonding between molecules such that the nearer co-ordinates for molecules are in this space the more likely they are to dissolve into each other [17].

5.3.2.3 Dispersants

Dispersing agents like polyacrylic acid are purely dispersant (by adsorption and charge stabilization). Others reduce the interfacial surface tension making it easier for dispersion to occur. Most are to some degree oleophilic and capable of surface tension depression in aqueous systems spreading over and wetting pigment surfaces. Zinc soaps have been widely used in alkyd resin coatings (generally as a grinding aid) to wet the surface and provide zinc to adsorb carboxylic resins and dispersants. *Speciality ranges of grinding aids and dispersants* include copolymers with pigment affinic groups (e.g. the Nuplex "Disperbyk" and "Byk"), low surface energy siloxane, perfluoro alkyls) and high molecular weight polymeric dispersants (Ciba "Efka" range) and others (Tego Dispers, Solsperse Hyperdispersants) to wet surfaces and disperse particles.

Most pigments are crystalline with sharp edges which have higher surface energy than the bulk of the crystals. These agents wet the crystals to benefit grinding and prevent *recombination.* For example, the Solsperse hyperdispersants demonstrate significant reduction in viscous inks processing owing to prevention of recombination.

The level of dispersant, to gain perfect colloidal dispersion, is a critical parameter which will vary according to particle size and surface area and is thus determined by specific weight per surface area which may typically be 0.005–0.01 g/m^2

5.3.3 Modes of Adsorption of Surfactants and Dispersants at Interfaces

When washing a piece of glass with ordinary soap (sodium palmitate) water wetting occurs because the soap is *surface-active* i.e. the soap concentrates at the interface between the glass and water. Because the palmitic hydrophobes (16 carbon alkane) are non-polar with little attraction between them, they are repelled by the highly polar and self attracting *water* and subsequently concentrate and spread at the surface or interface. Thus the soap *wets* the surface with water and this typically describes the normal adsorption process of soaps, surfactants and detergents [18].

5.3.3.1 Acid–Base (Lewis) Interaction' Adsorption/Chemisorption

If the piece of glass is washed in water with a cationic surfactant (e.g. lauryl trimethyl ammonium chloride), the water beads and the surface will remain non-wettable. This occurs because there is an acid–base (Lewis) interaction between the lauryl trimethyl ammonium cation and the silicic acid groups of the glass. Here the adsorption is the reverse of the above with the ammonium cation on the glass and the hydrophobes oriented towards the water hence beading will occur.

5.3.3.2 Anionic Surfactants and Dispersants

Anionic stabilizers are used in aqueous compositions. In the US the *de-facto* standard dispersant has been Tamol 731 (also known as Orotan) which is a sodium (or ammonium) salt of a maleic anhydride copolymer, an anionic polyelectrolyte which is highly carboxylated and has some wetting/surface tension depression properties at 1 % aqueous solution. There are many other varieties of ionic stabilizers: anionic such as alkyl sulphates, sulphosuccinates, sodium hexametaphosphate, and "ether" (ethylene oxide) *sulphates* or "ether" (ethylene oxide) *phosphates* of alkyl or aryl polyethoxylates.

5.3.3.3 Cationic Surfactants

Lauryl trimethyl ammonium chloride and cetyl trimethyl ammonium bromide are highly effective in comminution and stabilization of fine pigment particles. One particular application of cationic surfactants is in bitumen/asphalt emulsions for coating aggregate on roads.

5.3.3.4 Steric Enthalpic Stabilizers

"Enthalpic" is the classification for non-ionic surfactants (ethoxylates) because their thermodynamic stability comes from heat of hydration necessary to remove water on collision with the hydrated polyethylene oxide sheaths. Steric stabilizers for aqueous compositions are most particularly the ethylene oxide condensates of a variety of hydrophobes. The most popular were the ethoxylates of nonyl phenol and of the less biodegradable octyl phenol. Environmental concerns about the lack of sufficient biodegradability and thus toxicity has prohibited the use of ethoxylates of alkyl phenols in detergents. Fatty alcohol (alkyl) ethoxylates, particularly lauryl and decyl, may be used as they are biodegradable, but have diminished effectiveness. Ethoxylates are also notable for their *inverse solubility* with temperature and the "cloud point temperature" is used to control the endpoint of ethoxylation due to temperature dependent hydrogen bonding.

5.3.3.5 Entropic Stabilizers

Solsperse Hyperdispersants are an effective invention of predominately *entropic* stabilizers. They were invented within what was ICI Dyestuffs Division and are excellent for low to moderately polar systems including aliphatic, aromatic (particularly solvent gravure), esters, ketones, and glycol ethers. They comprised a variety of solvatable polymeric chains combined with a variety of 'anchor'/'pigment-affinic' groups which adsorb by Acid–Base (Lewis) Interaction (chemisorption).

The first member was Solsperse 3000 which has a carboxyl group suitable for attaching to basic pigments, particularly Calcium Rubine 4B Toner, Pigment Red 57:1, Colour Index 15,850:1, the predominant magenta pigment for trichromatic 'process' printing (Cyan, Yellow and Magenta). In the range there are a variety of anchor groups and a variety of solvatable groups which are suitable for most combinations of pigment and solvent. There are also two 'synergists' to precoat specific pigments for the Acid–Base (Lewis) Interaction of other appropriate Solsperses.

5.3.3.6 Polymer Resin Dispersancy

In solvent borne-coatings the polymeric resins themselves may have adequate wetting and dispersion properties for the colloidal stability of the suspended pigments. Alkyd resins have free hydroxyl and carboxyl groups and solvatable chains of fatty acids on polyester. Straight polymethyl methacryate (PMMA) has poor dispersancy. Acrylic resins, both solvent borne (thermoplastic and thermoset) and aqueous acrylic dispersions, are made with small proportions of carboxylic acid co-monomer to assist *adsorption* onto the pigment surface and provide *colloidal stability*.

5.4 Developing a Formulation

Factorial experiments are efficient because they simultaneously measure the effect of several factors, *and their interactions, with* replication and *without* wasteful repetition. In developing a formulation, *Two level*, multi factor experiments can be very useful; the number of experiments being 2^n where n is the number of factors. Factorial experiments may be used for optimizing the formulation for performance and cost and for discovering preferred ranges for exemplification in patents or for troubleshooting performance problems (e.g. solvent boil—bubbles in hot weather on sprayed automotive lacquers). There are many textbooks on Experimental Method and Process Optimization from [19–23].

Typically, experimental formulations are advantageously drafted in a spreadsheet where the left hand *vertical columns* list ingredients with the right hand columns contain the quantities of the ingredients and pertinent formulation descriptors or properties (e.g. v/v ratio, density etc.). This format conveniently allows all kinds of variations based on different raw materials across the pages, and is suitable for factorial experiments.

Coating formulae list ingredients with quantities by weight fraction or percentage adding up to unity or 100 respectively. This format may be easily 'extended' by multiplying by a factor to yield the required scale up of quantities for preparation or manufacture. *Master spread sheets* can be used to calculate the *volume* relationships: volume solids and pigment volume concentration, with costing, manufacturing instructions and testing instructions.

Relative volumes of the continuous and dispersed phases are most importantly calculated, both as liquid and dried solid, by spreadsheet. Experimental variation can proceed informed by the observable effect of pigment *volume* concentrations on properties. From volumes, and measurement of particle size, surface areas may be calculated which are important to determine (1) the level of dispersant and (2) the rheology of the dispersion via the thickness of the adsorbed barrier.

Formulations may be usefully managed and recorded in the following of manner:

A spread sheet should be able to calculate and provide:

- 100 % total by weight
- Volumes of ingredients (from density and weight)
- Density of the liquid formula determined from those of the ingredients (kg/L, Weight per Litre test)
- The *total* weight of solids and volume of solids of the liquid dispersion. The *pigment* weight content and pigment volume content of the liquid dispersion;
- *Th*e levels of additives and of binder on the pigment in pph w/w and v/v;
- Pigment weight content of the solid coating (PWC) and
- Pigment volume content of the solid coating (PVC)
- Costs and Values Columns of raw materials to calculate total cost may be included.
- Manufacturing instructions.
- Testing/characterisation instructions
  "Extensions" multiplying by a factor the 100 % or unit total by weight, to required batch sizes.

Table 5.1 presents an illustrative example of an aqueous based formulation containing two *dense pigments* (Pigment 1 density = 4.8 g/cc, Pigment 2 density = 2.94 g/cc) have differing concentrations by *weight* and *volume* (30.4 *and* 17.7 %) and which contains a dispersant, a stabiliser (alkaline), the resin base (40 % non volatiles) and make up water.

Table 5.1 Illustrative example of formulation

			Dispersion	w/w	SG		
	pph w/w	pph v/v	As made	Solids	Density	Volume	Surface area SA
Prep No:	On pigment	On pigment	Weight **w**	or **NV**	**d**	**v**	m^2
Ingredients							
Pigment 1			4.80	4.80	4.80	1.00	100
Pigment 2			11.20	11.20	2.94	3.81	381
Dispersant	30	89	4.80	4.80	1.12	4.29	
Resin 41 %NV	60	178	23.40	9.59	1.12	8.57	
Alkaline stabiliser NH3 25 % aq			6.00		1.00	6.00	
WATER (Hold out 5 % for final washout and makeup)			49.80		1.00	49.80	
Totals	**90**	**267**	**100.00**	**30.4**	**1.36**	**73.5**	**481**
Constants							
Total pigment				**16.00**	4.96	**4.81**	
Total weight of dispersion			100.00				
Total volume of dispersion						73.5	
Weight Solids or NV of dispersion w/w				30.4			
Volume Solids or NV of dispersion v/w						17.7	
Pigment weight content of dispersion w/w				16.0 %			
Pigment volume content of dispersion v/v						6.5 %	
Pigment weight concentration PWC of coating				52.6 %			
Pigment volume concentration PVC of coating						27.2 %	
Dispersant per square metre at surface area, g							**0.010**
(Dispersant + resin) per square metre at surface area, g							**0.030**

(continued)

Table 5.1 (continued)

			Dispersion	w/w	SG		
	pph w/w	pph v/v	As made	Solids	Density	Volume	Surface area SA
Prep No:	On pigment	On pigment	Weight **w**	or **NV**	**d**	**v**	m^2
Ingredients							
Manufacturing instructions							
Charge water and NH_3(aq) to premix vessel							
Charge and dissolve dispersant and aqueous resin							
Charge and premix pigments.							
Feed mill base via the high recirculation rate mill to the receiving vessel:-							
W.O.Bachofen Dyno 'ECM', or Netzsch 'Zeta', or Drais 'DCP'							
When premix tank is empty—mill with recirculation at high rate from and with return to receiving vessel.							
Progressively test against standard sample for colouristic strength until end point is reached							
Mill batch to an empty vessel ('pass' or 'passage' mill) to complete milling.							
Tests: CIELab colour, Gloss, Specific gravity, Viscosity.							

NV non volatiles
pph parts per hundred
pph w/w on pigment parts by weight of dispersant and resin on 100 parts by weight of pigment
= wt of dispersant/[(wt of pigment 1 + wt of pigment 2)/100] = 4.8/[(4.8 + 11.2)/100] = 30
pph v/v on pigment parts by volume of dispersant and resin on 100 parts by volume of pigment
= vol of dispersant/[(vol of pigment 1 + vol of pigment 2)/100]
= (wt of dispersant/density of dispersant)/[(wt of pigment 1/density of pigment 1) + (wt of pigment 2/density of pigment2)}/100
= (4.8/1.12)/[((4.8/4.8) + (11.2/2.94))/100] = 4.28/[(1 + 3.8)/100] = 428/4.8 = 89
Surface Area may be calculated from theory (based on size of particle) or determined from empirical or reported data
PWC pigment weight concentration; *PVC* pigment volume concentration

5.5 Summary

Self healing coatings have considerable technological and industrial potential for application across a number of sectors. To assist easy entry into the market place, self healing function needs to be incorporated or formulated into a coating in a manner typical to the coatings industry. This requires an understanding of what material considerations are important in formulating a commercial coating for easy deployment and application performance. This chapter has provided a brief guide to some of these considerations particularly in relation to component selection, formulation and production. It hopefully has provided the reader with an appreciation of the complexity that may arise when producing a formulation and reinforced the notion that it is not simply a matter of adding one component to another to produce the coating. It has not focused on any specific self healing agents but rather taken an overview using pigmented coatings as a model illustrative system and that subsequently the reader is better equipped to address the challenges in taking a self healing coating concept from the laboratory to a higher level of technology readiness.

To this end the main points of the chapter are that:

- Paints are blends of components and formulating and producing them requires the selection and use from a large number of raw materials.
- Polymer resins provide the essential properties and functionality for the utility of the coating.
- Polymerization propagates via either (i) 'step growth' mechanisms (condensation) with all molecules steadily reacting (to full conversion) gradually increasing molecular weight or (ii) a 'chain reaction' mechanism (free radical or addition) with the rapid propagation of few initial monomers up to termination and a *final molecular weight*.
- Pigments may be inorganic, used for hiding/opacity, as fillers, and for anti-corrosive properties or they may be organic used for colour.
- Solvents and additives, in addition to polymer resins and pigments, are major components of coatings. Dispersant additives are particularly important for providing, in both liquid and solid phases, colloidal stability.
- Formulation may be assisted by preliminary factorial experiments.

A formulation is better understood and managed when the *volume* relationships (such as pigment volume concentration PVC) are calculated on a spread sheet.

References

1. *McCutcheon's Emulsifiers & Detergents*, McCutcheon Division, vol. 1 (MC Publishing Company, 2014)
2. *The Raw Materials Index RMI*, vol. RMI 8, (Surface Coating Association of Australia, 2008)
3. *Colour Index International*, http://www.colour-index.com/

4. G.B.A.G. Paff (ed.), *Industrial Inorganic Pigments*, 3rd edn. (Wiley, Germany, 2005)
5. *Industrial Organic Pigments*, 3rd edn. (Wiley, 2006)
6. *High Performance Pigments*, 2nd edn. Revised and Expanded (Wiley, 2009)
7. A. Ostman, *Reducing the use of Dangerous Glycol Ethers* (Nordic Council of Ministers, Chemical Group, Copenhagen, 1993)
8. B.K. Blaiszik, S.L.B. Olugebefola, S.C. Moore, J.S. Sottos, N.R. White, *Self Healing Polymers and Composites*, in *Annual Review of Materials Research* (Annual Reviews, USA, 2010) p. 32
9. A. Szegvari, M. Yang, in *Seminar on Dispersion of Pigments and Resins in Fluid Media. Attritor Grinding and Dispersing Equipment* (Union Process, Kent State University, 1999), p. 7
10. S. Hochberg, *Process for Dispersing Pigments in Film-Forming Materials* (Du Pont, USA, 1952)
11. K. Engels, *Stirring Mill* (Draiswerke Gmbh, USA, 1978)
12. H. Herman, H.Z. Hockmeyer, *Basket Media Mill with Stirring Rods and Counterpart Stators*, in *US Patents Office*, USPTO ed. (Hockmeyer Equipment Corp., USA, 1998), p. 9
13. S.X. Jinli Zhang, L. Wei, High shear mixers: a review of typical applications and studies on power draw, flow pattern, energy dissipation and transfer properties. Chem. Eng. Process. **57–58**, 16 (2012)
14. K. Robin, J.L.K. Connelly, 3D numerical simulation of the flow of viscous newtonian and shear thinning fluids in a twin sigma blade mixer. Adv. Polym. Technol. **25**(3), 12 (2006)
15. *Fineness of Grind Gauges*, http://www.gardco.com/pages/dispersion/fg/finenessofgrind.cfm
16. B. Hansen, *Solubility Parameters for Film Formers.* Official Digest of Federation of Society of Paint Technoogy (the former name of J. Coat. Technol. Res.), **27**(369), 32 (1972)
17. C.M. Hansen, *Hansen Solubility Parameters: A User's Handbook*, 2nd edn. (CRC Press, 2007), p. 544
18. D.A.B. Ward, Orientation of surfactants adsorbed on a hydrophobic surface. J. Phys. Chem **97**, 2 (1993)
19. Fisher, R.A., *The Design of Experiments* (Macmillan Pub Co., 1971)
20. E.P. George, J.S.H. Box, G. William, *Hunter Statistics for Experimenters: Design, Innovation, and Discovery*, 2nd edn. Probability and Statistics (Wiley, 2005)
22. D.C.W. Abdullah Ekin, J.W. Daniels, S.J. Stafslien, F. Cassé, A. James, M.E. Callow, Callow, Synthesis, formulation, and characterization of siloxane–polyurethane coatings for underwater marine applications using combinatorial high-throughput experimentation. J. Coat. Technol. Res. **4**(4), 6 (2007)
23. L.R.R.A. Price, X.H. Li et al., Investigation of the effects of formulation on selected properties of UV curable IPN coatings. J. Coat. Technol. Res. **70**(885), 5 (1996)

Part II
Advances in Active Protective Coatings

Chapter 6
Fostering Green Inhibitors for Corrosion Prevention

M.F. Montemor

6.1 Introduction

Inhibitors are chemical species that can interact with a metallic surface in a way that the result is the decrease of the corrosion activity and corrosion rate. Classification of corrosion inhibitors have been the subject of some controversy. However, nowadays, inhibitors are commonly grouped according to their mechanism of inhibition as follows: anodic (or passivating) inhibitors, cathodic inhibitors, mixed inhibitors and volatile corrosion inhibitors.

Passivating inhibitors (also classified as anodic inhibitors) are in general oxidizing substances that when above a certain critical value, provide corrosion protection due to one (or more) mechanisms: (i) ability to shift the corrosion potential towards more positive values, forcing the metallic surface into the passive range; (ii) stabilization of passive oxide layers; (iii) repair of damaged protective (namely oxide) layers; (iv) Enhancement of the passive ability by incorporation of species into the passive layer; (v) decrease of the probability of absorption of aggressive ions, such as chloride ions. Well known examples of passivating inhibitors are chromates, nitrites and molybdates. These inhibitors are stable in various pH ranges, being quite effective in near-neutral solutions, forming sparingly soluble corrosion products, such as oxides, hydroxides, or salts, which are protective in various surfaces.

Cathodic inhibitors either slow down the rate of the cathodic reaction or, selectively, precipitate at the cathodic areas, where the pH increases due to hydroxyl release, raising the local impedance and hindering the diffusion of reducible species into these areas. These inhibitors work by different mechanisms that include: (i) decreasing of the reduction reaction rates (cathodic poisons); (ii) selective precipitation on the cathodic areas (cathodic precipitators). Oxygen scavengers are

M.F. Montemor (✉)
CQE, DEQ—Instituto Superior Técnico, Universidade de Lisboa,
Av. Rovisco Pais, 1049-001 Lisboa, Portugal
e-mail: mfmontemor@tecnico.ulisboa.pt

A.E. Hughes et al. (eds.), *Active Protective Coatings*,
Springer Series in Materials Science 233, DOI 10.1007/978-94-017-7540-3_6

often classified as cathodic inhibitors. Well known example are rare-earths, namely the cerium cation that is able to form hydroxides or stable oxides on the cathodic areas [1, 2].

Mixed corrosion inhibitors are mainly organic compounds that can be neither classified as cathodic nor as anodic inhibitors, because they affect both processes. However, many of these are more effective in hindering one redox reaction, comparatively to the other. The effectiveness of mixed corrosion inhibitors is controlled by their ability to adsorb on metal surface and by the subsequent degree of coverage. This process strongly depends on the chemical composition and structure of the inhibitor and metallic surface morphology, roughness and composition. Typically mixed corrosion inhibitors work by physical adsorption, chemisorption and formation of stable surface layers. Whatever may be the mechanism of adsorption, the electron density of the functional groups, polarizing ability and electronegativity are important parameters that control the inhibition efficiency. Several organic inhibitors show ability to inhibit both the anodic and cathodic processes. For example 8-hydroxyquinoline, salicylaldoxime and quinaldic acid form insoluble chelate layers that impede chloride adsorption on the metallic surface. Concerning the localized corrosion of AA2024 in the presence of chlorides, these inhibitors act by chemisorption and formation of complexes on the Al surface, including over the reactive intermetallics and hinder the corrosion process, reducing the rate of the anodic and cathodic processes [3]. Volatile corrosion inhibitors (VCIs), also called vapor-phase inhibitors (VPIs), are normally used in closed environments. They are expected to disperse and, when in contact with the metal surface, the vapor of these inhibitors (typically salts) condenses and reacts with moisture producing protective species. Amine based compounds are commonly used as VCIs. Over the years, the field of usage of VCIs has widened, now covering electronics, packaging, process industries, reinforced concrete, coatings and metalworking fluids.

Corrosion inhibitors can be applied to hinder all corrosion forms in a wide array of environments. However, the selection and application of a corrosion inhibitor is an extremely complex process, greatly influenced by the environmental parameters, aggressive agents, forms and mechanism of corrosion and other factors, that altogether govern the corrosion process and determine the inhibitor efficiency. Moreover, the choice is also greatly influenced by external factors such as regulations that limit the choice because of toxicity risks, disposal, health issues and environmental impact. Well-known is the actual legislation that has been created to replace inhibitive species containing hexavalent chromium compounds, which are known to be carcinogenic and non-environmentally friendly. The same trend is currently applied to some conventional organic inhibitors and, therefore there is an increasing pressure on finding more environmentally friendly and effective corrosion inhibitors.

This chapter aims at reviewing the most recent trends in corrosion inhibitors research for corrosion prevention of metallic parts. Thus, corrosion inhibitors of different natures, with emphasis on inhibitors having a clear “green” label are reviewed in this chapter.

6.1.1 Importance of Corrosion Inhibitors

A particular advantage of corrosion inhibition is that it can be easily implemented in situ, most of the times without disrupting the metallic application, and can be used to control a wide range of corrosion processes. The major industries applying corrosion inhibitors are the oil and gas industry, the chemical and petrochemical industry, heavy industrial manufacturing steel industry, water treatment facilities, mining, coatings industry and the product additive industries. The largest consumption of corrosion inhibitors world over is allocated to the oil industry, particularly primary and secondary production of oil and subsequent refining and transport processes. Different types of corrosion can be found in oilfield onshore and offshore under extremely aggressive environments that include high temperatures, high pressures, high contents of CO_2, H_2S and chlorides. Therefore very effective mixtures, working efficiently in a wide range of environmental conditions are required. In many cases, tons of inhibitors are used, and this demands very rigid regulations concerning disposal and waste treatment. In the end, there is an important economic impact in the operations cost. Therefore, there is a strong economic pressure driving the search for more effective and environmentally friendly corrosion inhibitors.

In a wide array of technical applications, inhibitors are rarely used in the form of single compounds. Currently, it is quite usual to employ formulations made up from two, three or even more inhibitors. This need is justified by the fact that protection provided by single inhibitors is only for a limited number of metals and that protection of multi-metal systems requires the presence of more than one inhibitor. Another reason is that more effectiveness can be achieved if, for example, both anodic and cathodic reactions are inhibited, requiring inhibitors of the anodic and cathodic types. Although this is a common procedure in many applications there is still a lack of knowledge related to the inhibition efficiency, detrimental interactions between inhibitors and optimal ratios in inhibitors mixtures. Some attempts are being made to overcome this gap [4, 5]. The forecasts indicate that the inhibitors market will increase in the near future mainly boosted by the economic growth of emerging economies. In this sense, all knowledge contributing to new generations of safe and effective inhibition systems is of utmost relevance.

6.1.2 New Trends Towards "Green" Inhibitors

Nowadays there is a growing search and demand for and the application of corrosion inhibitors that are less toxic and biodegradable compared to traditional formulations. Hence, "green" labeled corrosion inhibitors are attracting a lot of attention. These inhibitors must display high effectiveness and improved environmental compatibility. Undoubtedly, these will be the ones most widely used in the future.

The employment of green eco-friendly natural molecules, extracted from leaves, seeds, plants, biological systems and even from natural and synthetic drugs for example, are gaining large preference as corrosion inhibitors. If by one side these molecules are safe, easy to apply and biodegradable, they also need to present low cost and easy extraction from renewable sources. Thus, these compounds need to be effective, but also competitive both from the economic point of view and environmental friendliness. Abundance, renewal and bio disposal are the key for the implementation of such inhibitors.

The action mechanism of a corrosion inhibitor depends upon its chemical nature. For example, organic corrosion inhibitors (OI), when added in solution, adsorb on the metallic surface by displacing the water molecules (and eventually aggressive agents such as chloride ions) according to the reaction scheme depicted below.

$$OI_{(electrolyte)} + nH_2O_{(ads.Surf.)} \Leftrightarrow OI_{(ads.Surf.)} + nH_2O_{(electrolyte)}$$

In organic molecules, the presence of heteroatoms such as O, N, and S, which possess higher electron density and act as active centers for physical or chemical adsorption on the surface are the main reason for the inhibition ability of the molecule. Other parameters governing the effectiveness of the corrosion inhibitor include: the chemical structure of functional groups; the electron density at the various donor atoms, the presence of π bonds, non-bonding p-orbitals and the electronic properties of the molecule. All these factors contribute to the formation of homogeneous chemically or physically adsorbed layers, or stable inhibiting compounds, which are the key for an effective corrosion inhibition process. Therefore the selection of the inhibitors, either natural or synthetic has to be carefully defined in function of all the parameters that affect the inhibitor effectiveness. Most of the time, in natural or biological inhibitors the challenge is to identify and to isolate the active agents that present higher inhibition potential.

6.2 Biological Inhibitors

Biological inhibitors are species that are present in biological tissues and that are being probed and evaluated as corrosion inhibitors. The border between biological inhibitors and vegetable inhibitors is not precise. However, in this review, inhibitors extracted from plants, seeds, roots and shells are classified as vegetables, and thus discussed in the following section. The ones related with biological processes are treated in this section.

6.2.1 Chitosan

Chitosan (CS), which is derived from the polysaccharide chitin, is well known as a low cost, renewable marine polymer. The main source of chitin is the structural components of the shells of crustaceans, such as shrimps, lobsters and crabs.

Chitosan is one of the most abundant natural polymers next to cellulose. From a chemical point of view, chitosan is essentially a linear copolymer of (1-4)-2-amido-2-de-oxyd-glucan (glucosamine) and (1-4)-2-acetamido-deoxy-d-glucan (N-acetylglucosamine). It can be produced from chitin by means of alkaline N-deacetylation. Chitosan possesses unique physicochemical properties namely biocompatibility, antimicrobial activity, biodegradability and excellent film-forming ability. Indeed, it can be used to form thin coatings to protect metallic surfaces. When chitosan is dissolved in dilute acidic electrolytes, the amine groups are easily protonated and the additional positive charge confers a behavior that is similar to that of some polyelectrolyte macromolecules. This chemical functionality is extremely relevant when chitosan is used as corrosion inhibitor.

Another derivative of chitin is carboxymethyl chitosan that was tested as corrosion inhibitor for mild steel. It was shown that the amino groups and the hydroxyl ends can displace the chloride ions from the iron surface, thus providing corrosion protection [6]. Moreover, it was demonstrated that this inhibitor forms a complex in the presence of Cu^{2+}. This compound provides additional corrosion inhibition of mild steel in acidic environments like HCl electrolytes [7].

Fekry and Mohamed [8] synthesized acetyl thiourea chitosan in the form of a polymer that was used to inhibit the corrosion processes on mild steel exposed to acidic media. The results show that the corrosion rate increases when the temperature increases and when the polymer concentration decreases. The inhibition efficiency determined in a 0.5 M sulphuric acid solution reached 94.5 % for an inhibitor concentration of 0.76 mM. The chitosan quaternary ammonium was also tested as corrosion inhibitor for carbon steel in the same electrolyte with positive results [9].

The corrosion inhibition ability of chitosan was also tested on stainless steel in acidic media with relevant efficiency and acting mainly as an anodic inhibitor as reported elsewhere [10].

Chitosan can inhibit corrosion of copper in acidic media, namely in HCl electrolytes, and literature reports that chitosan acts essentially as a mixed-type inhibitor, which follows a Langmuir isotherm. Moreover, by chemical modelling it was found that the nitrogen and oxygen atoms in the chitosan molecule are the main active sites that result in adsorption of chitosan on the copper surface as depicted in Fig. 6.1 [11].

These works demonstrate that chitosan based compounds can be effectively used to mitigate the corrosion activity on steel and cooper substrates exposed to acidic environments. However, the mechanism is not clear. In some cases it is reported as anodic inhibitor and in other cases as mixed corrosion inhibitor.

Chitosan, because of its molecular structure, has been used to prepare coatings for corrosion protection of various metallic substrates. A protective coating for AA2024 was composed of a chitosan-based pre-layer deposited onto the metal surface followed by a barrier hybrid film. The chitosan layer could serve as reservoir, where other corrosion inhibitors, such as cerium ions could be stored for enhanced protection of AA2014. It was demonstrated that cerium ions form a complex with the functional groups of the chitosan macromolecules, providing a

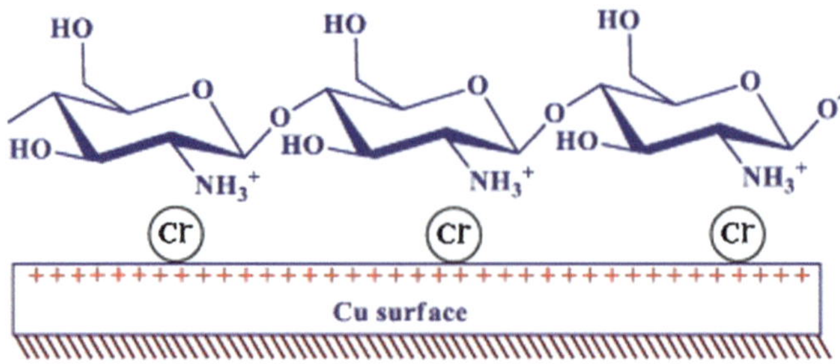

Fig. 6.1 The adsorption mode of chitosan onto the copper surface [11]

prolonged release of the inhibiting species, on demand, as reported by Zheludkevich et al. [12].

Anodized aluminium samples were dip coated into a chitosan–acetic acid solution in order to form a surface coating with good adhesion to the anodized layer. The film formation was successful and it presents protective behavior. A reduction in the permeability of the chitosan film could be observed after modification with copper cations. The increased inhibition efficiency is likely to be associated with copper cross linking within the chitosan structure [13]. Chitosan was used to synthesize an epoxy/chitosan cement slurry for cement oil wells [14]. The results underline the properties, including the corrosion protective features, of the epoxy/chitosan-modified cement slurry and its potential to be used as an environmental-friendly system for cementation of oil wells.

Given its biological nature, chitosan has been tested in the protection of metallic materials for biomedical applications. A poly-ethyleneimine (PEI)/heparin coating and chitosan/heparin coating was applied on NiTi alloy substrates using a layer-by-layer method. Both the corrosion behavior and biocompatibility of the coatings were studied. The results revealed that the biocompatibility of the PEI/heparin coating was better than that of the chitosan/heparin coating as well as that of the substrate [15].

Hydroxyapatite (HA)–chitosan composite coatings were deposited on AZ31 Mg alloy substrate using aerosol deposition (AD) in order to improve both the corrosion resistance and biocompatibility of bioresorbable AZ31 Mg alloys. The coating was uniform and provided a dense barrier layer. The corrosion resistance of AZ31 Mg alloy was increased by the HA–chitosan composite coating. In addition, the biocompatibility of the alloy was remarkably improved by the presence of the HA–Chitosan composite coating [16].

Chitosan based coatings were also studied over biomedical stainless steel grades. Electrophoretic deposition (EPD) was applied to deposit halloysite nanotubes (HNT) and to fabricate composites of chitosan–HNT and chitosan–HNT–hydroxyapatite films on 316 stainless steels. The resulting composite films, revealed stable properties and good biocompatibility, thus being promising materials for the fabrication of biomedical implants with advanced functional properties [17]. The literature demonstrates that chitosan has a high potential as corrosion inhibitor either in acidic media or neutral media in various metallic substrates and in various applications. It can be used by direct addition to the electrolyte or to fabricate functional or composite coatings. Furthermore it can be used to confine other inhibitors or functional agents through a layer by layer approach in which the protective species are stored between chitosan layers. It can also be used as surface modifier of natural clays for improved corrosion protection.

Undoubtedly the potential of chitosan as a corrosion inhibitor is very high and its abundance, low prices and availability, combined with its biological nature will attract attention in the near future.

6.2.2 *Aminoacids*

Aminoacids are a class of molecules that can be classified either as biological or natural. In this review it was decided to include them as biological molecules due to its role in the biological processes. Aminoacids are abundant, not expensive and easy to produce with a purity near 100 %. They contain heteroatoms and π-type electrons that make these molecules potential corrosion inhibitors. Moreover, the presence of amino groups that can be protonated, make the molecule able to interact with metallic surfaces. In fact, the inhibition efficiency of different aminoacids was compared and simulated by using computational tools. Quantum chemical studies revealed that amino acids can interact with nickel surface in the protonated rather than in the neutral state—Fig. 6.2 [18].

Aminoacids were tested as corrosion inhibitors for Cu and Ni alloys in neutral electrolytes. These alloys are of extremely high industrial relevance since they are used in power stations, boilers and heat exchangers. The problem is that they are sensitive to corrosion mainly because of the circulating water. Aminoacids, especially glycine, are able to inhibit the corrosion processes of these alloys. The mechanism of inhibition involves the adsorption of the amino acid on the active corrosion sites and/or the deposition of corrosion products with the formation of a 3-D layer on the alloy surface as reported by Badawy et al. [19].

Copper is also susceptible to corrosion in the presence of sulphates. Therefore, N-(5,6-diphenyl-4,5-dihydro-[1, 2, 4] triazin-3-yl)-guanidine (NTG) was evaluated as a corrosion inhibitor for copper in the presence of sulphates. The aminoacid derived compound provides corrosion protection of copper surfaces, acting by a cathodic inhibition mechanism. The process was concentration dependent because the thickness of the inhibiting layer increases as the concentration increases [20]. Effective corrosion inhibition was also observed for copper surfaces exposed to 3 % NaCl. However, in such environment, the process was associated with a mixed inhibitor mechanism [21]. The corrosion inhibition ability of various aminoacids was assessed in different metallic substrates used in various technological applications. Aminoacids were studied as corrosion inhibitors of vanadium substrates. In the presence of chlorides, the corrosion processes prevent the adsorption of the

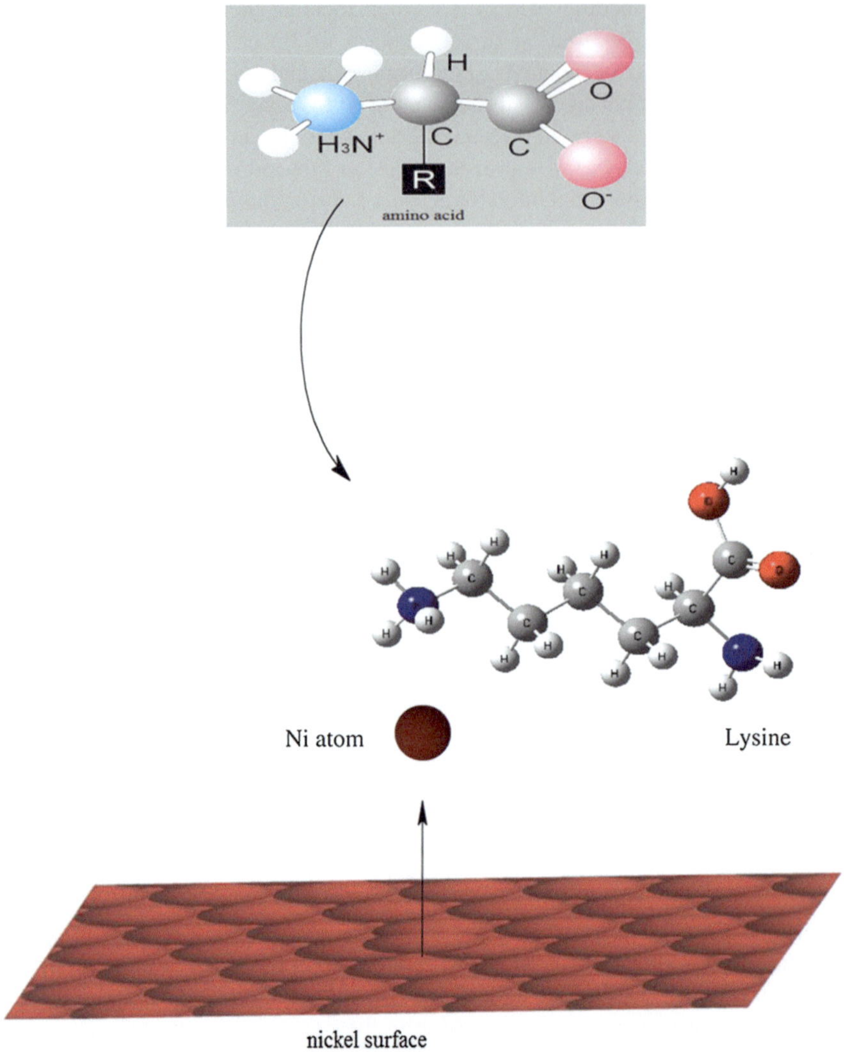

Fig. 6.2 Mechanism of inhibition by aminoacids and interaction with the Ni surface [18]

inhibitor molecules on the metallic surface and the corrosion processes could not be effectively inhibited [22].

Lead alloys found several applications in energy storage (for example in traditional lead acid batteries) being exposed to acidic (H_2SO_4) electrolytes. In general these alloys present significant corrosion resistance because lead can form thick, homogeneous and insoluble surface films. However, the composition of these products depend on the composition of the electrolyte and on the exposing environment that may include several species such as sulphates, chlorides or oxides and carbonates, among others. Aminoacids such as cysteine, methionine and alanine reveal good inhibition efficiency for lead alloys exposed to sulfuric acid. The inhibiting effect is reported as dependent on the aminoacid concentration, temperature and chemical structure of the inhibitor molecule. The inhibition effect increased with concentration, but decreased with temperature, and involves the heterogroups present in these molecules [23].

Aminoacids can be used to protect various metals but copper has been one of the most studied. However, the existing literature is somewhat inconclusive and the inhibition mechanism is controversial. In some cases the mechanism is attributed to an anodic inhibition effect, whereas in other cases, the mechanism is described as a mixed type one. However, it is conclusive that the process of inhibition involves strong adsorption of the inhibitor and protection efficiency increases with the inhibitor concentration.

Corrosion inhibition of mild steel in HCl environments was studied and synthetic aminoacids were proposed as corrosion inhibitors. Olivares-Xometl et al. applied different electrochemical, spectroscopic and microscopic techniques to study the inhibition process. The study ranks the efficiency of various aminoacids and discriminates specific features about the inhibition process of each one. The results highlight some important findings that include: the presence of a phenyl ring in tyrosine molecule supported electron retro-donation and consequently enhances the chemical adsorption of molecules; the inhibition efficiency depends on the length of the carbon chain, but decreases with temperature [24]. The role of the heteroatoms is determinant on the inhibitor efficiency. XPS measurements revealed that the N-containing groups present in some aminoacids are shifted to higher binding energies as consequence of protonation and interactions with the steel surface [25]. These findings support a mechanism involving a protonation, and proves the formation of an adsorbed film on the metallic surface.

Cystein is one aminoacid that can be oxidized to cysteic acid. It can be used as corrosion inhibitor in various substrates, mainly copper, but Silva et al. also studied its efficiency against corrosion of 304 stainless steel in acidic media. However it was found that in increased concentrations it activates the steel surface, whereas in low concentrations the passive film was not disturbed [26].

Aluminium alloys are susceptible to pitting corrosion. The corrosion inhibition ability of aminoacids to protect the AA 7075 in NaCl was investigated in detail by Bereket and Yurt [27]. The number and size of pits was decreased in the presence of glycine. A protonated form of the aminoacid molecule might be co-adsorbed on $AlCl_4^-$ based compounds, which were formed by the complexing of aluminium cation in oxide lattice with Cl^- anions. The consequence is that it prevents diffusion of $AlCl_4^-$, which is the main responsible for corrosion.

The inhibiting role of aminoacids is associated with the protonation of amino groups and to the presence of heteroatoms. The mechanisms are still controversial and divergent but it has been reported that after physical adsorption of the inhibitor molecule, there is an additional stage in which chemisorption occurs due to chelation over the metallic surface. The heteroatoms present in the aminoacid molecules may offer electrons to unoccupied d-orbitals of metals and accept the electrons in the d-orbitals of metals by using antibonding orbitals to form a feed-back bond as reported elsewhere [28].

6.2.3 Other Biological Inhibitors

There are several biological molecules that have been tested as corrosion inhibitors. Ascorbic acid is a biological like inhibitor, which has been reported as very effective for inhibition of steel corrosion. The mechanism is based on the ability of this molecule to induce early dissolution of Fe and formation of stable chelates, which delay the pitting initiation phase. At sufficiently high concentrations the inhibitor also delays the pit propagation phase [29]. One alkaloid called atropine can be found in some seeds and can be used for medicinal purposes. However, because its molecule contains some structural elements, such as one aromatic ring, carbonyl group and oxygen atom, it gathers some of the basic requirements to be used as a corrosion inhibitor. Raja et al. have tested it as corrosion inhibitor of mild steel in conditions simulating H_2SO_4 production. The inhibition ability depends upon the concentration and the mechanism involves adsorption and formation of a complex with the steel [30]. Caulerpin is also an alkaloid present in some marine algae. The corrosion inhibition effect studied in steel, in HCl electrolytes revealed that the efficiency is dependent upon concentration and temperature [31]. Uracil is a compound that plays important roles in biological activities and it is an N-heterocyclic compound. Condensed uracil compounds, resulting from the fusion of a uracil moiety with a pyran ring were studied as corrosion inhibitors for mild steel exposed to hydrochloric acid. By using quantum chemical simulation tools it was possible to establish the corrosion mechanism of this inhibitor (Fig. 6.3). The uracil compounds are protonated in acidic environment. The results show the formation of a surface film that follows a Langmuir isotherm that inhibits the corrosion process. The mechanism involves the formation of complexes with steel. When H_2 is released, the inhibitor returns to its neutral formula and the heteroatoms, which possess a free lone pair of electrons, promote chemical adsorption over the steel surface. The accumulation of electrons on the metallic surface turns it more negative, forcing the metal to transfer electrons from the d-orbital of Fe to a vacant π^* (antibonding) orbital of uracil (backdonation). This process strongly enhances the process of adsorption on the Fe surface and the formation of the protective layer [32].

A mussel protein obtained from the marine mussel *Mytilus edulis*, has drawn attention for its adhesive and corrosion protective properties as reported by Zhang et al. [33]. The adsorption of the protein was studied by AFM and the corrosion inhibition efficiency by electrochemical methods on steel surfaces exposed to NaCl electrolytes. The AFM observations showed that the protein film formed on the carbon steel surface is not homogeneous, presenting defects and/or large aggregates. It was expected that the mussel protein could form a surface layer by complexing with Fe^{3+} ions. Thus, in the defect sites there is formation of agglomerates, where the protein is able to complex with Fe^{2+} and Fe^{3+}. This process

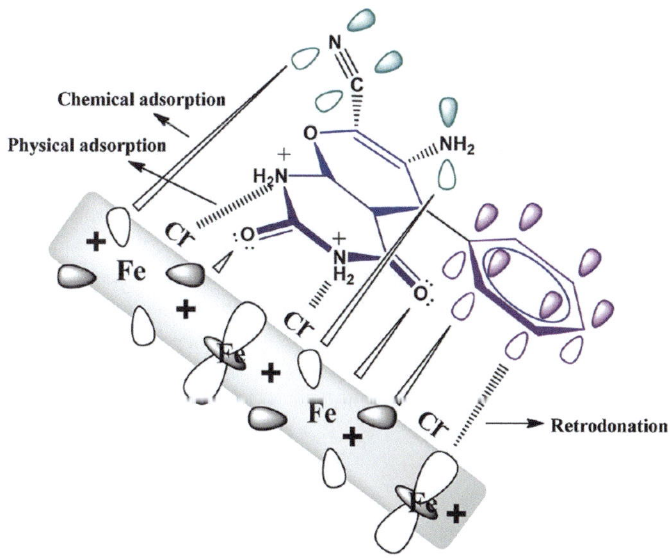

Fig. 6.3 Scheme showing the adsorption and inhibition mode of uracils on mild steel plates immersed in 1 M HCl [32]

slows down the inhibition efficiency at early stages of exposure in the aggressive solutions. The efficiency is similar to that of Bovine serum albumin.

Another class of biological inhibitors worthy of notice are the peptides (short chains of amino acids linked by amides). Like the aminoacids these compounds have interesting functionalities that make them suitable to be used as corrosion inhibitors. Abed et al. reports the use of Boc–Phe–Met–OCH_3 (where Boc is a blockage agent, *tert* butoxycarbonyl; Phe is phenylalanine and Met is methionine) as corrosion inhibitor for brass in 0.5 M nitric acid solution. The inhibition efficiency increases with the inhibitor concentration but decreased with temperature. The mechanism is essentially cathodic and involves the adsorption of the molecule on the brass surface, forming a barrier film [34].

The wide range of biological species that can be used as corrosion inhibitors is too wide to be fully covered. However, the most important features for increased inhibition efficiency and the main aspects governing the corrosion inhibition mechanism of the most common biological inhibitors are highlighted in the various works referred above.

6.2.4 *Bacteria and Fungi*

Living organisms are generally considered harmful when in contact with metallic substrates because they are likely to provoke corrosion. Microbiological induced corrosion is a well-known problem, well documented and with terrible implications in structural materials. However, there have been a few cases in which living organisms can provide corrosion protection. For example, a biological patina was reported to preserve copper artefacts. The capacity of a fungal strain, *Beauveria bassiana*, to precipitate copper oxalates was exploited for the stabilization of soluble patinas (copper hydroxysulfates) or the conversion of active corrosion products (copper hydroxychlorides) into protective compounds. In fact, copper oxalates form green compact patinas, showing a high degree of insolubility and chemical stability even in acid atmospheres close to pH 3 [35].

Stainless steel is very susceptible to corrosion in the presence of living organisms in inorganic media. Indeed, this is a severe phenomenon in marine environments and bio corrosion control is of the utmost relevance in ensuring the reliability of steel structures exposed to seawater. However, there are interesting findings in which some bacteria are reported to inhibit the corrosion process of stainless steel. Two bacterial species (*Pseudomonas sp.* and *B. megaterium*) were found to inhibit stainless steel 304 (SS 304) pitting corrosion in inorganic medium, containing nitrates and phosphates. The bacteria increase the production of extracellular polysaccharides and, hence, these promote the passivity of SS 304. The inhibition process could be related to the increased production of bacterial metabolites and formation of a passive film, composed of a dense and thick Fe_2O_3 and Cr_2O_3 layers. However, in organic media, the effect is the opposite and the same bacteria stimulate the pitting process [36].

The composition of the environment, the metal nature and the composition of interfacial electrolytes (e.g. pH, ions and their concentration, the oxygen amount) define the chemical composition and the properties of the protective films formed under biological activity. Therefore, it is very likely that in some specific conditions as the ones reported above, the microbial activity may induce protection of the metallic surfaces. The process generally involves either stabilization of protective surface layers or anodic inhibition.

6.3 Vegetable Inhibitors

Plant extracts, seeds and natural shell extracts are gaining importance nowadays as corrosion inhibitors. The reason is that these are classified as "green", biodegradable and in most of the cases very abundant and renewable. In general, vegetable extracts contain several constituents that may present functional groups that allow the formation of protective species or adsorbed layers on the metallic surface.

6.3.1 Extracts of Plants

Henna extracts (*Lawsonia inermis*) have been used, for centuries, in many applications, with different functions. The main constituents of Henna are lawsone, gallic acid, glucose and tannic acid. These extracts were investigated as corrosion inhibitors for mild steel coupons exposed to 1 M HCl. The inhibition efficiency depends upon the concentration of the henna extract. Electrochemical impedance measurements showed that the polarization resistance increased by about one order of magnitude when the concentration of the extract increased from zero up to 1.2 g/L. In addition in the presence of henna extract, the pitting potential is shifted towards more positive values. The inhibitor adsorbs on the iron surface and a chemisorption mechanism is likely to occur. However, the inhibition mechanism is dependent on the henna constituents. For example the inhibition by lawsone was attributed to the formation of sparingly insoluble complex compounds with the Fe^{2+} cations. Moreover, this compound, which is the most effective amongst the henna constituents, also shows good oxygen scavenging properties [37].

Another work reports that the extract of *Lawsonia* leafs can work as an effective inhibitor against corrosion of carbon steel, nickel and zinc in acidic, neutral and alkaline media. This inhibitor acts by a mixed mechanism, because it adsorbs on both the anodic and cathodic sites. When used with zinc substrates, it shifts the corrosion potential towards the anodic direction and decreases the corrosion current density in various media. However, the inhibition of zinc dissolution decreases in the order: neutral > alkaline > acid. The dependence of the corrosion inhibition efficiency on the electrolyte pH is quite different from that observed for the corrosion inhibition of carbon steel and nickel. For these two metals, the inhibitor forms stable and insoluble complexes on the surface that hinder the dissolution of the metal [38].

Acalypha torta is an extract that is used for medicinal purposes. However, it was also tested as corrosion inhibitor for mild steel in acidic media. The EIS results revealed improved protection with increased concentration of inhibitor—Fig. 6.4.

The extract behaves as a mixed inhibitor and its efficiency decreases with temperature. The inhibition mechanism proposes that some constituents of the molecule can be protonated, being able to form a protective surface film on the steel. The inhibition process follows the conventional mechanism, which involves

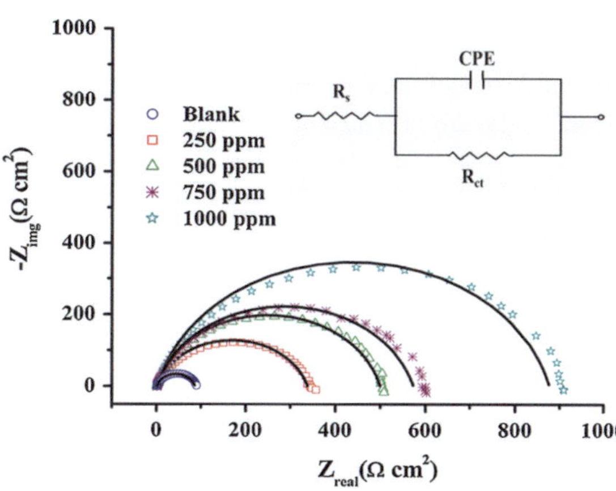

Fig. 6.4 Electrochemical impedance spectroscopy Nyquist plots (*solid line* shows fitted results) obtained on mild steel in 1 M HCl in the absence and presence of different concentrations of inhibitor [39]

the π-electrons of aromatic rings and lone pairs of electrons of heteroatoms [39]. Date palm (*Phoenix dactylifera L.*) is one of the most abundant palms. Extracts of their leaves can be used as corrosion inhibitors. Gerengi [40, 41] investigated its effectiveness in protecting aluminum alloys (AA) 7075 exposed to NaCl solutions. As it is well known this alloy is highly susceptible to pitting attack in the presence of chloride ions. The alloy contains zinc, as a primary alloying element along with copper and magnesium that create numerous types of intermetallics and precipitates on the surface, serving either as anodic or cathodic sites relatively to the metallic matrix. The extract of palm was found to act as a mixed corrosion inhibitor, with predominance of the cathodic effect. It was also demonstrated that the polarization resistance increased with the concentration of inhibitor. The extract forms a physically adsorbed surface layer, a process that seems to be closely associated with glucose compounds, which are present in the extract [40, 41].

The corrosion of aluminum plates in electrolytes of different pH and in the presence of different halides as aggressive anions was studied in the presence of *S. trifasciata* extract, which is obtained from the leafs of a plant. The results show that this extract inhibits aluminum corrosion by an adsorption process that follows a Freundlich isotherm. Moreover, it was found that the inhibition efficiency increased with the extract concentration and decreased with temperature, suggesting that a physical adsorption process is likely to explain the corrosion inhibition process. The presence of halide salts affects the inhibition efficiency of the extract [42]. Extracts of olive leaf displayed good inhibition ability for carbon steel exposed to acidic (HCl) electrolytes. The inhibition process was studied by El-Etre [43]. The results report that phenolic compounds, mainly oleuropein and hydroxytyrosol present in the olive leaf extract can physically adsorb on the steel surface, leading to an effective protection effect. The extract behaves essentially as a mixed inhibitor and it was found that it decreased the current density in the transpassive region. Another inhibitor tested on steel exposed to acid electrolytes (H_3PO_4) was artemisia oil. It contains various constituents, namely beta-thujone and works as cathodic inhibitor affecting the hydrogen reduction reaction [44].

Aloe vera is well known because of its medicinal properties. However, the literature reports that it can be an effective inhibitor for mild steel and aluminium in acidic environments [45]. *Aloe vera* contains heterocyclic compounds of high molecular weight and its molecular structure also contains oxygen and nitrogen atoms. Its inhibition efficiency was also tested on zinc exposed to HCl electrolytes. The maximum efficiency reported was 67 %, decreasing with an increase in temperature and the mechanism follows a Langmuir isotherm [46].

6.3.2 Shells and Seeds

The linseed oil molecule (see Fig. 6.5) is one of the most popular seed extracts for corrosion inhibition. This compound is compatible with various surface treatments and coatings and can also be encapsulated to prepare self-healing coatings. Thin

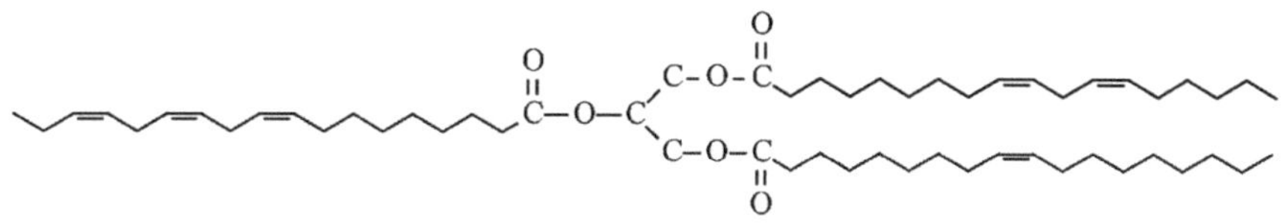

Fig. 6.5 Molecular structure of the linseed oil [47]

silane pre-treatments modified with linseed oil display better tribological properties and increased corrosion protection as reported by Bexell et al. [47].

Linseed oil is an effective inhibitor for steel. This effect was tested by encapsulating linseed oil in urea-formaldehyde capsules (also see Chap. 8). The incorporation of inhibitor was made during the synthesis of the capsules. The linseed oil is expected to be released from the filled microcapsules through the mechanical action of cracks that propagate in the coating in which the capsules are embedded. The released inhibitor can form a surface a film, resulting from oxidative polymerization of the molecules, that is able to prevent the ingress of dissolved aggressive agents [48]. The anti-corrosion behavior of encapsulated linseed oil in microcapsules for steel applications was also reported by Pilbáth et al. [49].

The corrosion inhibition of AA6061 in seawater in the presence of tapioca starch was studied by Rosliza and Wan Nik [50]. This molecule presents several hydroxyl groups as can be seen in the spectra depicted in Fig. 6.6. This spectrum reveals that the compound contains two main constituents, both polymers of glucose. One of these is amylpectin, which contains phosphate groups that establish stable bonds with the aluminum surface. The efficiency of the starch was reported to increase with the concentration of extract and the corrosion inhibition process involves the formation of stable precipitates on the metallic surface. The electrochemical measurements revealed that the extract works mainly as mixed inhibitor, however with predominance of cathodic inhibition.

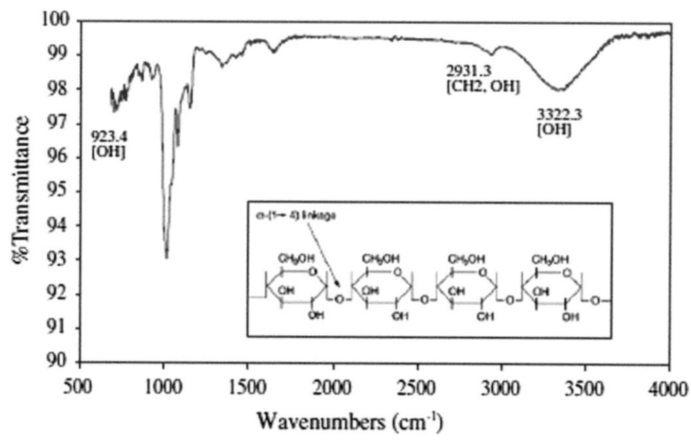

Fig. 6.6 FTIR spectra for the tapioca starch [50]

Caffeic acid (3-(3,4-dihydroxyphenyl)-2-propenoic acid, is a molecule extracted from the coffee grains and possessed properties and functional characteristics that suggest that it can work as corrosion inhibitor. Moreover, these compounds are soluble in water, in a wide pH range, being easy to use in different aggressive electrolytes. The compound was tested on iron coupons exposed in acidic media. The inhibitor works by adsorption on the steel surface, leading to a reduction of the anodic and cathodic current densities. Thus, by one side the inhibitor blocks the cathodic active sites and, on the other hand, it modifies the activation energy of the anodic reaction [51]. Caffeine was also tested as corrosion inhibitor for copper plates exposed to nitrate containing electrolytes. Voltammetric scans show that the anodic current decreased because of the formation of an adsorbed surface layer. The effect was observed even in the presence of KCl in the testing electrolyte [52]. Inhibitive effects were also observed on zinc surfaces exposed to caffeine [53]. Coconut derivatives were tested as corrosion inhibitors. An interesting performance was observed for mild steel in acidic media [54]. The *Coconut nucifera L.* is another extract that can be used as corrosion inhibitor. It shows good performance on aluminum coupons exposed to HCl electrolytes. The maximum inhibition efficiency was 93 % at 7.5 % (v/v) concentration. The inhibition mechanism follows an adsorption process that obeys a Langmuir adsorption isotherm [55]. There are numerous extracts of seeds with potential to be used as corrosion inhibitors. In general the mechanism depends upon the functional groups present in the active molecules. There are many instances in which the effectiveness of the extracts are reported, but little is known about either the active components or chemical composition and therefore, the mechanisms of corrosion inhibition are not fully explained and understood. These inhibitors are abundant and can be a renewable source of molecules with corrosion inhibition effectiveness.

6.3.3 Tannins

Tannins and polyphenolic compounds are very abundant and easy to extract from different sources. These molecules are well characterized and possess oxygen groups that can easily form chelates with metallic surfaces such as iron. The hydroxyl groups can form tannates with metallic ions, giving a dark product that stabilizes the surface oxides.

Tannins can be classified in two different families: (i) hydrolyzable tannins and (ii) condensed tannins. The corrosion efficiency of tannic acid depends on the metal or alloy to be protected and also on the characteristics of the environment.

Tannins can be used in a wide range of environments since the molecules are stable under different pHs. Mimosa tannin inhibits the corrosion of steel in alkaline media simulating reinforcing steel, in the presence of chlorides. The mechanism is explained by the formation of chelates with Fe(III) species that stabilize the oxide films [56]. The same inhibitor was also tested in acidic media (pH 1–2) and behaves as a mixed corrosion inhibitor, leading to the formation of adsorbed species on the steel surface. The mechanism of inhibition includes the formation of physically adsorbed ferric-tannate multilayers when the pH increased, as reported by Martinez and Stem [57]. The same authors reported that the inhibition efficiency decreased from pH 3–1.

Tannic acid was tested against mild steel corrosion in seawater under wet/dry cyclic conditions. The inhibition process was studied by weight loss measurements, DC polarization and electrochemical impedance spectroscopy. The inhibition effect of the compound was attributed to the formation of a protective ferric tannate film, which is relatively stable during the wet/dry cycles [58].

Mangrove tannins were studied as corrosion inhibitors for carbon steel exposed to HCl solutions. All the compounds tested are reported to act as cathodic inhibitors and their inhibitive efficiency was dependent on the concentration. Moreover, the efficiency is also pH dependent, being more effective in more acidic electrolytes [59].

The fundamentals of the inhibitive mechanism of chestnut tannin were studied by means of the frontier orbital and the inhibitor adsorption theories. Results obtained on cold rolled steel exposed to acidic electrolytes, evidenced that the various tannin constituents are most likely to adsorb as neutral molecules by means of electrostatic forces that are established between the dipoles of the molecule and the iron charged surface. The study highlights the fact that the chemisorption mechanism, via sharing of the oxygen lone-pair electrons or the aromatic ring π-electrons, between the inhibitor molecule and the vacant d-orbitals of iron, is not likely to occur. A plausible explanation resides on the unfavorable orientation of the molecules due to their complex geometric and structural properties. This result reveals that other mechanisms may be important to describe the inhibitive action of tannins [60]. Although tannins are established as effective corrosion inhibitors in various metallic substrates and, mainly, on steel, the action mechanism is not fully understood.

6.4 Pharmaceutical Drugs

Although very controversial, pharmaceutical drugs are attracting attention as corrosion inhibitors. There is some divergence in what concerns the "green" label of these molecules as corrosion inhibitors. Nevertheless, some relevant findings regarding their potential in corrosion inhibition are reviewed in this section.

Carbocyclic and heterocyclic compounds, with 5–6 rings are commonly found in many drugs. Moreover, benzene rings containing different substitutes are also commonly present in the molecular structure. The corrosion potential of numerous compounds were reviewed by Gece et al. in a recent review [61]. According to the literature, the most promising drugs are the ones that include in their molecular structure reactive centers like O, N and/or S atoms, containing lone pairs of electrons and aromatic rings with delocalized π-electrons and sufficiently high molecular weight. These molecular features are necessary to ensure that adsorption occurs on the metallic surface and that the metallic surface is effectively covered by the inhibitor film. Only in that way is the corrosion activity efficiently inhibited. However, the mechanism of action is dependent on many other parameters and although several works report the use of these compounds as corrosion inhibitors, the mechanisms of inhibition are still worthy of investigation.

Ketoconazole (KCZ) also known as Nizoral® is an antifungal drug. This molecule is a heterocyclic compound containing N, O, and aromatic rings, containing π bonds, which are expected to serve as active sites for the adsorption process (Fig. 6.7). This makes the compound attractive as a corrosion inhibitor. In addition, the molecule is quite heavy (molecular weight 508.4) and likely to effectively cover larger areas by an adsorption process. These molecules are also cheap and easily available [62].

Barbituric acids and its derivatives can be used as corrosion inhibitors and have been classified as compatible with health and environment concerns. These families of compounds were tested as corrosion inhibitors for mild steel exposed to acidic electrolytes and it was possible to understand its inhibition mechanism and to rank them regarding inhibiting efficiency. The inhibition mechanism involves the adsorption of the molecules on the steel surface, mainly by electrostatic interaction, following a Langmuir adsorption isotherm [63].

The corrosion inhibition mechanism of four antibacterial drugs compounds, namely, ampicillin (Amp), cloxacillin (Clox), flucloxacillin (Fluclox) and amoxycillin (Amox) were tested concerning corrosion of **aluminium** in 2 M HCl solution, taking into consideration the effect of temperature. The corrosion inhibition process was studied by combining different electrochemical techniques and weight loss measurements. The results of this study revealed important insights into the corrosion inhibition mechanism: (i) the inhibition efficiency increases by increasing the electron donor characteristics of the substituted groups; (ii) the efficiency increases when the temperature decreases; (iii) the mechanism involves the blocking of the electrode surface by adsorption of the inhibitor molecules through the active centers present in their structure; (iv) the adsorption obeys Langmuir adsorption [64].

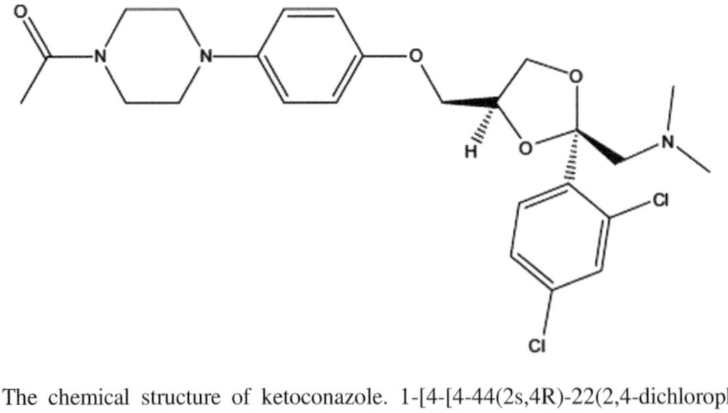

Fig. 6.7 The chemical structure of ketoconazole. 1-[4-[4-44(2s,4R)-22(2,4-dichlorophenyl)-22 (imidazol-1-ylmethyl)-1,3-dioxolan-4-yl]methoxy]phenyl]piperazin-1-yl]-ethanone [62]

The inhibiting action of some antihypertensive drugs (Enalapril maleate, Atenolol, Etilefrine hydrochloride) was studied in Al and three alloys of Al–Si exposed to hydrochloric acid. These compounds are expected to inhibit both uniform and pitting corrosion. The inhibition process is believed to be controlled by the formation of insoluble complexes on the aluminium surface. As the concentration of inhibitor increases, the efficiency also increases and the pitting potential is shifted towards more positive values. The inhibitor molecules can react with Al^{3+} ions; the process being dependent upon details of the molecular structure [65].

Xanthone is an organic compound with the molecular formula $C_{13}H_8O_2$ that can be used as an insecticide and to mitigate some bacterial activity. The molecule has high molecular weight and contains a carbonyl group and aromatic rings with delocalized bonds that can serve as initiator sites for the adsorption processes. Mild steel was exposed to sulphuric acid media and the corrosion inhibition was studied. The results reveal that the inhibitor efficiency increases when the concentration increases, but decreases when the temperature rises, effects that are likely to be associated with physical adsorption mechanisms [66].

Antibiotic molecules are also reported as potential corrosion inhibitors. For example, Cefatrexyl the commercial name of the sodium salt 7-[a-(4-pyridylthio)-acetamido]-cephalosporanic acid or simply, cephapirin sodium was tested as corrosion inhibitor for steel exposed to different acids. The inhibition efficiency depends upon the acid nature and temperature. Adsorption of cefatrexyl on iron surface follows a Langmuir isotherm in HCl solution, whereas in H_2SO_4, H_3PO_4 and $HClO_4$ electrolytes, the adsorption process follows Temkin's isotherm. The efficiency was found to be significantly decreased for HCl, especially at increased temperatures. The best performance was found in H_2SO_4 and H_3PO_4. In H_2SO_4 the neutral cefatrexyl molecules adsorb preferentially at the anodic areas, through a chemisorption process with the Fe surface. The process involves the unshared pairs of electrons existing on the N, S and O atoms and the vacant d-orbitals of Fe atoms. The process is more effective, if the inhibitor molecules follow an orientation parallel to the iron surface. A similar process was observed for H_3PO_4 [67]. An identical mechanism corrosion inhibition was reported for Streptomycin. This compound protonates in acidic media and behaves essentially as mixed corrosion inhibitor. In the process, the protonated species adsorb at the cathodic sites and inhibit H_2 evolution. Concomitantly, the adsorption on anodic sites occurs through π-electrons of the aromatic ring and the lone pairs of electrons present in the nitrogen atom, hindering the anodic dissolution of iron [68]. The effect of rhodanine drugs was evaluated in stainless steel 304 exposed to HCl electrolytes. These molecules are likely to provide corrosion protection, working by a mixed mechanism. The presence of methoxy groups in the inhibiting molecule seems to improve the inhibitor efficiency. The inhibition mechanism follows a Langmuir adsorption isotherm and the efficiency decreases with temperature [69].

It is clear that several drugs can be used as corrosion inhibitors and can be used to protect various metallic substrates. As concluded for the previous inhibitors (biological and vegetable) the mechanism depends on specific features of the inhibiting molecule, nature of the metallic surface and environmental parameters (pH, aggressive agents, temperature) and concentration of inhibitor.

6.5 Rare-Earth Inhibitors

Amongst the rare-earth corrosion inhibitors, cerium and lanthanum salts have been widely studied. Others such as yttrium and praseodymium have also been studied but to a much less extent. Its inhibition efficiency has been reported for aluminium (including various aluminium alloys), steel, galvanized steel and stainless steels. The mechanisms have been widely discussed but are strongly dependent on the metallic surface chemistry, local pH variations and nature of the rare-earth salt ion. For example cerium ions seem quite effective over aluminium and zinc, but much less on iron substrates. Rare-earths have been tested as direct additives in solution, to prepare conversion coatings or as additives in protective coatings.

6.5.1 Rare-Earth Cations

Water soluble rare-earth salts are one of the most effective sources of rare earth cations. Blin et al. reported the corrosion inhibition mechanism of rare earth cinnamate compounds, REM-4 hydroxy cinnamate (4-OHCin) (REM being lanthanum and mischmetal), on steel exposed to NaCl solutions. The electrochemical results revealed that $La(4\text{-}OHCin)_3$ affects both the rate of the anodic and cathodic reactions. However, there is a predominance of anodic inhibition, which reduces pitting activity on steel. Electrochemical impedance results revealed that the corrosion inhibition processes were accompanied by the build-up of a protective film over time. Furthermore, this film seems to have regions with different roles. Thus, the outer surface layers seem to offer less resistance to the diffusion of electrochemically active species than the highly resistive inner layer at the film/metal interface [70].

Steel corrosion induced by CO_2 and in NaCl in aerated electrolytes was studied in the presence of Praseodymium 4-hydroxycinnamate ($Pr(4OHCin)_3$). In naturally-aerated systems the rare earth inhibitor leads to the formation of a continuous and protective surface film, while in CO_2-saturated systems, more heterogeneous surface films are formed thus being less effective. The electrochemical data revealed that the $Pr(4OHCin)_3$ provides anodic and cathodic inhibition in the presence of CO_2 (in contrast to the predominantly anodic inhibition in aerated NaCl media) [71].

Cerium and other rare-earths have been widely investigated as inhibitors for aluminum alloys, namely the AA2024. Yasakau et al. [72] report that the addition of lanthanum(III) or cerium(III) cations leads to formation of rare-earth hydroxides over the active intermetallics, specifically the S-phase (Al_2CuMg), hindering both the anodic and the cathodic reactions, occurring at these corrosion sensitive locations. The hydroxide precipitation is a consequence of the pH increase that occurs in the neighborhood of the active S-phase. The presence of these precipitates slows down the copper re-deposition process and therefore the overall corrosion rate. The results also account for a linear time dependence of the rare-earth hydroxides growth, revealing that the corrosion processes require a balance between diffusion and electrochemical reactions. The work also demonstrates that cerium nitrate shows superior corrosion inhibition properties in comparison with lanthanum nitrate [72]. Cerium cinnamate is also reported as an effective inhibitor for AA2024, behaving as an anodic inhibitor at early stages. However, at later stages after pitting onset it works as a mixed inhibitor, delaying pitting propagation [73]. The corrosion inhibition of various AA 7xxx in the presence of cerium diphenyl phosphate was investigated by Hill et al. [74] and the results demonstrate that the surface of the alloys are protected because it is covered by a fine surface film, composed of distributed discrete Ce containing particles, either on the matrix or over the cathodic Cu containing intermetallics. Moreover, through TOF-SIMS analysis it was demonstrated that there was the formation of a complex involving Al cations and the cerium organophosphate.

A comparison between the inhibition efficiency of $LaCl_3$ and $CeCl_3$ in pure aluminium exposed to NaCl solutions was reported by Mishra and Balasubramaniam [1]. Both species lead to the formation of stable oxides and hydroxides precipitates over the cathodic regions. By increasing the $LaCl_3$ concentration up to 1000 ppm it was demonstrated that the Al surface was covered with a fine and relatively homogeneous surface film made of globular precipitates. The inhibition efficiency was found to be slightly better for $LaCl_3$ and the effects seem to be controlled by the more uniform distribution of precipitates.

Recently Yasakau et al. [72] reported that cerium molybdate nanowires inhibit the corrosion process in AA2024. It was demonstrated that the S phase was more stable in NaCl, when nanowires were added as anti-corrosion additive. The corrosion mechanism is related with the release of cerium ions from the cerium nanowires that precipitate preferentially over the anodic sites, especially at later stages of the corrosion process. The molybdate cations also play a relevant role, suppressing dissolution of Al and Mg.

6.5.2 Rare-Earth Conversion Coatings

Rare earth metals have been preferred to deposit corrosion protective conversion coatings, aiming to replace chromate based surface pre-treatments. These conversion coatings are being applied to various substrates. The corrosion protective behavior depends on several factors such as the nature of the anion in the rare-earth salt. This effect was investigated by Montemor et al. [75] on Mg AZ31 alloys and it was found that the anion has a great influence on the inhibition of corrosion activity of the bare material. Potentiodynamic polarization measurements revealed that the

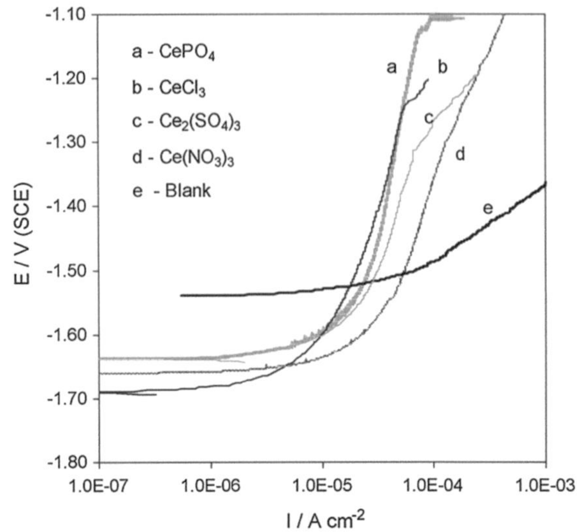

Fig. 6.8 Anodic polarization curves for the AZ31 coupons immersed in NaCl 0.005 M. The coupons were treated for 3600 s in the conversion baths

corrosion protection mechanism involves a clear displacement of the corrosion potential towards more cathodic values and a significant decrease of both anodic and cathodic current densities (Fig. 6.8). Moreover, XPS and Auger measurements provided some relevant highlights concerning the protection of the conversion films formed under the various salts. Thus, the thicker films were the ones obtained from $CeCl_3$, whereas the thinner ones were obtained from $CePO_4$. The films formed in $CeCl_3$ solutions are richer in Ce(IV) when compared to those formed in $CePO_4$ and $Ce_2(SO_4)_3$. However, the composition depends upon the time and, for longer exposure times, Ce(IV) oxides predominate over cerium hydroxides in all the conversion films.

The corrosion mechanism of AZ31 Mg alloys involves the dissolution of Mg and Al ions. When the AZ31 is exposed to Ce ion containing solutions, Ce^{3+} and Ce^{4+} ions can precipitate as hydroxides on the surface, mainly at the cathodic areas, where hydroxyl ions are made available. The E-pH diagram for Cerium suggest that Ce^{4+} hydroxides precipitate at lower pH values and that these compounds are expected to be formed in a first stage [76]. With increasing exposure time, the presence of cerium oxides starts to be important as well. This can be attributed to the conversion of $Ce(OH)_4$ into CeO_2. Hayes et al. [77] propose a mechanism that involves a solid phase transformation according to the scheme below:

$$Ce(OH)_4 \rightarrow CeO_2 + 2H_2O$$

Taking these results into consideration, the following mechanism was postulated to describe the corrosion protection of conversion coatings applied on AZ31 [2]: (i) dissolution of the native oxide, accompanied by formation of hydroxyl ions and pH rise; (ii) growth of a first layer composed of Ce(IV) hydroxides and Ce(III) hydroxides mixed with Mg and Al hydroxides during the first instants of immersion; (iii) thickening of the surface film and weaker pH changes with preferential deposition of $Ce(OH)_4$ and its conversion into CeO_2, forming an outer layer, richer in Ce(IV) species. Identical mechanism can also be applied to lanthanum conversion layers.

The mechanism of corrosion inhibition of rare-earth conversion coatings was observed in various metallic substrates. For example, conversion layers on galvanized steel from cerium, yttrium and lanthanum nitrate solutions result in a surface film that constrains the cathodic reduction of oxygen and the anodic dissolution of zinc, leading to a decrease of the corrosion rate. The effectiveness is dependent upon the cation, with lanthanum nitrate being more effective than cerium nitrate [78]. Corrosion inhibition of galvanized steel corrosion by lanthanum conversion layers was studied in the presence of citric acid. The mechanism involves inhibition of the anodic and cathodic activity [79].

6.5.3 *Rare-Earth Compounds Additives in Coatings*

The inhibition ability of rare earth compounds make them attractive to be used as additives in organic, hybrid and inorganic coatings. Montemor et al. reported the inhibition ability of Ce modified silane coatings and it was found that the presence of cerium provides corrosion protection due to the formation of protective species at the active corrosion sites and by increasing the crosslinking of hybrid silane films. The protective role of Ce modified coatings was confirmed to various substrates: galvanized steel, cold rolled steel and Mg alloys [80–83]. The effects of corrosion inhibition on AA 2024-T3 provided by cerium cations and cinnamate anions added to epoxy coatings were investigated by Shi et al. [84]. The corrosion activity in artificial defects was investigated by the scanning vibrating electrode technique. The results revealed a cathodic inhibition mechanism, in which Ce(III) hydroxides are converted into CeO_2 precipitates over the cathodic sites. The mechanism of inhibition was supported by XPS measurements that revealed the formation of Ce containing compounds at the metal-coating interface.

Incorporation of cerium nitrate in smart reservoirs or on the surface of pH sensitive nanoparticles, such as hydroxyapatite, calcium carbonate microbeads and nanoclays is a very interesting strategy for developing self-healing coatings and has been investigated by various authors. It was shown that cerium loaded carriers can provide enhanced corrosion inhibition [85, 86]. Effective corrosion inhibition effects were also observed for silica and ceria particles loaded with Ce(III) ions. These cerium modified additives are compatible with sol-gel coatings and provide corrosion protection of galvanized steel substrates. In these systems the corrosion protection mechanisms present some particularities. It was demonstrated that ceria particles, by themselves could shift the corrosion potential of zinc towards more anodic values, with a simultaneous reduction of the anodic current density. However, the loading of the ceria nanoparticles with cerium ions induced relevant changes in the corrosion protection mechanism that included a significant reduction of the cathodic current density [81]. The corrosion inhibition mechanism of cerium loaded inorganic nanoparticles is linked with the formation of stable and protective layers of metallic oxides and hydroxides formed under the alkaline environment that characterizes the cathodic processes. Recently it was reported that this process is effective only if the inhibitor is able to penetrate in the alkaline region. This effect was demonstrated by simulation results, which reveal a threshold for the corrosion inhibitor solubility below which no passivation of the cathode occurs [87].

The literature points out that rare-earth compounds fulfill various requirements that classify them as environmentally friendly since they have low toxicity and very good efficiency. However, being rare-earths there are some concerns regarding their price and availability as corrosion inhibitors for massive application [88]. Nevertheless, taking into account the corrosion inhibition effectiveness of rare earths in various metallic substrates and the possibility of using them in various formulations, namely the possibility of encapsulation, makes these compounds very attractive to solve several corrosion problems.

6.6 Mixtures and Synergies of New Inhibitors

Synergistic combination of corrosion inhibitors is a strategy that presents high potential in the advancement of novel corrosion protection strategies. Through this process it is possible to combine different inhibitors, allowing them to work together in a synergistic way. This ensures higher inhibition efficiency and, eventually the use of smaller amounts of corrosion inhibitors, which might have important implications, for example, in coatings design.

Mixtures of rare earth cations can provide synergistic effects for corrosion inhibition of AA2024 immersed in NaCl. A combinatorial matrix of rare earth chloride mixtures was studied recently and it allowed the definition of the best combinations of inhibitors. Figure 6.9 displays modelling results that show how the corrosion current density is expected to change for various combinations of rare earth inhibitors. The results show that the highest inhibition efficiency is obtained with a ratio of Ce:(La, Pr) of around 3:1. It is demonstrated that as the amount of Ce decreased from levels of 1E-3 M, the corrosion inhibition effect is expected to decrease, regardless of the concentration of other inhibitors present in solution (La, Pr and Nd) and regardless of the fact that they are added alone or combined. Moreover, the individual rare-earth inhibitors, La, Pr and Nd display levels of efficiency that are in general lower than that of Cerium ions. In the absence of Ce, the most promising mixture is based on approximately 50 % Pr and 50 % Nd [89].

The corrosion protective synergies between cerium(IV) ions and vanillin was assessed on cold rolled steel (CRS) in 1.0 M HCl. The individual inhibitors provided some corrosion protection, but when combined there was an important improvement of the process of corrosion inhibition. Vanillin, individually, behaves as mixed corrosion inhibitor, but with predominance of cathodic inhibition. On the other hand, Ce(IV) by itself shows little effects on the corrosion of steel, but the

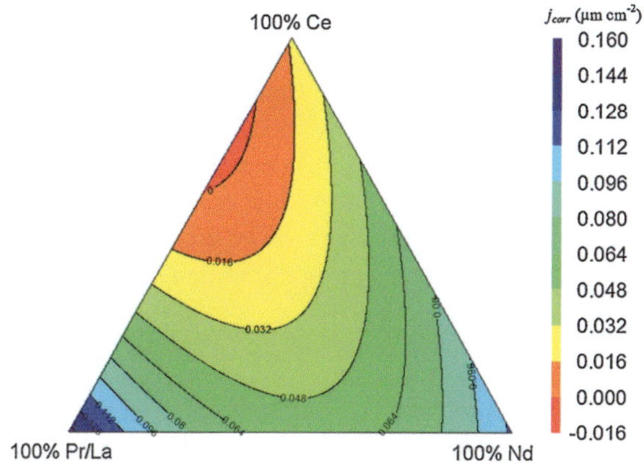

Fig. 6.9 Response curve of the quadratic model for corrosion current (A cm^{-2}) as determined by extrapolation from cathodic polarisation for mixtures. The *scale* on the *right hand side* shows the *colour* coding used, with *red* representing the lowest currents (and thus best corrosion inhibition) and *violet* representing the highest [89]

mixture of both shows ability to inhibit both anodic and cathodic reactions. The corrosion mechanism is proposed to involve a Ce(IV)-Vanillin complex [90].

Catubig et al. [91] investigated the synergistic behavior of cerium mercapto acetate ($Ce(MAcet)_3$) and praseodymium mercapto acetate ($Pr(MAcet)_3$) on the corrosion process of AA2024. The mechanism of individual and combined inhibitors was studied by various electrochemical techniques and it was found that there is an important reduction of the corrosion activity. The effectiveness depends both on the immersion time and inhibitor concentration. Increased concentration of inhibitors led to dominance of cathodic inhibition, while anodic inhibition was more significant after longer immersion times. A protective film, consisting of various $RE(MAcet)_3$ inhibitor species is expected to be formed on the metallic surface, limiting the cathodic reactions. The presence of thiol groups is expected to favor the deposition of the inhibitors over the cathodic sites [91].

Sodium oleate and Ce^{4+} cations revealed a synergistic effect on the corrosion process of cold rolled steel in acidic media. The corrosion protection mechanism involves the formation of a complex between Ce^{4+} and the oleate that promotes the inhibition process. Oleate acts essentially as a cathodic inhibitor, while the mixture provides a mixed corrosion inhibition process [92]. The synergistic effect of mixtures of various inorganic salts that work as corrosion inhibitors was studied by Chambers and Taylor [93]. The authors report various possible combinations of inhibitors and found a dependence both on the concentration ratio of inhibitors and pH. Some inhibitor mixtures revealed performances comparable to chromates. Mixtures of $CeCl_3/Na_2MoO_4$, $LaCl_3/Na_2MoO_4$, and $NaVO_3/Na_3PO_4$ demonstrated very effective inhibition ability comparatively to conventional Na_2CrO_4 [93].

The synergistic combination of inhibitors is a promising route to increase the protective efficiency of inhibitors. On the one hand, it is expected to achieve higher efficiency with the combination, but on the other hand it is expected to reduce the usage of the inhibitors, as efficiency above the sum of the individuals is expected. However, it is essential to ensure that no detrimental interactions occur between distinct inhibitor compounds, with negative results. One way to achieve successful results consists on combining inhibitors that have different inhibition mechanisms or different kinetics of inhibition. In general dual combinations are reported, but the combinative options can be wider and mixtures of three or even more inhibitors can be applied to force synergistic corrosion inhibition efficiency.

6.7 Future Trends

The search for corrosion inhibitors that can be classified as "green" and that can be easily found and extracted from renewable sources is undoubtedly one research area where several developments are continually evolving. This chapter overviewed some of the most relevant insights concerning the latest developments in this field. If on the one hand several compounds have been established as promising corrosion inhibitors, on the other hand the corrosion inhibition mechanisms of several corrosion inhibitors from "green" generations are still unknown. Thus, this uncertainty opens a wide array of possibilities to explore in the future.

Effective corrosion inhibitors can be found in biological species and vegetable mixtures or even in synthetic molecules that aim at mirroring natural compounds. Several organic compounds can be proposed as green and natural corrosion inhibitors. Although numerous studies have been reported in the literature, the corrosion inhibition mechanisms are still unknown in many cases. Generally the corrosion inhibition process proceeds by one (or more) mechanisms: (i) physical and/or chemical adsorption at the metal/electrolyte interface; (ii) electrostatic interactions that involve protonated groups that exist in the inhibitor molecule and metallic cations; (iii) chemical interaction between the unshared electrons (like those present on N, S and O atoms) and vacant d-orbitals present in the electronic structure of metallic atoms; (iii) formation of stable inhibitor-metallic complexes. However, corrosion efficiency is strongly controlled by the formation of chemically stable, homogeneous and well organized protective surfaces. The stability of protective species is also controlled by the physical and chemical properties of the inhibiting compounds, the presence of functional groups such as hydroxyl, carbonyl, amino and aromatic rings. Other relevant parameters include the density of electronic donors, composition of the aggressive electrolyte, local and bulk pH and oxygen contents, morphology, chemical composition and charge characteristics of the metal surface.

Rare-earth compounds play a major role in inhibiting the corrosion activity in various substrates. Generally for these compounds a cathodic inhibition process that involves the formation of stable hydroxides and oxides is reported. However, some other mechanisms that include anodic inhibition, mixed inhibition and formation of complexes are necessary to explain the effective role of rare-earths.

Undoubtedly, synergistic mixtures of corrosion inhibitors, presenting different mechanisms of action and kinetics of release, will become important routes to mitigate corrosion activity in various technical domains. The exploration of the various possibilities open an avenue of new research lines with a high technical and economic impact.

Of utmost relevance, is the use of mixtures of compounds that both provide synergistic corrosion inhibition as well as the beneficial effects over other properties required for the metallic substrate such as wear, lubricating and hydrophilic or hydrophobic properties, among others. These multifunctional corrosion inhibitors, of green label and natural origin, especially if able to form functional layers on the metallic surface to be protected can lead to disruptive discoveries in the near future.

Acknowledgement FCT for financial support to CQE- UID/QUI/00100/2013.

References

1. A.K. Mishra, R. Balasubramaniam, Corrosion inhibition of aluminium by rare earth chlorides. Mater. Chem. Phys. **103**(2–3), 385–393 (2007)
2. M.F. Montemor, A.M. Simoes, M.J. Carmezim, Characterization of rare-earth conversion films formed on the AZ31 magnesium alloy and its relation with corrosion protection. Appl. Surf. Sci. **253**(16), 6922–6931 (2007)
3. S.V. Lamaka, M.L. Zheludkevich, K.A. Yasakau, M.F. Montemor, M.G.S. Ferreira, High effective organic corrosion inhibitors for 2024 aluminium alloy. Electrochim. Acta **52**(25), 7231–7247 (2007)
4. M.R. Arshadi, M. Lashgari, G.A. Parsafar, Cluster approach to corrosion inhibition problems: interaction studies. Mater. Chem. Phys. **86**(2–3), 311–314 (2004)
5. A.E. Hughes, F.H. Scholes, A.M. Glenn, D. Lau, T.H. Muster, S.G. Hardin, Factors influencing the deposition of Ce-based conversion coatings, Part I: The role of Al_{3+} ions. Surf. Coat. Technol. **203**(19), 2927–2936 (2009)
6. S. Cheng, S. Chen, T. Liu, X. Chang, Y. Yin, Carboxymenthylchitosan as an ecofriendly inhibitor for mild steel in 1 M HCl. Mater. Lett. **61**(14–15), 3276–3280 (2007)
7. S. Cheng, S. Chen, T. Liu, X. Chang, Y. Yin, Carboxymethylchitosan + Cu^{2+} mixture as an inhibitor used for mild steel in 1 M HCl. Electrochim. Acta **52**(19), 5932–5938 (2007)
8. A.M. Fekry, R.R. Mohamed, Acetyl thiourea chitosan as an eco-friendly inhibitor for mild steel in sulphuric acid medium. Electrochim. Acta **55**(6), 1933–1939 (2010)
9. X.-H. Yang, S.-Q. Liao, J.-H. Liao, Study on corrosion inhibition of chitosan quaternary ammonium for carbon steel in sulfuric acid. Corros. Sci. Prot. Technol. **19**, 255–258 (2007)
10. A. Eddib, Y.A. Albrimi, A.A. Addi, J. Douch, R.M. Souto, M. Hamdani, Inhibitory action of non toxic compounds on the corrosion behaviour of 316 austenitic stainless steel in hydrochloric acid solution: comparison of chitosan and cyclodextrin. Int. J. Electrochem. Sci. **7**, 6599–6610 (2012)
11. M.N. El-Haddad, Chitosan as a green inhibitor for copper corrosion in acidic medium. Int. J. Biol. Macromol. **55**, 142–149 (2013)
12. M.L. Zheludkevich, J. Tedim, C.S.R. Freire, S.C.M. Fernandes, S. Kallip, A. Lisenkov, A. Gandini, M.G.S. Ferreira, Self-healing protective coatings with "green" chitosan based pre-layer reservoir of corrosion inhibitor. J. Mater. Chem. **21**(13), 4805–4812 (2011)

13. O. Lundvall, M. Gulppi, M.A. Paez, E. Gonzalez, J.H. Zagal, J. Pavez, G.E. Thompson, Copper modified chitosan for protection of AA-2024. Surf. Coat. Technol. **201**(12), 5973–5978 (2007)
14. A.R. Cestari, E.F. Vieira, F.J. Alves, E.C. Silva, M.A. Andrade Jr., A novel and efficient epoxy/chitosan cement slurry for use in severe acidic environments of oil wells-structural characterization and kinetic modeling. J. Hazard. Mater. **213–214**, 109–116 (2012)
15. P. Dong, W. Hao, Y. Xia, G. Da, T. Wang, Comparison study of corrosion behavior and biocompatibility of polyethyleneimine (PEI)/heparin and chitosan/heparin coatings on NiTi alloy. J. Mater. Sci. Technol. **26**(11), 1027–1031 (2010)
16. B.-D. Hahn, D.-S. Park, J.-J. Choi, J. Ryu, W.-H. Yoon, J.-H. Choi, H.-E. Kim, S.-G. Kim, Aerosol deposition of hydroxyapatite–chitosan composite coatings on biodegradable magnesium alloy. Surf. Coat. Technol. **205**(8–9), 3112–3118 (2011)
17. I. Deen, X. Pang, I. Zhitomirsky, Electrophoretic deposition of composite chitosan–halloysite nanotube–hydroxyapatite films. Colloids Surf. A **410**, 38–44 (2012)
18. G. Gece, S. Bilgiç, A theoretical study on the inhibition efficiencies of some amino acids as corrosion inhibitors of nickel. Corros. Sci. **52**(10), 3435–3443 (2010)
19. W.A. Badawy, K.M. Ismail, A.M. Fathi, Corrosion control of Cu–Ni alloys in neutral chloride solutions by amino acids. Electrochim. Acta **51**(20), 4182–4189 (2006)
20. K.F. Khaled, Adsorption and inhibitive properties of a new synthesized guanidine derivative on corrosion of copper in 0.5 M H2SO4. Appl. Surf. Sci. **255**(5), 1811–1818 (2008)
21. K.F. Khaled, Guanidine derivative as a new corrosion inhibitor for copper in 3 % NaCl solution. Mater. Chem. Phys. **112**(1), 104–111 (2008)
22. M.M. El-Rabiee, N.H. Helal, G.M.A. El-Hafez, W.A. Badawy, Corrosion control of vanadium in aqueous solutions by amino acids. J. Alloy. Compd. **459**(1–2), 466–471 (2008)
23. M.A. Kiani, M.F. Mousavi, S. Ghasemi, M. Shamsipur, S.H. Kazemi, Inhibitory effect of some amino acids on corrosion of Pb–Ca–Sn alloy in sulfuric acid solution. Corros. Sci. **50**(4), 1035–1045 (2008)
24. O. Olivares-Xometl, N.V. Likhanova, M.A. Domínguez-Aguilar, E. Arce, H. Dorantes, P. Arellanes-Lozada, Synthesis and corrosion inhibition of α-amino acids alkylamides for mild steel in acidic environment. Mater. Chem. Phys. **110**(2–3), 344–351 (2008)
25. O. Olivares, N.V. Likhanova, B. Gómez, J. Navarrete, M.E. Llanos-Serrano, E. Arce, J.M. Hallen, Electrochemical and XPS studies of decylamides of α-amino acids adsorption on carbon steel in acidic environment. Appl. Surf. Sci. **252**(8), 2894–2909 (2006)
26. A.B. Silva, S.M.L. Agostinho, O.E. Barcia, G.G.O. Cordeiro, E. D'Elia, The effect of cysteine on the corrosion of 304L stainless steel in sulphuric acid. Corros. Sci. **48**(11), 3668–3674 (2006)
27. G. Bereket, A. Yurt, The inhibition effect of aminoacids and hydroxy carboxylic acids on pitting corrosion of aluminum alloy 7075. Corros. Sci. **43**, 1179–1195 (2011)
28. A. Yurt, G. Bereket, C. Ogretir, Quantum chemical studies on inhibition effect of amino acids and hydroxy carboxylic acids on pitting corrosion of aluminium alloy 7075 in NaCl solution. J. Mol. Struct. (Thoechem) **725**(1–3), 215–221 (2005)
29. L. Valek, S. Martinez, D. Mikulić, I. Brnardić, The inhibition activity of ascorbic acid towards corrosion of steel in alkaline media containing chloride ions. Corros. Sci. **50**(9), 2705–2709 (2008)
30. P. Bothi Raja, M.G. Sethuraman, Atropine sulphate as corrosion inhibitor for mild steel in sulphuric acid medium. Mater. Lett. **62**(10–11), 1602–1604 (2008)
31. C. Kamal, M.G. Sethuraman, Caulerpin—a bis-indole alkaloid as a green inhibitor for the corrosion of mild steel in 1 M HCl solution from the marine Alga Caulerpa racemosa. Ind. Eng. Chem. Res. **51**(31), 10399–10407 (2012)
32. D.K. Yadav, M.A. Quraishi, Application of some condensed uracils as corrosion inhibitors for mild steel: gravimetric, electrochemical, surface morphological, UV–visible, and theoretical investigations. Ind. Eng. Chem. Res. **51**(46), 14966–14979 (2012)
33. F. Zhang, J. Pan, P.M. Claesson, Electrochemical and AFM studies of mussel adhesive protein (Mefp-1) as corrosion inhibitor for carbon steel. Electrochim. Acta **56**(3), 1636–1645 (2011)

34. Y. Abed, M. Kissi, B. Hammouti, M. Taleb, S. Kertit, Peptidic compound as corrosion inhibitor for brass in nitric acid solution. Prog. Org. Coat. **50**(2), 144–147 (2004)
35. E. Joseph, A. Simon, R. Mazzeo, D. Job, M. Wörle, Spectroscopic characterization of an innovative biological treatment for corroded metal artefacts. J. Raman Spectrosc. **43**(11), 1612–1616 (2012)
36. A. Rajasekar, Y.-P. Ting, Role of inorganic and organic medium in the corrosion behavior of Bacillus megaterium and Pseudomonas sp. in stainless steel SS 304. Ind. Eng. Chem. Res. **50** (22), 12534–12541 (2011)
37. A. Ostovari, S.M. Hoseinieh, M. Peikari, S.R. Shadizadeh, S.J. Hashemi, Corrosion inhibition of mild steel in 1 M HCl solution by henna extract: a comparative study of the inhibition by henna and its constituents (Lawsone, Gallic acid, α-d-Glucose and Tannic acid). Corros. Sci. **51**(9), 1935–1949 (2009)
38. A.Y. El-Etre, M. Abdallah, Z.E. El-Tantawy, Corrosion inhibition of some metals using lawsonia extract. Corros. Sci. **47**(2), 385–395 (2005)
39. P.M. Krishnegowda, V.T. Venkatesha, P.K.M. Krishnegowda, S.B. Shivayogiraju, Acalypha torta leaf extract as green corrosion inhibitor for mild steel in hydrochloric acid solution. Ind. Eng. Chem. Res. **52**(2), 722–728 (2013)
40. H. Gerengi, Anticorrosive properties of date palm (Phoenix dactylifera L.) fruit juice on 7075 type aluminum alloy in 3.5 % NaCl solution. Ind. Eng. Chem. Res. **51**(39), 12835–12843 (2012)
41. H. Gerengi, H.I. Sahin, Schinopsis lorentzii extract as a green corrosion inhibitor for low carbon steel in 1 M HCl solution. Ind. Eng. Chem. Res. **51**(2), 780–787 (2012)
42. E.E. Oguzie, Corrosion inhibition of aluminium in acidic and alkaline media by Sansevieria trifasciata extract. Corros. Sci. **49**(3), 1527–1539 (2007)
43. A.Y. El-Etre, Inhibition of acid corrosion of carbon steel using aqueous extract of olive leaves. J. Colloid Interface Sci. **314**(2), 578–583 (2007)
44. M. Benabdellah, M. Benkaddour, B. Hammouti, M. Bendahhou, A. Aouniti, Inhibition of steel corrosion in 2 M H3PO4 by artemisia oil. Appl. Surf. Sci. **252**(18), 6212–6217 (2006)
45. A.O. James, A.O. Atela, Aloe vera: an inhibitor of aluminium corrosion in hydrochloric acid solution. Int. J. Pure Appl. Chem. **3**, 159–163 (2008)
46. O.K. Abiola, A.O. James, The effects of Aloe vera extract on corrosion and kinetics of corrosion process of zinc in HCl solution. Corros. Sci. **52**(2), 661–664 (2010)
47. U. Bexell, M. Olsson, M. Johansson, J. Samuelsson, P.E. Sundell, Preparation and characterization of microcapsules containing linseed oil and its use in self-healing coatings. Prog. Org. Coat. **63**, 72–78 (2008)
48. C. Suryanarayana, K.C. Rao, D. Kumar, Preparation and characterization of microcapsules containing linseed oil and its use in self-healing coatings. Prog. Org. Coat. **63**(1), 72–78 (2008)
49. A. Pilbáth, T. Szabó, J. Telegdi, L. Nyikos, SECM study of steel corrosion under scratched microencapsulated epoxy resin. Prog. Org. Coat. **75**(4), 480–485 (2012)
50. R. Rosliza, W.B. Wan Nik, Improvement of corrosion resistance of AA6061 alloy by tapioca starch in seawater. Curr. Appl. Phys. **10**(1), 221–229 (2010)
51. F.S. de Souza, A. Spinelli, Caffeic acid as a green corrosion inhibitor for mild steel. Corros. Sci. **51**(3), 642–649 (2009)
52. T. Fallavena, M. Antonow, R.S. Gonçalves, Caffeine as non-toxic corrosion inhibitor for copper in aqueous solutions of potassium nitrate. Appl. Surf. Sci. **253**(2), 566–571 (2006)
53. F.N. Grosser, R.S. Gonçalves, Electrochemical evidence of caffeine adsorption on zinc surface in ethanol. Corros. Sci. **50**(10), 2934–2938 (2008)
54. S.A. Umoren, I.B. Obot, A.U. Israel, P.O. Asuquo, M.M. Solomon, U.M. Eduok, A.P. Udoh, Inhibition of mild steel corrosion in acidic medium using coconut coir dust extracted from water and methanol as solvents. J. Ind. Eng. Chem. (2013)
55. O.K. Abiola, Y. Tobun, Cocos nucifera L. water as green corrosion inhibitor for acid corrosion of aluminium in HCl solution. Chin. Chem. Lett. **21**(12), 1449–1452 (2010)
56. J. Flis, T. Zakroczymski, Impedance study of reinforcing steel in simulated pore solution with tannin. J. Electrochem. Soc. **143**, 2458–2464 (1998)

57. S. Martinez, I. Stem, Inhibitory mechanism of low-carbon steel corrosion by mimosa tannin in sulphuric acid solutions. J. Appl. Electrochem. **31**, 973–978 (2001)
58. B. Qian, B. Hou, M. Zheng, The inhibition effect of tannic acid on mild steel corrosion in seawater wet/dry cyclic conditions. Corros. Sci. **72**, 1–9 (2013)
59. A.A. Rahim, E. Rocca, J. Steinmetz, M.J. Kassim, R. Adnan, Sani Ibrahim, M., Mangrove tannins and their flavonoid monomers as alternative steel corrosion inhibitors in acidic medium. Corros. Sci. **49**(2), 402–417 (2007)
60. S. Martinez, I. Štagljar, Correlation between the molecular structure and the corrosion inhibition efficiency of chestnut tannin in acidic solutions. J. Mol. Struct. (Thoechem) **640**(1–3), 167–174 (2003)
61. G. Gece, Drugs: a review of promising novel corrosion inhibitors. Corros. Sci. **53**(12), 3873–3898 (2011)
62. I.B. Obot, N.O. Obi-Egbedi, Adsorption properties and inhibition of mild steel corrosion in sulphuric acid solution by ketoconazole: experimental and theoretical investigation. Corros. Sci. **52**(1), 198–204 (2010)
63. M. Özcan, R. Solmaz, G. Kardaş, İ. Dehri, Adsorption properties of barbiturates as green corrosion inhibitors on mild steel in phosphoric acid. Colloids Surf. A **325**(1–2), 57–63 (2008)
64. M. Abdallah, Antibacterial drugs as corrosion inhibitors for corrosion of aluminium in hydrochloric solution. Corros. Sci. **46**(8), 1981–1996 (2004)
65. M. Abdallah, I. Zaafarany, S.O. Al-Karanee, A.A. Abd El-Fattah, Antihypertensive drugs as an inhibitors for corrosion of aluminum and aluminum silicon alloys in aqueous solutions. Arab. J. Chem. **5**(2), 225–234 (2012)
66. I.B. Obot, N.O. Obi-Egbedi, Anti-corrosive properties of xanthone on mild steel corrosion in sulphuric acid: experimental and theoretical investigations. Curr. Appl. Phys. **11**(3), 382–392 (2011)
67. M.S. Morad, Inhibition of iron corrosion in acid solutions by Cefatrexyl: behaviour near and at the corrosion potential. Corros. Sci. **50**(2), 436–448 (2008)
68. S.K. Shukla, A.K. Singh, I. Ahamad, M.A. Quraishi, Streptomycin: a commercially available drug as corrosion inhibitor for mild steel in hydrochloric acid solution. Mater. Lett. **63**(9–10), 819–822 (2009)
69. M. Abdallah, Rhodanine azosulpha drugs as corrosion inhibitors for corrosion of 304 stainless steel in hydrochloric acid solution. Corros. Sci. **44**, 717–728 (2002)
70. F. Blin, P. Koutsoukos, P. Klepetsianis, M. Forsyth, The corrosion inhibition mechanism of new rare earth cinnamate compounds—electrochemical studies. Electrochim. Acta **52**(21), 6212–6220 (2007)
71. N.D. Nam, A. Somers, M. Mathesh, M. Seter, B. Hinton, M. Forsyth, M.Y.J. Tan, The behaviour of praseodymium 4-hydroxycinnamate as an inhibitor for carbon dioxide corrosion and oxygen corrosion of steel in NaCl solutions. Corros. Sci. **80**, 128–138 (2014)
72. K.A. Yasakau, J. Tedim, M.F. Montemor, A.N. Salak, M.L. Zheludkevich, M.G.S. Ferreira, Mechanisms of localized corrosion inhibition of AA2024 by cerium molybdate nanowires. J. Phys. Chem. C **117**(11), 5811–5823 (2013)
73. H. Shi, E.-H. Han, F. Liu, Corrosion protection of aluminium alloy 2024-T3 in 0.05 M NaCl by cerium cinnamate. Corros. Sci. **53**(7), 2374–2384 (2011)
74. J.-A. Hill, T. Markley, M. Forsyth, P.C. Howlett, B.R.W. Hinton, Corrosion inhibition of 7000 series aluminium alloys with cerium diphenyl phosphate. J. Alloy Compd. **509**(5), 1683–1690 (2011)
75. M.F. Montemor, A.M. Simoes, M.G.S. Ferreira, M.J. Carmezim, Composition and corrosion resistance of cerium conversion films on the AZ31 magnesium alloy and its relation to the salt anion. Appl. Surf. Sci. **254**(6), 1806–1814 (2008)
76. A.E. Hughes, T.G. Harvey, N. Birbilis, A. Kumar, R.G. Buchheit, Coatings for corrosion prevention based on rare earths, in *Rare Earth-Based Corrosion Inhibitors,* pp. 186–232 (2014); A.E. Hughes, Cerium-based corrosion inhibitors, in *Cerium: Molecular Structure, Technological Applications and Health Effects*, pp. 43–72 (2012)

77. S.A. Hayes, P. Yu, T.J. O'Keefe, M.J. O'Keefe, J.O. Stoffer, The phase stability of Cerium species in aqueous systems: II. The Ce(III IV)-H_2O-$H_2O_2O_2$ systems. Equilibrium considerations and Pourbaix diagram calculations. J. Electrochem. Soc. **153**, C74 (2006)
78. M.F. Montemor, A.M. Simões, M.G. Ferreira, Composition and corrosion behaviour of galvanised steel treated with rare-earth salts: the effect of the cation. Prog. Org. Coat. **44**, 111–120 (2002)
79. G. Kong, L. Lingyan, J. Lu, C. Che, Z. Zhong, Corrosion behavior of lanthanum-based conversion coating modified with citric acid on hot dip galvanized steel in aerated 1 M NaCl solution. Corros. Sci. **53**(4), 1621–1626 (2011)
80. M.F. Montemor, M.G.S. Ferreira, *A Review on the Use of Nanostructures and Functional Organosilane Coatings Modified with Corrosion Inhibitors as Environmentally Friendly Pre-treatments for Metallic Substrates*. European Federation of Corrosion Series: UK, 2011; Vol. 58
81. M.F. Montemor, R. Pinto, M.G.S. Ferreira, Chemical composition and corrosion protection of silane films modified with CeO_2 nanoparticles. Electrochim. Acta **54**, 5179–5189 (2009)
82. W. Trabelsi, P. Cecilio, M.G.S. Ferreira, M.F. Montemor, Electrochemical assessment of the self-healing properties of Ce-doped silane solutions for the pre-treatment of galvanised steel substrates. Prog. Org. Coat. **54**(4), 276–284 (2005)
83. M.F. Montemor, M.G.S. Ferreira, Analytical characterization of silane films modified with cerium activated nanoparticles and its relation with the corrosion protection of galvanised steel substrates. Prog. Org. Coat. **63**, 330–337 (2008)
84. H. Shi, E.-H. Han, S.V. Lamaka, L.F. Zheludkevich, M.G.S. Ferreira, Cerium cinnamate as an environmentally benign inhibitor pigment for epoxy coatings on AA 2024-T3. Prog. Org. Coat. **77**, 765–773 (2014)
85. D. Snihirova, S.V. Lamaka, M. Taryba, A.M. Salak, S. Kallip, M.L. Zheludkevich, M.G.S. Ferreira, M.F. Montemor, Hydroxyapatite microparticles as feedback-active reservoirs of corrosion inhibitors. ACS Appl. Mater. Interfaces **2**(11), 3011–3022 (2010)
86. C. Motte, M. Poelman, A. Roobroeck, M. Fedel, F. Deflorian, M.G. Olivier, Improvement of corrosion protection offered to galvanized steel by incorporation of lanthanide modified nanoclays in silane layer. Prog. Org. Coat. **74**(2), 326–333 (2012)
87. M. van Soestbergen, S.J.F. Erich, H.P. Huinink, O.C.G. Adan, Inhibition of pH fronts in corrosion cells due to the formation of cerium hydroxide. Electrochim. Acta **110**, 491–500 (2013)
88. A.E. Hughes, J.M.C. Mol, I.S. Cole, The cost and availability of rare earth-based corrosion inhibitors, in *Rare Earth-Based Corrosion Inhibitors*, ed. by M. Forsyth, B.R.W. Hinton, vol. 61 (Woohead Publishing, Cambridge, 2014), pp. 292–305
89. T.H. Muster, H. Sullivan, D. Lau, D.L.J. Alexander, N. Sherman, S.J. Garcia, T.G. Harvey, T. A. Markley, A.E. Hughes, P.A. Corrigan, A.M. Glenn, P.A. White, S.G. Hardin, J. Mardel, J. M.C. Mol, A combinatorial matrix of rare earth chloride mixtures as corrosion inhibitors of AA2024-T3: optimisation using potentiodynamic polarisation and EIS. Electrochim. Acta **67**, 95–103 (2012)
90. X. Li, S. Deng, H. Fu, G. Mu, Synergism between rare earth cerium(IV) ion and vanillin on the corrosion of cold rolled steel in 1.0 M HCl solution. Corros. Sci. **50**(12), 3599–3609 (2008)
91. R. Catubig, A.E. Hughes, I.S. Cole, B.R.W. Hinton, M. Forsyth, The use of cerium and praseodymium mercaptoacetate as thiol-containing inhibitors for AA2024-T3. Corros. Sci. **81**, 45–53 (2014)
92. X. Li, S. Deng, H. Fu, G. Mu, Synergistic inhibition effect of rare earth cerium(IV) ion and sodium oleate on the corrosion of cold rolled steel in phosphoric acid solution. Corros. Sci. **52** (4), 1167–1178 (2010)
93. B.D. Chambers, S.R. Taylor, The high throughput assessment of aluminium alloy corrosion using fluorometric methods. Part II—A combinatorial study of corrosion inhibitors and synergistic combinations. Corros. Sci. **49**(3), 1597–1609 (2007)

Chapter 7
Active Protective Coatings: Sense and Heal Concepts for Organic Coatings

H.R. Fischer and S.J. García

7.1 Introduction

Smart engineering materials able to mimic sense-response principles found in nature are a long dream of engineers, scientists and constructors [1]. In nature, this concept of sense-response ensures the survival of organisms and increases their chances of survival, reproduction and further evolution. Many very complex examples of smart materials or systems can be found in nature as integral parts or as complete smart structures and organisms. The common ground for all natural responsive structures is that they are composed of numerous elements including sensors, actuators, control algorithms, processing units, and structural parts able to perform the necessary corrective actions required in processes such as healing (e.g. skin system). Some natural systems are thus capable of initiating a response depending on the level of impact/damage of its own system by translating mechanical, physical or chemical changes into readable signals. Engineering smart materials aim thus at implementing the capability of sensing and monitoring their status during their service lifetime in order to be able to promote the restoration (healing) of a lost functionality when needed.

Damage during the preparation and operation of material systems is practically unavoidable with the consequence of a partial or total loss of a certain function of the original system. The traditional approach to deal with damage is the use of better, stronger and more resistant materials and systems resulting in systems with initially much higher performance than required for a certain application leading to

H.R. Fischer (✉)
TNO Technical Sciences, Materials for Integrated Products,
De Rondom 1, 5612, AP Eindhoven, The Netherlands
e-mail: hartmut.fischer@tno.nl

S.J. García
Delft University of Technology, Novel Aerospace Materials,
Kluyverweg 1, 2629, HS Delft, The Netherlands

A.E. Hughes et al. (eds.), *Active Protective Coatings*,
Springer Series in Materials Science 233, DOI 10.1007/978-94-017-7540-3_7

the so called of the safety factor. This traditional approach is nevertheless not the most preferred approach anymore. As materials become scarce, these need to be used as efficiently as possible during the longest possible time while maintaining certain possibilities for re-use and recycling. An alternative to the traditional approach are strategies using the concept of "damage management". Such strategies are based on the acceptance that damage will eventually occur as opposite to the classical strategy of "damage avoidance" where materials are stronger to try to delay the damaging event. The more preferable strategy follows a more dynamic approach, compared to the classical static one. Under this concept, known as self-healing or self-repair, the initial system properties may be just sufficient to meet the requirements. In return, the system needs to have incorporated functionalities to sense damage, process the obtained signals and to initiate single or multiple healing events or processes. The healing process is fundamentally based on (local) mobility resulting in a full or partial recovery of the original properties leading to an extension of the service life-time of the system. Only the very parts of the system, in which damage is to be expected, need to be equipped with such functionalities which saves again resources and limits complexity. Depending on the dimensions or scale of damage, different strategies of repair may be possible [2]; however, detection of damage should start preferably during or immediately after the occurrence of damage. Figure 7.1 shows the implementation scheme of such a strategy. Route "a" in Fig. 7.1 shows the concept damage-sense-monitor-manual repair as has been implemented so far, while route "b" shows the ideal damage-sense-healing or damage-sense-monitor-healing routes. Despite route "a" being the most common one; route "b" represents the ideal concept to be implemented in order to develop the full potential of the self-healing concept.

This chapter will focus on different concepts for damage sensing and healing of anticorrosive coatings not aiming at showing all existing concepts but more aiming at inspiring the reader in the concept of sense and repair while at the same time offering the reader an overview of systems and concepts in use or ready to be used.

7.2 Organic Polymeric Coatings

(Organic) coatings can be understood as thin (organic) solid phases mechanically or/and chemically attached to a much thicker substrate or structure in order to add a certain functionality such as mechanical and chemical (corrosion) protection, wear resistance, thermal protection, electrical and thermal conductivity or isolation, colour, gloss or a specific roughness, light reflectance, hydrophobicity or hydrophilicity, self-cleaning, or other functionalities not obtainable from the substrate material itself [2]. Such systems are commonly known as paints [3], although the words "coating" and "paint" are interchanged in daily use. Major components of all paints are a binder (resin and plasticizers), and fillers (colouring and protective pigments) combined with other substances added in low quantities like surface-active agents, biocides or thickeners (this is discussed in detail in Chap. 5). The binder forms a continuous

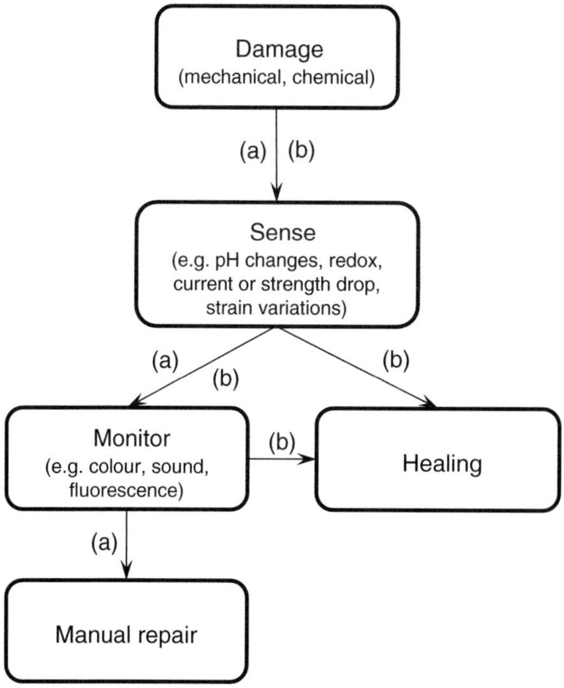

Fig. 7.1 Process of damage-sense-monitor-heal/repair found in natural systems and implemented in engineering materials. Route (**a**) shows the process as it has been implemented so far, while route (**b**) represents the new concept of (self)-healing

phase that gives the main characteristics (e.g. mechanical, chemical) to the coating, while the fillers are the discontinuous phase giving additional or improved properties to the coating. Clearly, the selection of the specific components will affect the general performance of the coating and particularly the interaction of the phases at the three main interfaces: coating-substrate, coating-environment, and binder-filler. Firstly, the compatibility between coating and substrate has a direct impact on the adhesion properties and thus on the long term protection of the underlying material (e.g. corrosion protection). Secondly, the compatibility between fillers and binder is a key factor influencing mainly the mechanical properties. And thirdly, the interface coating-environment (see also Chap. 3) will have a direct impact on film formation and densification of the protective layer as well as on the degradation resistance to external agents such as water/humidity and other components coming in direct contact to the coating surface [2].

The continuous phase in a coating is usually a three-dimensional network of covalently interconnected polymer segments, known as thermoset (e.g. epoxy, polyurethane), or a system of entangled polymeric chains known as thermoplastic (e.g. acrylic resins, polyesters). Thermosets are the most widely used binders in paints technology because of their more favourite properties before, during and after application and especially their final network architecture which offers higher resistance to chemicals, solvents and mechanical stresses compared to thermoplastics.

Coatings in general, can be applied as a single paint (coating) layer for certain applications, but most generally they are applied as layered systems, where each layer has been optimized to maximize a different functionality [2]. For example, an anticorrosive paint system in the aerospace industry has three main layers: (1) a conversion coating of 100 nm (e.g. chromate conversion coating) up to a few microns (e.g. anodizing layer) just above the metallic surface to offer good adhesion of the subsequent polymeric layer and to offer active corrosion protection; (2) a primer, which is normally an epoxy layer of around 20–25 μm containing anti-corrosion inhibitors to be leached out when damage takes place; and (3) a topcoat, generally a polyurethane, which can have different thicknesses from 50 to 100 μm and offer functional properties as colour, barrier, hydrophobicity, gloss or UV resistance.

7.3 Sensing Mechanisms Incorporated into Organic Coatings

Ideally, paints offer their intended functionalities (e.g. colour, corrosion protection of the underlying substrate) during the whole lifetime of the structure. Unfortunately, and despite the great advances in coatings technology, this is not what usually happens. Coatings degrade due to mechanical factors such as impacts, scratches, thermal cycles, and fatigue related to the use of the structure, as well as chemical factors as UV radiation, acid and alkali aggressions, moisture and oxygen influences, and thermo-oxidation; processes globally known as weathering and ageing. In the case of a sudden damage event like a key scratching the paint of a car door, the damage is directly at the macro scale, although it implies damage at the nano-scale (i.e. bonds breaking). However, in the case of a time-related degradation process (fatigue damage), all scales are involved. For example, the time-degradation of a coating will start at the nano-scale due to hydrolysis combined with polymer oxidation amongst others, followed by the micro scale because aging and weathering leads to crazing, finally involving the macro scale where the coating breaks down due to the progressive embrittlement of the polymer, all being a combination of degradation processes related to ambient humidity, and mechanical forces related to the normal use.

This means, that sensing of a possible damage should start ideally at a molecular level, and immediately alter the occurrence of damage relying on a material undergoing a transformation through its interaction with the damaging environment. On the other side, the ability to sense damage should be present or preserved during the whole service life-time of the coating.

7.3.1 Sensor Systems Based on Colour Response

Such systems respond to an (external) damage by a visible colour change or change in fluorescence/phosphorescence as a result of a variety of stimuli including mechanical action, redox reactions, pH changes, presence of ions, electrical current etc. These response principles can be employed to detect early stages of corrosion, where the protected material reacts with the environment (water, oxygen) after damage of the passive barrier, or even indicate when a corrosion inhibition function has started.

Several approaches can be found such as: (i) molecules that change colour by chemical complexation with corrosion products or pH changes; and (ii) systems that upon mechanical or chemical rupture release a colouring dye.

There is a large variety of compounds able to change colour upon local changes of pH. The use of pH indicators goes back to the 1926, when Evans [4] used indicators to locate the presence of anodic and cathodic corrosion sites on corroding iron. The application of a broad range indicator in a gel or saturated saccharose solutions allows the simultaneous study of the corrosion of aluminium and the localization and monitoring of pH changes at cathodic and anodic sites. One of the most used compounds is phenolphthalein, changing its appearance from colourless to red while reaching a pH of 9.4 as shown in Fig. 7.2. When corrosion stops, the pH level returns to normal and the coloration vanishes.

Other compounds used to indicate the beginning of corrosion even under virtually undamaged coatings, are fluorescing organic compounds such as 8-hydroxyquinoline sulfonic acid hydrate [5], lumogallion [6], or 9-phenyl-3-fluorone [7]. These systems rely on changes upon oxidation or complexion with metal cations generated during the first stages of the corrosion process.

Finally, other colour sensoric functionalities (e.g. dyes) are released upon reaction to mechanical damage or chemical variations (e.g. pH variations). One example of this are microcapsules incorporated into the coating that contain a dye [8]. In this case, marker dyes are part of the capsule load and a leaking will be initiated by capsule damage.

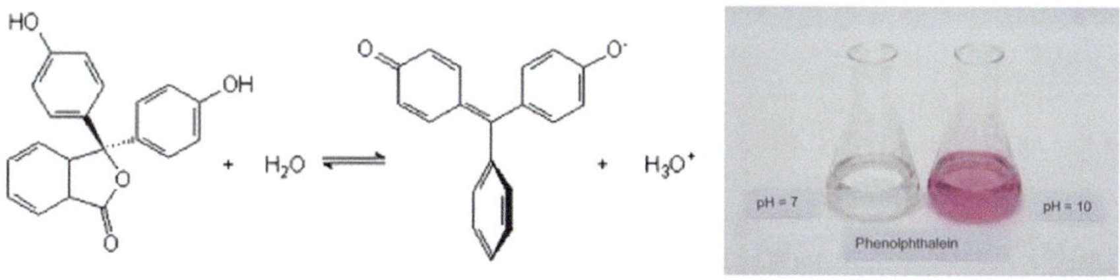

Fig. 7.2 Phenolphtalein colour change at given pH

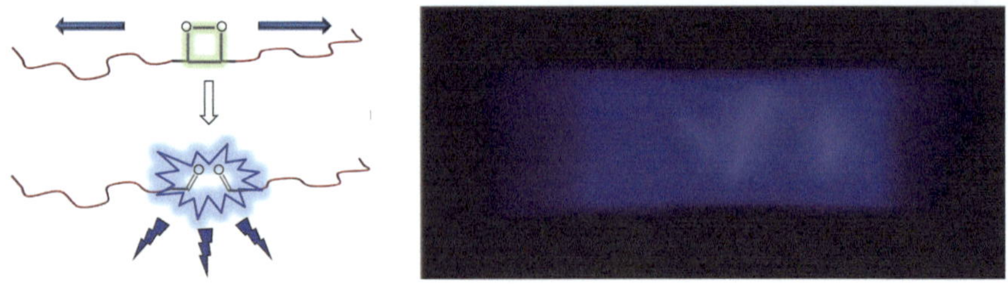

Fig. 7.3 Chemiluminiscence from 1–2 dioxetane units

7.3.2 Sensor Systems Based on Non-colour Response

Response to mechanical damage can also lead to effects other than colour. Very recently, a nice system has been developed, which transduces local mechanical forces into chemiluminescence. A bright blue luminescence occurs on straining polymer networks containing 1,2-dioxetane units as shown by Chen et al. [9] (Fig. 7.3).

The potential of this approach is that it is feasible to increase the signal strength and the sensitivity by energy transfer to suitable acceptors that allows a change in wavelength of the emitted light. This would thus enable the sensing and discrimination of strains indicating failure at different locations.

Other type of sensors act based on external/internal monitoring of the system. For instance dielectric sensors embedded in a coating could indicate molecular mobility while fibre optic strain sensors based on fiber Bragg gratings (FBG) could detect strain variations via resistance measurements. Moreover, external devices such as flash thermography or ultrasound could help to monitor at each time the state of the system.

7.4 Responsive Mechanisms Incorporated into Organic Coatings

Besides the pure detection or sensing of damage of the coating and/or the progress of corrosion, a response to damage depending on the level of damage and the status of the system is needed to ensure the envisaged performance and life-time. Such a response can either occur without or by external intervention and aims at restoring the original functional state as much as possible. The response is not always immediately required; however, repair actions are more successful when initiated at very early stages of damage. It seems obvious that any useful response or repair approach highly depends on a number of points [2]: (a) initial requirements of the coating and description of the most critical functionalities to be restored; (b) the components of the coating system (binder and eventually also fillers); (c) size scale of damage; (d) damage type; (e) and effect of the implemented response/healing mechanism on the general performance of the coating meaning that the implemented response should not affect other required properties of the coating beyond an unacceptable level.

A response leading to a cessation of corrosion or even to repair requires at least some part of the coating system being mobile, either stored in containers or intrinsically incorporated into the coating matrix. Such an agent should only be mobile where damage occurs and when needed meaning a local temporary release of the agent. As mentioned earlier, a supply of a modest amount of externally applied energy (e.g. local temperature, UV, high kinetic energy in high speed impacts) may also be required to the increase the local mobility involved in the response action and to minimize the recovery period to acceptable time scales as well as an external supply of agents/materials (such as moisture or oxygen) in case the restoration of the damage site requires the supply of new material to fill the initially empty damage volume. A period of reduced loading conditions (the recovery period) to allow healing to take place is preferential. Most important, and an essential part of an active protective casting, is a sensing system able to determine the occurrence of damage and to activate the response action.

The existing self-healing approaches applied or potentially applicable to coatings have recently been published and discussed [2]. Table 7.1 sums up the different existing concepts based on different categories which may be useful for the design and implementation of a system with self-repairing properties: *concept, architecture, healing mechanism, and damage size-scale.*

Table 7.1 Classification of self-healing strategies that have been applied or are applicable to organic coatings (adapted from [2])

Concept (level 1)	Architecture (level 2)	Healing mechanism (level 3)	Damage scale (level 4)
Additives or extrinsic (discontinuous)	Containers (capsules, fibers, networks)	Liquid spreads and reacts (flow)	μm–mm (micro–macro)
	Expansive phases	Reacts and expands (expansion)	μm (micro)
	Corrosion inhibitors	Release and reacts with metallic surface (flow)	nm–μm (nano–micro)
Intrinsic (continuous)	No architecture	Reversible chemical bonds (re-flow)	nm–mm (nano–macro)
		Reversible non covalent bonds (re-flow)	nm–mm (nano–macro)
		Delayed elasticity (re-flow)	nm–μm (nano–micro)
		Unreacted groups (re-flow)	
		Surface stratification (re-flow)	

Another design criterion for such systems and often used refers to the mode of repair action, whether it is autonomous or non-autonomous. However such differentiation may be difficult, since systems that are considered non-autonomous under certain circumstances can be autonomous under others. For instance, the use of isocyanate based coatings with retarded elasticity from Bayer [10] and AkzoNobel-Sikkens [11] are autonomous in a sunny day, but in cold and cloudy weathers heat needs to be applied directly to the damaged site. Strictly speaking a non-autonomous system would be one that would require external interference, a direct action of an agent that is not in the normal environment of use of the system or a supply of energy of any kind.

7.4.1 Corrosion Triggered Protective Response: Underlying Metal Protection

Despite not being named as such until the nineties, the most traditional way to response and to cease corrosion at damaged coatings sites is by the use of corrosion inhibitors that are released from the coatings via leaching [12, 13]. In such systems, the corrosion inhibitor leaches upon interaction with water at the corrosion site and by the ingress of water at the periphery of the damaged coating. Once leached out the corrosion inhibitor subsequently coordinates with the metallic surface to create a passive layer that replaces the original corrosion protection (Fig. 7.4).

This response mechanism works already in intact organic coatings, since water and oxygen ingress through coatings is always present to some extent. Therefore, an active protection is always needed as well. So far, the most efficient corrosion inhibitors are based on hexavalent chromium [14], but the toxicity issues and carcinogenic effects associated with their use, despite being known since the eighties [15], have become increasingly highlighted in recent years [16–18]. This fact has created, in recent years, a high effort into the development of new environmentally friendly corrosion inhibitors [19–24] which can offer equal or better corrosion protection to Cr^{VI} based ones, as well as on the development of high-throughput techniques to develop and evaluate corrosion inhibitors in reduced times [19, 25–28]. The incorporation of such new corrosion inhibitors into coatings to offer an effective and long term protection independently of the polymer matrix

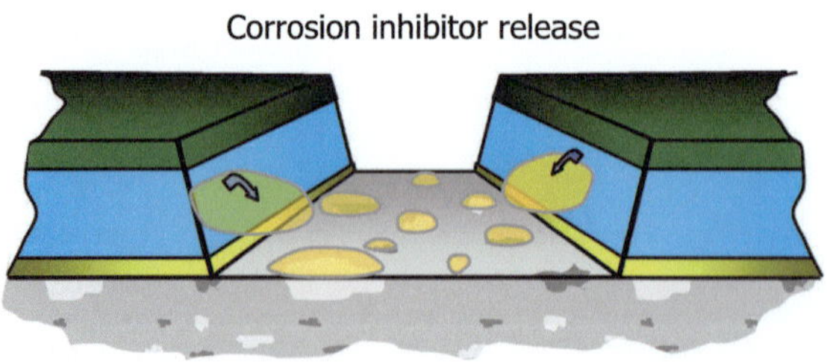

Fig. 7.4 Release of corrosion inhibitors by desorption [2]

has also been object of extensive studies; thus creating a whole area of research under the umbrella of self-healing materials. Several new concepts aiming at a controlled release from carriers have thus been developed. So far most the studied and successful concepts include delivery of inhibitors from inorganic carriers by ion exchange and/or desorption and the release of inhibitors from organic carriers by pH environmental changes.

7.4.1.1 Controlled Release by Desorption and Ion Exchange

Changes and increase in local ion concentration and mobility due to the start of the corrosion process can be used as trigger mechanism to release corrosion inhibitors on demand. For such concept nano and micro inorganic minerals can be used as carriers and support of the inhibitors. Such materials are either cation exchangers such as smectite type alumosilicate materials (e.g. montmorillonite) but also anion-exchangers such as the layered double hydroxides (LDH's) or hydrotalcites (see also Chap. 8). These carriers can be doped with a single corrosion inhibitor (cationic or anionic) that is released by desorption or ion exchange once the corrosion process has started as introduced in the very extensive works done by, amongst others, Shchukin, Zheludkevich and Buchheit [29–39]. As an alternative to single inhibitor-carrier concept, Garcia et al. [40] recently introduced the concept of dual kinetic release by a dual inhibitor doping of nanocarriers (a cathodic and an anionic inhibitor). This new concept was demonstrated with zeolites and proposes a fast release of one inhibitor by desorption followed by a slow release of the second inhibitor by ion exchange whereby one inhibitor can be cathodic inhibitor and the other anodic. This promising concept opens the path for the development of more efficient and multifunctional carriers using the traditional idea of nanocarrier doped systems.

It is worth mentioning that LDH's have not only the potential to release anionic corrosion inhibitors upon ion exchange but also upon pH-changes, being thus possible the release of active metal ions due to decomposition of the carrier themselves [41, 42]. A follow up of the doped nanocarrier concept is the incorporation of inhibiting cations like Ce^{3+} into amorphous oxide nanoparticles (e.g. ZrO_2) which act as reinforcement of a sol-gel matrix and at the same time as reservoir of the inhibitor [43]. Finally also microgels as storage entities for corrosion inhibitors were exploited showing a similar controlled desorption [44].

7.4.1.2 Controlled Release by Environmental pH Changes

Different from a desorption or ion controlled release is the controlled release of inhibitors initiated by local changes of pH during the corrosion process. To increase the control over the release of inhibitors, inorganic nanocarriers based on SiO_2 were coated with polyelectrolyte shells using the Layer-by-Layer (LbL) assembly technique [32, 45, 46]. The corrosion inhibitor benzotriazole was incorporated between oppositely charged polyelectrolyte layers. The aim was to use the permeability dependence of polyelectrolyte shells on pH to provide release of corrosion inhibitors, only when corrosion processes were about to start or have already started.

Alternatively, cerium molybdate containers loaded with 2-mercaptobenzothiazole [47] or porous calcium carbonate beads loaded with cerium nitrate, salicyldioxime or 2,5-dimercapto-1,3,4 thiadiazolate [48] are described as smart pH sensitive particulate reservoirs. When compared to a desorption concept the controlled release by pH or ion exchange are more effective for protection applications as the ion controlled release of species will provide a control not only in time but also spatially because only the nanoreservoirs close to the defects where corrosion activity initiates will sense local pH changes or accumulation of ions necessary for the inhibitor release.

An example of a complete smart active system, combining corrosion indicators, corrosion inhibitors and healing agents working in an autonomous way has been developed by Lutz et al. [49]. Here all three components were incorporated via microcapsules into coating systems. The shell of the microcapsules opens up by ester-hydrolysis, initiated and catalyzed by high pH and naturally also on rupture. In such a way, completely autonomous and active response is guaranteed upon corrosion. This process is being initiated either by mechanical damage and/or by the start of a corrosion reaction due to the degradation/ageing of the passive protective coating. In the complete concept the reaction does not only sense and signal the damage, but also is stopped by the use of corrosion inhibitors and the reformation of a passive layer by the siloxane healing agent (Fig. 7.5).

All these approaches are used or considered at industrial levels and provide obviously a good corrosion protection for undamaged coatings or small damages. However, a system that only recovers the active protection by corrosion inhibitors reacting with the metallic substrate lacks the recovery of the barrier protection and the recovery of some aesthetic properties in case of serious damage (cracks, scratches etc.). At such circumstances, the release is difficult to control locally and

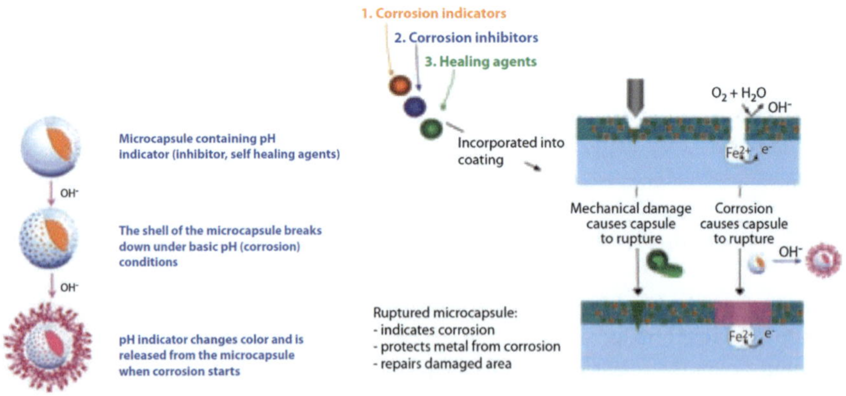

Fig. 7.5 Release of corrosion inhibitors by pH change

in time and a recovery of the passive protection is difficult. For this reason the combination of such concepts with gap closure concepts hereafter introduced allow a complete protective healing system.

7.4.2 *Response Aiming a Closure of the Passive Protective Surface*

7.4.2.1 Surface Coverage

A serious damage of the coating system is related to a loss of material leading to a gap in the passive protective coating system. In order to react adequately to avoid further damage of the underlying surface, an active protective system must be able to:

1. Fill a gap in a corrosive environment. In an ideal case, this should occur without trapping the corrosive agents under the repaired coating layer and avoiding severe pitting corrosion and undercutting corrosion (corrosion under the delaminated film) and
2. Protect the damaged area of the substrate as well as that of the coating system to avoid ingress of corrosive substances and water through the side-wall of the scratches/damages and producing under-film corrosion promoted by the lack of good adhesion and coverage between the repaired coating and the substrate (Fig. 7.6).

Both extrinsic and intrinsic approaches have been employed to restore barrier protection in parallel with corrosion protection. Several works can be found in the literature using encapsulated agents for restoration of barrier and corrosion protection. Cho et al. [50] used an encapsulated-catalyst system to protect steel from corrosion at scribed areas, showing relevant protection by salt fog spray tests. Kumar et al. [51] and Samadzadeh [52] used the encapsulation of tung oil. Despite these systems showing good corrosion protection, several doubts about the long term protection have been posed due to the non-full coverage of the underlying metallic surface and possible entrapment of corrosive species during the healing process, key issues in corrosion protection [53]. Substantial progress and a very elegant solution was introduced by Garcia et al. [53, 54]. This research team

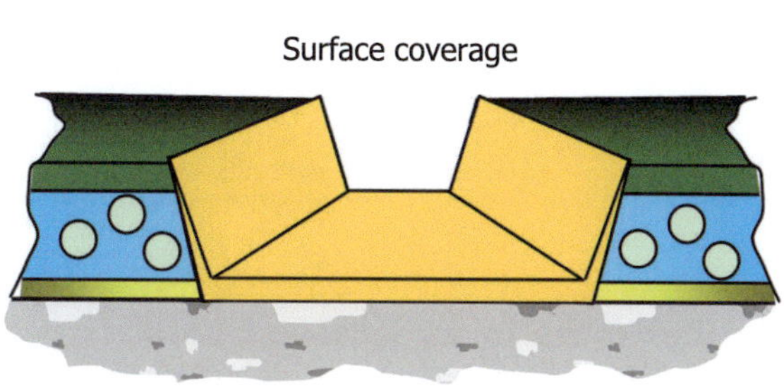

Fig. 7.6 Surface coverage [2]

proposed a new single reactive healing agent, specifically designed for corrosion protection of aluminium substrates but extendable to other metallic surfaces. The new concept (known as silyl ester concept) showed excellent long term protection by electrochemical impedance spectroscopy (EIS), scanning vibrating electrode technique (SVET) and scanning electrochemical microscopy (SECM). The concept is based on capsules containing a single component healing agent providing after rupture and reaction with moisture a complete coverage of the metallic surface and the damaged coating edges. This approach was proven with a newly synthesized healing agent, namely a silyl ester, that showed excellent quick response and corrosion protection under immersion in corrosive solution, The new concept is based on wetting, reaction with the metallic surface and ambient humidity, adhesion to metal and polymer, and formation of a hydrophobic layer to extend the corrosion protection. Based on these protective principles the concept can be applied not only for active corrosion protection and barrier functionality but also for the recovery of adhesion between polymer and metal. Several advantages with respect to previous concepts exist: (i) the system is based on surface healing instead of volumetric-gap filling; (ii) it does not require catalyst or crosslinker as surface hydroxides and environmental humidity suffice for reaction to occur; and (iii) the reaction takes place only in case of corrosion damage

7.4.2.2 Gap Filling

A concept different from surface coverage is gap filling (Fig. 7.7). This concept aims at the full closure of the damage area and several approaches can be used.

One of the most known and efficient ways of closing wide and/or deep damage is by the use of self-healing polymers based on reversible chemistries where the most common healing trigger is induced temperature. The retro-Diels-Alder reaction is particularly useful for that, especially if the active components are used in powder coating systems [55, 56], where reversibility was found at temperatures around 150 °C. Also the earlier mentioned disulfide based systems [57, 58] as well the photo-induced reversible crosslinking of antracene containing networks [59] show repeated healing possibilities for surface damages. Supra-molecular approaches may be also applied in the restoration of damaged coatings as proposed by Bosman et al. [60]. Although the previous systems have been mostly evaluated in terms of

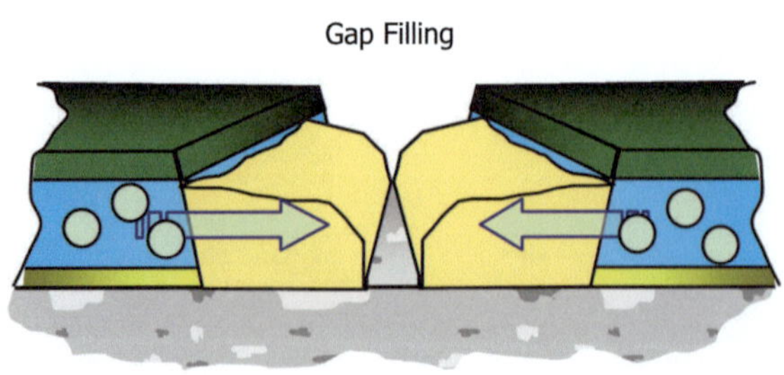

Fig. 7.7 Gap filling [2]

crack or scratch closure in free standing films and the quality of the formed scar has not been addressed or quantified in terms of sealing level, some studies have shown the potential of intrinsic approaches for corrosion protection using shape memory polymers [61].

An alternative to the use of reversible chemistries is the use of external components that can react with a coating in order to promote damage closure. Such components need to be transported via the gas phase or via contact with a liquid carrier, and have the ability to form new or to modify (repair) the matrix material with an increase in volume due to the incorporation of additional material and as such to fill possible cracks completely. The healing is autonomous during presence of the external component, as the ingredients to be used are naturally present in the application foreseen. Typical examples of this route are the oxidative healing of high temperature ceramic coatings or thermal barrier coatings on turbines in which local oxidation (oxygen is the external component) is used to fill the crack with a load bearing deposit. The external components easily accessable and useable are H_2O, O_2 and CO_2. This concept is very interesting for coatings in general although to be applied in organic coatings it would require the introduction of different approaches.

One successful example of this concept is the use of external components for the self-healing of insulating PPS coatings used in hydrothermal environments [62]. Hydraulic inorganic grains, e.g. calcium silicate and calcium aluminate are especially attractive for a use as self-repairing fillers since their crystals grow after being in contact with a hot water which acts as the external component for autonomous self-healing. The expanding crystals then are able fill the open cracks densely with a reconstruction of the damaged coatings. As a result of the decalcification–hydration reactions of the CaO, Al_2O_3 and CaO $2Al_2O_3$ filler, block-like bohemite crystals are formed which fill and seal the open cracks.

Much easier than the use of a liquid as external self-healing system component is the use of gases like water vapour or oxygen as being present anywhere. Water can be efficiently absorbed and leads even without a chemical reaction involved to a volume increase which can be used to close scratches in coatings [63, 64]. The clay-based systems demonstrate a very quick water absorption accompanied by overexpansion of the layer leading to the scratch closure and removal of the water from the surface (Fig. 7.8).

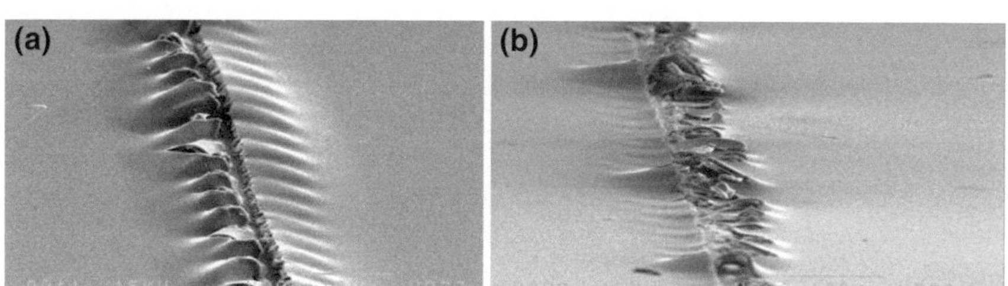

Fig. 7.8 Crack-healing of multilayer coatings due to phase expansion upon moisture absorption before (**a**) and after 21 h of healing (**b**) in wet atmospheric conditions [64]

This expansive layer system has recently been studied by numerical methods in order to establish the key parameters to be considered when trying to improve the healing efficiency while decreasing the side negative effects as the levelling at the healed area [65]. In the study it was found that the scratch closing behaviour was dependent on relative thickness of the layers in the system, relative stiffness of the same, but also on the length of initial interfacial debonding, finding that some initial lateral debonding was necessary to obtain the healing of the system.

Another more straight forward approach of the use of humidity as the healing agent comes from the work done by South and Lyon [66] on hydrogel thin films, who show how damage at the surface of the microgel polyelectrolyte multilayers exhibit repeatable self-healing behaviour upon multiple mechanical deformations, which could be an important advance in the application of hydrogel biomaterial coatings likely to be damaged by routine surgical handling.

The obvious advantages of such systems are their efficiency in repair action via the volume expansion providing a complete filling of cracks and scratches and the option to perform a multiple repair action, their flexibility in design and the absence of fluids as repairing components, whereas the main drawback is their obvious dependency on the external components. Furthermore, the ability of these compounds to react with species present in the environment makes the isolation of the reactive agents in intact coatings a key priority to avoid reactions occurring prior to the occurrence of damage.

7.4.3 Recovery of Other Functionalities

So far, in organic coatings, the active response concept has been applied, mainly, for the detection and recovery of macro and micro damages, as exposed in the previous points. This section evaluates the attempts to recover specific functionalities such as interfacial adhesion/bonding. The recovery of the interfacial bonding between coatings and metal substrates has attracted some attention already [19], although much less than other concepts due to the complexity of the adhesion itself and the lack of perfect understanding of the underlying mechanisms. So far, and up to the knowledge of the authors, no clear successful attempts has been reported in this field, although Mardel et al. [67] demonstrated that the incorporation of certain corrosion inhibitors (i.e. $Ce(dbp)_3$) improved the resistance to filiform corrosion by the formation of an adhesive oxide between the metal and the coating once exposed to humid environments. Other alternative approaches to this could use nano-encapsulated or compartmented adhesion promoters that upon damage are released and wet both the coating and metal interface reacting with either materials. Interestingly, Lane et al. [68] showed a proof of concept for this approach for the repair of dielectric interfaces, although the concept has not been further developed.

7.5 Concluding Remarks

This work presents several concepts for sense, monitor and heal that have been implemented or can be implemented into coatings. Clearly many different approaches can be applied to organic coatings although the level of success depends on the application of the coatings themselves and response wanted and needed. Sensing and signalling/monitoring of mechanical failure and/or corrosion initiation is very useful, however, in most cases a manual intervention to stop the corrosion and to prevent excessive damage is needed but not possible. This opens the need to develop coating systems that do not only sense but also act (heal) damages, thus developing implemented concepts of sense and heal. Encapsulated approaches have already been implemented under different concepts and seem to be already at the stage of production and real applications, although certain issues such as the entrapment of corrosive species under the healed area and the lack of complete coating-metal adhesion recovery should be revised as they can be more detrimental to the system than the damage itself. A combination of sensing/signalling, corrosion inhibition and repair (heal) of the barrier seems up to now the most advanced system. The use of expansive phases seems to be very appropriate for certain applications but issues like the adhesion to the underlying surface should be taken into account. For such systems, mathematical models have shown the strong influence of the coating architecture and damage topology.

Furthermore, functionalities of coatings other than mechanical and electrochemical substrate protection (e.g. adhesion, color, surface tension) have not yet received enough attention to be sufficiently developed and further efforts are necessary.

Acknowledgments The authors gratefully acknowledge the sustained financial support from the Innovative Research Program (IOP) in The Netherlands.

References

1. H. Fischer, Self-repairing material systems-a dream or a reality? Nat. Sci. **2**, 873–901 (2010)
2. S.J. García, H.R. Fischer, S. Van Der Zwaag, A critical appraisal of the potential of self healing polymeric coatings. Prog. Org. Coat. **72**, 211–221 (2011)
3. J. Boxall, J.A. von Fraunhofer, *Paint Formulation: Principles and Practice* (George Godwin Limited, London, UK, 1980)
4. U.R. Evans, The fundamental principles of corrosion. J. Soc. Chem. Ind. **45**, 504–516 (1926)
5. D.E. Bryant, D. Greenfield, The use of fluorescent probes for the detection of under-film corrosion. Prog. Org. Coat. **57**, 416–420 (2006)
6. M.P. Sibi, Z. Zong, Determination of corrosion on aluminum alloy under protective coatings using fluorescent probes. Prog. Org. Coat. **47**, 8–15 (2003)
7. S.-M. Li, H.-R. Zhang, J.-H. Liu, Preparation and performance of fluorescent sensing coating for monitoring corrosion of Al alloy 2024. Trans. Nonferrous Met. Soc. China (English Edition) 16 (SUPPL.), 159–s164 (2006)

8. C. Suryanarayana, K.C. Rao, D. Kumar, Preparation and characterization of microcapsules containing linseed oil and its use in self-healing coatings. Prog. Org. Coat. **63**, 72–78 (2008)
9. Y. Chen, A.J.H., Spiering, S. Karthikeyan, G.W.M. Peters, E.W. Meijer, R.P. Sijbesma, Mechanically induced chemiluminescence from polymers incorporating a 1,2-dioxetane unit in the main chain. Nat. Chem. **4**, 559–562 (2012)
10. www.bayer.com
11. www.akzonobel.com
12. S. van der Zwaag (ed.) in *Self-Healing Materials: An Alternative Approach to 20 Centuries of Materials Science*. Springer Series in Materials Science, vol 100. (Springer, Dordrecht, The Netherlands, 2007)
13. J. Sinko, Prog. Org. Coat. **42**, 267 (2001)
14. J. Zhao, G.S. Frankel, R.L. McCreery, J. Electrochem. Soc. **145**, 2258 (1998)
15. US Public Health Service, Toxicological Profile for Chromium, Report No. ATSDR/TP-88/10, July, 1989
16. M. Costa, C.B. Klein, Crit. Rev. Toxicol. **36**, 155 (2006)
17. P.T. La Puma, J.M. Fox, E.C. Kimmel, Regul. Toxicol. Pharmacol. **33**, 343 (2001)
18. N.A. Dalager, T.J. Mason, J.F. Fraumeni, R. Hoover, W.W. Payne, J. Occup. Med. Toxicol. **22**, 25 (1980)
19. S.J. García, J.M.C. Mol, T.H. Muster, A.E. Hughes, J. Mardel, T. Miller, T. Markely, H. Terryn, J.H.W. de Wit, Advances in the selection and use of Rare-Earth-Based Inhibitors for Self Healing Organic Coatings, in *European Federation of Corrosion Series: Self-Healing Properties of New Surface Treatments* ed. by L. Fedrizzi, vol. 58 (ISBN: 978 1906540364, Maney Publishing, March 2011)
20. J. Mardel, S.J. García, P.A. Corrigan, T. Markley, A.E. Hughes, T.H. Muster, D. Lau, T.G. Harvey, A.M. Glenn, P.A. White, S.G. Hardin, C. Luo, X. Zhou, G.E. Thompson, J.M.C. Mol, Prog. Org. Coat. **70**, 90–101 (2011)
21. M. Forsyth, T. Markley, D. Ho, G.B. Deacon, P. Junk, B. Hinton, A. Hughes, Corrosion **64**, 191 (2008)
22. N. Birbilis, R.G. Buchheit, D.L. Ho, M. Forsyth, Electrochem. Solid. State. Lett. **8** C180 (2005)
23. D. Ho, N. Brack, J. Scully, T. Markley, M. Forsyth, B. Hinton, J. Electrochem. Soc. **153** (B392) (2006)
24. M. Forsyth, C.M. Forsyth, K. Wilson, T. Behrsing, G.B. Deacon, Corros. Sci. **44**, 2651 (2002)
25. B.D. Chambers, S.R. Taylor, M.W. Kendig, Corrosion **61**, 480 (2005)
26. T.H. Muster, A.E. Hughes, S.A. Furman, T. Harvey, N. Sherman, S. Hardin, P. Corrigan, D. Lau, F.H. Scholes, P.A. White, M. Glenn, J. Mardel, S.J. García, J.M.C. Mol, Electrochim. Acta **54**, 3402 (2009)
27. S.J. García, T.H. Muster, Ö. Özkanat, N. Sherman, A.E. Hughes, H. Terryn, J.H.W. de Wit, J. M.C. Mol, Electrochim. Acta **55**, 2457 (2010)
28. P.A. White, A.E. Hughes, S.A. Furman, N. Sherman, P.A. Corrigan, M.A. Glenn, D. Lau, S.G. Hardin, T.G. Harvey, J. Mardel, T.H. Muster, S.J. García, C. Kwakernaak, J.M.C. Mol, Corros. Sci. **51**, 2279 (2009)
29. D.G. Shchukin, D.O. Grigoriev, H. Mohwald, Soft Matter **6**, 720 (2010)
30. D.V. Andreeva, D. Fix, H. Mohwald et al., Adv. Mater. **20**, 2789 (2008)
31. S.V. Lamaka, M.L. Zheludkevich, K.A. Yasakau, M.F. Montemor, P. Cecílio, M.G.S. Ferreira, Electrochem. Commun. **8**, 421 (2006)
32. D.G. Shchukin, M.L. Zheludkevich, K.A. Yasakau, S.V. Lamaka, H. Möhwald, M.G.S. Ferreira, Adv. Mater. **18**, 1672 (2006)
33. S.V. Lamaka, M.L. Zheludkevich, K.A. Yasakau, R. Serra, S.K. Poznyak, M.G.S. Ferreira, Prog. Org. Coat. **58**, 127 (2007)
34. S.V. Lamaka, M.L. Zheludkevich, K.A. Yasakau, M.F. Montemor, M.G.S. Ferreira, Electrochim. Acta **52**, 7231 (2007)
35. D.G. Shchukin, S.V. Lamaka, K.A. Yasakau, M.L. Zheludkevich, H. Möhwald, M.G.S. Ferreira, J. Phys. Chem. **112**, 958 (2007)

36. M.L. Zheludkevich, D.G. Shchukin, K.A. Yasakau, H. Möhwald, M.G.S. Ferreira, Chem. Mater. **19**, 402 (2007)
37. M. Zheludkevich, Self healing anticorrosion coatings, Ch. 4, in *Self-Healing Materials: Fundamentals, Design Strategies, and Applications* ed. by S.K. Ghosh (Wiley-VCH, 2009), pp. 101–139
38. S.P.V. Mahajanarn, R.G. Buchheit, Corrosion **64**, 230 (2008)
39. R.G. Buchcheit, H. Guan, JCT Res. **1**, 277 (2004)
40. E.L. Ferrer, A.P. Rollon, H.D. Mendoza, U. Lafont, S.J. Garcia, Double-doped zeolites for corrosion protection of aluminium alloys. Microporous Mesoporous Mater. **188**, 8–15 (2014)
41. Z. Yang, H. Fischer, J. Cerezo, J.M.C. Mol, R. Polder, Aminobenzoate modified MgAl hydrotalcites as a novel smart additive of reinforced concrete for anticorrosion applications. Constr. Build. Mater. **47**, 1436–1443 (2013)
42. Z. Yang, H. Fischer, R. Polder, Synthesis and characterization of modified hydrotalcites and their ion exchange characteristics in chloride-rich simulated concrete pore solution. Cement Concr. Compos. **47**, 87–93 (2014)
43. M.L. Zheludkevich, J. Tedim, M.G.S. Ferreira, "Smart" coatings for active corrosion protection based on multi-functional micro and nanocontainers. Electrochim. Acta **82**, 314–323 (2012)
44. A. Latnikova, D. Grigoriev, M. Schenderlein, H. Möhwald, D. Shchukin, A new approach towards "active" self-healing coatings: exploitation of microgels. Soft Matter **8**, 10837–10844 (2012)
45. D.O. Grigoriev, T. Bukreeva, H. Möhwald, D.G. Shchukin, New method for fabrication of loaded micro- and nanocontainers: emulsion encapsulation by polyelectrolyte layer-by-layer deposition on the liquid core. Langmuir **24**, 999–1004 (2008)
46. D.G. Shchukin, H. Möhwald, Surface-engineered nanocontainers for entrapment of corrosion inhibitors. Adv. Funct. Mater. **17**, 1451–1458 (2007)
47. I.A. Kartsonakis, A.C. Balaskas, E.P. Koumoulos, C.A. Charitidis, G. Kordas, Evaluation of corrosion resistance of magnesium alloy ZK10 coated with hybrid organic-inorganic film including containers. Corros. Sci. **65**, 481–493 (2012)
48. D. Snihirova, S.V. Lamaka, M.F. Montemor, "SMART" protective ability of water based epoxy coatings loaded with $CaCO_3$ microbeads impregnated with corrosion inhibitors applied on AA2024 substrates. Electrochim. Acta **83**, 439–447 (2012)
49. http://ntrs.nasa.gov/archive/nasa/casi.ntrs.nasa.gov/20120003396.pdf
50. S.H. Cho, S.R. White, P.V. Braun, Adv. Mater. **21**, 645 (2009)
51. A. Kumar, L.D. Stephenson, J.N. Murray, Prog. Org. Coat. **55**, 244–253 (2006)
52. M. Samadzadeh, S. Hatami Boura, M. Peikari, A. Ashrafi, S.M. Kasiriha, Prog. Org. Coat. **70**, 383–387 (2011)
53. S.J. García, H.R. Fischer, P.A. White, J. Mardel, Y. González-García, J.M.C. Mol, A.E. Hughes, Prog. Org. Coat. **70**, 142–149 (2011)
54. Y. González-García, S.J. García, H.R. Fischer, A.E. Hughes, J.M.C. Mol, Local electrochemical evaluation of a self-healing coating based on encapsulated healing-agent. in *18th International Corrosion Congress*, vol. 1, 2011. pp. 408–413
55. M. Wouters, E. Craenmehr, K. Tempelaars, H. Fischer, N. Stroeks, J. van Zanten, Prog. Org. Coat. **64**, 156–162 (2009)
56. M. Wouters, M. Burghoorn, B. Ingenhut, K. Timmer, C. Rentrop, T. Bots, G. Oosterhuis, H. Fischer, Tuneable adhesion through novel binder technologies. Prog. Org. Coat. **72**, 152–158 (2011)
57. J. Canadell, H. Fischer, G. de With, R.A.T.M. van Benthem, J. Polym. Sci., Part A: Polym. Chem. **48**, 3456–3467 (2010)
58. J.A. Yoon, J. Kamada, K. Koynov, J. Mohin, R. Nicolaÿ, Y. Zhang, A.C. Balazs, T. Kowalewski, K. Matyjaszewski, Self-healing polymer films based on thiol-disulfide exchange reactions and self-healing kinetics measured using atomic force microscopy. Macromolecules **45**, 142–149 (2012)

59. P. Froimowicz, H. Frey, K. Landfester, Towards the generation of self-healing materials by means of a reversible photo-induced approach. Macromol. Rapid Commun. **32**, 468–473 (2011)
60. R.P. Bosman, R.P. Sijbesma, E.W. Meijer, Supramolecular polymers at work. Mater. Today 34–39 (2004)
61. J.-B. Jorcin, G. Scheltjens, Y. Van Ingelgem, E. Tourwé, G. Van Assche, I. De Graeve, B. Van Mele, H. Terryn, A. Hubin, Investigation of the self-healing properties of shape memory polyurethane coatings with the 'odd random phase multisine' electrochemical impedance spectroscopy. Electrochim. Acta **55**, 6195–6203 (2010)
62. T. Sugama, K. Gawlik, Mater. Lett. **57**, 4282–4290 (2003)
63. A. Hikasa, T. Sekino, Y. Hayashi, R. Rajagopalan, K. Niihara, Mater. Res. Innovations **8**, 84–88 (2004)
64. F. Miccichè, H. Fischer, R. Varley, S. van der Zwaag, Surf. Coat. Technol. **202**, 3346–3353 (2008)
65. R. Rey, E. Javierre, S.J. Garcia, S. van der Zwaag, J.M. Garcia-Aznar, Numerical study of the scratch-closing behaviour of coatings containing an expansive layer. Surf. Coat. Technol. **206**, 2220–2225 (2012)
66. A.B. South, L.A. Lyon, Angew. Chem. Int. Ed. **49**, 767–771 (2010)
67. J. Mardel, S.J. García, P.A. Corrigan, T. Markley, A.E. Hughes, T.H. Muster, D. Lau, T.G. Harvey, A.M. Glenn, P.A. White, S.G. Hardin, C. Luo, X. Zhou, G.E. Thompson, J.M.C. Mol, Prog. Org. Coat. **70**, 90–101 (2011)
68. M.W. Lane, A. Roush, S.E. Callahan, in *Tenth International Workshop on Stress-Induced Phenomena in Metallization. AIP Conference Proceedings*, vol. 1143, 2009, pp. 71-86

Chapter 8
Delivery Systems for Self Healing Protective Coatings

M.L. Zheludkevich and A.E. Hughes

8.1 Introduction

Application of protective coatings is the most common and cost effective method of corrosion protection for a wide range of engineering structures, from cars to aircrafts, from heavy industry, energy industries and chemical factories to household equipment. The main role of coating in corrosion protection is to provide a dense barrier against corrosive species. Additionally the protective coatings can confer other important functions such as desired aesthetic appearance, improved mechanical properties, and self-cleaning just to mention a few. However, defects are inevitable during the lifetime of coated structures leading to local disruption of the barrier function. The corrosion processes develop faster after failure of the protective barrier. Therefore active self-healing of defects in coatings is necessary in order to provide effective long-term protection.

The term self-healing in materials science was actively introduced by White only in beginning of 21st century [1]. The main definition of self-healing is based on self-recovery of the initial properties of the material after degradation due to the external environment. The same definition can be also used for protective coatings. The classical

M.L. Zheludkevich (✉)
Institute of Materials Research, Helmholtz-Zentrum Geesthacht,
Zentrum für Material- und Küstenforschung GmbH, Max-Planck-Straße 1,
21502 Geesthacht, Germany
e-mail: Mikhail.Zheludkevich@hzg.de

M.L. Zheludkevich
Department of Materials and Ceramic Engineering, CICECO-Aveiro Institute of Materials,
University of Aveiro, 3810-193 Aveiro, Portugal

A.E. Hughes
Institute for Frontier Materials, Deakin University, Waurn Ponds Geelong, Australia

A.E. Hughes
CSIRO Mineral Resources, Clayton, VIC 3169, Australia

A.E. Hughes et al. (eds.), *Active Protective Coatings*,
Springer Series in Materials Science 233, DOI 10.1007/978-94-017-7540-3_8

understanding of self-healing is based on the complete recovery of the functionalities of the coating due to a real healing of the defect based on the recovery of the coating integrity. However, a partial recovery of the main functionality of the coating can be also considered as a self-healing ability. Thus, in the case of corrosion protective coatings the term "self-healing" can be interpreted in a different way [2]. As pointed out in the introduction to this book, the main function of anticorrosion coatings is the protection of a metallic substrate against an environment-induced corrosion attack. Thus, hindering of corrosion activity in a defect through agents in the coating itself, is already considered as self-healing. If the corrosion protective system recovers its main function, namely the corrosion protection, it can be considered as self-healing. A temporary self-healing functionality can also be achieved by incorporation of encapsulated water-displacing agents which can repeal aqueous electrolytes from the defects shortly after damage appears in the coating. Moreover active preventive mechanisms can also be integrated. One of the examples is entrapment of aggressive species reducing their penetration through the coating system down to the metallic substrate. Of course changes to the current strategy in coating design are required in order to implement these new strategies. Some strategies will require changes to the design of the polymer system whereas others will require the addition of active agents and their delivery systems [3]. This chapter examines aspects of these changes in more detail than other chapters in this book. However, prior to delving into the details of these areas this chapter begins with an overview of strategies for self healing.

8.2 Overview of Self Healing Coatings

Several self-healing or active protection mechanisms can also be combined in the same coating system being introduced to the different layers or even mixed together in a one layer coating. Recently the concept of multi-level self-healing protective coatings was introduced by Zheludevich et al. [4]. This multilevel protection draws on much of the underlying research in this field and from that perspective is worth a review. The multi-level self-healing concept is based on gradual active feed-back of the protective systems to the environmental conditions as illustrated in Fig. 8.1. Different embedded active protection mechanisms can function depending on the aggressiveness of the environment and the defectiveness of the coating. In the case of an intact coating exposed to an aggressive environment the preventive entrapment of chloride anions can result in a reduction in the permeation of these corrosive species through the coating. The water displacement mechanism can be temporarily activated by mechanical disruption of capsules when defects are formed. The formed defects can be sealed by expandable phases or reflow based approaches. Finally, the corrosion inhibition mechanism can be in play when corrosion processes start at the metallic surface. The combination of several mechanisms in the same coating system can result in synergistic effect and efficient long-lasting active protection. Different self-healing mechanisms involved in the multi-level protection concept will be briefly addressed in the present chapter.

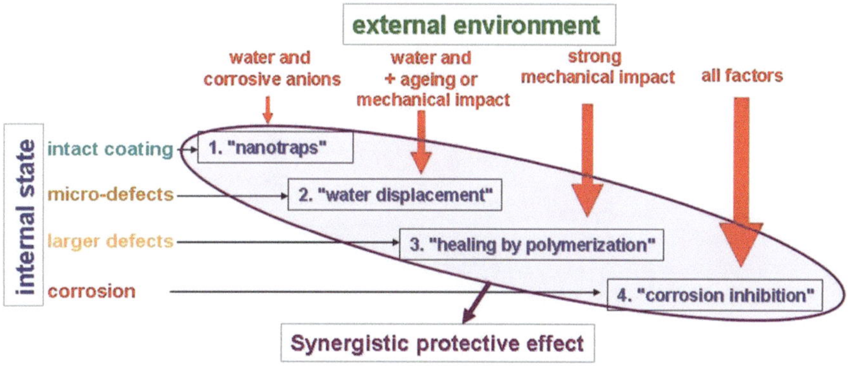

Fig. 8.1 Schematic representation of multi-level active protection concept [5]

8.2.1 *Nanotraps*

The first level of active protection can be achieved by the incorporation of nanotraps for chloride ions into the coating. Tedim et al. [6, 7] has demonstrated a promising chloride-entrapment effect by the Layered Double Hydroxide (LDH) nanoparticles. Zn–Al LDH intercalated with nitrate anions can serve as chloride nanotraps because of the high ion-exchange ability of the LDH. The addition of such particles to a polymer layer drastically reduces the permeability of corrosive chloride anions through the protective coating. Figure 8.2 clearly shows that a coating modified with LDH–NO_3 exhibits significantly lower permeability to chlorides in comparison to both blank and LDH–Cl-containing coatings. Immersion of the coating modified with Zn(2)–Al–NO_3 results in a nitrate-to-chloride anion exchange. By the end of the measurements (1 month), about 70 % of nitrate anions have been substituted by chlorides as followed from using XRD measurements. This demonstrates the potential of LDHs in delaying coating degradation and corrosion initiation. Moreover, LDH nanotraps loaded with corrosion inhibitor can play a double function capturing aggressive anions in the coating and releasing corrosion inhibitor on demand when aggressiveness of environment is high [8].

An improvement of barrier properties of polymer coatings via addition of particles which can trap water is another active corrosion prevention mechanism. The possibility of decreasing the water permeability through polymer coatings by adding pigments with high affinity to water is still disputed. However, the recent work presented by Krzak et al. [9] show a significant decrease of permeability when water microtraps (cross-linked poly(methacrylic acid) sodium salt spherical particles) are added to epoxy-based coating [9]. The water diffusion coefficient for the best trap-containing coating was three times lower than for the pure epoxy coating. The modelling of such system predicted that it is beneficial if the water-traps are positioned in the middle of the coating system. This effect was also proven by the experimental permeability measurements.

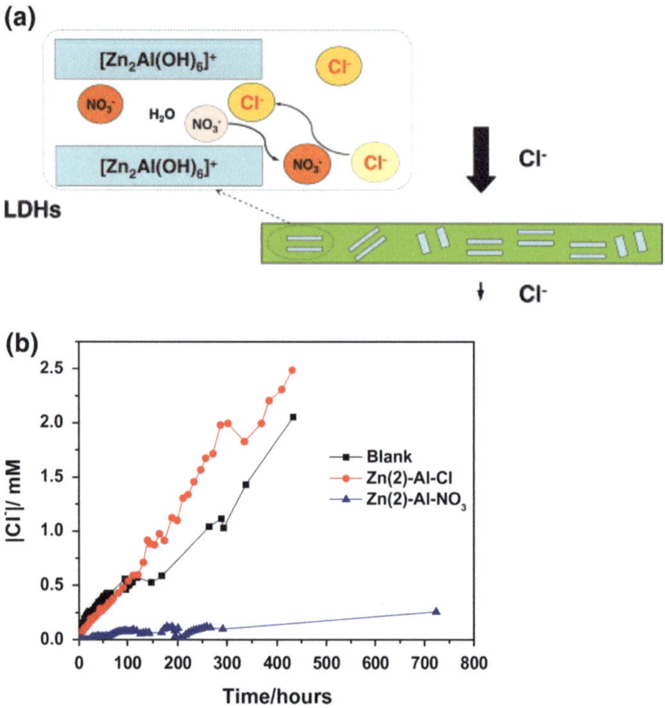

Fig. 8.2 Scheme of LDH nanotrap effect in a polymer coating: **a** Permeability of chloride anions through the coatings (blank, coating with full Zn(2)–Al–Cl traps, and coating with 'empty' Zn(2)–Al–NO_3 traps) **b** as a function of time (adapted from [6])

8.2.2 *Water Displacing from Defects*

The demonstrated effect of nanotraps is important as a preventive action when the coating is still intact. However, it will not be very efficient when the first defects appear and corrosive agents penetrate through the defects rather than by the diffusion though polymer. In this case different healing mechanisms should be activated.

Several groups have proposed the use of encapsulated hydrophobic materials for water displacement ranging from drying oils (alkyds) to silanes [10–16]. As an example, Latnikova et al. [12] have proposed the introduction of hydrophobic silane in encapsulated form into protective polymer paint. The capsules can be disrupted upon a scratch formation or under mechanical stresses caused by a strong deformation. The content of the capsules flows into the defects and spreads on the substrate surface. Simultaneously stretching the coated substrate and measurement of the dynamic contact angle can provide information about water displacement properties from induced defects that are formed [17]. The results indicated that the advancing contact angle is increasing over a long period of time after their

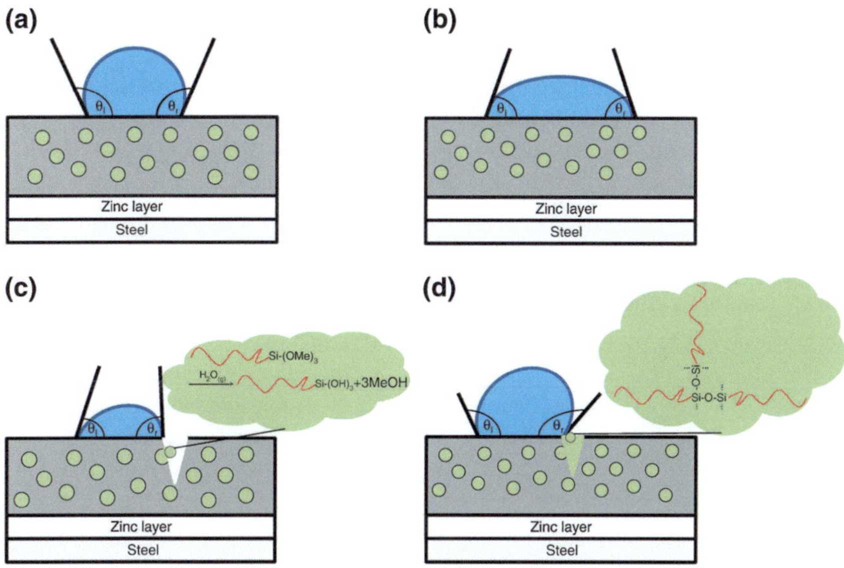

Fig. 8.3 Schema of hydrophobization occurring at the defect induced by stretch forming [17]

formation. With high probability a release of the hydrophobizing agent during coating stretching causes the surface to become more hydrophobic Fig. 8.3. Thus not only the scratches but also defects introduced by environmental stressors can be hydrophobized providing temporary corrosion protection.

8.2.3 Self-sealing

The preventive entrapment of corrosive species and a temporary protection in the formed defects are not sufficient when long term healing of induced defects is needed. More permanent defect healing approaches should be considered. For example the phenomenon of autogenous crack healing in concretes has been known for centuries [18]. The defects in ancient structures like Roman aqueducts are self-repaired by unhydrated cement particles that are in contact with penetrating water. Crack healing occurs due to carbonation of leached calcium cations when both water and CO_2 are available inside the defect. Following the idea of concretes the self-sealing protective coatings can be designed. The barrier properties of a damaged coating can be recuperated by a simple blocking of the defects with insoluble precipitates. These deposits in cracks can originate from reaction of the corrosive medium or corrosion products with components of self-sealing coating.

The high-temperature self-sealing protective coatings were reported several decades ago [19]. Plasma sprayed Cr/Cr_2O_3 layers with closed porosity are able to recover the diffusion barrier against oxidizing species because of Cr oxidation and closing the defects when exposed to an oxidizing environment at temperatures above 900 °C. Sugama and Gawlik [20] have reported a poly (phenylenesulfide) (PPS) self-sealing coating for protection of carbon steel heat exchanger tubes operating at brine temperatures up to 200 °C. The calcium aluminate fillers dispersed in the coating matrix undergo the decalcification–hydration reactions close to the defect promoting a fast growth of boehmite crystals in the cracks. The block-like boehmite crystals (~4 μm in size) fill the cracks increasing the pore resistance of the damaged coating about two orders of magnitude after 24 h. Extension of the exposure time to 20 days results in a stable value of pore resistance meaning that the sealing of the cracks by boehmite crystals plays an essential role in the recovery of the protective function of a PPS coating.

The self-sealing coatings operating at lower temperatures have been also reported [21]. The two layer coating consisting of an expandable clay layer and a polymer top coat was designed. The idea is based on the expansion of the clay phase in the defects leading to its partial sealing. Figure 8.4 demonstrates the schematics of the coating and micrographs of scratched zone before and after sealing. Fast filling of cracks occurred mainly within the first two hours of exposure resulting in a good restoration of the surface flatness. A similar approach was also suggested by Hikasa et al. [22], where a sodium-clay (hectorite) and silica multilayers were deposited using spin coating technique. The hectorite is a swelling clay which expands cubically due to the reaction with water. When water penetrates to

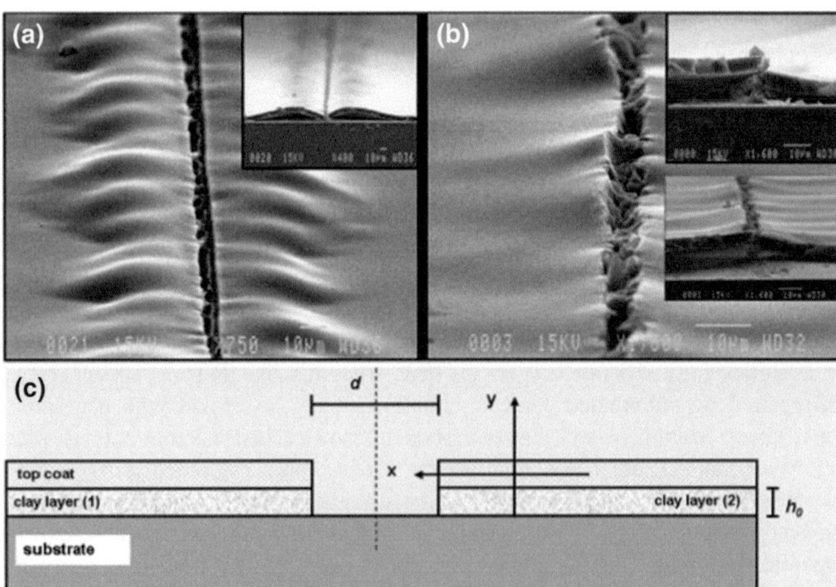

Fig. 8.4 SEM images of a scratch on a multi-layer coating obtained by a 0.15 N load before (**a**) and after 21 h of healing (**b**); (**c**) graphical representation of the two-layer coating used to model the crack-filling effect; d = scratch width and h_0 = clay layer height. Adapted from [21]

the crack in clay/silica composite film it reacts with hectorite causing its swelling and consequent blocking of the defect.

The swelling function can also be achieved by addition of superabsorbent polymers as active pigments [23]. The protective vinyl ester polymer coating containing superabsorbent polymer particles were developed to ensure active protection of steel substrates. Electrochemical impedance study demonstrates that barrier properties of the coating damaged by a knife increase remarkably upon immersion in electrolyte. The swelled polymer particles create a layer which prevents the diffusion of oxygen from the bulk solution to the scratched surface of the substrate.

In spite of the proof of several self-sealing concepts the real application of self-sealing coatings for low-temperature corrosion protection has not yet been demonstrated. One of the important restrictions is the limited degree of barrier recovery. The sealing often is not sufficient to fully prevent ingress of electrolyte and oxygen into the defect.

8.3 Self Healing in Polymer Networks

Another approach widely discussed in self-healing polymer or composite bulk materials is based on reversible formation of polymer bonds under different conditions. This type of self-healing is often called as intrinsic healing. In this case the healing capability is intrinsically connected to the polymer matrix. The two parts of the damaged polymer should be brought together at the first stage in order to ensure the contact. Only then the reconstruction of polymer bonds and recovery of the polymer matrix integrity can be achieved [24]. There are two main chemical approaches based on supramolecular interactions (e.g., hydrogen bonding [25], ionic interactions [26], π–π interactions, or host-guest interactions)) or reversible covalent bond formation (e.g., based on the Diels-Alder (DA) reaction) [27]. A comprehensive review on the different self-healing polymer based materials is out of scope for this chapter since most of the approaches are not applicable to protective coatings. In the case of coating the geometrical limitations can play a significant role. A relative high mobility of the polymer matrix is needed to close the gap by relatively thin layer of polymer. In many cases the width of the defect is larger than the thickness of the polymer coating. Nevertheless several intrinsic self-healing polymer coatings were recently developed as briefly overviewed below.

An example of supramolecular copolymer used to create self-healing coating was recently presented by Bode et al. [28]. The chains forming the copolymer network are connected by metal-ligand interactions in this case. The healing process occurs on the molecular level via formation of ionic clusters which consist of doubly positively charged iron center and two terpyridines as ligands (Fig. 8.5). However, certain energy is required to reach sufficient mobility of polymer network for the reformation of ionic clusters.

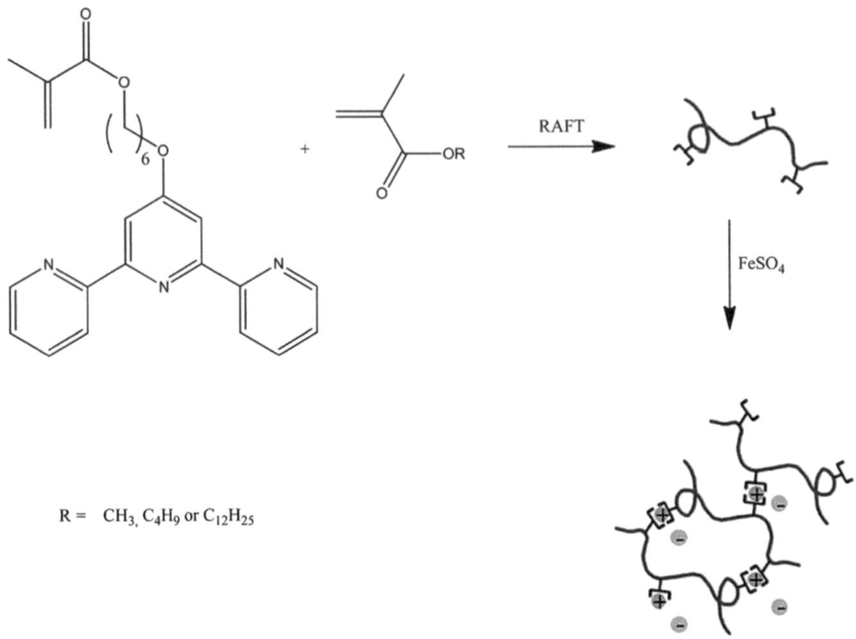

Fig. 8.5 Schematic representation of the iron(II) crosslinked polymer network with ionic clusters [28]

The Diels-Alder (DA) reaction, widely explored in the self-healing bulk materials, can also be considered as a promising approach when designing active coatings with intrinsic self-repair function. The healing action in the case of DA based polymers is relying on reversible formation of covalent bonds under different temperature conditions. Thus the defects can be healed by reformation of polymer bonds upon heating to moderate temperatures. One of the most typical systems is the furan/maleimide couple. Kötteritzsch et al. [29] have synthesized novel terpolymers with MIMA (2-(1, 3-Dioxo-3a, 4, 7, 7atetrahydro-1H-4, 7-epoxyisoindol-2(3H)-yl)ethyl methacrylate) and FMA (furfuryl methacrylate) as functional units for DA reactions. As co-mono-mers MMA (methyl methacrylate), BMA (butyl methacrylate), and LMA (lauryl methacrylate) were used. The most flexible co-monomer is (LMA), enabling an effective reflow of the material in the range of the DA temperature. While BMA-based co-polymer provided the healing effect only on the nanometer range, the LMA-containing system was also able to heal millimetre-wide defects. It was demonstrated that the scratches can be repaired within a short time (2 min) at 160 °C or over longer periods using lower temperatures (110 °C) (Fig. 8.6).

The mobility of the polymer networks plays a crucial role in the healing of defects using the DA reaction approach. Higher flexibility can be achieved not only by choosing long chain co-monomers but also via introduction of proper plasticizers. The healing performance of the polymer based on thermoreversible DA

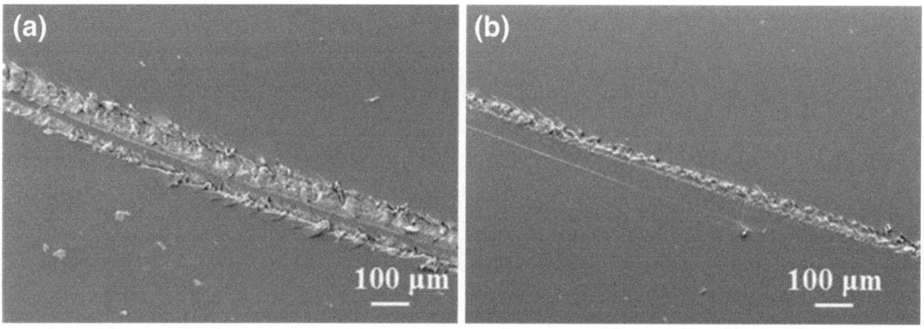

Fig. 8.6 Self-healing experiment of LMA-copolymer using SEM for visualization. **a** Scratch before annealing, **b** scratch after annealing at 110 °C for 30 min [29]

reaction was significantly improved by the addition of 10 % w/w of benzyl alcohol as plasticizer [30]. Benzyl alcohol is thermally stable in the temperature range needed to activate the retro DA reaction. The plasticizer-containing polymer has demonstrated an excellent recovery of aesthetic appearance of the scratched surface.

Shape memory polymers can also be used to formulate protective coating capable healing of mechanical defects using the viscoelastic reflow effect. The shape memory polymers are usually double-phase materials with high flexibility capacity. Under the action of the external trigger such as UV or temperature one of the phases becomes very mobile allowing a limited reflow of the polymer. In this way the scratch can be filled again with the polymer. González-García has used a shape-memory polyurethane (SMPU) film as a protective coating on AA2024 [31]. The shape memory polymer film consisted of a two-segmented block co-polymer of poly-(ε-caprolactone) (PCL) as the soft domains and polyurethane segments (PUs) as the hard domains. This material, at a temperature above the melting point of the soft part, offers enough mobility and sufficient mechanical stability to avoid excessive flow of the coating.

The idea of reflow-healing of protective coatings has already found its commercial realization. Nissan has recently implemented the "Scratch Guard Coat" painting system which contains a newly developed high-elastic resin providing reflow in artificial scratches [32]. The new coating system is effective for about 3 years and is five times more resistant to abrasions caused by car-washing machine than a conventional clear paint. However the cost effect is still a limiting factor for the mass cars.

8.4 Coating Systems with Encapsulated Inhibitors

As described above self healing or responsive coatings, require the incorporation of responsive mechanisms or agents to produce the healing effect. In the previous section a number of mechanisms were reviewed where the response is built into the polymer itself. However, there is a very broad range of self healing applications where it is either desirable or necessary to isolate the healing agent from the polymer matrix using a variety of encapsulation approaches. There are good reasons to isolate the healing agents because (i) it may react with the polymer matrix material, (ii) it may be easily leached from the polymer or (iii) it may be desired to have a triggered release and so the capsule must respond to the triggering mechanism.

The term *encapsulation* itself encapsulates a wide variety of approaches to incorporating agents into coatings. It can range from inert capsules which play no other role than to isolate the active agent from the polymer resin [16], to ion exchange materials where the healing agent (e.g. chromate) that is released is exchanged for an agent that causes degradation (e.g. the chloride ion) [33]. There are several good reviews on this subject [34, 35]. There is no straight forward classification system for encapsulated, self healing agents since they could be classified based on capsule geometry, size, material, mechanism (e.g. ion exchange capacity mentioned above), or application to cite just a few possible classification systems. The approach adopted here is to classify the capsules based on the material from which they are made.

Before we discuss in detail the different capsule types, it is worth noting that many of the encapsulation approaches adopted by researchers in the field are borrowed from either the food or pharmaceutical industries where encapsulation is used extensively. In the food industry one of the major uses is for encapsulating flavours [36]. In the pharmaceutical industry it is used for targeted delivery of drugs [37–41]. It is interesting to note that for many oral pharmaceuticals, the acid environment of the stomach is a key target environment for the design of encapsulation systems. Different design strategies may vary from slow release to complete inertness in the stomach to allow the capsule to survive the stomach environment to release "downstream" in the intestines. The reason why this application is of interest is that the stomach is a hydrochloric acid environment of pH 3 which is similar to the anolyte solution in an active pit [42, 43], which can be used as a trigger mechanism.

Encapsulating containers for coating applications have different requirements than for other applications. The most obvious limitation is that the width of the container must be smaller than the thickness of the coating as noted above. Convoluted with this issue is the amount of material that is required for healing. Approaches to this issue involve hollow tubes that can form 2-D structures [44] to methods for the manufacture of elongated capsules [45].

Prior to further description of encapsulation, particularly where an active agent is absorbed into a porous material, it is important to note that there are IUPAC definitions for porosity [46]. These are:

(i) *micropores*: less than 2 nm
(ii) *mesopores*: between 2 and 50 nm
(iii) *macropores*: greater than 50 nm

These definitions are based on the size of the pore diameter (from one side of the pore to the other), but as can be seen from the definitions both microporous and mesoporous materials have pores that are nano-dimensioned. Further, the smallest macropores are also in the nanometer range. The term nanoporous is commonly used to describe the porosity in materials, but if this term refers to pore size then the porosity could be as large as a macropore and not particularly small by the definitions above. It is also worth noting that microporous materials (<2 nm) may be good for small oxyanions such as a number of inorganic inhibitors, e.g. chromate, but may not be so useful for larger organic inhibitors or polymer healing agents.

8.4.1 Volume of Self Healing Agent Required

Before examining the detail of the types of containers that could be used for self healing, it is important to consider the quantity of material required for healing. The simplest case is a single, once-only dose, the most complicated case is ongoing delivery of a healing agent in repeated doses. The types of approaches for each situation could be vastly different. In the former case numerous small independent (non-interacting) containers may be sufficient. On the other hand, if ongoing, triggered release is required then one can imagine a complicated system with supply networks for the self healing agents along with sensors to detect the need for repair, and actuators to stimulate release of the healing agent.

Some insight into the volumes of healing material required can be gleaned from the use of multiple tubes to deliver the agent for repair of a defect as presented in Fig. 8.7. Here the defect is a scratch with a length of 1 cm through a 25 μm thick coating. The x-axis represents the width of the defect (scratch width) and the y-axis reflects the volume of materials required for repair. The volume is expressed as the

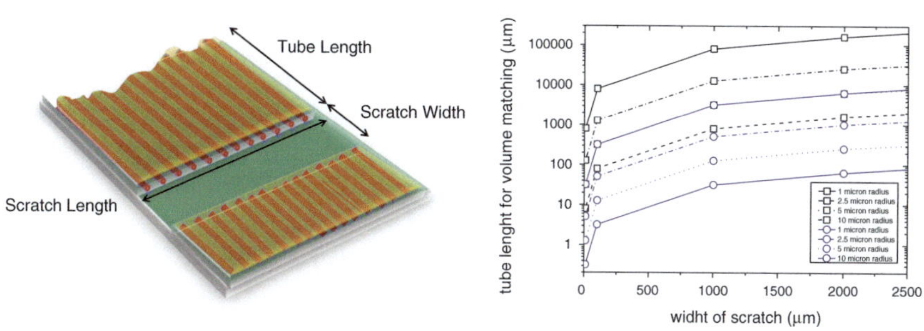

Fig. 8.7 Conceptual scenarios for volume of material required for self healing of a defect in a paint film. The model chosen here is a series of hollow tubes embedded into a 25 μm thick film. For the purposes of calculation of the volume of material released, two radii of tube are considered; 1 and 10 μm. The volume of material that is required for repair is expressed in terms of the length of tube for either of the two radii

length of tube (tube length) that is required for four different radii of tubes (1–10 μm) that are embedded into the film. A 10 μm radius for the tube is close the thickness of the film. The calculation for the data presented in Fig. 8.7 assumes that there are 1000 tubes supplying the defect. In the case of the tubes with a 10 μm radius the tubes will be nearly touching each other along the walls of the scratch. It is immediately clear that as the scratch widens there is a significant increase (orders of magnitude) in the volume of material required for volume filling of the scratch which is equivalent to full functional and cosmetic repair as represented by the data in black. As an example a 1 mm (1000 μm) scratch would require all 1000 tubes to release healing agent from around 10 cm of their length which will limit the availability of healing agent for further repairs. On the other hand, if the objective is to simply provide a functional repair then, represented by a 1 μm thick film in Fig. 8.7, then only 1 mm length of tube is required to release enough material from all the tubes to provide a functional repair thus reducing the design demands for the coating system. This reasoning is what has driven a lot of self healing activity in the coatings area for a functional, as opposed to full cosmetic, repair.

The strategies based on encapsulation of self-healing agents in single capsules confront the same issues related to the amount of healing agent available at the defects. The number of capsules and their diameter play the major role. Recent theoretical analysis has demonstrated that the size of the microcapsule and the geometry of the crack are very important in coating healing [47]. In the case of small cracks the optimal diameter of microcapsule has to be used. The probability of healing significantly decreases with an increase in diameter and volume of available healing agent in a single capsule since the probability of rupturing sparsely located large microcapsules by a small crack is very low. In the case of a large crack the maximum possible microcapsule diameter has to be used for efficient healing. It is related to the fact that the probability of hitting the capsules by the crack is close to 1 and repair depends only on the amount of healing agent available. The dependence of the theoretical healing efficiency on the capsule dimensions for different loading levels and different crack sizes is demonstrated in Fig. 8.8. It can be clearly seen that even at high loading levels such as 15 % the first remarkable signs of a self-healing effect can be expected only in the case of capsules larger than 20 μm. However, as stated, capsules of this size approach the film thickness for many potential applications. Therefore the self-healing coatings with polymerizable agents can be only efficient for coatings with much higher thickness used in oil and gas or marine industries for example.

The only possible way to achieve the recovery of mechanical integrity is application of polymerizable agents which form expanding phases where the volume of the polymer is significantly higher than the volume of original monomers. It can be achieved for example by use of polyepoxide or polyurethane foams [48].

Beyond these conceptual ideas for addressing the issue of the volume of required healing agents that is required, are the mechanics and materials science of incorporating these concepts into real coatings. Small isolated containers will have an influence on the mechanical properties of the coating on the local scale, whereas the incorporation of flow networks, sensors and actuators will have an obvious

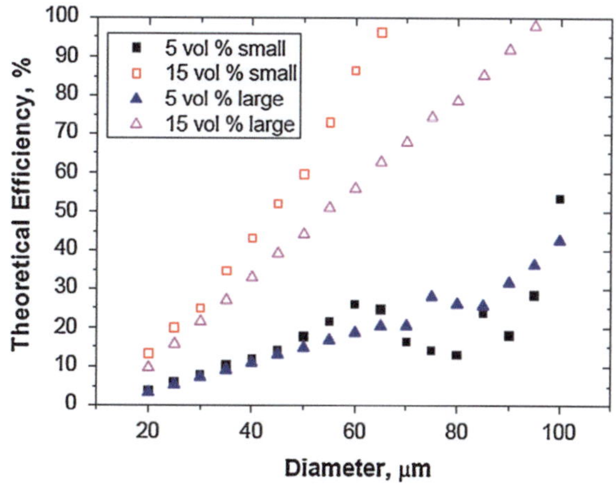

Fig. 8.8 Dependence of theoretical healing efficiency on diameter of the capsules for different levels of loading and different crack sizes [47]

influence on the local properties of the coatings but also on the properties and engineering at the scale of the painted structure, which may introduce a whole new set of issues.

8.5 Encapsulation Systems

The following sections describe various methods for encapsulating either inhibitors or polymer healing agents. Not all approaches are suitable for both types of healing applications. For example, monomers for polymer healing are unlikely to be suited to the mesoporous inorganic capsules described below since in many instances they will react with the carrier.

In terms of capsule technology, there are two types of chemical approaches that are broadly used in preparation of capsules that will be discussed prior to a more detailed description of encapsulation systems that have been used to date. These are (i) sol-gel and (ii) emulsification approaches.

8.5.1 Sol-Gel: Transition Metals

Sol-gel processes are used extensively in modern chemistry [49]. There are two fundamental classes of reaction that are important for sol-gel processing for the manufacture of inorganic capsules which include (i) hydrolysis and (ii) condensation reactions. These reactions are used extensively in coating formulations and also in making porous materials for a range of applications from catalysis to drug delivery.

Metals ions dissolved in water are solvated by water molecules. Hydrolysis of these metal ions can be generally defined by the following equilibria:

$$[M(OH_2)]^{z+} \leftrightarrow [M-OH]^{(z-1)+} + H^+ \leftrightarrow [M{=}O)]^{(z-2)+} + 2H^+ \qquad (8.1)$$

Where there are three types of non-complexing ligands including the *aquo* ligand $M(OH_2)$, the *hydroxo* ligand (M–OH) and the *oxo* ligand (M = O). The equilibrium species will depend on the co-ordination number (N), the charge (z), the electronegativity and the pH. A general representation of these precursor species is $[MO_NH_{2N\text{-}h}]^{(z-h)+}$ where h is the molar ratio of hydrolysis. In terms of the species, if h = 0 then it is an *aquo* ion $[MO_NH_{2N}]^{(z)+}$, and if h = 2 N then it is an *oxo* ion $[MO_N]^{(2N-z)-}$. At intermediate values of h there are *oxy-hydroxo* complexes $[MO_x(OH)_{N\text{-}x}]^{(N+x-z)-}$ or *hydroxo-aquo* complexes $[M(OH)_x(OH_2)_{N\text{-}x}]^{(z-x)+}$. In the case of the transition metal cations, there is charge transfer into the d orbitals from the $3a_1$ bonding orbital of water. In the case of the rare earths the f orbitals will also be involved.

These species can then interact to form structures that eventually result in precipitates; a process called condensation. This can proceed via two nucleophylic mechanisms either nucleophylic substitution where the preferred co-odination is satisfied [49]:

$$M_1\text{-}OX + M_2\text{-}OY \rightarrow M_1\text{-}\overset{\overset{\displaystyle X}{|}}{O}\text{-}M_2 + OY \qquad (8.2)$$

or nucleophylic addition where the preferred co-ordination is not satisfied:

$$M_1\text{-}OX + M_2\text{-}OY \rightarrow M_1\text{-}\overset{\overset{\displaystyle X}{|}}{O}\text{-}M_2\ \text{-}\ OY \qquad (8.3)$$

Condensation occurs via *olation* or *oxolation* (which includes *alcoxolation*) as described below. During *olation* a hydroxyl bridge(s) is formed between two metal centres whereas *oxolation* is a process where an oxo bridge(s) is formed between two metal centres with one type of oxo-bridge depicted in the schema below. Where the metal is co-ordinatively unsaturated then oxolation occurs by nucleophylic addition and leads to edge sharing or face sharing polyhedral [49]. For co-ordinatively saturated metal ions then oxolation occurs by 2 step nucleophylic substitution leading to a M–O–M bond and water elimination. Oxolation occurs via:

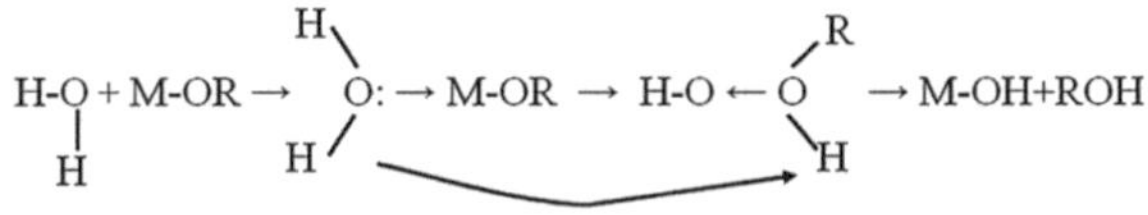

(8.4)

Whereas alcoxolation occurs via:

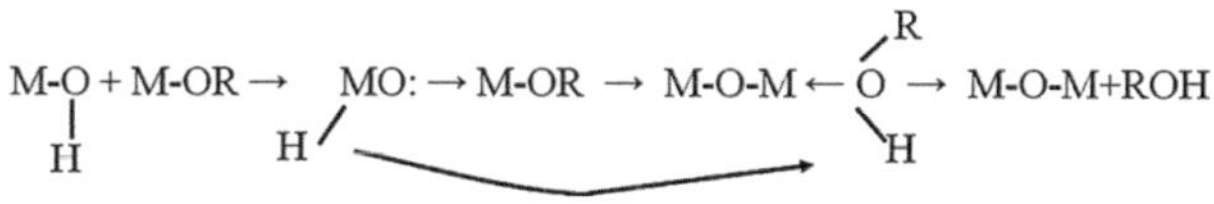
(8.5)

Olation occurs via the formation of a hydroxyl bridge between the two metal centres as depicted in one schema below. There are numerous other types of bridging configurations that can occur as discussed by Brinker and Scherer [49]. Nucleophylic substitution reactions for transition metal ions result in dimmers or trimers in certain pH regimes since the net charge diminishes with each substation to the point where there is no further driving force for nucleophylic addition.

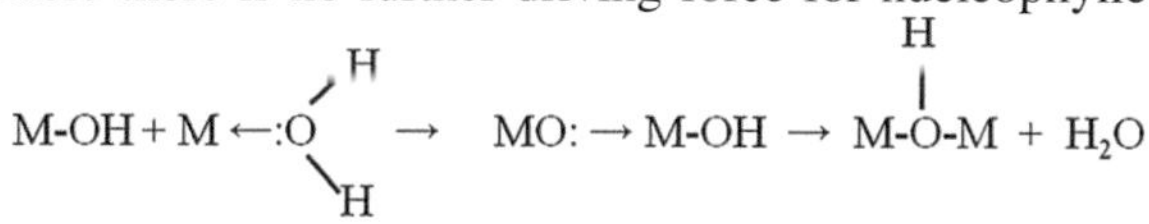
(8.6)

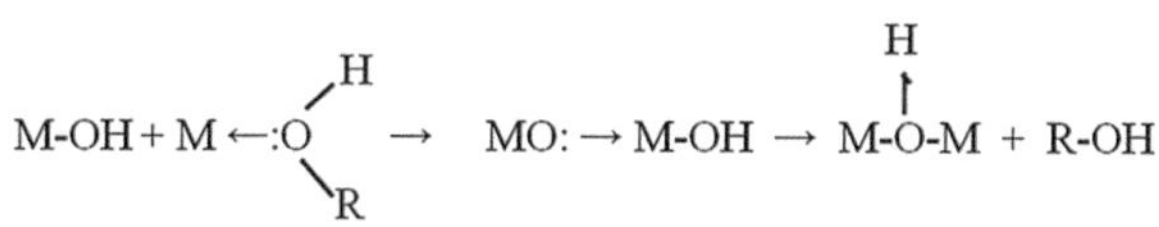
(8.7)

Close to the iso-electric point neutral precursors are able to condense indefinitely to form either hydroxides or oxy-hydroxides. Polyanions are formed at high pH with the chromate reaction being a familiar example in the world of inhibitors:

$$[Cr_2O_5(OH)_2]^0 \rightarrow [Cr_2O_6(OH)]^- + H^+ \tag{8.8}$$

$$[Cr_2O_6(OH)]^- \rightarrow [Cr_2O_7]^{2-} + H^+ \tag{8.9}$$

8.5.2 Sol-Gel: Silicon

Silicon alkoxides are the most commonly used alkoxides and have some distinct differences to the transition metal alkoxides. As summarised by Brinker these include [49]:

1. lower electronegativity of Si compared to the transition metals,
2. In polar solvents neither oligomerisation nor alcohol association are observed with Si-alkoxides whereas they are observed with the transition metal alkoxides,
3. Processing for Si-alkoxides is less strict than for transition metal alkoxides which can react extremely quickly.

Silicon is hydrolysed in H_2O to form $Si(OH)_4$ (silicic acid). Above pH 7 further hydrolysis occurs resulting in species such as $SiO_x(OH)_{4-x}^{4-}$ and protons. The same reactions occur with silicon alkoxides ($Si(OR)_4$). Polynuclear species are characteristic of silicic acid condensation reactions particularly below pH 7.

The formation of silicate gels generally proceeds via the hydrolysis of the tetra-alkoxides facilitated by the presence of either acid or base [49] and is described by the following three reactions, beginning with hydrolysis:

$$\equiv \mathrm{Si{-}OR} + \mathrm{H_2O} \leftrightarrow \equiv \mathrm{Si{-}OH} + \mathrm{ROH} \tag{8.10}$$

Followed by esterification via alcoholysis:

$$\equiv \mathrm{Si{-}OR} + \mathrm{HO{-}Si} \equiv \leftrightarrow \equiv \mathrm{Si{-}O{-}Si} + \mathrm{ROH} \tag{8.11}$$

Or hydrolysis:

$$\equiv \mathrm{Si{-}OH} + \mathrm{HO{-}Si} \equiv \leftrightarrow \equiv \mathrm{Si{-}O{-}Si} + \mathrm{H_2O} \tag{8.12}$$

Of course the R group need not be the same for all ligands and can be substituted by a hydroxyl (–OH) group. These basic reactions underpin our understanding for sol and gel formation for the transition metals as well as silicon. Of course there will be differences between different alkoxides as well as different metals which will determine the exact parameters for processing individual alkoxides.

8.5.3 Emulsification and Sols

Emulsions and sols are commonly used in the preparation of capsules. Emulsions are formed by the dispersion of one liquid into a second liquid where two phases are immiscible and have a tendency to separate. The two will work to minimise their surface free energy which, in the absence of surfactants, leads to phase separation and the formation of two layers, but in the presence of surfactants and agitation an emulsion can be formed. The advantage of the emulsion approach is that it increases the surface area and dispersion of one phase with respect to the second. Thus for equivalent volumes, a spheres with 100 nm radius will have over a thousand times more surface area compared to a single sphere with a 1 μm radius.

Emulsions are generally stabilised by molecules that have end-groups of a different nature to each other. Thus a fat/water emulsion can be stabilised by fatty acids, that have lyophobic and lyophillic end groups, which sit at the interface between the two phases. The presence of this interfacial layer provides an opportunity for further reaction as discussed in a number of encapsulation processes below. Inverse micro-emulsions have also been used to control particle size of inorganic delivery systems since the precipitation process is confined to the inverse emulsion droplet volume [50].

Sols (solids suspended in a liquid) are also potentially useful for encapsulation via processes which involve preferential interaction at the surfaces. Sols are often stabilised via charge which means that molecules of opposite charge can react with the surface. Examples of this include the polyanion/polycation approaches to encapsulation. In this respect, Hughes et al. also report an oleic acid encapsulation of La-dibutylphosphate inhibitor in epoxy [3].

8.5.4 Inorganic Capsules

Inorganic chemistry is a vast field of chemistry and from within this field there is an equally vast number of possibilities for capsule structure and chemistry derived from both the natural and man-made world. Many of these materials are used in other industries such as in pharmaceuticals [51], others are already used in the paint industry. For example porous aluminosilicates formed from flyash left after the combustion of coal (geopolymers) have been used as starting materials for delivery systems in the pharmaceutical industry [52, 53]. Furthermore, natural capsules are formed during coal combustion and captured from the gas effluent stream such as that presented in Fig. 8.9, and are already used as fillers in paint simply to reduce the volume of more expensive components of the paint such as the polymer and pigments [54].

8.5.4.1 Natural Materials

Among natural materials there are three broad classes that are used as containers for inhibitors including silicates, polyphosphates and layered double hydroxides (LDH). Our understanding of these materials has led to many man-made structures, some of which only have a curiosity value while others have very practical applications; these will be discussed below. Some of the most commonly used natural minerals include alumino-silicates such as natural zeolites and layered

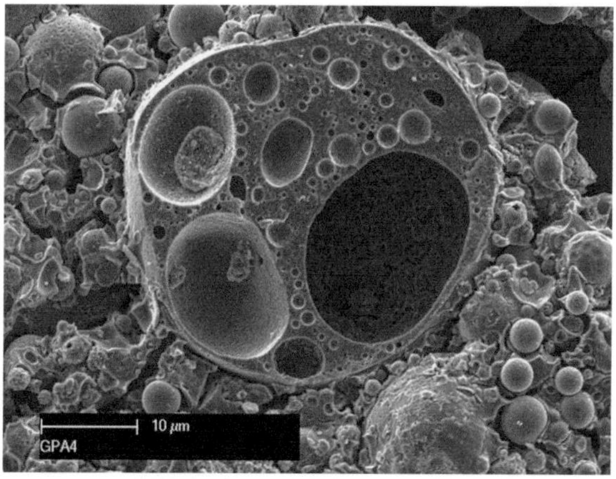

Fig. 8.9 Example of alumino-silicate hollow-sphere formed during coal combustion used as filler in paint, sealants, plasterboard and polymers (Courtesy of Dr. Damian Fullston, CSIRO)

double hydroxides which can be silicate-based (clays) or based on other elements (e.g. Hydrotalcites). For example, NaX Zeolite occurs both naturally but can also be synthesised and has been used as a container by loading it with Ce(III) [55, 56]. Even corrosion product itself may form LDHs having a porosity that may influence the corrosion process as well as provide opportunities for the design of self healing mechanisms [57]. A number of natural materials have been explored for application in the paint industry. For example Ce [58] and Ca-exchanged bentonite clays [59] have been tested in paints. Halloysite, which is a mineral clay that forms nanotubes, has been used to encapsulate organic inhibitors [51] with a particular focus on benzotriazole (BTA) as the inhibitor [51, 60, 61]. Endcaps for the Halloysite tubes were formed by exposure to a $CuSO_4$ and controlled release was achieved by treatment of the capsules with ammonia. Release of the active component from these materials is generally triggered by either ion exchange, pH and or solubility (or a combination thereof).

The materials mentioned above operate as "capsules" at the molecular level, i.e., largely as microporous materials. However, mesoporous or macroporous materials can also be of interest since the larger scale pore structure in these materials can be used to store healing agents. For example, there are many types of porous rocks that are of current interest because they contain gas or oil within their pore structures. The pore structures within these natural minerals are only beginning to be investigated [62, 63]. There may be as yet undiscovered opportunities to use these materials as delivery systems for inhibitors or healing agents where the pore size distribution is of an appropriate dimension.

8.5.4.2 Synthetic Materials

There are numerous approaches for synthesising materials for various types of capsules as a method of incorporating inhibitors in coatings. These types of capsules can vary from simple precipitates which may not have a well defined pore structure to materials with more ordered structures as discussed below.

Disordered Mesoporous Structures

The hydrolysis and condensation reactions described above can be used to make precipitates that may have a disordered mesoporous structure but still have useful properties for either adsorbing or absorbing inhibitors and other agents. These can be simple inorganic precipitates (e.g., oxides hydroxides, phosphates, sulphates) to precipitates that have some organic functionality such as might be produced when metal alkoxides are used to prepare the precipitate, e.g., when silanes are used (RSiOR'). A second category of inhibitor carriers includes inorganic (usually oxide) nano-particles with no microporosity onto which an active agent is adsorbed [5, 64–68]; the term nano-containers or nano-reservoirs are also used to describe these types of capsules [5, 65, 68–70] and even extended to boehmite ($Al_2O_3.nH_2O$, n = 1; boehmite, n = 2 or 3; pseudoboehmite) particles which have been loaded with Ce ions [71]. For example, Ce ions have been adsorbed onto the surface of ZrO_2

[67, 68], TiO_2 [5, 72, 73] and CeO_2 nano-particles [66, 74, 75] and shown good performance. This type of work follows on from putting inhibitors into sol-gel coatings such as Ce ions into ZrO_2 coatings [76].

3D Porous Structures and Templating Methods

As stated above inspiration has been drawn from natural structures to design man-made materials. Some of the most notable developments in this area are the development of the ZSM series of molecular sieves for catalysis in the petrochemical industry [77] and later the Santa Barbara materials (e.g. SBA) [78] and the Mobil Crystalline Materials (MCM) [79]. ZSM zeolites are formed using organic templating (see below) and hydrothermal synthesis whereas the SBA and MCM materials can be manufactured at room temperature. The approach used to manufacture these structures is based on an organic template which defines a porous structure within which the inorganic material precipitates. The template can be modified to provide different structures. The beauty of this approach is that if the template is stable then mixtures of alkoxides can be used to design the "chemistry" within the structure or introduce aliovalent ions that can be used to create charge in the framework. For example, Al ions or alkoxides can be introduced into silica gels of precipitates obtained from alkoxides to provide an Al^{3+} site thus introducing charge into the final structure. This was well demonstrated with the ZSM series of zeolites which can be made with a range of Si:Al ratios that allow for cation exchange into them and so the properties could be tuned. Of course some of the natural zeolites can be formed synthetically and have been used as inhibitor carriers. For example, Dias et al. [55, 56] loaded synthetic NaX zeolite with Ce(III) in a Ce $(NO_3)_3$ solution.

Organic templating (or scaffolding is another term used here) is a common technique these days for the preparation of microporous and mesoporous materials [80]. One particular class of organic templates are the Cubosomes™ which are nanostructured, cubic lyotropic liquid crystalline colloidal particles made from lipid bi-layers [81]. These cubic liquid crystals are most commonly formed from glyceride-based monoolein and phytantriol. They are stabilised by block co-polymers made from ethylene and propylene oxides. An example of the structure of such an organic template is displayed in Fig. 8.10. In this instance the organic structure, which is frozen for the purposes of imaging in the TEM, is formed from a myverol-based organic used for making mesoporous materials for drug delivery [82, 83]. While these types of structures have a high lipid content, they also have water channels for the incorporation of metal alkoxides for the manufacture of an inorganic "negative" of the template. Some common structures which are commerially manufactured through these approaches, but used in other applications such as catalysis, are the SBA [84] and the MCN structures mentioned above [79]. Inorganic particles can also be absorbed onto the internal surface structure thus stabilising it as a framework [85]. Unlike the molecular sieves such as the zeolites these material are mesoporous structures. SBA, for example, has uniaxial channels around 5 nm in diameter and smaller pores along the walls of the channels as displayed in Fig. 8.11.

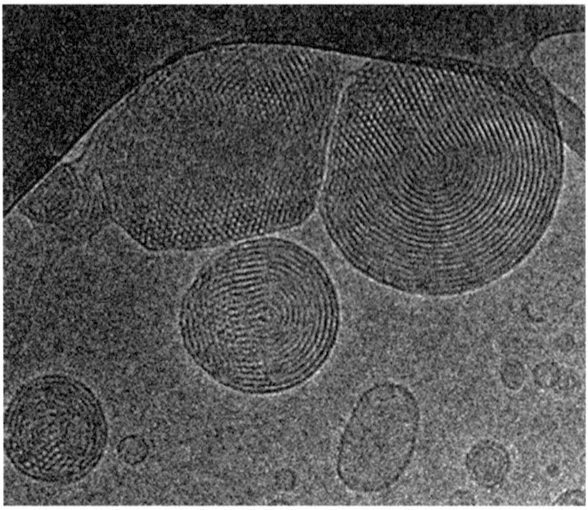

Fig. 8.10 Organic structure used as a precursor for templating. In this case it is Myverol-based organic. The organic has been cryogenically sectioned using diamond microtomy and examined cryogenically as well (Image provided courtesy of Dr. Lynne Waddington, CSIRO.)

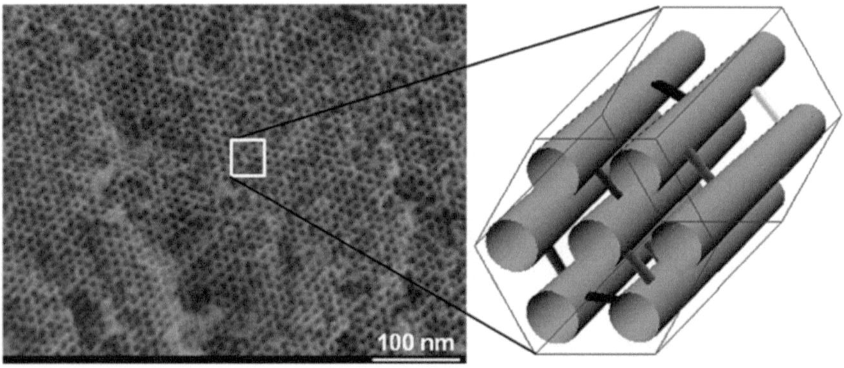

Fig. 8.11 TEM Image of the mouths of the channels in SBA-15 (*left*) and schematic of the structure (*right*). Images provided courtesy of Dr. Lynne Waddington, CSIRO

In principle, organic templating is simple and involves the following steps (i) formation of the organic template (e.g., Fig. 8.10), (ii) precipitation of the inorganic within the porosity of the template (within the bright rings of the structure in Fig. 8.10) and (iii) burning off the organic to leave an inorganic material which is the "negative" of the porous organic template (Fig. 8.11). Second generation materials can also be produced, by causing precipitation within the porous network of the inorganic micro or mesoporous material and then dissolving out the inorganic material. In the case of silicates using hydrofluoric acid or strong alkali results in dissolution of the structure. The second generation precipitation generally involves inorganic materials that are HF and alkali resistant. This results in a reproduction of the original organic framework. The inorganics are typically metal alkoxides, but silicon alkoxides are by far the most characterised and best understood as described in the sol-gel section.

The structure of the organic template is sensitive to the chemistry and can be easily destroyed, thus templating with other metal alkoxides may not be so straight forward as with silicon alkoxides. As discussed above, the rapidity of hydrolysis and condensation reactions for the transition metal alkoxides is one difficulty, but others include whether the optimum pH is achieved for hydrolysis and condensation reactions, or whether the rapid release of alcohols destabilises the templating structure. For the rapidity of hydrolysis, Zheludkevich et al. [86] rates the reactivity of a number of metal alkoxides with similar alkoxides groups as $Si(OR)_4 \ll Sn(OR)_4 < Ti(OR)_4 < Zr(OR)_4 < Ce(OR)_4$. Nevertheless examples based on using Ti propoxides [87] Zr-oxychlorides [88] and Ce acetate have been reported [89].

All these templating processes rely on sol-gel chemistry described above where metal alkoxides undergo full hydrolysis followed by polymerisation. A variant on this process is where traditional hydrolysis and condensation is performed in the presence of silanes. The species here are of the form $R_nSi(OR')_{4-n}$ where the R groups can perform some additional function such as providing linkers for cross linking and coating formation. These materials are called Ormosils after "organically modified silicate" [90]. This type of sol-gel chemistry is used widely in coatings research and there are many excellent reviews of this area [86, 90–92]. There are numerous examples of sol-gel coatings in the inhibitor and coatings literature including silica- [93–101], titania- [72, 102], zirconia- [67, 103–108], ceria- [76, 109] and molybdate- [110] based processes. Some examples include the loading of titania nanocontainers with 8-hydroxyquinilone (8-HQ) [72] or benzotriazole [102]. There is an equally large literature looking at the incorporation of nanoparticles or inhibiting ions into sol-gel coatings [64, 65, 68, 76, 95, 106, 111, 112]. Often sol-gel approaches are used in hybrid applications where native oxide thickening is also employed [109, 110, 113, 114]. Even conversion coatings such as the chromate conversion coating has been viewed from the perspective of sol-gel processes [115].

The most common silicon alkoxides that are used in this application is tetraethoxy silane (TEOS), tetramethoxy silane (TMOS) and glycidoxypropyltrimethoxysilane (GTMS). Bis-(triethoxysilylpropyl)-tetrasulfide silane has also been explored extensively [116–119]. The most common titanium and zirconium alkoxides that are used are the metal propoxides [120].

Triggering release from the pore structure in these materials is often via simple dissolution, but more sophisticated approaches can be designed. "Gates" and coatings sensitive to pH are some of the most obvious approaches and are borrowed from the pharmaceutical industry [121].

8.5.4.3 Sheet Materials

Layered Double Hydroxides

Layered double hydroxides (LDH) or hydrotalcite-like structures have been of great interest in recent years as carriers for inhibitors. As mentioned above HT materials exchange carbonate most efficiently in the presence of chloride ions [122] making this class of materials very suitable for use in corrosion since the presence of chloride can be used as a trigger. Thus the exchange of organic inhibitors such as benzotriazole [123] and inorganic inhibitors such as nitrate, chromate [33] or vanadates [124, 125] for chloride ions has been reported in the literature. Other types of LDH include "green rust" which is a natural occurring Fe(II)–Fe(III) LDH and a transient species in iron corrosion [126] and may delay the onset of corrosion. LDHs have the following composition:

$$\left[M^{2+}_{1-x}M^{3+}_{x}(OH)_2\right]^{x+}\left(A^{n-}_{x/n}\right).mH_2O,$$

where the divalent and trivalent metal ions can undergo isomorphic substitution. The structure consists of Brucite-like layers with edge $M(OH)_6$ octahedra. Methods for making synthetic hydrotalcites are well known and they can be manufactured from the desired chlorides (typically Mg or Zn with Al) under pH control.

LDHs can be used as coatings themselves [127–129] as well as pigments in coating formulations [130–132]. The synthesis of HTs is well known [133] and has been applied to the manufacture of vanadate- [131, 132] and molybdate [134] containing HT pigments. When these pigments have been incorporated into a paint coating they demonstrate a considerable improvement in filiform corrosion resistance due to chloride uptake [33]. As noted above, HT-based coatings themselves have also been explored as alternatives to chromate conversion coatings either grown using elements of the substrate metal [129, 135, 136] or cast onto a surface using spin coating [127].

Inhibitors that have been incorporated into LDHs include inorganic [124, 130, 137] and organic inhibitors [123, 137–139]. Buchheit et al. [131] reported that the replacement of decavanadate during the uptake of the chloride ion results in a change in the crystal structure of the host Zn(II)–Al(III) hydrotalcite which can be detected by X-ray diffraction. They further concluded that this could be used as a sensor for signalling chloride ingress. Zheludkevich et al. also noted significant changes in XRD patterns with inhibitor loading and treatment conditions for Mg (II)–Al(III) and Zn(II)–Al(III) hydrotalictes [132, 138]. Kendig and Hon [139] reported good performance for a 2, 5-dimercapto-1, 3, 4-thiadiazole/HT pigment on AA2024-T3 with the inhibitor itself showing good performance on Cu [140]. Trujillano et al. [141] prepared Cu(II)–Al(III) LDH with intercalated sulphonates which are of interest in catalysis and also for their magnetic properties, but may also be good containers for homo or heterocyclic thio inhibitors [142].

Phosphates

Similar to the zeolite molecular sieves are the phosphate based molecular sieves [143, 144] which were developed in the late 1980s and early 1990s. As with the ZSM series of zeolites, the small pore diameters in these materials makes them less attractive for inhibitor capture. Other phosphate structures exist within the apatite family with the best known being hydroxyapatite which is used as a scaffold for bone regeneration [145]. The preparation of high surface area boron phosphates [146] and titanium organo phosphates [147] has also been reported. The exfoliation of layered Zr-phosphates been investigated for coating formation but more from the perspective of better adhesion rather than for corrosion protection [148]. Exfoliation was achieved by loading the interlayer space with tetra-(n-buytlammonium) ion. To the authors knowledge these have not been used to encapsulation inhibitor compounds in coatings.

Many phosphates themselves have inhibitive properties as well solubilities that make them suitable for direct incorporation into paint systems although their performance is generally not considered to be as good as chromate [149–156], but exceptions have been reported. For example, Kalendova et al. [157, 158] examined a range of phosphate-based materials and found that Zn and Ca phosphates performed better than strontium chromate in epoxyester paints at a pigment volume concentration (PVC) of 10 % and a PVC/CPVC of 0.65. Pokhmurs'kyi et al. found that Ca-hyperphosphate inhibition approached that of strontium chromate [159]. Howard et al. [155] found that commercial phosphate pigment was similar in performance to chromate, but that performance was related more to release rate rather than intrinsic inhibition. Raps et al. [106] found that phosphates added during sol-gel formation improved the barrier functions of sol-gel coatings. On the other hand these materials are not being exploited as delivery systems for other inhibitors as is the case where other porous materials have been employed as containers for inhibitors, so there may be opportunities to develop new inhibitor delivery system using porous phosphates.

8.5.5 Exfoliation and Reassembly into Capsules—Pickering Particles

Exfoliation of layered structures is well known and extensively studied in recent decades. Delamination involves intercalation of fatty or amino acids followed by treatment in organic acids [160, 161] but treatment in formamide has also been reported [162]. Exfoliated layered materials are generally used to improve the mechanical properties of materials. For example, exfoliated clays are used extensively for mechanical reinforcement in polymers and laminates more commonly known as nano-composites [163–165]. They increase the strength of these polymers by arresting crack propagation. Exfoliated layered material is also used to increase fire resistance in polymers since the addition of an inorganic to the organic matrix retards combustion [166].

In encapsulation, the exfoliated layers are used as interfacial layers between two phases in emulsions to form shells [162]. Li et al. [162] described a process for coating polystyrene (PS) beads with HT exfoliated sheets. The PS was subsequently burnt out of the structure. Bon et al. described a process for coating using exfoliated Laponite clay sheets [167, 168] as well as a process using TiO_2 nanoparticles as the inorganic phase [169]. Bon et al. [170] also described a process for making non-spherical droplets to demonstrate the manufacture of nonspherical rods by forcing clay-shell stabilised oil emulsions through a tube. In principle, exfoliated phosphate layers (see above) could also be used for shell manufacture [148].

8.5.6 Polymer Capsules

Polymer capsules have been used in a number of examples of self healing in the coatings area. These are derived from the food industry where polymeric encapsulation is used for a range of applications but typically for extending the lifetime of flavours [171]. In both applications the mechanism of release is via breakage of the capsule. Thus they provide a once-only dose. There are other approaches to using polymers for encapsulation such as polyelectrolyes or conducting polymers which respond to pH changes and these are discussed below. Only the hard polymer capsule approaches will be discussed in this section.

Hard polymer capsule approaches have been used to encapsulate both inhibitors and polymeric healing agents [172–175]. There are a number of methods of making polymeric capsules but they all use the emulsion processes described earlier. There are both one-step and two-step processes described in the literature for the dispersion/polymerisation steps for the manufacture of the polymer shells, although all processes use an acid polymerisation step [13].

The polymer capsule approach for self healing in epoxy systems have been reported for both composite materials and coatings [34, 176]. For example in composites, epoxy and mercaptans were separately encapsulated and demonstrated in a number of studies [177–180] to provide healing in epoxy coatings. In this instance the mercaptan was used as the hardener and highlights the multiple functions that chemicals can have since mercaptans can also be used as inhibitors [142, 181–184]. It also highlights that good organic inhibitors, such as thio-compounds, can interfere with polymerisation and need to be isolated during film-formation. Nesterova et al. [13] examined the synthesis conditions for urea formaldehyde and melamine urea formaldehyde formulations when the encapsulant was a polymer healing agent. It was found that encapsulation procedures needed to be adjusted for each encapsulant and that viscous materials, such as bisphenol A or bishenol F epoxidised resins, were difficult to encapsulate. These led to sticky products rather than free flowing powders. An example of hard polymer shells incorporated into PMMA/PS is shown in Fig. 8.12. In this instance the polymer shell comprises ureaformaldehyde containing a solvent for solvent welding of the PMMA/PS. Other polymer capsule formulations have been reported that may avoid

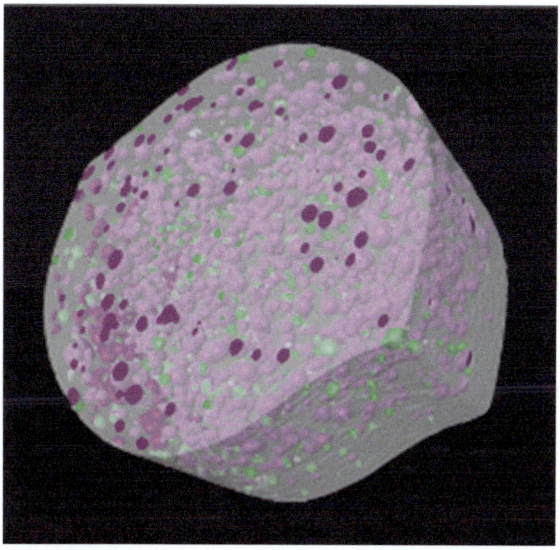

Fig. 8.12 Example of embedded polymer capsules in a section of a polymer tube. The image is a reconstruction from X-ray absorption data. Full capsules are in mauve whereas empty capsules are in green. Capsules were 15 wt%, ~60 μm and made from ureaforlaldehyde. They contained dichlorobenzene for solvent welding repair of PMMA/PS (after Mookhoek et al. [172])

this problem [175]. In a variant on the polymer encapsulation, He and Shi [185] reported a process where methylmethacrylate (MMA) and catalyst ($K_2S_2O_8$ or Na thiosulphate) were encapsulated in a shell formed from TEOS. This was achieved by forming an emulsion of the MMA in funtionlised (sulphonated) polystyrene spheres (PSS), then polymerising the TEOS and finally removing the PSS.

Apart from the hard capsules mentioned above, there are other methods for putting a polymer capsule onto an inhibitor. For example solids can be coated when in suspension in solvent/polymer mixture and spray dried. Plasma coating is another process used for coating particles and has been demonstrated for $BaCrO_4$, tolyltriazole, benzotriazole and 1, 2, 4-triazole where the plasma coating was perfluorohexane monomer [186].

Beyond polymer capsules, large organic molecules such as the dendrimers described below can be used to trap inhibitors. Cyclodextran has also been used as a type of capsule since it has a hollow cage structures that can be used to adsorb molecules [98]. Other polysaccharides have been used for drug delivery as reviewed by Aginhotri et al. [187] and poly- glycolic and lactic acids as well as polylactide-co-glycolide have been reviewed by Hans and Lowman [40]. Not all these materials will be applicable to coatings since some are highly soluble in an aqueous environment.

8.5.7 Hollow Tubes

The introduction of tubes within materials offers a primitive approach to building proto-vascular networks for the supply of inhibitors and healing agents. While inorganic and organic hollow tubes have been explored for concrete [188] and composites [189], the manufacture of hollow tubes in the size ranges required for coatings is still an active area of research. In addition it is uncertain how the incorporation of tubes into coatings modifies the properties of the coating particularly the mechanical properties and durability. Clearly the rupture of the tube would provide the triggering mechanism for release of the healing agent. Perhaps soluble tubes as well as porous tubes might also be options.

Inorganic Hollow Tubes

Inorganic hollow tubes can be formed by anodising metals. Anodising aluminium produces columnar structures which are hollow and often treated with inhibitors or dyes for colouring and then sealed in hot water [190]. Chromate sealing has been used for many years in anodising processes so in a sense filled hollow tubes have been around for a long time. Similarly titanium can also be anodised producing similar structures to aluminium. Kowalski et al. [191] looked at filling Ti anodised layer with polypyrrole and found a range of structures could be produced. While these tubes are not free tubes that can be manipulated and then incorporated into a coating, they can, in combination with a paint coating, be used to manufacture a self healing system. However, free tubes manufactured from anodised coatings, have been removed from their substrate and used as membranes or sensor arrays [192], so the potential is there is further explore these types of inorganic materials.

Organic Hollow Fibres

Porous polypropylene fibres have been used for self healing applications in concrete structures [188]. Capability now exists for building a range of internal structures in polymer fibres.

8.5.8 Self Assembly

Self assembly is a process where entities, typically molecules organise themselves into structures to minimise their free energy. As described above the driving force for mixtures of immiscible liquids is for them to separate into phases. "Suspensions" of one phase dispersed in a second phase can be stabilised as micelles or inverse micelles [92]. This requires "self assembly" of surfactants at the interface between the two phases as described above.

In the coatings area self assembly for corrosion prevention has been investigated for metal surfaces, particularly thiols and phosphonates for Fe and Cu surfaces [152] and phosphonates for aluminium [152, 193]. Many of these films that self assemble have protective properties and some are used commercially [194]. Nothing more than a summary of this work will be provided here because, apart from the micelle application, the topic of self assembly on surfaces is not really devoted to encapsulation. Knag [195], Knag et al. [196] found that docoyltriethylammonim bromide and 1-octadeanol formed monolyers reducing the corrosion by 60 and 5 % on Fe substrates respectively. Ramachandran et al. explored imidizole derivates adsorbed onto Fe_2O_3 single crystal surfaces and found that the long hydrophobic tail tilts to form a packed hydrophobic monolayer [197]. Felhősi et al. [198] found that 1-phosphono-alkane derivates rapidly adsorbed on Fe surfaces but underwent a longer term ordering process. From an encapsulation perspective ad-micelles are often observed on surfaces [199] and it may be possible to fill these with inhibitor and incorporate them into a coating. Shida et al. [200] reported a self assembling Zr-based coating on AA 1100 produced in a multistep process that provided some protection. Other surfaces can be coated as well such as the self assembly of Zr-phosphate layers in Si reported above [148].

Another form of self assembly that appears in the literature are the self-assembled nanophase coating process (SNAP) [94, 201, 202] coatings. The self assembly here refers to the production of nanoparticles based on $SiO_x(OR)_{4-x}$ on addition of TMOS and GPTMS to acidified solution. These functionlised particles are then connected by the addition of cross linking agents when coating formation is required. SNAP coatings themselves have good barrier properties but do not represent an encapsulation system. Encapsulated inhibitors need to be added to the coatings to further improve the performance. For example, encapsulation of organic inhibitors in inhibitor/β-cyclodextrin complexes has been reported for this system [98]. The inhibitors being mercaptobenzothiazole and mercaptobenzimidazole [98, 203].

8.5.9 *Dendrimers*

Dendrimers form a class of materials where branching occurs at the molecular level. There are several good reviews of dendrimers in the literature [39, 204]. As can be seen in Fig. 8.13, the key structural feature of the dendrimer is the generation step where branching occurs. Dendrimers are ideal for hosting substances at the molecular level and are used in drug delivery. According to Sahoo and Labhastewar [39] dendrimers can be manufactured either divergently (from the centre outwards) or convergently (from the exterior inwards). Agents can be incorporated either within the internal volume or attached to the outside of the dendrimer. There is little in the corrosion/inhibition literature on the use of dendrimers as carriers for healing agent, but these molecules seem to offer opportunities for further exploration, but generally these types of molecules would be suited to organic inhibitors [205, 206].

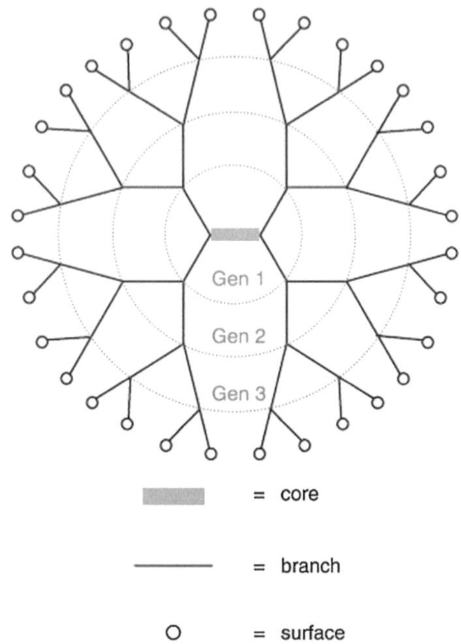

Fig. 8.13 3rd Generation dendrimer. Each generation is indicated by the circles marked with the "gen" labels and the level of branching is indicated by the *black branching links*

8.5.10 Conductive Polymers

Research into conductive polymers is certainly an area of emerging interest for a few decades now. Interest in this area seems to have arisen with the report by DeBerry [207] which examined the protection of ferritic stainless steels in acid media, by polyaniline. Generally, polyaniline (PANi), polypyrole (PPy), polythiophene (PTh) and polyphenylene vinylene [208] are the conducting polymers of interest [209–211]. The mechanism of protection relies on the ability of the conducting polymer to conduct electrons. In one model of the mechanism of conductive polymers, they at act as a an oxidising agent resulting in oxide formation on the underlying metal [212–214]. Alternatively, it has been proposed that cathodic reduction is moved from the surface into the coating [215]. Within the coating itself there has been further advancement in revealing the underlying structure. Currently, the "organic nanometal" model is used to explain the mechanism of protection [216, 217]. In this model small conducting polymer particles (nanometers in size) are separated by only a very thin less conducting layer through which electrons tunnel thus producing conductivity. The conductivity is facilitated through structural entities upon complexation with metal ions. DeBerry's paper also highlighted the need for a method of coating application which has been investigated by a number of authors [210, 218] Electrochemical deposition has been reported to provide good coatings, but this process is not considered generally suitable for industry. The issues associated with dispersion into a solvent has also been investigated [217, 219].

Much of the work in this area has focused on the corrosion protection offered by the coating itself as described above. Of course the interest in conducting polymers in this chapter is their role as delivery systems for self healing agents [220]. Release of species in response to external stimuli was first reported by Barisci et al. [221] for a range of applications including corrosion protection. Since then a number of different inhibitors have been incorporated into PANi including mercaptothiadiazole [140], ammonium salt, dithiocarbamate [140], molybdates [220] various transition metal ions [217] and oxyanions [222]. Conducting polymers have also been stored in nanowires for later release [191].

8.6 Summarizing Discussion on Triggering Mechanisms

As discussed in the introduction of this book there are many types and degrees of complexity of triggering that can be envisaged. In the end, the level of sophistication of the triggering will depend on the application area, which defines the triggers that are available, and the cost, which will determine the cost/benefit balance for the application.

8.6.1 Autonomic Triggering

The most obvious triggers are those that occur during normal operation and are provided by cycles in the operational environment. The triggers for autonomic release, listed in Table 8.1, are based on variables of the ambient environment that regularly go through cycles and could be used for self-responding systems. But as pointed out by Garcia et al. [35] a self healing mechanism such as the Diels-Alder method for repair of autopaint coatings requires that the ambient temperature rises above the transition temperature for repair to be effected. In the middle of winter in the northern hemisphere, this temperature may never be reached and the required

Table 8.1 Trigger classification

Autonomic triggering	External
pH	Field gradients
Water	Sensors
Temperature	Actuators
Mechanical damage	Centrally controlled delivery systems
UV radiation	
Corrosive ions	

temperature may need to be applied by an auto-technician making it an external rather than an autonomic trigger. So the classifications listed in Table 8.1 must be taken as guidelines rather than strict classifications.

In the examples of various encapsulation techniques presented above the most prominent triggers for release of the healing agent are mechanical damage for hard capsules [10, 14, 16, 172], pH for configurational change of polyelectrolytes [223] or conducting polymers [140], pH for dissolution of sparingly soluble inhibitors [55, 56], and ion-exchange into porous materials such as LDHs [33, 124, 137]. Temperature and UV radiation are more common as triggers for polymer healing than for inhibitor release [176].

Some authors discuss the use of internal signalling to internal agents to effect the release of the healing agent. The advantage of this approach for individual healing events is not immediately obvious since using the trigger directly to effect the healing event seems more efficient. On the other hand, if sequential healing events are required such as inhibitor release followed by polymer healing and the events interfere with each other then internal autonomic control of the release is necessary.

8.6.2 External Triggering and Transduction

Externally triggered systems require some level of added intervention above autonomic triggering. Furthermore, it also implies that signalling events occur, either within the coating or to an external agent. On detection of the signal, intervention could be delivered using sensors and monitoring systems which activate a response automatically or via human intervention. Of course these types of systems are likely to cost more than autonomic systems, therefore the benefit in the application area will need to outweigh the cost. Some applications where this type of technology might be cost effective include off-shore structures (oil rigs and windmills), onshore pipelines, bridges and aircraft, particularly where it is difficult to inspect. Specifically these could become cost effective where there are occluded spaces such as in aircraft wings, hazardous environments such as off-shore structures, bridges or chemically dangerous environments or extended structures such as oil pipelines where monitoring thousands of kilometres of pipeline is a challenging task. Additionally, external triggering, or signal transduction will need to be built into service equipment.

Sensor and signalling systems can include colour [224–227], changes in transmittance, or fluorescence [228–230] to name a few signals. Closely related to molecular level sensing and signalling are the nanosensors. The challenge here is to produce significant signal level from very small sensors that have very little effect on the mechanical properties when incorporated into a polymer matrix. One example of these includes Surface Enhanced Raman Spectroscopy (SERS) sensors which make use the enhancement of Raman signal produced as a result of surface plasmons. SERS sensors typically contain Ag or Cu nanoparticles and can be used as sensors to detect organic molecules and potentially inhibitor release. For example, SERS has been used to detect adsorbed citrate [231] suggesting the possibility of detecting other carboxylates. It has also been used to detect proteins, even in the presence of potentially interfering molecules, through surface functionalisation [232]. Quantum dots are also being explored for these types of application. Current generation CdSe/ZnS have been used to demonstrate how much chromate is left in an epoxy coating [233]. However, the use of Cd-containing sensors represents a challenge from a health and safety perspective and other types of Q-dots are being explored.

8.7 Practical Considerations

From this short overview of different approaches to designing self-healing protective coatings it is clear that intrinsic self-healing mechanisms require a complete change of current paradigms for coating chemistry. Moreover full cosmetic healing of large defects can only be achieved when using mobility of the polymeric chains at relatively high temperatures. Unfortunately, high mobility is often associated with relatively weak barrier properties. The recovery of mechanical integrity encapsulated polymerizable agents has the important physical limitation that it is not possible to provide a sufficient volume of polymerizable component to fill the defect in thin coatings. However, this type of healing mechanism is likely to be applicable to industries where relatively thick polymer coatings are applied: marine industry, oil and gas sector, and large infrastructures.

The concepts based on incorporation of corrosion inhibitors seem to be the most matured and close to applications at current stage. The introduction of self-healing functionality can be achieved with minimal modifications of existing coating formulations. The encapsulated inhibitors can be added in the form of pigments and combined in different proportions in order to achieve potentially synergistic combinations of different inhibiting species in the same coating. Moreover, some immobilization approaches are based on the nano-sized containers which can be added even to very thin coating layers without disrupting the barrier properties.

In spite of many encapsulation approaches developed during last decade most of the active coatings based on such reservoirs do not offer expected self-healing performance. Very often only a minor inhibition effect in artificial defects is observed. At the same time in majority of the cases the barrier properties of polymers are negatively influenced causing a total underperformance of protective layers.

The most critical factor is the local concentration of corrosion inhibitor which can be achieved in the defect close to metal surface. The local inhibitor concentration in the defect of nanoreservoir-containing coatings is defined by the number of factors:

- loading of nanocontainer with inhibitor;
- concentration of nanocontainers within the matrix;
- total concentration of available inhibitor;
- kinetics of inhibitor release from NCs;
- kinetics of delivery to the defect (diffusion through the coating or along the interface);
- electrolyte conditions (concentration, amount, flow, cycling).

When the combination of the abovementioned factors does not allow the inhibitor to reach a minimal necessary concentration at the metal surface the active protection will not be effected. Therefore often the inhibitors that are efficient in tests with direct addition to electrolyte do not perform well when used in the active protective coatings. The efficiency of inhibiting compounds is normally evaluated on bare metallic substrates adding the corrosion inhibitor to the corrosive electrolyte in mM concentration range. However such concentrations rarely can be achieved in the defects when thin coatings with relatively low amount of containers. For example, it is hard to expect that the addition of 0.01 wt% of benzotriazole-loaded (5 wt%) nanocontainers in the coating can ensure any efficient long-term inhibition in a defect [51]. The same is applied for the cases when very low loading of nanocontainers with corrosion inhibitors is used as in the case of work reported by Zhao et al. [234]. Some recent work on treating the polymer as the paint itself as the encapsulating medium and developing appropriate network of channels is also emerging as an alterntive method for packaging the inhibitor [235].

One of the main advantages of many different types of nanocontainers is their release on demand functionality. It is especially important from the standpoint of long-term corrosion protection since uncontrollable fast leaching of inhibitor can be prevented, keeping an acceptable level of inhibitor concentration in the system for a long time. Additionally, it ensures that no inhibitor is leached out of nanocontainers during preparation and shelf-life of the coating formulations. Combining the nanocontainers which release corrosion inhibitors at different rates under action of different triggers opens a wide range of possibilities to tailor the active protection coating for long service life. For example, one type of nanocontainer can deliver a high amount of inhibitor under extremely aggressive conditions while the second one can ensure very slow release of small amounts during long service [236].

Analysing different possibilities for self-healing coatings from standpoint of proximity to real applications, in the short term coatings containing encapsulated corrosion inhibitors seem to be the most promising. There are several types of nanocontainers which can be easily upscaled such as nano-clay based ion-exchangers. The nano-clays can also have a good compatibility with traditional polymer coating formulations and ensure an additional nano-trap role for corrosive species. The important next steps to ensure faster industrial implementation should bring more efficient inhibiting systems. Therefore the loading of existing nanocontainers has to be increased and more efficient corrosion inhibitors which confer healing even at very low concentration level should be explored.

From a medium-term perspective the systems which combine corrosion inhibition with another active protection mechanism might become feasible. The coatings which combine capsules with water-displacing agent mixed with corrosion inhibitor seem promising especially when exploited at conditions where there is no immersion. The nanocontainers of corrosion inhibitors which can also be traps for corrosive species are also of high interest.

The self-healing coatings which can provide a recovery of physical integrity of the polymer layers still need to mature more to be considered for real applications. This is especially true for polymer coatings with intrinsic healing properties since most of the basic polymer formulations currently used in industry have to be reconsidered. The self-healing coatings based on encapsulation of liquid polymerazable agents have important limitations in the case of relatively thin coatings. However this technology can be accepted for thick coatings already in a relatively short time if long term performance in service conditions is demonstrated. The economic aspects will play a decisive role here as well.

8.8 Conclusions

This chapter has covered a considerable body of work on the delivery systems that are being investigated for self healing coatings. These delivery systems cover two broad areas; (i) those used for polymer healing and (ii) those used for corrosion inhibition. Clearly there is a very broad range of chemical approaches being taken to the delivery of self healing in coatings which covers inorganic, polymer and organic chemistry. Given the breadth of these areas there is an almost unlimited range of possibilities. However, for uptake by industry, the practical considerations delineated in the previous section will probably dictate which technologies will eventually be taken up. In the polymer area there is probably scope for technology involving reversible bonds since these are multiply rehealable. For inhibitors, delivery systems based on silicon sol-gel chemistry, hard capsules (organic or inorganic) or using the polymer film itself are likely to be avenues of fruitful research. In the inhibitor case, because the critical inhibitor concentration [149] of chromate replacement inhibitors is likely to be higher than that of chromate (10^{-5} M [237]), then the search will be for the delivery which can deliver the largest amount of inhibitor.

References

1. S.R. White et al., Autonomic healing of polymer composites. Nature **409**(6822), 794–797 (2001)
2. M.L. Zheludkevich, in Self-Healing Materials, ed. by S.K. Ghosh. Self-Healing Anticorrosion Coatings (Wiley-VCH, 2008)

3. A.E. Hughes, I.S. Cole, T.M. Muster, R.J. Varley, Combining green and self healing for a new generation of coatings for metal protection. Nat. Asia Mater. **2**(4), 143–151 (2010)
4. M.L. Zheludevich, D. Raps, A.C. Bastos, T. Hack, M.G.S. Ferreira, *Self Healing Coatings with Multiple Level Protection Based on Active Nanoparticles, in NANOSPAIN-2008* (Braga, Portugal, 2008)
5. M. Zheludkevich, M. Ferreira, S. Poznyak, in *Self-Healing Corrosion Protection Coatings with Nanocontainers of Corrosion Inhibitors*. ed by V.E. Borisenko, S.V. Gaponenko, V.S. Gurin. Physics, Chemistry and Application of Nanostructures: Reviews and Short Notes (2007), pp. 380–383
6. J. Tedim et al., Enhancement of active corrosion protection via combination of inhibitor-loaded nanocontainers. ACS Appl. Mater. Interfaces **2**(5), 1528–1535 (2010)
7. J.K. Tedim, A.N. Salak, F. Montemor, D. Snihirova, M. Pilz, M.L. Zheludkevich, M.G.S. Ferreira, Zn–Al layered double hydroxides as chloride nanotraps in active protective coatings. Corros. Sci. **55**, 1–4 (2012)
8. T. Stimpfling, F. Leroux, H. Hintze-Bruening, Organo-modified layered double hydroxide in coating formulation to protect AA2024 from corrosion. Colloids Surf. A **458**, 147–154 (2014)
9. M. Krzak, Z. Tabor, P. Nowak, P. Warszyński, A. Karatzas, I.A. Kartsonakis, G.C. Kordas, Water diffusion in polymer coatings containing water-trapping particles. Part 2 experimental verification of the mathematical model. Prog. Org. Coat. **75**, 207–214 (2012)
10. S.J. Garcia et al., Self-healing anticorrosive organic coating based on an encapsulated water reactive silyl ester: synthesis and proof of concept. Prog. Org. Coat. **70**(2–3), 142–149 (2011)
11. Y. Gonzalez-Garcia et al., A combined redox-competition and negative-feedback SECM study of self-healing anticorrosive coatings. Electrochem. Commun. **13**(10), 1094–1097 (2011)
12. A. Latnikova et al., Polyfunctional active coatings with damage-triggered water-repelling effect. Soft Matter **7**(2), 369–372 (2011)
13. T. Nesterova, K. Dam-Johansen, S. Kiil, Synthesis of durable microcapsules for self-healing anticorrosive coatings: a comparison of selected methods. Prog. Org. Coat. **70**(4), 342–352 (2011)
14. C. Suryanarayana, K.C. Rao, D. Kumar, Preparation and characterization of microcapsules containing linseed oil and its use in self-healing coatings. Prog. Org. Coat. **63**(1), 72–78 (2008)
15. M. Samadzadeh et al., Tung oil: an autonomous repairing agent for self-healing epoxy coatings. Prog. Org. Coat. **70**(4), 383–387 (2011)
16. A. Kumar, L.D. Stephenson, J.N. Murray, Self-healing coatings for steel. Prog. Org. Coat. **55** (3), 244–253 (2006)
17. M. Wiesener et al., In-situ contact angle studies of the release of water displacing agents from capsule filled organic coatings. Surf. Coat. Technol. **206**(21), 4481–4487 (2012)
18. T.G. Nijland, J.A. Larbi, R.P.J. van Hees, B. Lubelli, M. de Rooij, *Self Healing Phenomena in Concretes and Masonry Mortars: a Microscopic Study, in Self Healing Materials* (Springer, Eindhoven, 2007), p. 31
19. G. Perugini, Plasma-sprayed self-sealing ceramic coatings: materials chemistry and high temperature protective properties. Thin Solid Films **108**(4), 415–425 (1983)
20. T. Sugama, K. Gawlik, Self-repairing poly(phenylenesulfide) coatings in hydrothermal environments at 200 °C. Mater. Lett. **57**(26–27), 4282–4290 (2003)
21. F. Miccichè et al., Moisture induced crack filling in barrier coatings containing montmorillonite as an expandable phase. Surf. Coat. Technol. **202**(14), 3346–3353 (2008)
22. A. Hikasa et al., Preparation and corrosion studies of self-healing multi-layered nano coatings of silica and swelling clay. Mater. Res. Innovations **8**(2), 84–88 (2004)
23. A. Yabuki, K. Okumura, Self-healing coatings using superabsorbent polymers for corrosion inhibition in carbon steel. Corros. Sci. **59**, 258–262 (2012)
24. S.J. Garcia, Effect of polymer architecture on the intrinsic self-healing character of polymers. Eur. Polymer J. **53**(1), 118–125 (2014)

25. P. Cordier et al., Self-healing and thermoreversible rubber from supramolecular assembly. Nature **451**(7181), 977–980 (2008)
26. S.J. Kalista Jr, T.C. Ward, Thermal characteristics of the self-healing response in poly (ethylene-co-methacrylic acid) copolymers. J. R. Soc. Interface **4**(13), 405–411 (2007)
27. X. Chen et al., A thermally re-mendable cross-linked polymeric material. Science **295**(5560), 1698–1702 (2002)
28. S. Bode et al., Self-healing metallopolymers based on cadmium bis (terpyridine) complex containing polymer networks. Polym. Chem. **4**(18), 4966–4973 (2013)
29. J. Kötteritzsch et al., One-component intrinsic self-healing coatings based on reversible crosslinking by diels-alder cycloadditions. Macromol. Chem. Phys. **214**(14), 1636–1649 (2013)
30. G. Postiglione, S. Turri, M. Levi, Effect of the plasticizer on the self-healing properties of a polymer coating based on the thermoreversible Diels-Alder reaction. Prog. Org. Coat. **78**, 526–531 (2015)
31. Y. Gonzalez-Garcia et al., A combined mechanical, microscopic and local electrochemical evaluation of self-healing properties of shape-memory polyurethane coatings. Electrochim. Acta **56**(26), 9619–9626 (2011)
32. Nissan Develops World's First Clear Paint that Repairs Scratches on Car Surfaces. JCNN News, 2005
33. H.N. McMurray, G. Williams, Inhibition of filiform corrosion on organic-coated aluminum alloy by hydrotalcite-like anion-exchange pigments. Corrosion **60**(3), 219–228 (2004)
34. H.R. Fischer, Self-repairing material systems—a dream or a reality? Nat. Sci. **2**, 873–901 (2010)
35. S.J. Garcia, H.R. Fischer, S. van der Zwaag, A critical appraisal of the potential of self healing polymeric coatings. Prog. Org. Coat. **72**(3), 211–221 (2011)
36. B.F. Gibbs et al., Encapsulation in the food industry: a review. Int. J. Food Sci. Nutr. **50**(3), 213–224 (1999)
37. I. Roy et al., Calcium phosphate nanoparticles as novel non-viral vectors for targeted gene delivery. Int. J. Pharm. **250**(1), 25–33 (2003)
38. M. Nishikawa et al., Hepatocyte-targeted in vivo gene expression by intravenous injection of plasmid DNA complexed with synthetic multi-functional gene delivery system. Gene Ther. **7** (7), 548–555 (2000)
39. S.K. Sahoo, V. Labhasetwar, Nanotech approaches to delivery and imaging drug. Drug Discovery Today **8**(24), 1112–1120 (2003)
40. M.L. Hans, A.M. Lowman, Biodegradable nanoparticles for drug delivery and targeting. Curr. Opin. Solid State Mater. Sci. **6**(4), 319–327 (2002)
41. O.G. Schramm et al., Polymeric nanocontainers with high loading capacity of hydrophobic drugs. Soft Matter **5**(8), 1662–1667 (2009)
42. J.R. Galvele, Transport processes and mechanism of pitting of metals. J. Electrochem. Soc. **123**(4), 464–474 (1976)
43. W.H. Slabaugh et al., Filiform corrosion of aluminum. J. Paint Technol. **44**(566), 76–& (1972)
44. S.D. Mookhoek, H.R. Fischer, S. van der Zwaag, Alginate fibres containing discrete liquid filled vacuoles for controlled delivery of healing agents in fibre reinforced composites. Compos. A Appl. Sci. Manuf. **43**(12), 2176–2182 (2012)
45. S.D. Mookhoek, H.R. Fischer, S. van der Zwaag, A numerical study into the effects of elongated capsules on the healing efficiency of liquid-based systems. Comput. Mater. Sci. **47** (2), 506–511 (2009)
46. J. Rouquerol et al., Recommendations for the characterization of porous solids. Pure Appl. Chem. **66**(8), 1739–1758 (1994)
47. T. Nesterova, *Self-Healing Anticorrosive Coatings, in Materials Science* (Technical University of Denmark, Copenhagen, 2012)

48. S.I. Rae, et al. in *Novel Self-Healing Systems: Expanding and Inhibited Healing Agents*. ASME 2014 Conference on Smart Materials, Adaptive Structures and Intelligent Systems, SMASIS 2014. 2014
49. C.J. Brinker, G.W. Scherer, *Sol-Gel Science*, 1st edn. (Academic Press, San Diego, 1990)
50. C. Lai et al., Formation of calcium phosphate nanoparticles in reverse microemulsions. Mater. Lett. **59**(2–3), 210–214 (2005)
51. Y.M. Lvov et al., Halloysite clay nanotubes for controlled release of protective agents. ACS Nano **2**(5), 814–820 (2008)
52. E. Jamstorp et al., Mechanically strong geopolymers offer new possibilities in treatment of chronic pain. J. Controlled Release **146**(3), 370–377 (2010)
53. J. Forsgren et al., A ceramic drug delivery vehicle for oral administration of highly potent opioids. J. Pharm. Sci. **99**(1), 219–226 (2010)
54. H. Ku et al., Fracture toughness of phenol formaldehyde composites reinforced with E-spheres. J. Compos. Mater. **43**(7), 741–754 (2009)
55. S.A.S. Dias et al., Sol–gel coatings modified with zeolite fillers for active corrosion protection of AA2024. Corros. Sci. **62**, 153–162 (2012)
56. S.A.S. Dias et al., The role of Ce(III)-enriched zeolites on the corrosion protection of AA2024-T3. Electrochim. Acta **112**, 549–556 (2013)
57. S. Thomas et al., Self-repairing oxides to protect zinc: review, discussion and prospects. Corros. Sci. **69**, 11–22 (2013)
58. S. Bohm et al., Novel environment friendly corrosion inhibitor pigments based on naturally occurring clay minerals. Materials and Corrosion-Werkstoffe Und Korrosion **52**(12), 896–903 (2001)
59. N. Granizo et al., Ion-exchange pigments in primer paints for anticorrosive protection of steel in atmospheric service: Cation-exchange pigments. Prog. Org. Coat. **75**(3), 147–161 (2012)
60. E. Abdullayev, Y. Lvov, Clay nanotubes for corrosion inhibitor encapsulation: release control with end stoppers. J. Mater. Chem. **20**(32), 6681–6687 (2010)
61. D.G. Shchukin et al., Halloysite nanotubes as biomimetic nanoreactors. Small **1**(5), 510–513 (2005)
62. H.P. Wang et al., Data-constrained modelling of an anthracite coal physical structure with multi-spectrum synchrotron X-ray CT. Fuel **106**, 219–225 (2013)
63. S. Yang, T.E. Gureyev, M.B. Tulloh, M.B. Clennell, M. Pervukhina, Feasibility of a data constrained prediction of hydrocarbon reservoir sandstone microstructures. Meas. Sci. Technol. **21**, (2010)
64. M.F. Montemor et al., The synergistic combination of bis-silane and $CeO_2.ZrO_2$ nanoparticles on the electrochemical behaviour of galvanised steel in NaCl solutions. Electrochim. Acta **53**, 5913–5922 (2008)
65. M.L. Zheludkevich et al., Anticorrosion coatings with self-healing effect based on nanocontainers impregnated with corrosion inhibitor. Chem. Mater. **19**(3), 402–411 (2007)
66. M.F. Montemor, R. Pinto, M.G.S. Ferreira, Chemical composition and corrosion protection of silane films modified with CeO(2) nanoparticles. Electrochim. Acta **54**(22), 5179–5189 (2009)
67. M.L. Zheludkevich et al., Nanostructured sol-gel coatings doped with cerium nitrate as pre-treatments for AA2024-T3-corrosion protection performance. Electrochim. Acta **51**(2), 208–217 (2005)
68. M.L. Zheludkevich et al., Oxide nanoparticle reservoirs for storage and prolonged release of the corrosion inhibitors. Electrochem. Commun. **7**(8), 836–840 (2005)
69. M.L. Zheludkevich et al., Corrosion protective properties of nanostructured sol-gel hybrid coatings to AA2024-T3. Surf. Coat. Technol. **200**(9), 3084–3094 (2006)
70. D.G. Shchukin, M. Zheludkevich, H. Moehwald, Feedback active coatings based on incorporated nanocontainers. J. Mater. Chem. **16**(47), 4561–4566 (2006)
71. N.P. Tavandashti, S. Sanjabi, Corrosion study of hybrid sol-gel coatings containing boehmite nanoparticles loaded with cerium nitrate corrosion inhibitor. Prog. Org. Coat. **69**, 384–389 (2010)

72. A.C. Balaskas et al., Improvement of anti-corrosive properties of epoxy-coated AA 2024-T3 with TiO2 nanocontainers loaded with 8-hydroxyquinoline. Prog. Org. Coat. **74**(3), 418–426 (2012)
73. E.D. Mekeridis, I.A. Kartsonakis, G.C. Kordas, Multilayer organic-inorganic coating incorporating TiO_2 nanocontainers loaded with inhibitors for corrosion protection of AA2024-T3. Prog. Org. Coat. **73**(2–3), 142–148 (2012)
74. M.F. Montemor, M.G.S. Ferreira, Analytical characterization of silane films modified with cerium activated nanoparticles and its relation with the corrosion protection of galvanised steel substrates. Prog. Org. Coat. **63**(3), 330–337 (2008)
75. E.D. Mekeridis et al., Release studies of corrosion inhibitors from cerium titanium oxide nanocontainers. J. Nanopart. Res. **13**(2), 541–554 (2011)
76. M. Schem et al., CeO_2-filled sol-gel coatings for corrosion protection of AA2024-T3 aluminium alloy. Corros. Sci. **51**(10), 2304–2315 (2009)
77. H. Lermer et al., Synthesis and structure refinement of ZSM-5 single-crystals. Zeolites **5**(3), 131–134 (1985)
78. F. Kleitz et al., Transformation of highly ordered large pore silica mesophases (Fm3 m, Im3 m and p6 mm) in a ternary triblock copolymer-butanol-water system. Chem. Commun. **13**, 1536–1537 (2004)
79. C.T. Kresge et al., Ordered mesoporous molecular-sieves synthesized by a liquid-crystal template mechanism. Nature **359**(6397), 710–712 (1992)
80. K.M. Ryan et al., Control of pore morphology in mesoporous silicas synthesized from triblock copolymer templates. Langmuir **18**(12), 4996–5001 (2002)
81. J.Y.T. Chong et al., Steric stabilisation of self-assembled cubic lyotropic liquid crystalline nanoparticles: high throughput evaluation of triblock polyethylene oxide-polypropylene oxide-polyethylene oxide copolymers. Soft Matter **7**(10), 4768–4777 (2011)
82. X. Mulet et al., High throughput preparation and characterisation of amphiphilic nanostructured nanoparticulate drug delivery vehicles. Int. J. Pharm. **395**(1–2), 290–297 (2010)
83. D. Ebrahimi et al., High throughput screening arrays of rhodium and iridium complexes as catalysts for intramolecular hydroamination using parallel factor analysis. Analyst **133**, 817–822 (2008)
84. D.Y. Zhao et al., Morphological control of highly ordered mesoporous silica SBA-15. Chem. Mater. **12**(2), 275–279 (2000)
85. P.J. Colver, S.A.F. Bon, Cellular polymer monoliths made via pickering high internal phase emulsions. Chem. Mater. **19**(7), 1537–1539 (2007)
86. M.L. Zheludkevich, I.M. Salvado, M.G.S. Ferreira, Sol-gel coatings for corrosion protection of metals. J. Mater. Chem. **15**(48), 5099–5111 (2005)
87. A. Bhaumik, Mesoporous titanium phosphates and related molecular sieves: synthesis, characterization and applications. Proc. Indian Acad. Sci. Chem. Sci. **114**(4), 451–460 (2002)
88. M. Mokhtar, S.N. Basahel, T.T. Ali, Effect of synthesis methods for mesoporous zirconia on its structural and textural properties. J. Mater. Sci. **48**(6), 2705–2713 (2013)
89. D.M. Lyons, K.M. Ryan, M.A. Morris, Preparation of ordered mesoporous ceria with enhanced thermal stability. J. Mater. Chem. **12**(4), 1207–1212 (2002)
90. T.L. Metroke, R.L. Parkhill, E.T. Knobbe, Passivation of metal alloys using sol-gel-derived materials—a review. Prog. Org. Coat. **41**(4), 233–238 (2001)
91. D. Wang, G.R. Bierwagen, Sol-gel coatings on metals for corrosion protection. Prog. Org. Coat. **64**(4), 327–338 (2009)
92. F.J. Rossier-Miranda, C. Schroen, R.M. Boom, Colloidosomes: versatile microcapsules in perspective. Colloids Surf., A **343**(1–3), 43–49 (2009)
93. S.E. Hornstrom et al., Paint adhesion and corrosion performance of chromium-free pretreatments of 55 % Al-Zn-coated steel. J. Adhes. Sci. Technol. **10**(9), 883–904 (1996)
94. A.N. Khramov et al., Nanostructured sol-gel derived conversion coatings based on epoxy- and amino-silanes. Prog. Org. Coat. **47**(3–4), 207–213 (2003)

95. L.S. Kasten et al., An XPS study of cerium dopants in sol–gel coatings for aluminum 2024-T3. Surf. Coat. Technol. **140**(1), 11–15 (2001)
96. L. Mascia et al., Molybdate doping of networks in epoxy-silica hybrids: domain structuring and corrosion inhibition. Prog. Org. Coat. **56**(1), 13–22 (2006)
97. N.C. Rosero-Navarro et al., Improved corrosion resistance of AA2024 alloys through hybrid organic–inorganic sol–gel coatings produced from sols with controlled polymerisation. Surf. Coat. Technol. **203**(13), 1897–1903 (2009)
98. A.N. Khramov et al., Hybrid organo-ceramic corrosion protection coatings with encapsulated organic corrosion inhibitors. Thin Solid Films **447**, 549–557 (2004)
99. A. Wittmar et al., Hybrid sol-gel coatings doped with transition metal ions for the protection of AA 2024-T3. J. Sol-Gel. Sci. Technol. **61**(3), 600–612 (2012)
100. S.M.A. Hosseini, A.H. Jafari, E. Jamalizadeh, Self-healing corrosion protection by nanostructure sol-gel impregnated with propargyl alcohol. Electrochim. Acta **54**(28), 7207–7213 (2009)
101. N.C. Rosero-Navarro et al., Effects of Ce-containing sol-gel coatings reinforced with SiO_2 nanoparticles on the protection of AA2024. Corros. Sci. **50**(5), 1283–1291 (2008)
102. D.G. Shchukin et al., Layer-by-layer assembled nanocontainers for self-healing corrosion protection. Adv. Mater. **18**(13), 1672–1678 (2006)
103. I.A. Kartsonakis et al., Hybrid organic–inorganic multilayer coatings including nanocontainers for corrosion protection of metal alloys. Corros. Sci. **57**, 56–66 (2012)
104. F. Andreatta et al., Water-based ZrO_2 pretreatment for AA2024 aluminum alloy. Surf. Interface Anal. **42**(4), 293–298 (2010)
105. F. Andreatta et al., Corrosion behaviour of sol–gel treated and painted AA2024 aluminium alloy. Prog. Org. Coat. **69**(2), 133–142 (2010)
106. D. Raps et al., Electrochemical study of inhibitor-containing organic-inorganic hybrid coatings on AA2024. Corros. Sci. **51**(5), 1012–1021 (2009)
107. N.N. Voevodin et al., An organically modified zirconate film as a corrosion-resistant treatment for aluminum 2024-T3. Prog. Org. Coat. **41**(4), 287–293 (2001)
108. R.B. Greegor et al., X-ray spectroscopic investigation of the Zr-site in thin film sol-gel surface preparations. J. Sol-Gel. Sci. Technol. **20**(1), 35–50 (2001)
109. A.S. Hamdy, Advanced nano-particles anti-corrosion ceria based sol gel coatings for aluminum alloys. Mater. Lett. **60**(21–22), 2633–2637 (2006)
110. A.S. Hamdy, A clean low cost anti-corrosion molybdate based nano-particles coating for aluminum alloys. Prog. Org. Coat. **56**(2–3), 146–150 (2006)
111. S.R. Taylor et al., Increasing the functionality of military coatings using nano-dimensioned materials. Corros. Rev. **25**(5–6), 491–522 (2007)
112. V. Palanivel, D.Q. Zhu, W.J. van Ooji, Nanoparticle-filled silane films as chromate replacements for aluminum alloys. Prog. Org. Coat. **47**(3–4), 384–392 (2003)
113. A.S. Hamdy, D.P. Butt, Novel anti-corrosion nano-sized vanadia-based thin films prepared by sol-gel method for aluminum alloys. J. Mater. Process. Technol. **181**(1–3), 76–80 (2007)
114. C. Sanchez et al., Designed hybrid organic-inorganic nanocomposites from functional nanobuilding blocks. Chem. Mater. **13**(10), 3061–3083 (2001)
115. J.H. Osborne, Observations on chromate conversion coatings from a sol-gel perspective. Prog. Org. Coat. **41**(4), 280–286 (2001)
116. A. Kumar et al., A pseudoboehmite-silane hybrid coating for enhanced corrosion protection of AA2024-T3. J. Electrochem. Soc. **157**(10), C346–C356 (2010)
117. A. Cabral et al., Analytical characterisation and corrosion behaviour of bis-triethoxysilylpropyl tetrasulphide pre-treated AA2024-T3. Corros. Sci. **47**(3), 869–881 (2005)
118. D.Q. Zhu, W.J. van Ooij, Corrosion protection of AA 2024-T3 by bis-3-(triethoxysilyl) propyl tetrasulfide in neutral sodium chloride solution. Part 1: corrosion of AA 2024-T3. Corros. Sci. **45**(10), 2163–2175 (2003)

119. D.Q. Zhu, W.J. van Ooij, Corrosion protection of AA 2024-T3 by bis-3-(triethoxysilyl) propyl tetrasulfide in sodium chloride solution. Part 2: mechanism for corrosion protection. Corros. Sci. **45**(10), 2177–2197 (2003)
120. S.V. Lamaka et al., Novel hybrid sol-gel coatings for corrosion protection of AZ31B magnesium alloy. Electrochim. Acta **53**(14), 4773–4783 (2008)
121. A. Bögershausen et al., Drug release from self-assembled inorganic—Organic hybrid gels and gated porosity detected by positron annihilation lifetime spectroscopy. Chem. Mater. **18** (3), 664–672 (2006)
122. N. Iyi et al., Deintercalation of carbonate ions from a hydrotalcite-like compound: enhanced decarbonation using acid-salt mixed solution. Chem. Mater. **16**(15), 2926–2932 (2004)
123. G. Williams, H.N. McMurray, Inhibition of filiform corrosion on polymer coated AA2024-T3 by hydrotalcite-like pigments incorporating organic anions. Electrochem. Solid State Lett. **7**(5), B13–B15 (2004)
124. A.N. Salak et al., Anion exchange in Zn-Al layered double hydroxides: in situ X-ray diffraction study. Chem. Phys. Lett. **495**(1–3), 73–76 (2010)
125. S.P.V. Mahajanam, R.G. Buchheit, in *Corrosion and Protection of Light Metal Alloys*, ed. by R.G. Buchheit et al. Characterization of Zn-Al-V10O286-Corrosion-Inhibiting Hydrotalcite Pigments in Epoxy Resins (2004), pp. 270–282
126. M. Mullet, V. Khare, C. Ruby, XPS study of Fe(II)-Fe(III) (oxy) hydroxycarbonate green rust compounds. Surf. Interface Anal. **40**(3–4), 323–328 (2008)
127. F.H. Zhang et al., Fabrication of oriented layered double hydroxide films by spin coating and their use in corrosion protection. Chem. Eng. J. **141**(1–3), 362–367 (2008)
128. R.B. Leggat et al., Performance of hydrotalcite conversion treatments on AA2024-T3 when used in a coating system. Corrosion **58**(4), 322–328 (2002)
129. R.G. Buchheit et al., Active corrosion protection in Ce-modified hydrotalcite conversion coatings. Corrosion **58**(1), 3–14 (2002)
130. S.P.V. Mahajanarn, R.G. Buchheit, Characterization of inhibitor release from Zn-Al-$[V_{10}O_{28}]^{6-}$hydrotalcite pigments and corrosion protection from hydrotalcite-pigmented epoxy coatings. Corrosion **64**(3), 230–240 (2008)
131. R.G. Buchheit et al., Active corrosion protection and corrosion sensing in chromate-free organic coatings. Prog. Org. Coat. **47**(3–4), 174–182 (2003)
132. M.L. Zheludkevich et al., Active protection coatings with layered double hydroxide nanocontainers of corrosion inhibitor. Corros. Sci. **52**(2), 602–611 (2010)
133. N. Granizo et al., Ion-exchange pigments in primer paints for anticorrosive protection of steel in atmospheric service: anion-exchange pigments. Prog. Org. Coat. **76**(2–3), 411–424 (2013)
134. X. Yu et al., One-step synthesis of lamellar molybdate pillared hydrotalcite and its application for AZ31 Mg alloy protection. Solid State Sci. **11**(2), 376–381 (2009)
135. R.G. Buchheit, H. Guan, Formation and characteristics of Al-Zn hydrotalcite coatings on galvanized steel. JCT Res. **1**(4), 277–290 (2004)
136. W. Zhang, R.G. Buchheit, Hydrotalcite coating formation on Al-Cu-Mg alloys from oxidizing bath chemistries. Corrosion **58**(7), 591–600 (2002)
137. G. Williams, H.N. McMurray, Anion-exchange inhibition of filiform corrosion on organic coated AA2024-T3 aluminum alloy by hydrotalcite-like pigments. Electrochem. Solid State Lett. **6**(3), B9–B11 (2003)
138. S.K. Poznyak et al., Novel inorganic host layered double hydroxides intercalated with guest organic inhibitors for anticorrosion applications. ACS Appl. Mater. Interfaces **1**(10), 2353–2362 (2009)
139. M. Kendig, M. Hon, A hydrotalcite-like pigment containing an organic anion corrosion inhibitor. Electrochem. Solid State Lett. **8**(3), B10–B11 (2005)
140. M. Kendig, M. Hon, Environmentally triggered release of oxygen-reduction inhibitors from inherently conducting polymers. Corrosion **60**(11), 1024–1030 (2004)
141. R. Trujillano et al., Preparation, physicochemical characterisation and magnetic properties of Cu-Al layered double hydroxides with CO_3^{2-} and anionic surfactants with different alkyl chains in the interlayer. Physica B **373**(2), 267–273 (2006)

142. T.G. Harvey et al., The effect of inhibitor structure on the corrosion of AA2024 and AA7075. Corros. Sci. **53**(6), 2184–2190 (2011)
143. D. Hasha et al., Studies of silicoaluminophosphates with the sodalite structure. J. Am. Chem. Soc. **110**(7), 2127–2135 (1988)
144. M.E. Davis et al., A molecular-sieve with 18-membered rings. Nature **331**(6158), 698–699 (1988)
145. H.K. Varma et al., Porous calcium phosphate coating over phosphorylated chitosan film by a biomimetic method. Biomaterials **20**(9), 879–884 (1999)
146. J.B. Moffat, J.F. Neeleman, Preparation and some properties of high surface-area boron phosphate. J. Catal. **31**(2), 274–277 (1973)
147. K. Sarkar et al., A porous open-framework titanium oxophenylphosphate. J. Solid State Chem. **181**(8), 2065–2072 (2008)
148. H.N. Kim et al., Characterization of zirconium phosphate polycation thin films grown by sequential adsorption reactions. Chem. Mater. **9**(6), 1414–1421 (1997)
149. J. Sinko, Challenges of chromate inhibitor pigments replacement in organic coatings. Prog. Org. Coat. **42**(3–4), 267–282 (2001)
150. H.F. Clay, J.H. Cox, Chromate and phosphate pigments in anti-corrosive primers. J. Oil Colour Chem. Assoc. **56**(1), 13–16 (1973)
151. A.C. Bastos, M.G.S. Ferreira, A.M. Simoes, Comparative electrochemical studies of zinc chromate and zinc phosphate as corrosion inhibitors for zinc. Prog. Org. Coat. **52**(4), 339–350 (2005)
152. N.S. McIntyre et al., XPS studies of octadecylphosphonic acid (OPA) monolayer interactions with some metal and mineral surfaces. Surf. Interface Anal. **37**(9), 749–754 (2005)
153. X.F. Liu, S.J. Huang, H.C. Gu, Crack growth behaviour of high strength aluminium alloy in 3.5 %NaCl solution with corrosion inhibiting pigments. Int. J. Fatigue **24**(7), 803–809 (2002)
154. B. del Amo et al., High performance water-based paints with non-toxic anticorrosive pigments. Prog. Org. Coat. **45**(4), 389–397 (2002)
155. R.L. Howard et al., Inhibition of cut edge corrosion of coil-coated architectural cladding. Prog. Org. Coat. **37**(1–2), 83–90 (1999)
156. R.L. Twite, G.P. Bierwagen, Review of alternatives to chromate for corrosion protection of aluminum aerospace alloys. Prog. Org. Coat. **33**(2), 91–100 (1998)
157. A. Kalendova, D. Vesely, Study of the anticorrosive efficiency of zincite and periclase-based core-shell pigments in organic coatings. Prog. Org. Coat. **64**(1), 5–19 (2009)
158. A. Kalendova et al., Anticorrosion properties of inorganic pigments surface-modified with a polyaniline phosphate layer. Prog. Org. Coat. **63**(2), 209–221 (2008)
159. V.I. Pokhmurs'kyi et al., Corrosion protection of aluminum alloys by inhibiting pigments. Mater. Sci. **42**(5), 573–578 (2006)
160. M. Adachi-Pagano, C. Forano, J.P. Besse, Delamination of layered double hydroxides by use of surfactants. Chem. Commun. **1**, 91–92 (2000)
161. T. Hibino, W. Jones, New approach to the delamination of layered double hydroxides. J. Mater. Chem. **11**(5), 1321–1323 (2001)
162. L. Li et al., Hollow nanoshell of layered double hydroxide. Chem. Commun. **29**, 3125–3127 (2006)
163. P.C. Lebaron, Z. Wang, T.J. Pinnavaia, Polymer-layered silicate nanocomposites: an overview. Appl. Clay Sci. **15**(1–2), 11–29 (1999)
164. K. Haraguchi, Synthesis and properties of soft nanocomposite materials with novel organic/inorganic network structures. Polym. J. **43**(3), 223–241 (2011)
165. J.J. Luo, I.M. Daniel, Characterization and modeling of mechanical behavior of polymer/clay nanocomposites. Compos. Sci. Technol. **63**(11), 1607–1616 (2003)
166. A. Okada, A. Usuki, Twenty years of polymer-clay nanocomposites. Macromol. Mater. Eng. **291**(12), 1449–1476 (2006)
167. S.A.F. Bon, T. Chen, Pickering stabilization as a tool in the fabrication of complex nanopatterned silica microcapsules. Langmuir **23**(19), 9527–9530 (2007)

168. S.A.F. Bon, P.J. Colver, Pickering miniemulsion polymerization using Laponite clay as a stabilizer. Langmuir **23**(16), 8316–8322 (2007)
169. T. Chen, P.J. Colver, S.A.F. Bon, Organic-inorganic hybrid hollow spberes prepared from TiO_2-stabilized pickering emulsion polymerization. Adv. Mater. **19**(17), 2286–2289 (2007)
170. S.A.F. Bon et al., Route to stable non-spherical emulsion droplets. Eur. Polymer J. **43**(11), 4839–4842 (2007)
171. F.J. Rossier-Miranda, K. Schroen, R. Boom, Mechanical characterization and pH response of fibril-reinforced microcapsules prepared by layer-by-layer adsorption. Langmuir **26**(24), 19106–19113 (2010)
172. S.D. Mookhoek et al., Applying SEM-based X-ray microtomography to observe self-healing in solvent encapsulated thermoplastic materials. Adv. Eng. Mater. **12**(3), 228–234 (2010)
173. S.H. Cho, S.R. White, P.V. Braun, Self-healing polymer coatings. Adv. Mater. **21**(6), 645–649 (2009)
174. D.S. Xiao et al., Self-healing epoxy based on cationic chain polymerization. Polymer **50**(13), 2967–2975 (2009)
175. D.S. Xiao, M.Z. Rong, M.Q. Zhang, A novel method for preparing epoxy-containing microcapsules via UV irradiation-induced interfacial copolymerization in emulsions. Polymer **48**(16), 4765–4776 (2007)
176. D.Y. Wu, S. Meure, D. Solomon, Self-healing polymeric materials: a review of recent developments. Prog. Polym. Sci. **33**(5), 479–522 (2008)
177. Y.C. Yuan et al., Self-healing polymeric materials using epoxy/mercaptan as the healant. Macromolecules **41**(14), 5197–5202 (2008)
178. T. Yin et al., Self-healing epoxy composites—preparation and effect of the healant consisting of microencapsulated epoxy and latent curing agent. Compos. Sci. Technol. **67**(2), 201–212 (2007)
179. G. Bierwagen et al., Active metal-based corrosion protective coating systems for aircraft requiring no-chromate pretreatment. Prog. Org. Coat. **68**(1–2), 48–61 (2010)
180. E.N. Brown, S.R. White, N.R. Sottos, Microcapsule induced toughening in a self-healing polymer composite. J. Mater. Sci. **39**(5), 1703–1710 (2004)
181. M.L. Zheludkevich et al., Triazole and thiazole derivatives as corrosion inhibitors for AA2024 aluminium alloy. Corros. Sci. **47**(12), 3368–3383 (2005)
182. M. Benmessaoud et al., Inhibiting effect of 2-mercaptobenzimidazole on the corrosion of Cu-30Ni alloy in aerated 3 % NaCl in presence of ammonia. Corros. Sci. **49**(10), 3880–3888 (2007)
183. H.F. Yang et al., pH-dependent 2-amino-5-mercapto-1, 3, 4-thiadiazole monolayers at the silver surface: surface enhanced Raman scattering spectroscopic and electrochemical observations. Appl. Surf. Sci. **255**(5), 2994–2999 (2008)
184. L.J. Zhang et al., 2-Amino-5-(4-pyridinyl)-1,3,4-thiadiazole film at the silver surface: observation by Raman spectroscopy and electrochemical methods. Appl. Surf. Sci. **257**(15), 6347–6352 (2011)
185. X.D. He, X.M. Shi, Self-repairing coating for corrosion protection of aluminum alloys. Prog. Org. Coat. **65**(1), 37–43 (2009)
186. H. Yang, W.J. van Ooij, Plasma deposition of polymeric thin films on organic corrosion-inhibiting paint pigments: a novel method to achieve slow release. Plasmas Polym. **8**(4), 297–323 (2003)
187. S.A. Agnihotri, N.N. Mallikarjuna, T.M. Aminabhavi, Recent advances on chitosan-based micro- and nanoparticles in drug delivery. J. Controlled Release **100**(1), 5–28 (2004)
188. C.M. Dry, M.J.T. Corsaw, A time-release technique for corrosion prevention. Cem. Concr. Res. **28**(8), 1133–1140 (1998)
189. K.S. Toohey et al., Self-healing materials with microvascular networks. Nat. Mater. **6**(8), 581–585 (2007)
190. S. Wernick, R. Pinner, P.G. Sheasby, *The Surface Treatment and Finishing of Aluminium and its Alloys*, 5th edn. (Finishing Publications and ASM International, Teddington, 1987)

191. D. Kowalski, A. Tighineanu, P. Schmuki, Polymer nanowires or nanopores? site selective filling of titania nanotubes with polypyrrole. J. Mater. Chem. **21**(44), 17909–17915 (2011)
192. C.Y. Han et al., Anodized aluminum oxide membranes as templates for nanoscale structures. Plat. Surf. Finish. **91**(7), 40–45 (2004)
193. M. Rohwerder, M. Stratmann, Surface modification by ordered monolayers: new ways of protecting materials against corrosion. MRS Bull. **24**(7), 43–47 (1999)
194. T. Schmidt-Hansberg, P. Schubach, A comparative study of innovative aluminium pretreatments. ATB Metallurgie **43**(1–2), 9–14 (2003)
195. M. Knag, Fundamental behavior of model corrosion inhibitors. J. Dispersion Sci. Technol. **27**(5), 587–597 (2006)
196. M. Knag et al., Langmuir-Blodgett films of dococyltriethylammonium bromide and octadecanol on iron. Deposition and corrosion inhibitor performance in CO_2 containing brine. Corros. Sci. **48**(9), 2592–2613 (2006)
197. S. Ramachandran et al., Self-assembled monolayer mechanism for corrosion inhibition of iron by imidazolines. Langmuir **12**(26), 6419–6428 (1996)
198. I. Felhősi et al., Kinetics of self-assembled layer formation on iron. Electrochim. Acta **47**(13–14), 2335–2340 (2002)
199. V. Subramanian, W.A. Ducker, Counterion effects on adsorbed micellar shape: experimental study of the role of polarizability and charge. Langmuir **16**(10), 4447–4454 (2000)
200. A. Shida et al., Zirconium-phosphate films self-assembled on aluminum substrate toward corrosion protection. Surf. Coat. Technol. **169**, 686–690 (2003)
201. M.S. Donley, V.N. Balbyshev, N.N. Voevodin, Self-assembled NAnophase Particle (SNAP) surface treatments for corrosion protection of AA2024-T3. Prog. Org. Coat. **52**(1), 34–38 (2005)
202. N.N. Voevodin et al., Nanostructured coatings approach for corrosion protection. Prog. Org. Coat. **47**(3–4), 416–423 (2003)
203. A. Khramov et al., Sol-gel-derived corrosion-protective coatings with controllable release of incorporated organic corrosion inhibitors. Thin Solid Films **483**(1–2), 191–196 (2005)
204. A. Trinchi, T.H. Muster, A review of surface functionalized amine terminated dendrimers for application in biological and molecular sensing. Supramol. Chem. **19**(7), 431–445 (2007)
205. T. Wang, G. Zhang, F. Fei, Synthesis and application of imidazoline dendrimer corrosion inhibitor. Speciality Petrochemicals **27**(5), 1–5 (2010)
206. G. Fan, J. Ge, L. Chai, Synthesis and evaluation of new kind of imidazoline corrosion inhibitor. Speciality Petrochemicals **28**(1), 73–76 (2011)
207. D.W. DeBerry, Modification of the electrochemical and corrosion behavior of stainless steels with an electroactive coating. J. Electrochem. Soc. **132**(5), 1022–1026 (1985)
208. P. Zarras, J. He, D.E. Tallman, N. Anderson, A. Guenthner, C. Webber, J.D. Stenger-Smith, J.M. Pentony, S. Hawkins, L. Baldwin (ed.), in *Chapter 10 Electroactive Polymer Coatings as replacements for Chromate Conversion Coatings. Smart Coatings, ACS Symposium Series*, ed. by T.E.A. Prover (American Chemical Society, Washington, 2007)
209. J.O. Iroh, Conducting polymer coating: a viable alternative to chromate conversion coating. Surf. Eng. **17**(4), 265–267 (2001)
210. M. Rohwerder, A. Michalik, Conducting polymers for corrosion protection: What makes the difference between failure and success? Electrochim. Acta **53**(3), 1300–1313 (2007)
211. M. Rizzi, M. Trueba, S.P. Trasatti, Polypyrrole films on Al alloys: the role of structural changes on protection performance. Synth. Met. **161**(1–2), 23–31 (2011)
212. B. Wessling, Scientific engineering of anti-corrosion coating systems based on organic metals (polyaniline). J. Corros. Sci. Eng. **1**, (1999)
213. J. Reut, A. Opik, K. Idla, Corrosion behavior of polypyrrole coated mild steel. Synth. Met. **102**(1–3), 1392–1393 (1999)
214. M.M. Attar, J.D. Scantlebury, Polyaniline as a possible inhibitor for the corrosion of mild steel. J. Corros. Sci. Eng. **1**, (1997)
215. A. Michalik, M. Rohwerder, Conducting polymers for corrosion protection: a critical view. Zeitschrift fur Physikalische Chemie **219**(11), 1547–1559 (2005)

216. B. Wessling, Polymers to organic metals. Chem. Innov. **31**(1), 35–40 (2001)
217. B. Wessling, New insight into organic metal polyaniline morphology and structure. Polymers **2**(4), 786–798 (2010)
218. M. Rohwerder, Conducting polymers for corrosion protection: a review. Int. J. Mater. Res. **100**(10), 1331–1342 (2009)
219. J. He et al., Conducting polymers and corrosion III. A scanning vibrating electrode study of poly (3-octyl pyrrole) on steel and aluminum. J. Electrochem. Soc. **147**(10), 3667–3672 (2000)
220. M. Rohwerder, in *ACS Symposium Series*. Intelligent corrosion protection by conducting polymers (2009), pp. 274–287
221. J.N. Barisci et al., Conducting polymers as a basis for responsive materials systems. J. Intell. Mater. Syst. Struct. **9**(9), 723–731 (1998)
222. N. Jadhav, M.B. Jensen, V. Gelling, Tungstate and vanadate-doped polypyrrole/aluminum flake composite coatings for the corrosion protection of aluminum 2024-T3. J. Coat. Technol. Res. **12**(2), 259–276 (2015)
223. D.V. Andreeva, E.V. Skorb, D.G. Shchukin, Layer-by-layer polyelectrolyte/inhibitor nanostructures for metal corrosion protection. ACS Appl. Mater. Interfaces **2**(7), 1954–1962 (2010)
224. A.J. Vreugdenhil, M.E. Woods, Triggered release of molecular additives from epoxy-amine sol-gel coatings. Prog. Org. Coat. **53**(2), 119–125 (2005)
225. C. Rottman, G. Grader, D. Avnir, Polarities of Sol-Gel-derived ormosils and of their interfaces with solvents. Chem. Mater. **13**(10), 3631–3634 (2001)
226. E.V. Skorb et al., Light responsive protective coatings. Chem. Commun. **40**, 6041–6043 (2009)
227. J. Zhang, G.S. Frankel, Corrosion-sensing behavior of an acrylic-based coating system. Corrosion **55**(10), 957–967 (1999)
228. M. Buchler, T. Watari, W.H. Smyrl, Investigation of the initiation of localized corrosion on aluminum alloys by using fluorescence microscopy. Corros. Sci. **42**(9), 1661–1668 (2000)
229. G. Liu, H.G. Wheat, in *Coatings for Early Corrosion Detection, in Corrosion*, ed. by D.C. Hansen, A. Alfantazi, V.J. Gelling (2010), pp. 239–247
230. H.G. Wheat, G. Liu, in *Proceedings of the Eighteenth*. ed. by H.W. Jin, Y.Y. Wang, D.B. Lillig. Fluorescent Coatings for Corrosion Detection (2008), pp. 217–223
231. D. Fornasiero, F. Grieser, Analysis of the visible absorption and SERS excitation spectra of silver sols. J. Chem. Phys. **87**(5), 3213–3217 (1987)
232. F.H. Scholes et al., Silica-overcoated substrates for detection of proteins by surface-enhanced Raman spectroscopy. J. Raman Spectrosc. **39**(5), 673–678 (2008)
233. A. Trinchi et al., in *2012 IEEE Aerospace Conference*. Distributed Quantum Dot Sensors for Monitoring the Integrity of Protective Aerospace Coatings (2012)
234. D. Zhao et al., The fabrication and corrosion resistance of benzotriazole-loaded raspberry-like hollow polymeric microspheres. Surf. Coat. Technol. **238**, 15–26 (2014)
235. A.E. Hughes et al., Revelation of intertwining organic and inorganic fractal structures in polymer coatings. Adv. Mater. **26**, 4504–4508(2014)
236. M.F. Montemor et al., Evaluation of self-healing ability in protective coatings modified with combinations of layered double hydroxides and cerium molibdate nanocontainers filled with corrosion inhibitors. Electrochim. Acta **60**, 31–40 (2012)
237. M.W. Kendig, R.G. Buchheit, Corrosion inhibition of aluminum and aluminum alloys by soluble chromates, chromate coatings, and chromate-free coatings. Corrosion **59**(5), 379–400 (2003)

Part III
Characterisation Techniques—Measuring Self Healing or Repair

Chapter 9
Electrochemical Techniques for the Study of Self Healing Coatings

Y. Gonzalez-Garcia, S.J. Garcia and J.M.C. Mol

9.1 Electrochemistry in the Context of Corrosion

As discussed in chapter one, corrosion can be defined as the degradation of the material properties due to its interaction with the environment [1]. In general this interaction takes place by an electrochemical reaction. While in chemical reactions elements are added or removed from a chemical species in electrochemical reactions electrons are also involved in the process. In electrochemical reactions elements might be added or removed from a chemical species but also at least one species undergoes a change in the number of valence electrons (oxidation or reduction). Electrochemistry is thus the branch of chemistry focused mainly on the study of chemical changes caused by the passage of an electric current and the production of electrical energy by chemical reactions.

There are four necessary prerequisites for electrochemical reactions to take place thereby triggering the corrosion process: an anode (where oxidation occurs), a cathode (where reduction takes place consuming the electrons released from the anode), an electrolytic path for ionic conduction between the two reaction sites, and an electrical path for electron conduction between the reaction sites. These requirements are illustrated schematically in Fig. 9.1.

Y. Gonzalez-Garcia (✉) · J.M.C. Mol
Materials Science & Engineering Department, Delft University of Technology,
Mekelweg 2 2628 CD, Delft, The Netherlands
e-mail: y.gonzalezgarcia@tudelft.nl

S.J. Garcia
Novel Aerospace Materials, Delft University of Technology,
Kluyverweg 1 2629 HS, Delft, The Netherlands

A.E. Hughes et al. (eds.), *Active Protective Coatings*,
Springer Series in Materials Science 233, DOI 10.1007/978-94-017-7540-3_9

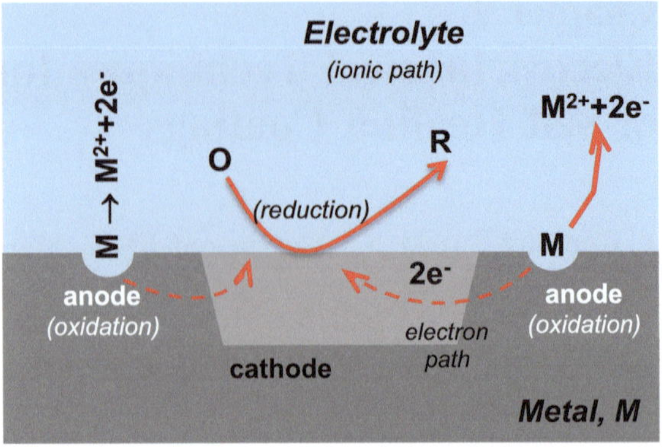

Fig. 9.1 Diagram displaying the main elements required for a corrosion process to occur: anode, cathode, electron path and ionic path

Corrosion research requires a thorough understanding of the electrochemical events within a particular metal-electrolyte system. The driving force for electrochemical events must be known in order to classify the type of electrochemical cell to be characterized to prevent corrosion.

9.2 Electrochemical Evaluation of Self-healing/Self-repair Coatings

In most metal applications, coatings are required to prevent the systems from fast corrosion or degradation. The underlying protection of coatings relies on barrier, adhesion, and active protection. The active protection mechanism is being improved using active and responsive materials (i.e. "smart coatings"). Such systems provide on-demand resistance to corrosion upon mechanical or chemical damage [1]. Such coatings inhibit corrosion processes of the underlying metal by active release of inhibiting species or by the on-demand or autonomous closure/repair of the damage. Significant research effort has focused on finding alternatives to introduce corrosion inhibitors into the organic coating. The traditional approach of using inhibitors is by their direct mixing with the coating binder, however problems such as dispersion and interaction with the polymer matrix affecting the release kinetics and inhibiting efficiency are likely to occur. In order to avoid such interactions encapsulation strategies of the corrosion inhibitor or healing agent have been proposed [2, 3] (see also Chap. 8). The incorporation of corrosion inhibitors into capsules is intended for controlled and effective release of the inhibitor. Therefore, an ideal performance of these encapsulated systems, would release the inhibiting species only when they are needed. The triggering of the inhibitor release can be established according to different mechanisms: breakdown by mechanical damage, pH, dissolution, exchange, humidity or moisture, to name a few. In the end, the ability to suppress or decrease corrosion to tolerable levels will depend on efficiency of the healing behavior of the coating system. It is at this point where the evaluation methods are relevant. The techniques used for evaluation of (self-healing) anticorrosive coatings can be divided in three main groups:

- accelerated tests used in industry
- conventional (global) electrochemical methods, and
- localized electrochemical techniques.

Accelerated testing is possibly the most widespread methodology for corrosion resistance evaluation at an industrial scale and presents many different variants and adaptations. These tests are typically based on exposure of the material to controlled (atmospheric) severe conditions: humidity, salinity, pollutants, temperature, etc. Alternating wet/dry cycles changes the thickness of electrolyte layer. The presence of water, temperature, and chloride ions promotes the corrosion processes at the scribe (degree of corrosion) as well as at the intact coating (delamination, blisters). The degradation of coated metals (or materials in general) is assessed by visual inspection. Accelerated tests only provide information that allows a ranking of the specimen from "unacceptable" to "excellent". They are screening methods to differentiate corrosion performance of the specimen under study. However, accelerated test results form no basis for the understanding of the corrosion (or protection) mechanism. These types of evaluation can accumulate sometimes large amounts of error due to the extreme aggressive conditions, the long duration of the test and a subjective, user-dependent evaluation. Moreover, despite all its drawbacks, since it is used broadly in industry, this is an almost mandatory technique before any system is introduced in the market. More about the evaluation and application of self-healing systems from an industrial point of view is discussed in Chaps. 12–14.

To gain insight into the healing-process behind the behaviour of the system other techniques have to be used. These techniques include electrochemical techniques described in this chapter and physico-chemical characterization described in Chap. 10.

Electrochemical methods are versatile and they can provide a wide range of information that can include the understanding of multi-step reactions, the kinetics of heterogeneous electron-transfer reactions, coupled chemical reactions and adsorption processes. For corrosion studies, electrochemical evaluation provides valuable information such as kinetics of the corrosion or protection processes, formation of protective films, evolution of barrier properties, identification and quantification of redox species.

Potential measurements, potentiodynamic experiments (including polarization curves and cyclic voltammetry) and electrochemical impedance spectroscopy (EIS) are the most common electrochemical methods used for corrosion and self-healing research. In this chapter they are classified as global methods. More advanced techniques that provide electrochemical properties with spatial resolution are considered localized techniques. Within this group micro-capillary cell, scanning vibrating electrode technique (SVET) and scanning electrochemical microscope (SECM) are included.

It is very important to identify the information that we can obtain from the electrochemical evaluation as function of the self-healing system under investigation. A summary of the information to be retrieved from electrochemical evaluation is presented in the Table 9.1. Three main types of healing-mechanism are considered: active and physical healing.

Table 9.1 Summary of the electrochemical techniques used to evaluate self-healing processes

Type of healing	System	Information	Electrochemical technique
Active	Corrosion inhibitor in solution	Screening of inhibitors	Potentiodynamic measurements
			High-throughput (multi-electrode) approach
		Mechanistic information	Electrochemical Impedance Spectroscopy (EIS)
		Kinetics of protection mechanism	
		Protection efficiency	Localized information: SVET
		Inhibition mechanism (localized)	Electrochemical microcapillary cell
Active-smart release	Inhibitor or healing agent incorporated into coating (directly in matrix or with micro-carriers)	Barrier properties of the coating	Electrochemical Impedance Spectroscopy (EIS)
		Kinetics of healing process (release)	
		Efficiency of the healing (protection) for long immersion time	
		Efficiency of healing (localized in the defect) for short immersion time	Scanning Vibrating Electrode Technique (SVET)
			Scanning Electrochemical Microscope (SECM)
		Properties of the coating around defect and healing activity in the defect	Local Electrochemical Impedance Spectroscopy (LEIS)

(continued)

Table 9.1 (continued)

Type of healing	System	Information	Electrochemical technique
Active-physical recovery	Smart polymeric coatings	Healing of barrier properties of the coating	Electrochemical Impedance Spectroscopy (EIS)
		Fast and quantitative information about the healing degree	AC/DC/AC technique
		Localized evaluation of the defective zone after healing process	Local Electrochemical Impedance Spectroscopy (LEIS)
		Confirmation of hindered corrosion activity after (localized) defect healing	Scanning Vibrating Electrode Technique (SVET)
			Scanning Electrochemical Microscope (SECM)

9.3 Conventional Electrochemical Techniques

In conventional electrochemical experiments, the electrode (metal) response to a perturbation signal corresponds to a surface-averaged measurement attributable to the behaviour of the whole electrode surface.

9.3.1 Open Circuit Potential

Measurements of the open circuit potential (OCP) as function of time is commonly used to study healing recovery of corroding metals.

When using this technique complementary tests are recommended to confirm the healing efficiency nevertheless it can be used as a first order approximation technique.

9.3.2 Potentiodynamic Measurements

Potentiodynamic methods consist of forcing the electrode potential to follow a known potential scan. The potential may be held constant or may be varied with time in a predetermined manner as the current is measured as a function of time or potential. In general, for these methods, systems in which the mass transport of electroactive species occurs only by diffusion should be considered.

Linear sweep voltammetry, alternatively named polarization curve measurement, is the most common potentiodynamic measurement. In Fig. 9.2 a scheme of polarization curve for a metal M during immersion in aqueous solution.

From polarization curves, the corrosion current density (i_{corr}) can be extrapolated from the linear parts of the Tafel plots (blue lines in Fig. 9.2). The intersection of the lines for cathodic and anodic processes correspond to the values of the corrosion current density. The linear parts of the curves should be in the range of potential ±150 mV above the corrosion potential (E_{corr}) value.

Another alternative to obtain the corrosion current is the linear polarization resistance (LPR) method. The polarization resistance of a material is defined as the slope of the potential-current density ($\Delta E/\Delta i$) curve at the free corrosion potential. Resistance polarization refers to the potential drop due to either the high resistivity of the electrolyte surrounding the electrode or an insulation effect of the film on the electrode surface formed by the reaction products.

Polarization resistance R_p is related to the corrosion current by the Stern-Geary equation:

$$R_p = \frac{B}{i_{corr}} = \frac{(\Delta E)}{(\Delta i)}_{\Delta E \to 0} \tag{9.1}$$

where, ΔE is a variation of the applied potential around the corrosion potential and Δi is the resulting polarization current. The proportionality constant, B, for a particular system can he determined empirically (calibrated from separate weight loss measurements) or, as shown by Stern and Geary, can be calculated from b_a and b_c, the slopes of the anodic and cathodic Tafel slopes:

$$B = \frac{b_a b_c}{2.3(b_a + b_c)} \tag{9.2}$$

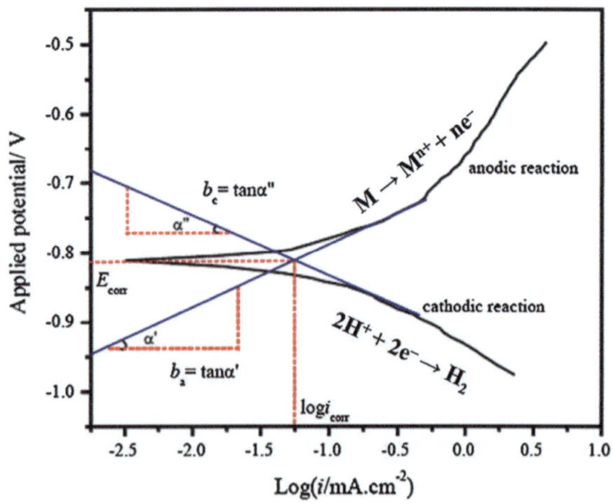

Fig. 9.2 Example of polarization *curve* for a metal M in an acidic environment. E_{corr} and i_{corr} are estimated by the intersection of the linear section of the *curves* corresponding to the Tafel *slope*

The Tafel slopes themselves can be evaluated experimentally using real polarization plots. A high R_p for a metal implies high corrosion resistance and a low R_p implies low corrosion resistance.

The extrapolation from the Tafel curves is only valid when the anodic and cathodic reactions are under activation control. In practice, determining the linear section of the Tafel plot is not always straightforward. Experimental polarization curves sometimes can present peculiarities that bring doubts about where to choose the linear section in the curve. This issue is a regular cause of errors when estimating the corrosion current.

Corrosion current is correlated to corrosion rate (r) of the process through Faraday's law by the following equation:

$$r = \frac{a i_{corr}}{nFD} \tag{9.3}$$

where a is the atomic number of the metal, n number of electrons, F the Faraday constant and D the density of the metal.

Within the context of self-healing materials, polarization methods are commonly used to evaluate and compare corrosion inhibitor performance in solution. The test consists of analyzing the response of the metal while it is immersed in solution containing dissolved corrosion inhibitor [4]. The resulting polarization curves give access to the corrosion rate. Furthermore it visualizes the anodic, cathodic or mixed protection performance of the inhibitor (Fig. 9.2). The analysis of the curves and interpretation of the results is well explained in Chap. 4.

Conventional potentiodynamic methods are usually not applicable for coated substrates due to important limitations arising from the high IR drop over the coating. This method cannot give any useful information on the active corrosion protection and the self-healing ability of protective coatings.

An alternative is the direct current (DC) measurement. For this case the coated sample is kept at high constant potential (potentiostatic method) and the current that flows through the system is measured. The electrochemical measurements are carried out in an electrochemical cell with typically a three electrode system (working, reference and counter electrodes) immersed in an electrolyte. The application of this constant potential can fully reduce or oxidize species in a given solution depending on the potential values, the substrate, pH of the electrolyte and other factors. It moves the systems out of electrochemical equilibrium possibly leading to (i) accelerated degradation of the system; or (ii) forced deposition-reaction of healing agents.

This technique has been used to study the healing capabilities of epoxy-amine coating doped with organic fillers containing inorganic corrosion inhibitors on steel [1] by monitoring the current density values at fixed potential (-0.85 $V_{Ag/AgCl}$ and -1.1 $V_{Ag/AgCl}$) with time. In this study, healing was detected by a decrease of the current density in a scratched doped-system, while the non-doped one showed an increase. The same system evaluated by electrochemical impedance spectroscopy (EIS) only showed a slight improvement with respect to the reference one. A similar method has been also employed to assess healing of a steel plate coated with a self-healing coating using microencapsulation of liquid healing agents [5]. In this case, the experiment was performed at 3 V polarization measuring the current density evolution with time of the scratched samples. Healed samples showed two orders of magnitude lower current density than the reference systems indicating the formation of a barrier layer at the scribe (gap filling of the damage).

The evaluation of healing properties under high potentiostatic polarization is useful mostly when the final application of the coating requires similar conditions as the applied experimental parameters.

9.3.3 Electrochemical Impedance Spectroscopy (EIS)

Electrochemical impedance spectroscopy (EIS) is one of the most used techniques at scientific level in the fields of corrosion and coatings technology and thus its extension and importance in the evaluation of self-healing (anticorrosive) coatings. EIS is a non-destructive test that allows assessment of the degradation (or healing) of the system with time. But more importantly it provides quantitative information of the processes taking place within the system.

Electrochemical impedance spectroscopy is based on electric circuit theory and the description of the different electrochemical processes occurring in the coating-oxide-metal system by a combination of electrical parameters such as resistors and capacitors. The technique uses the description of the behaviour of an electric circuit when an alternating current or voltage is applied as a function of the frequency.

The concept of electrical resistance is well known and is defined by Ohm's law. Resistance is the ability of a circuit to resist the flow of current, mathematically expressed as:

$$R = V/I \tag{9.4}$$

where R is resistance in ohms, V is voltage in volts, and I is current in amperes. However, this simple relationship is limited to one circuit element, the resistor, and it does not represent more complex behaviour that many systems exhibit in the real world. Therefore, the concept of impedance (Z) is introduced. The impedance is a measure of the tendency of a circuit to resist the flow of an alternating electrical current. The equivalent mathematical expression of Eq. (3) for the case of impedance is:

$$Z = V_{ac}/I_{ac} \tag{9.5}$$

This simple and basic equation actually hides some more complicated assumptions and concepts. First, it applies only to the time-varying, alternating, or ac components of the current and voltage. Secondly, it is not sufficient to just indicate the magnitude of the voltage and the current signals, it must also be expressed how they are related in time, since they are both time-varying ac quantities. Finally, the size and time-relationships nearly always depend on the frequency of the alternating current and voltage.

Figure 9.3 shows a sine wave voltage applied to an electrochemical cell. The response is obtained as a sinusoidal current with certain phase shifting with respect to the perturbation. The response is shifted in time due to the response of this system depending on the combination of resistors and capacitors affecting the kinetics of the process. This time shift is expressed as an angle, the phase angle shift of the current response, or simply the phase angle, φ.

The size of the impedance of this system can be expressed by taking the ratio of the size between the voltage sine wave (in volts) and the current sine wave (in amperes). The signal obtained from the test is the complex impedance ($Z = Z_{re} - j \cdot Z_{im}$). To characterize impedance, Z, both its magnitude or modulus (in ohms), |Z|, and phase, φ, as well as the frequency, *f* (in cycles per second, or Hertz), at which it is measured must be specified together with the amplitude used in the signal input. These three parameters are often plotted on what is known as a Bode plot, shown in Fig. 9.4a, or in the complex impedance form, which is the Nyquist plot (Fig. 9.4b).

The experimental set-up for EIS experiments normally consists of a three-electrode cell: the specimen plate (working electrode), a counter electrode (e.g. Pt wire or mesh, carbon) and a reference electrode (e.g. SCE, Ag/AgCl). The sample is kept under immersion conditions and the impedance test is performed in time. In Fig. 9.5 a scheme of a typical electrochemical cell for EIS measurements is shown.

Commonly the sinusoidal voltage perturbation applied to the electrochemical system is of small amplitude, typically in the range of ±10 mV. The voltage is applied at a given frequency range, usually from 10^5 Hz to 10^{-2} Hz. The instrumentation required to measure the impedance of an electrochemical cell includes a waveform generator to produce the sine wave electrical signal, and a potentiostat to control the potential (Fig. 9.6). These components are controlled by a computer used to run the experiment and to display the results in real time.

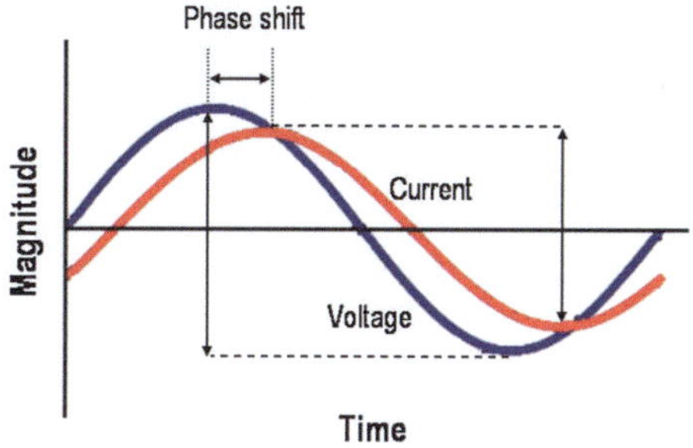

Fig. 9.3 Graphic representation of the current and voltage as a function of time

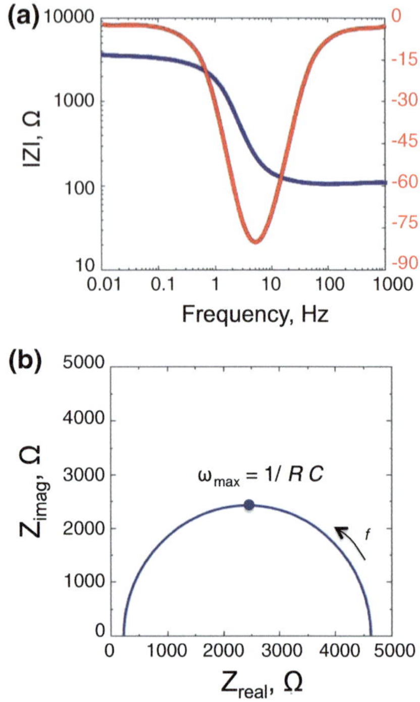

Fig. 9.4 Data representation for impedance data: **a** Bode plot and **b** Nyquist plot

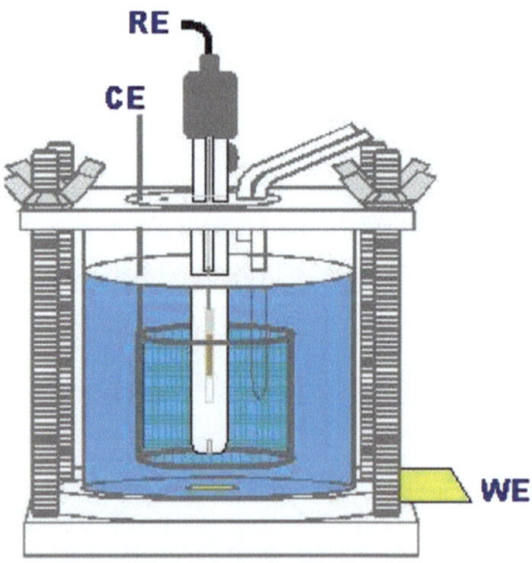

Fig. 9.5 Electrochemical cell used in EIS experiments: *WE*, working electrode; *RE*, reference electrode; and *CE*, auxiliary electrode

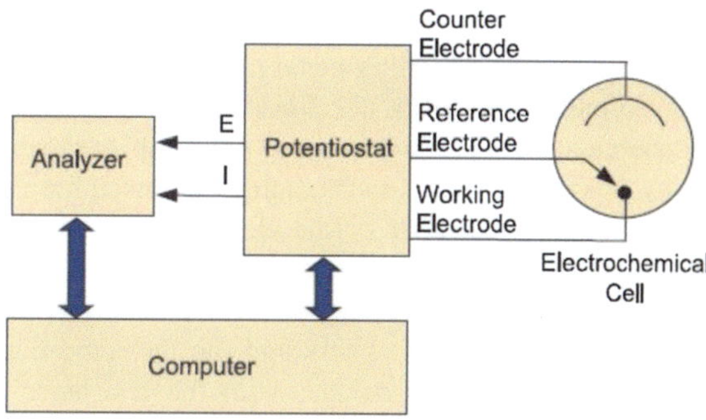

Fig. 9.6 A block diagram of the instrumentation used to conduct EIS measurements

As mentioned at the beginning of this section, EIS is based on electric circuit theory. Therefore the fitting of experimental data with electric equivalent circuits can give extra quantitative information about the system under study. Electrochemical systems such as coated surfaces or corroding metals often behave as simple electronic circuits when an alternating voltage is applied. And, they can be simulated by a set of electric circuit elements such as resistors, capacitors and inductors, which will constitute the *equivalent electric circuit*. This equivalence allows to associate a real chemical-physical process to each of the circuit components.

For example, a simple electrochemical reaction that involves a step of electron transfer such as:

$$O^{n+} + ne- \rightarrow R \tag{9.6}$$

The process can be represented by a simple equivalent circuit which is called Randles' circuit [6]. This circuit is depicted in Fig. 9.7, and consists of a combination of a capacitor and two resistors.

It must be noticed that when an electrochemical reaction occurs at an electrode, at least two associated interfacial processes must be considered, namely:

- the formation of the double layer in the interface, which behaves as a capacitor; and
- the electronic transfer through the interface, which is a faradaic process (resistor).

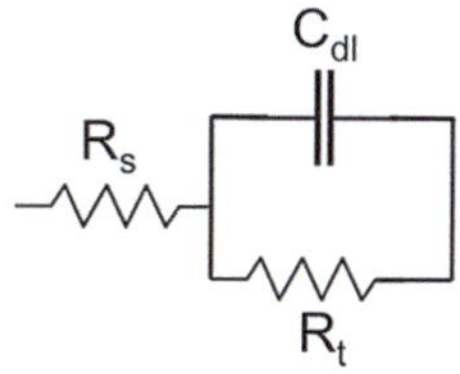

Fig. 9.7 Randles equivalent circuit for a charge transfer-controlled simple electrochemical reaction

This circuit can be used to either represent a simplification of a non-defective coating or a bare corroding metal immersed in an electrolyte solution although such EC simplification rarely describes inhibiting systems [7]. Depending on the electrochemical system the values and meanings of the circuit components are different. Thus, in the case of an intact coating immersed for a short time in an electrolyte, R_S represents the resistance of the electrolyte solution, or uncompensated resistance, between the reference electrode and the surface of the coating. The capacitor, C_{dl}, represents the coating capacitance and is normally represented as C_{coat}, and can be characterized by the thickness and the dielectric constant of the coating material. The resistor, R_t, is associated with the resistance of the coating and generally referenced as R_{pore} (pore resistance). It is also a property of the coating material and varies with its thickness and composition as well as defects.

Corrosion processes are more complex phenomena than just a simple electrochemical reaction. For example in the case of a coating-metal systems immersed for long times, besides the electrochemical reactions (oxidation and reduction), the existence of the coating-metal interface and the simultaneous occurrence of diffusion and adsorption processes should also be taken into consideration. Then a more complex equivalent electric circuit is required to represent the system. For example in the case of corrosion processes occurring underneath the coating, a more specific circuit that can describe this case is shown in Fig. 9.8 [8–12]. The elements of the circuit represent the following characteristics of the system [13]; R_S, the resistance of the electrolyte; R_{pore}, the resistance of the conducting pores in the coating layer; R_t, the charge transfer resistance of the corrosion reaction; C_{coat}, the capacitance of the coating layer; and C_{dl}, the double layer capacitance at the metal/electrolyte interface.

Often better fitting of the equivalent circuit model is obtained by replacing the capacitances in the equivalent circuit presented in Fig. 9.8 with constant phase elements (CPE) [14]. In reality, when studying coated-metal systems, it is difficult to find a pure capacitor behaviour. The deviation from the ideal capacitor can be due to:

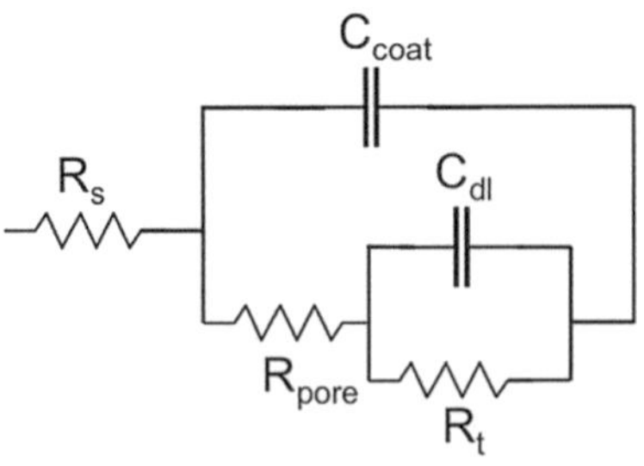

Fig. 9.8 Equivalent circuit of intact coated metal showing corrosion processes underneath the film

- gradients in properties within the coating (laterally and vertically). Such gradients can be caused by resistivity changes [15–17], differences in water takeup [18], porous electrodes [19, 20] to name a few.
- changes in structure such as occurs with layers within a coating or nanoporosity as shown for thin silica layers where the n parameter in the CPE has a linear relation to porosity [21].

However as pointed out by Tribollet and co-workers [15–18] there needs to be a good physical reason for the introduction of CPEs.

Additionally, water uptake, which is calculated through the coating capacitance, is complicated when CPE's are used because of the need to calculate the capacitance from the fitted parameters for the CPE. One equation that has broad acceptance is [22]:

$$C = A\left(\omega_m''\right)^{\alpha-1} \tag{9.7}$$

where A is a constant, ω_m'' is the frequency where the imaginary component of the impedance is a maximum and α is a measure of the capacitive behavior.

One other component worth mentioning is the Warburg impedance which is used to model diffusive behavior at the interface. The use of a Warburg component is generally justified on the basis of the value of the phase angle, where a phase angle of 22.5° or 45° is indicative of a Warburg process [19].

In conclusion, provided a good analysis is performed, information given by EIS can be directly linked to water absorption, kinetics of reaction or absorption, delamination, porosity and adhesion amongst other processes (e.g. diffusion). It is a powerful method for evaluation of (self-healing) anticorrosive coatings since it can provide information about different processes occurring on the system almost simultaneously. For example:

- evolution of the barrier properties of the coatings (water uptake, blistering, delamination)
- corrosion rate at the metal-coating interface
- surface changes due to action of corrosion inhibitors (passivation of the metal surface)

EIS has already been used to assess self-healing anticorrosive (organic) coatings, mainly using total impedance and phase angle as parameter. The main application of this technique has been in extrinsic self-healing coatings using corrosion inhibitors (e.g. loaded nanocontainers with corrosion inhibitors into a sol-gel film [3]). In these tests the normal procedure is the application of an artificial defect reaching the substrate followed by immersion in a corrosive medium (electrolyte). During the experiment it is expected that the corrosion inhibitors will leach out of the coating allowing them to react with the metallic surface at the damage site, decreasing the electrochemical activity, which is observed as a significant increase of the impedance of the system.

EIS has also been successfully employed for liquid encapsulated extrinsic self-healing coatings using two component microencapsulated liquid healing agents [23] or one component microencapsulated healing agent [24] in both cases with excellent results of the use of the technique. In the case of systems using micro-encapsulation, the technique generally shows a more significant increase in impedance than when used for corrosion inhibitor systems due to the restoration of the barrier effect offered by gap filling or full surface coverage [22] in encapsulated concepts and thus major hindering of the electrochemical reactions. EIS has also been recently combined with X-ray microtomography to further understand the mechanisms of protection of single reactive healing agents such as the silyl ester involving delamination and underfilm pitting processes [25].

9.3.4 Odd-Random Phase Multisine Electrochemical Impedance Spectroscopy

As an extension of use of EIS, the odd random phase multisine electrochemical impedance spectroscopy (ORP-EIS) has also been used as a characterization tool for diverse electrochemical processes [26, 27]. ORP-EIS cannot be considered as conventional electrochemical test, but it is included in this section as it provides global (averaged) information of the specimen.

In the previous section the capacity of EIS for the study and evaluation of diverse self-healing systems was presented. A reliable model that describes the system under investigation can only be obtained if the measurement fulfills the conditions of causality, linearity and time-invariance. However, the two last conditions can be difficult to satisfy. EIS measurements are often performed with very small amplitude excitation signals (for reasons of linearity) in the steady state regime of the process (for reasons of time-invariance). As a consequence, the measurements (i) can suffer from poor signal-to-noise ratios and (ii) cannot describe the initial, mostly rapidly evolving, stages of electrochemical phenomena of corrosion processes. Moreover proper fitting of the model to experimental data needs to be achieved, by minimising the residuals between the data and the fit across the whole frequency range [7, 28, 29].

Apart for shorter time measurements ORP-EIS allows measuring the level of disturbing noise, the level of the non-linear distortions and the level of the non-stationary behaviour. The additional information given by this methodology is useful to verify the quality of the measurement. In Fig. 9.9 a typical ORP-EIS spectrum is shown, in which, additional to the conventional impedance data, it is possible to observe the noise level and non-stationary behaviour of the system [30]. As happens with traditional EIS, ORP-EIS becomes a powerful method when combined with the fitting of the data. The additional information helps to reject, accept or improve electrical equivalent circuit models used to fit the conventional impedance spectra. The advantage of the technique has been demonstrated on an electric circuit [27] and the study of corrosion of coated metal [31]. However, it is still scarcely used on the study of self-healing corrosion systems.

Fig. 9.9 ODR-EIS Impedance spectra plotted on Bode coordinates showing the impedance value (*red*), noise level (*green*), non-stationary behaviour (*black*) and the non-linear behaviour (*blue*) of the corroding metal system

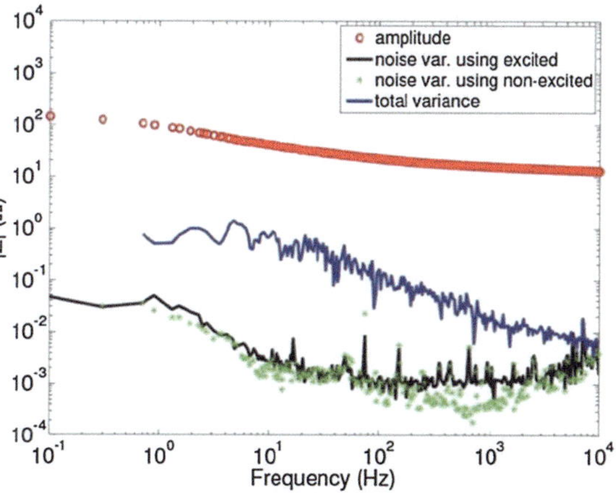

Jorcin et al. [30, 32] applied for the first time this methodology to the study self-healing coatings based on a shape memory polyturethane coating leading to partial physical recovery of the film after macroscale damage. By applying the odd-random phase multisine methodology in contrast to the conventional EIS, it was possible to propose a new equivalent circuit with a more accurate representation of the processes taking place in the coating after damaged. Even though a careful fitting of traditional EIS can also lead to highly accurate equivalent circuits, the ORP-EIS facilitates this task and increases the certainty in the selected EC.

The EC most commonly used circuit to fit damaged coatings (Fig. 9.10) assumes that the impedance data is mainly driven by the defect, representing the coating only as a capacitor C_{coat}. The use of this EC is not based on the overall error

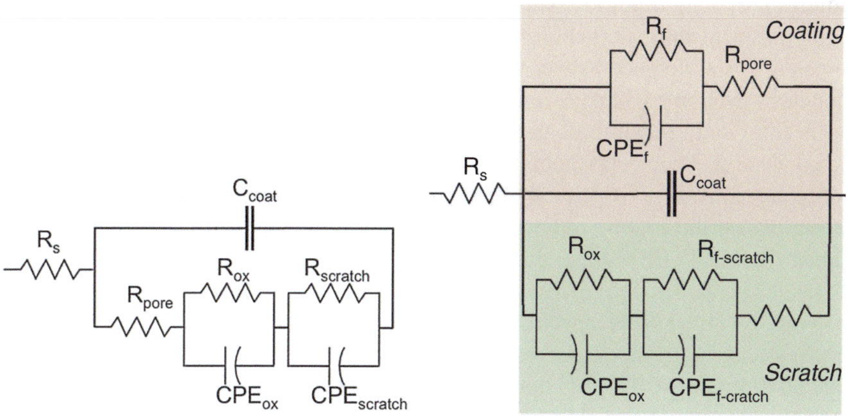

Fig. 9.10 (*left*) Initial equivalent electrical circuit used for the data analysis of the damaged coatings proposed by Bonora et al. [33]. (*right*) New proposition for an equivalent electrical circuit used for the odd-random phase data analysis of the damaged coatings

minimization [7, 24] but on the chi-square, however, it is generally accepted without questioning its validity. When this circuit is applied to fit the odd-random EIS data, a residue (difference between model and experimental data) higher than the stochastic noise of the measurement is observed. This is a clear indication that the equivalent circuit is not well adapted to this electrochemical system probably as a result of a simplification of the physical phenomenon in a damaged coating. Based on this lack of correlation it was possible to propose a new EC consisting in two parts [30]: one related to the organic coating and one related to the phenomena occurring in the artificial defect. The fitting of the data with this new model presented a residue at the level of stochastic noise of the measurement, indicating that all information in the spectrum is taken into account.

The odd-random phase impedance spectroscopy (ORP-EIS) appears as a valuable alternative to conventional impedance methodology for the study of corrosion and self-healing processes. The main advantages are: rapid acquisition of the data and more accurate information to model the system of interest. Nevertheless, analysis of the spectra and the fitting procedure is still tedious and requires a certain level of expertise that makes it not accessible yet for routine research.

9.3.5 AC/DC/AC Accelerated Electrochemical Protocol

Despite the fact that EIS and ORP-EIS can lead to very reliable information about the properties of intact and damaged coatings, they require very long times to assess intact high impedance coatings. In the case of intrinsic self-healing coatings, EIS and ORP-EIS can discern quickly the level of healing when the healed interface is only weakly healed (closed). When the damage interface of the healed coating is of high quality and close to the original network properties non-destructive EIS requires very long times to assess the healing of the scribe. A very recent study with highly efficient intrinsic healing coatings demonstrated that EIS is not capable of detecting any difference between the intact coating and healed coating (macroscale scratches) even after 30 days immersion in corrosive media [34]. In order to detect the healing state (healing degree) the researchers introduced the use of an accelerated destructive electrochemical protocol named AC/DC/AC or ACET. Such technique consists of cycles of EIS measurements/polarization to −2 or −4 V depending on the coating quality/potential relaxation/new EIS to measure the after damage state. [35–38]. When this protocol was applied to self-healing coatings [34] the healed scratches (interfaces) were detected after a number of cycles by a sudden drop of the impedance spectrum. While EIS indicated that the healed coating behaved as a high impedance coating (same behavior as the intact coating) for at least 30 days the AC/DC/AC indicated in three cycles (approximately 12 h) that the damages do act as heterogeneities in the coating even when the healing degree is high (Fig. 9.11). Such a result suggests that healed interfaces can protect damages for long times (at least 1 year as shown in the same article) but that the polymer network at the healed interface remains different from the bulk coating. This first

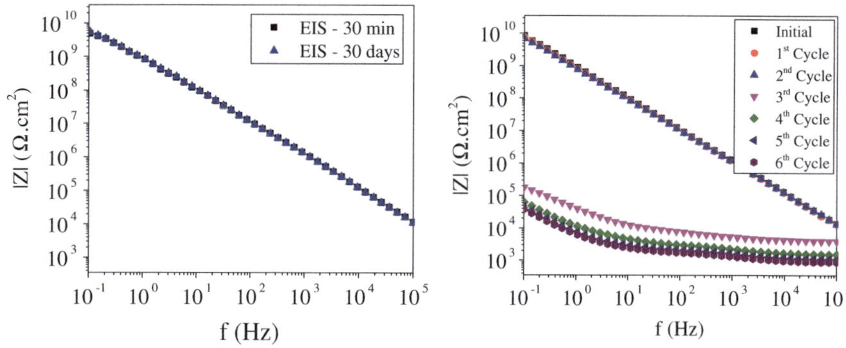

Fig. 9.11 Healed coating after 30 days exposure to electrolyte measured by EIS (*left*) and after 6 cycles of the AC/DC/AC procedure (*right*). (adapted from [34])

time results suggest that the AC/DC/AC technique can be a powerful technique to evaluate healing in intrinsic coatings and potentially even quantify healing degree based on the number of cycles to failure as well as polarization potential applied.

9.4 High-Throughput Screening Techniques

In the previous section the most common electrochemical methods to evaluate the efficiency of self-healing systems were introduced. In general, these techniques require testing times ranging from hours (in the case of electrochemical studies) to weeks (for accelerated laboratory testing, mass loss testing) or even years (in the case of outdoor exposure testing) and produce data on only one system of interest at a time. Hence, if a broad range of systems is to be investigated, more rapid means of evaluating their performance need to be found. This is typically an issue for the research focused on development of new corrosion inhibitive coatings. Restrictions on using Cr(VI) compounds for corrosion protection has forced an increase in the effort to search for equally efficient substitute inhibitors. Up to now the research of alternative corrosion inhibitors for all sorts of metals is still an ongoing topic. Within the development of new protective coatings we can identify five stages or milestones (Fig. 9.12):

- selection of alternative inhibitor candidates
- evaluation of their intrinsic efficiency
- incorporation into the coating
- total performance evaluation
- final optimization of delivery kinetics

In all these steps the main problem, apart from finding the perfect inhibitor, is that it is a very time consuming process, especially when thousands of inhibitor candidates have to be tested. For this reason the introduction of high throughput

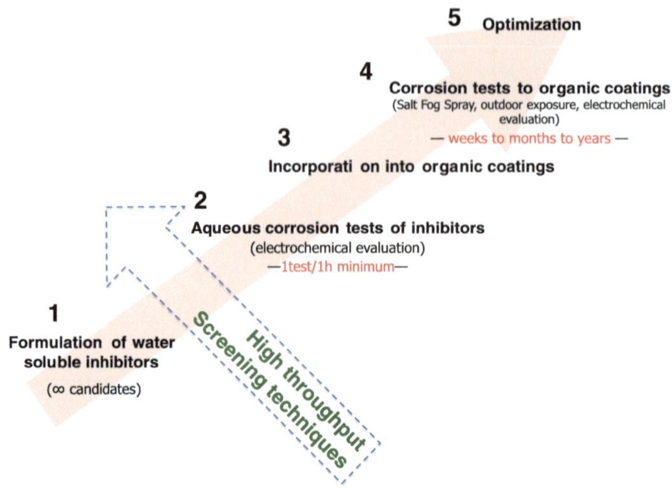

Fig. 9.12 Scheme presenting the milestones to develop corrosion inhibitive coatings [41]

techniques for inhibitor selection is a very necessary step previous to traditional aqueous corrosion tests. It is needed to test the efficiency of a large number of chromate-free alternative inhibitors and to be able to provide a shortlist of candidates in a reasonable time. Furthermore, due to the reliable and efficient performance of chromate-based inhibiting technologies, attempts to find a 'direct replacement' have largely failed, and therefore, present day approaches aim to develop 'multifunctional' corrosion inhibitors, which have the ability to passivate both anodic and cathodic reaction sites at the metal surface. Inhibitor development and evaluation approaches therefore require a large number of experiments if a systematic survey of potential cathodic, anodic and combined (multifunctional) inhibitors is to be performed [39, 40]. Then, high-throughput or rapid screening methods need to be used for assessing inhibiting efficiency of a broad range of potential inhibitors.

In this section the development of rapid screening techniques for high-throughput testing of inhibitors is described

Multi-electrode arrays have been introduced in recent times as valuable methods to carry out corrosion measurements [42–44] and high-throughput testing of corrosion inhibitors [45–49].

In the approach proposed by Taylor et al. [46], identical pairs of metal wires were placed into a large number of separate reaction wells. The experiment consisted of polarizing one of the electrodes with respect to the second pair which is held at a potential corresponding to the open-circuit potential of the metal. The current between each pair was monitored and used to measure the linear polarisation resistance (LPR) from which the corrosion rates can be calculated.

Muster et al. [41, 50] presented an adaptation of the multiple-electrode method that allowed the simultaneous investigation of several metals in one solution. Pairs of metals were immersed into various inhibitor solutions at relatively low polarization. The current flowing between the individual electrodes of each metal pair was measured. The current exchange between the individual electrodes of the same metal then was expressed in inhibitor efficiency with respect to that in uninhibited solutions.

Other main differences with respect to Taylor and Chambers approach are:

- no reference electrode is used to fix electrode potentials of the metals
- a blank sample is used as a control for each inhibitor evaluation

Electrochemical methods are usually not carried out in the absence of a reference electrode. However, for the rapid screening of multifunctional corrosion inhibitors, the omission of the reference electrode offers the benefit of simplicity whilst retaining the key information regarding inhibition.

High-throughput screening methods reduce the experimental time significantly. For example, in the work of Taylor and Chambers [45, 46] potential corrosion inhibitors and their synergistic combinations for aluminium alloy AA2024 were studied. The inhibition characteristics of 50 separate chemistries were assessed simultaneously. The test using direct current (DC) polarization in the multiple-electrode system was performed within 9 h while for electrochemical impedance experiments for the same number of inhibitors at least 10 days of measurements was needed.

In addition to electrochemical techniques that aim to determine the corrosion current, or potentiostatic current as above, high throughput studies of corrosion inhibition have also been applied to the detection of chemical changes. These techniques are non-electrochemical, high-throughput screening methods. These methods are based on fluorimetric [45, 51] and spectroscopic measurements [52]. They are used to determine the concentration of chemical species in solution.

A different type of method for high-throughput screening is based on micro-fluidic approaches [53]. These differ from the multielectrode in that only one metal can be studied in the one test. The metal is exposed to a range of aqueous corrosion inhibitors and concentrations via micro-fluidic channels. The sample is visually inspected and solutions are chemically analyzed. Another non-electrochemical high-throughput test employs arrays of wells which are filled with inhibitor-doped solutions and after exposure the bottom wells are then examined by digital micrograph analysis [54].

9.5 Local Electrochemical Techniques

In Sect. 9.3, general electrochemical methods were presented. These methods are characterized by having the sample of interest as working electrode. The obtained results provide averaged information of the electrochemical processes taking place on the entire surface of the working electrode. In this section we examine localised electrochemical techniques.

9.5.1 Electrochemical Microcapillary Cell

The electrochemical microcapillary cell is a technique used to perform conventional electrochemical measurements (open circuit potential measurements, potentiodynamic polarization and EIS) on areas with dimensions in the micrometer range [55, 56]. In general the microcapillary cell is a powerful technique since it allows performing electrochemical measurements on very small surface areas, in the micrometer range.

The microcapillary cell is based on the miniaturization of the standard three-electrode cell setup. For this, the area of the working electrode is reduced by using a glass microcapillary which holds a droplet of the electrolyte on the sample surface. The most common sizes of the tip of the capillary are found between 10–100 μm for corrosion research, but can be even less. The droplet of electrolyte can be held in position in two different ways: free droplet and silicone rubber gasket [57]. The silicon gasket provides a cushion while approaching and reaching the sample and it avoids leakage of electrolyte. The silicone with its high deformability shapes to the roughness of the surface. In any case, the wetted area on the sample surface defines the working electrode.

In Fig. 9.13 a scheme of the components of the microcapillary cell technique is shown. The setup consists of micro-electrochemical cell (3-electrode cell), microscope, camera, potentiostat and computerized controller. The microscope and camera enable a fast and precise positioning of the microcapillary on the desired spot of the sample surface. The components that will determine the resolution and quality of the electrochemical measurements are mainly:

- The potentiostat. A high resolution potentiostat is required. The current detection limit of the potentiostat will determine the resolution of the measurements. Currently modern instruments have a lower detection limit for the current down

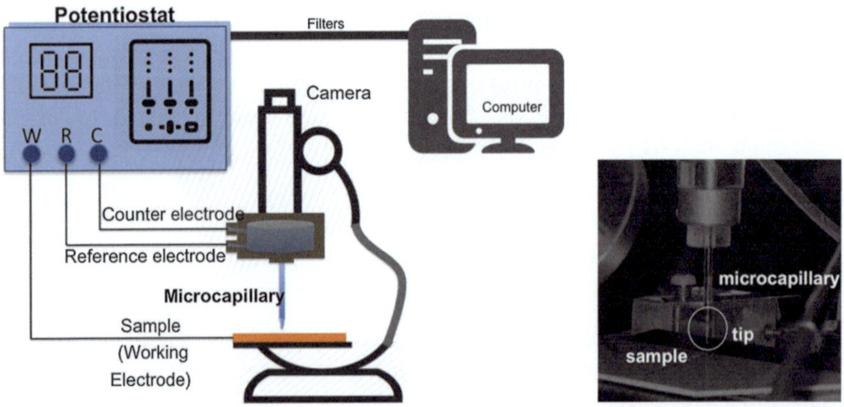

Fig. 9.13 (*left*) Schematic of the electrochemical microcapillary cell setup. (*right*) Detail of the microcapillary approaching the sample surface

to pA and fA levels. Furthermore, adequate isolation and filters to reduce noise influencing the measurements is required.

- Microcapillary. Preparation of the microcapillary consists of "pulling" a glass capillary until the desired shape and diameter is obtained. The final size of the capillary is achieved by careful grounding and polishing of the capillary tip. Afterwards, the silicone gasket is applied. The final inner diameter of the capillary will influence the limiting current density during the measurements. It is important to take into account this factor when planning a experiment. For example, a current density of 10 $\mu A/cm^2$ corresponds to a capillary of 1 μm diameter (exposed area). This is equivalent to corrosion currents in the order of 10 fA [57].

The electrochemical microcapillary cell is a very attractive technique to study the effect of microstructural features on the corrosion mechanisms of metal alloys. The microcapillary cells have been successfully used to the study of the local behaviour of intermetallics in aluminium alloys [58–60] and pit initiation of steels [61–63].

In this sense, the microcapillary cell can be used to study the protection mechanism of corrosion inhibitors for individual phases or intermetallics from the metal alloy. Some works have been done focusing on rare-earth corrosion inhibitors and their cathodic-type behaviour for AA2024 alloy.

Birbilis et al. [64] studied the protection mechanism of AA2024 alloy offered by cerium dibutyl phosphate ($Ce(dbp)_3$). Taking advantage of the capabilities of the microcapillary cell, it was possible to perform potentiodynamic measurements on individual intermetallics and to study their behaviour in presence and absence of the inhibitor. Cathodic polarization curves on the main cathodic phases (Al_2Cu, Al_7Cu_2Fe, $Al_{20}Cu_2Mn_3$, and Al_3Fe) showed that the $Ce(dbp)_3$ effectively inhibits oxygen reduction on each of them. Interestingly, anodic polarization curves measured for anodic phases (Al_7Cu_2Fe and Al_2CuMg) demonstrated that the inhibitor is able to stabilize the passivity of these phases. This work suggested that $Ce(dbp)_3$ acts as a mixed inhibitor offering cathodic and anodic inhibition for the AA2024 alloy. It may effectively inhibit localized corrosion owing to its mixed nature, providing retardation of the oxygen reduction reaction upon noble intermetallics and concomitant retardation of dissolution of phases that may be prone to attack.

In the work of Andreatta et al. [65] the localized corrosion inhibition of $CeCl_3$ was investigated for clad AA2024. In this case the research focused on the behaviour of areas containing Fe-rich intermetallic compounds and areas without inclusions (matrix). Microcapillary based potentiodynamic polarization evidenced that the deposition of Ce-species occurs on the entire sample surface being affected by the alloy microstructure. Moreover, it is activated by alkaline etching of the substrate. The deposition takes place preferentially at the sites of intermetallics due to their cathodic behaviour and to local alkalinization associated with cathodic reactions. It was possible to demonstrate that the amount of Ce detected at the sites of intermetallics is larger than on the matrix.

More recently, the microcapillary cell has been used to characterize the protective layer formed in a scratch by inhibitor released from the coating after it is damaged. In this work by Recloux et al. [66], benzotriazole (BTA) inhibitor was incorporated into a silica mesoporous thin film as pretreatment applied on an AA2024 alloy sample. The acquisition of local anodic polarization curves with the microcapillary cell on the bare metal exposed in the scribed coating was used to monitor the formation of a passive film by the BTA within the defect. The results demonstrated that the inhibitor was directly released from the mesoporous pretreatment once the top coat was damaged and acted rapidly to protect the bare metal and to delay the initial stage of corrosion phenomena.

It is clear that this type of information and conclusions are not accessible from conventional (macroscopic) potentiodynamic experiments. The microcapillary cell is a powerful method for localised electrochemical measurements and to establish correlation between corrosion/inhibition mechanisms and microstructural features at metal surfaces.

Nevertheless, the electrochemical microcapillary cell technique also presents several drawbacks or precautions:

- risk of crevice corrosion at the silicon periphery of the capillary
- requires a large number of measurements to assure the reproducibility of the results
- possibility of contamination of the solution during the measurement
- small size of the microcapillary can affect the transport of species to the substrate, such as oxygen and protons resulting in diffusion limited corrosion behaviour
- problems with high ohmic resistance between working and counter electrode can occur. The choice of an appropriate geometry of the cell will minimize this issue.

9.5.2 Scanning Vibrating Electrode Technique (SVET)

The SVET technique produces in situ measurements with high spatial resolution and provides valuable information on the behaviour of the corroding system at a microscopic level [67, 68]. The SVET offers the possibility of mapping variations in the ionic current densities in the micrometric range facilitating the location of anodic and cathodic zones (related to corrosion processes). This technique is based on the detection of electric fields generated in a solution due to ionic concentration gradients. The electric field is zero when the solution is at rest, but if there is a gradient of concentration caused by a source of ions, a variation of potential in the solution occurs. Ionic flows can arise from corrosion processes on a metal. The oxidation reactions occurring at anodic sites on a metal surface in contact with an electrolyte cause electrons to flow through the metal substrate to adjacent cathodic areas. This flow of electrons through the metal is supported by a flow of ionic current in the electrolyte, which in turn causes potential gradients to exist in the solution close to sites of localized corrosion. Figure 9.14 depicts a scheme of the

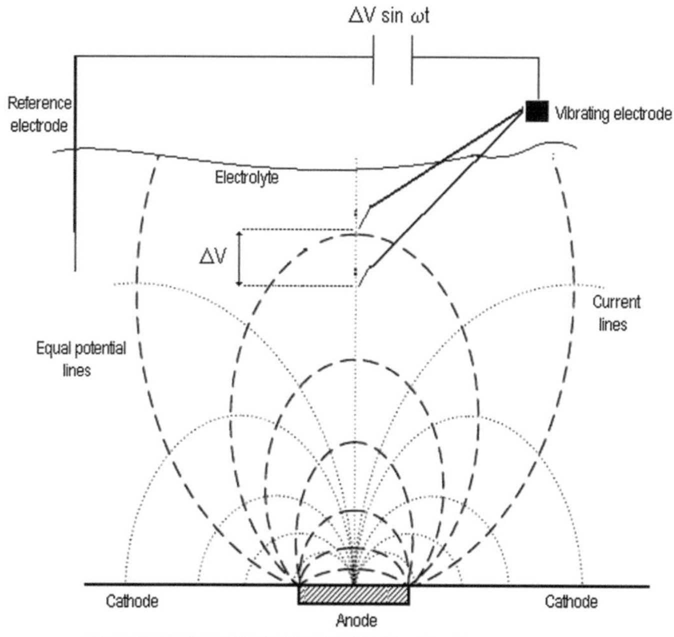

Fig. 9.14 Schematic representation of the operation principle of the SVET technique

distribution of equal-potential and current lines during a corrosion process and the operation principle of the SVET.

The SVET uses a vibrating microelectrode that scans the surface to measure these gradients in situ. Measurement is made by vibrating a fine tip microelectrode a few hundred microns above the sample, usually in a plane perpendicular to the surface. The electrochemical potential of the microelectrode is recorded at the extremes of the vibration amplitude, resulting in the generation of a sinusoidal AC signal (cf. Fig. 9.14). Then this signal is measured using a lock-in amplifier, which is tuned to the frequency of probe vibration. The measured potential variation, ΔV, can be related with the ionic currents (I) by use of the equation [7, 69]:

$$I = \frac{E}{\rho} = \frac{1}{\rho}\frac{\Delta V}{\Delta r}. \tag{9.8}$$

where E is the electric field measured between two points of the solution, ρ is the resistivity of the solution, and Δr the distance between the two extremes or vibration amplitude of the microelectrode.

The resulting signal, which is a measure of the DC potential gradients in solution, can be converted to current density by a calibration procedure. This consists of placing the vibrating microelectrode at a known distance from another electrode that acts as source of a given constant current. Then the potential measured by the

SVET probe at determined distance from the source corresponds to a known current. The potential (*V*) measured in each point can be related to the current intensities (*I*) of the processes occurring at the metal surface by Ohm's Law, using the following equation:

$$V = \frac{I\rho}{2\pi d} \tag{9.9}$$

where ρ is the resistivity of the solution, and d the distance between the point of measurement and the source of the current. The calibration is valid for the solution used and while the amplitude and the frequency of the vibration remain unaltered [70].

The scanning vibrating electrode instrument is basically constituted by:

- The electrochemical cell (detail in Fig. 9.15)
- The piezo-oscillator system which produces the vibration of the microelectrode
- Two lock-in amplifiers that measure and filter the signal produced in the probe.
- Tri-axial piezoelectric motors to control with precision the movement and the position of the vibrating probe
- Computer, interface and display system

The SVET microelectrode consists of a platinum/iridium (Pt/Ir: 80 %/20 %) wire insulated with paralene C® and arced at the tip to face the metal. The tip is platinized to form a small spherical platinum black deposit of 10–20 μm diameter. An image of the tip after platinization is shown in Fig. 9.15.

A map of current distribution can be obtained by making a grid of points and measuring at each point the potential difference while scanning in a plane parallel to the surface of the sample. The SVET is therefore a technique able to make in situ measurements of the localised corrosion activity occurring at the surface of the sample. Figure 9.16 shows an example of the application of the SVET for measuring the distribution of anodic and cathodic activities on AA2024 alloy sample during immersion in chloride solution.

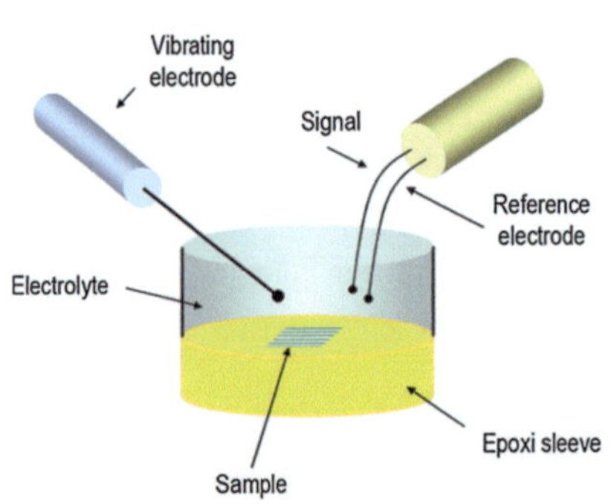

Fig. 9.15 (*left*) Configuration of the electrochemical cell employed in the SVET experiments (*right*) Image of the SVET microelectrode tip after platinization

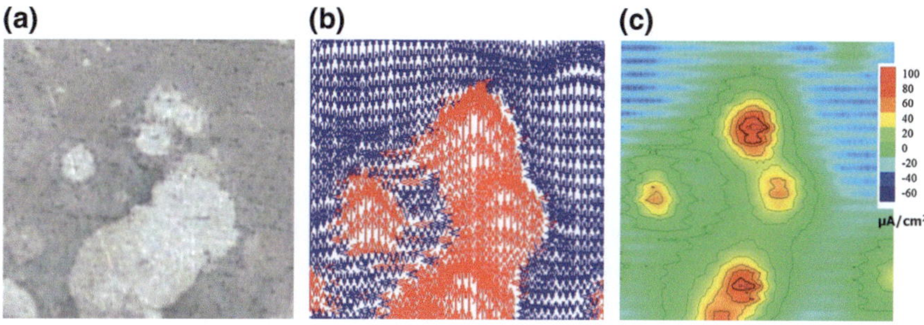

Fig. 9.16 **a** Image of AA2024-T3 sample after 2 days immersed in 0.05 M NaCl solution (image obtained while in solution). Scale 1 mm × 1 mm. **b** Vector map of the SVET data representing the direction of ionic currents. **c** Current density map of the surface. Anodic and cathodic currents can be correlated with the aspect of the sample in (**a**)

One of the important advantages of the SVET is that it is not a destructive method, in the sense that it allows to visualize corrosion activity occurring on the same sample surface during immersion [71]. The SVET is currently one of the most extended scanning micro-electrochemical techniques applied for corrosion research.

SVET measurement is widely used for analysis of corrosion inhibitor performance [72, 73]. In this case the bare metal substrate is immersed in the solution containing the inhibitor under study. Sequential SVET maps are obtained with immersion time to monitor the corrosion activity on the surface in situ.

Inspired by the high-throughput concept, a fast screening of the corrosion inhibitors by SVET measurements has also been proposed [74, 75]. It consists of a multi-electrode cell with a range of metal wires embedded in insulating material. Then the inhibition performance of the inhibitor for each metal wire is monitored all at the same time by using SVET. It can be used as preliminary testing when a large number of metals and corrosion inhibitors have to be tested.

With the emerging interest on finding new, environmentally friendly inhibitors and novel inhibitor controlled release concepts, the number of publications using the SVET has increased significantly in the last years [76–78]. With this technique it is possible to monitor the corrosion or inhibition activity in a coating defect. Therefore it is a useful and straightforward method to evaluate the performance of self-healing polymeric coatings [24, 79].

9.5.3 Selective-Ion Electrode Technique (SIET)

The SIET consists of a potentiometric micro-probe that measures the concentration of specific ions at a quasi-constant distance (at micrometer range) over the surface of interest. Potentiometric micro-probes have been used for biological applications for a long time now. Profiles and gradients of concentrations of Ca^{2+}, Mg^{2+}, K^+, Na^+, NH_4^+, Cl^- determined by potentiometric micro-sensors above biological specimens have been reported in numerous publications [80–83]. It is not until recent years that the SIET technique has been used to monitor specific ions in real corroding systems. The first time, it was applied for mapping local activity of Mg^{2+} and H^+ over the surface of a Mg-based alloy in chloride-containing solution [84].

The main components of a SIET device are: an ion-selective microelectrode, a reference electrode (e.g. Ag/AgCl mini-electrode), 3D computerized stepper-motors systems, a camera to control the location of the microelectrode over the sample.

The microelectrode consists of a glass-capillary filled with a selective ionophore-based oil-like membrane. The microelectrode also includes an inner reference electrolyte and the reference electrode Ag/AgCl wire. The diameter of the glass-capillary microelectrode varies from 0.1 to 5 μm. This type of electrode is highly fragile and have limited life time (<1 day). Furthermore, the detection limit is biased by the flux of primary ions from the ion-selective membrane and inner reference electrolyte. Recently, more robust designs have been introduced. They are based on solid-contact ion-selective microelectrodes which do not require an inner reference solution [85–89].

The potential measurements are correlated to the ion concentration (activity) by the Nernst equation:

$$E_i = E_i^0 + \frac{RT}{Fz_i} lna_i \tag{9.10}$$

where F is the Faraday constant, R is the universal gas constant, and T is the absolute temperature.

Before and after measuring a sample the ion-selective microelectrode has to be calibrated. The calibration consists of recording the potential E_i versus time dependence with sequential increase of ion activity, a_i. This also allows to check the stability of the potential measurements, the drift and the time response. More details about potentiometric measurements can be found in several reviews [90, 91].

SIET mapping has been used in several studies to evaluate corrosion inhibitor efficiency and mechanism [92, 93]. For example, the corrosion processes on AZ31 alloy coated with hybrid sol-gel was studied in NaCl solution containing three different potential inhibitors by Karavai et al. [94]. Localized measurements of pH and Mg^{2+} concentration at microdefects of the sample showed that it was possible to obtain complementary information of electrochemical reactions inside the defect.

9.5.4 Scanning Electrochemical Microscope (SECM)

The SECM is a non-optical scanning microscopic instrument which is included in the group of scanning probe microscopies (SPMs) [95]. The SECM is based on the reaction that occurs on the surface of a microelectrode immersed in the electrolyte solution. The microelectrode is scanned in close proximity to the sample surface to characterize the topography and/or redox activity of the microelectrode/sample interface [96]. This latter feature is very important, because it allows SECM to gain information about reactions that take place in the solution space between the tip and the sample, as well as on those occurring on the surface of the scanned sample. The SECM has become a very powerful technique for probing a great variety of electrochemical reactions, due to the combination of its high spatial resolution and its electrochemical sensitivity. Its capability for the direct identification of chemical species with high lateral resolution is of great interest in the study of corrosion processes [97].

Prior to explaining the operation and the response of the SECM, it is necessary to understand the behaviour of the microelectrode inside the electrochemical cell. Let's consider the case where the microelectrode is immersed in a solution containing an electrolyte and a reducible species, *R*. When a potential sufficiently negative is applied to the microelectrode, the reduction of the specie occurs at the surface of the microelectrode:

$$R^{n+} + ne- \rightarrow O \tag{9.11}$$

If this reaction is kinetically controlled by the diffusion of *R* from the bulk of the solution to the electrode surface, the current decays due to the formation of a diffusion layer of *R* around the electrode, and attains a steady-state value rapidly given by:

$$i_{lim} = 4nFDca \tag{9.12}$$

where *F*, is the Faraday constant; *a*, the microelectrode radius; *D*, the diffusion coefficient of the reducible species; and, *c*, its concentration.

A steady-state current results from the constant flux of *R* to the electrode surface due to an expanding hemispherical diffusion layer around the microelectrode (Fig. 9.17).

The measurements at a SECM microelectrode are not affected by stirring or other convective effects. In SECM measurements, the proximity of the tip to the substrate is the perturbation that constitutes the microelectrode response [96]. If the microelectrode is brought to the vicinity of an insulating substrate, the steady-state current that flows through the tip, tends to be smaller than i_{lim} (Fig. 9.18). This is a result of the insulating substrate partially blocking the diffusion of A^{n+} towards the tip. The current at the tip becomes smaller when the tip is closer to the substrate, and tends to zero when the distance between tip and substrate, d, approaches zero. This effect is known as *negative feedback*. In contrast, if the tip is close to a conductive substrate at which the oxidation reaction of *A* can occur (Fig. 9.18). Then a flux of A^{n+} from the substrate to the tip occurs, in addition to some flux from the bulk solution towards the tip. This effect leads to an enhancement of the current at the tip, i_{tip}, which is higher than i_{lim}. This effect is known as *positive feedback*.

Then by using A^{n+} as electrochemical mediator, the nature of the substrate (conductive or insulating) can be established. The SECM technique operating in

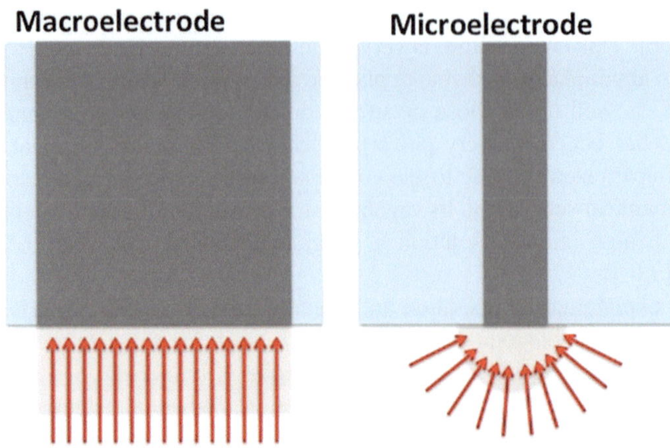

Fig. 9.17 Scheme of the geometry and diffusion field for (*left*) conventional macro-electrode, and (*right*) microelectrode. They present linear and hemispherical diffusion layer respectively

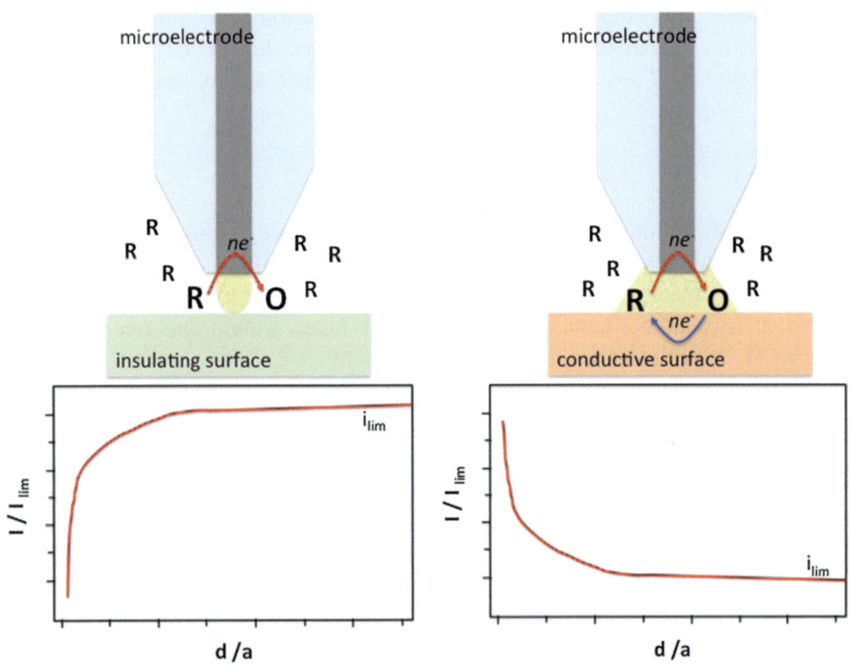

Fig. 9.18 Basic principle of SECM: (*left*) Negative feedback: The tip is placed near an insulating substrate which hinders the diffusion of species A^{n+}. (*right*) Positive feedback: The tip is located near a conductive substrate where the oxidation of species A occurs

this mode provides an excellent tool for the in situ study of film growth on various materials. Such is the case for the formation of a passive layer by corrosion inhibitors. Several works have been published in the study of organic inhibitors for the protection of copper [98, 99]. SECM was used to follow the transition of the copper surface from conducting to insulating as film growth proceeds. A scheme showing this effect is presented in Fig. 9.19. A sequence of approaching curves of the SECM microelectrode towards the metal (copper) is recorded. It represents the transition of the metal from active to passive with the time of immersion in the solution containing the inhibitor.

When studying with the SECM, organic corrosion inhibitors such as mercatobenzimidazole, benzotriazole or similar, it should be noted that often contamination of the microelectrode by the deposition of the inhibitor on it, may occur. Measurements should be performed with caution and regular checking of the state of the electrode is needed. There is no report so far of using the SECM for the evaluation of inorganic chromate-free inhibitors for the protection of aluminium and steel alloys.

Another way of operating the SECM is the generator-collector mode. In this case the tip current is used to monitor the flux of electroactive species generated or consumed by the metal substrate. The microelectrode is polarized at certain potentials where a redox reaction of one of the species takes places. This configuration has been applied to determine concentration profiles of species involved in the corrosion processes [100]. Iron, zinc and oxygen profiles have been monitored when corrosion is taking place on bare metals and defective coatings [101–103].

Recently this SECM operating mode has been applied for the evaluation of self-healing processes. In this case measurements are based on monitoring changes of dissolved oxygen concentration in the solution at the proximity of the substrate. The reduction of oxygen is commonly the cathodic reaction of corrosion processes in neutral/alkaline solutions. A consequence of this is the local depletion of

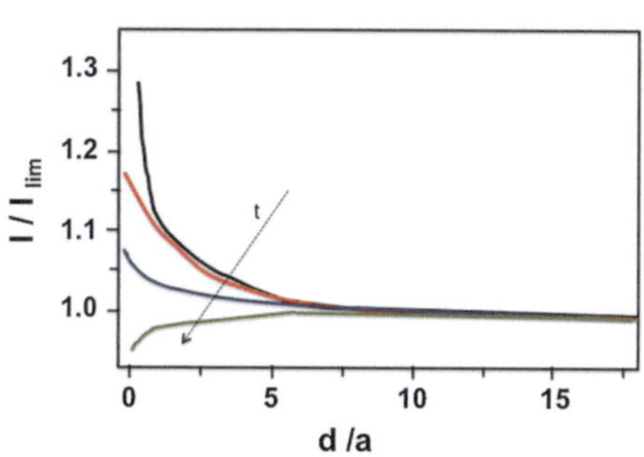

Fig. 9.19 Plot representing approaching curves of the microelectrode toward the metal surface at different immersion times. By the formation of a passive layer by the inhibitor action, a change of the response of the metal from conductive to insulating is detected. X axis: normalized distance (d/a = distance/electrode radius). Y axis: normalized current (I/I_{lim} = measured current/limiting current)

dissolved oxygen concentration in the solution near the cathode area (Fig. 9.20). This is an indirect way to follow the spontaneous corrosion processes occurring on a metal system and their depletion by the action of the healing process. This mode was used to evaluate the healing efficiency of polymer healing in a shape-memory type of coating for protection of an aluminium substrate [104]. Also this operation mode was used to study an autonomous healing system consisting of encapsulated healing-agent incorporated into organic coating for protection of the underlying metal substrate [105, 106].

The SECM oxygen sensing measurements are very interesting for corrosion research, however the results (profile concentration of oxygen) should be carefully discussed. While reducing oxygen at the SECM tip, two undesired processes will take place at the proximities of the microelectrode: consumption of dissolved oxygen and increase of pH by the generation of OH^- ions. These processes can have an impact on the corrosion processes taking place on the metal substrate, affecting the kinetics and the formation of corrosion products. The SECM measurement by itself can influence the natural corrosion mechanism under study. This undesired effect has been pinpointed by electrochemical simulation [107] and the analysis of the corrosion products [108]. The magnitude and impact of the undesired effect depends on: the proximity of the tip to the substrate and the duration of the experiments. When performing the experiments, short-time measurements and a

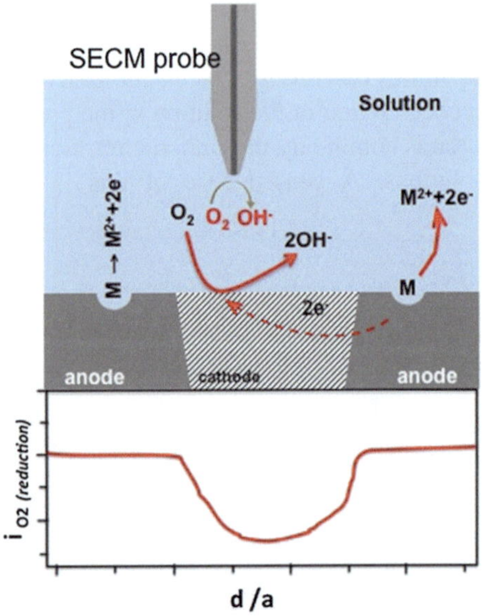

Fig. 9.20 (*top*) Diagram showing the anodic and cathodic reactions (oxygen reduction) on a corroding metal. The SECM microelectrode is polarized at a potential corresponding to the reduction potential of oxygen. (*bottom*) SECM line scan of the microelectrode across the metal surface. It shows the depletion of current when scanning over the cathodic region where less oxygen is available for the reduction reaction at the microelectrode

large tip-substrate distance is desired. The SECM oxygen sensing mode is a valuable method for *qualitative* indication of corrosion activity on the metal surface, but quantitative data should be carefully interpreted.

In general the main components of the scanning electrochemical microscope are:

- The tip movement and position controller, which is the SPM component.
- The electrochemical cell, which is formed by the tip, the counter electrode, the reference electrode and the substrate that, occasionally, can act as second working electrode.
- The bipotentiostat, which constitutes the SECM electrochemical setup together with the electrochemical cell. The bipotentiostat allows to polarize microelectrode and sample.
- Computer, interface and display system.

The most common microelectrode used for corrosion research is a platinum wire embedded in a glass capillary, resulting in a microdisc electrode. The diameter size of the electrode can vary from 1 to 25 μm.

SECM is a more versatile technique when compared to SVET or the micro-capillary cell since it provides higher resolution (wide range of microelectrode sizes) and chemical selectivity at the same time to monitor electrochemical processes occurring on the surface. When performing SECM measurements several facts needs to be considered:

- resolution of the measurement depends on the microelectrode dimensions and tip-sample distance,
- possible contamination or poisoning of the microelectrode. It is important to check the state of the tip after experiments,
- undesired influence of the SECM measurement on the system to be studied, The redox processes occurring on the microelectrode can affect the electrochemical processes taking place on the sample.
- select the adequate scan rate to assure limiting steady-state during measurement

9.5.5 *Local Electrochemical Impedance Spectroscopy (LEIS)*

SVET and SECM are based on the measurements of events or processes taking place between the probe and the specimen. A disadvantage of both these techniques is that this information may be disturbed by noise or limited response from the substrate, and the impossibility to obtain information of the processes underneath a coating. Local Electrochemical Impedance Spectroscopy (LEIS) appears an alternative where it is important to have direct information of the response of the substrate under different conditions such as during a healing event.

The LEIS technique is based on the hypothesis that the local impedance can be generated by measuring the AC-local current density in the vicinity of the working electrode in a three-electrode cell configuration. This was introduced by the Isaacs group in the 1990s [109]. Quantitative information was achieved by using a dual microelectrode for sensing the local AC-potential gradient, and therefore the local current by direct application of Ohm's law. The use of a multichannel frequency response analyzer also allows for the simultaneous measurement of global and local response of the specimens. The resolution of the technique depends on the size of the electrodes used in the bi-electrode probe to sense the local potential and the spacing between the electrodes. More detailed information about the fundaments of LEIS can be found in the recent review by Huang et al. [110].

The first applications of LEIS method were in the study of corrosion processes and focused on the investigation of localized delamination of coatings. As an example, Jorcin et al. in 2006 [32] mapped local electrochemical impedance which allowed them to observe and quantify the delamination process beneath the steel/epoxy-vinyl primer interface. An interesting observation was that the delaminated surface area measured by LEIS was significantly higher than that observed by visual examination of the systems. These results pointed out that LEIS is a powerful technique for visualizing, with high accuracy, the zones of coatings where adherence to the metal substrate has been lost.

The use of LEIS to evaluate the local efficiency of a healing mechanism against corrosion is becoming more common in recent years. Several works have demonstrated the capabilities of LEIS method for evaluation of corrosion inhibitor performance [111, 112]. Snihirova et al. [113] presented a new smart anticorrosion system for AA2024, based on the incorporation of encapsulated inhibitor into a coating. Controlled delivery was triggered by local changes of pH induced by the corrosion processes. Conventional EIS and local impedance spectroscopy were used to evaluate the efficiency of the systems as function of the active inhibitor employed and their incorporation into the coating.

The resolution of the LEIS technique depends on the size of the electrode/sensor and the distance between the electrode and the sample. The ultimate resolution achievable is constrained by the sensitivity of the potential measuring circuitry. Measurement sensitivity in the order of 1 nV limits the resolution to the micrometer range [110].

9.6 Conclusions

This chapter describes several global and local electrochemical techniques used for the evaluation and characterization of novel protection approaches based on extrinsic and intrinsic self-healing concepts.

Potential measurements, potentiodynamic experiments and electrochemical impedance spectroscopy (EIS) are the most commonly used methods used for corrosion and self-healing research. These global methods provide kinetic and mechanistic information of the self-healing processes. Apart from these traditional techniques some newer methods based on EIS have been proposed to obtain faster and more accurate information about the healing degrees and process such as the odd-random EIS and the AC/DC/AC protocol.

High-throughput techniques have also attracted considerable interest due to their powerful use for fast screening of a large number of samples. This is tremendously useful for evaluation of potential corrosion inhibitors of potential use in self-healing concepts.

In order to obtain local information about the evolution of damage and protected or healed damage local electrochemical techniques with spatial resolution have been proposed. Within this group micro-capillary cell, scanning vibrating electrode technique (SVET), selective ion-electrode technique (SIET), scanning electro-chemical microscope (SECM), and local electrochemical impedance spectroscopy (LEIS) are the most common. Even though these techniques have so far only been used to obtain pseudo-quantitative information about the healing degree they have the potential to be used for more quantitative studies of the healing processes themselves.

There is no doubt that electrochemical evaluation of self-healing processes is considered a key requirement to understand the healing mechanisms and to evaluate the corrosion performance of new developments. Nevertheless, these studies should always be supported by dedicated complementary surface, optical and/or physical chemical analysis.

References

1. V. Sauvant-Moynot, S. Gonzalez, J. Kittel, Self-healing coatings: an alternative route for anticorrosion protection. Prog. Org. Coat. **63**(3), 307–315 (2008)
2. S.R. White et al., Autonomic healing of polymer composites. Nature **409**(6822), 794–797 (2001)
3. D.G. Shchukin et al., Layer-by-layer assembled nanocontainers for self-healing corrosion protection. Adv. Mater. **18**(13), 1672–1678 (2006)
4. K. Aramaki, The inhibition effects of cation inhibitors on corrosion of zinc in aerated 0.5 M NaCl. Corros. Sci. **43**(8), 1573–1588 (2001)
5. S.H. Cho, S.R. White, P.V. Braun, Self-healing polymer coatings. Adv. Mater. **21**(6), 645–649 (2009)
6. J.E.B. Randles, K.W. Somerton, Kinetics of rapid electrode reactions. Part 3. - Electron exchange reactions. Trans. Faraday Soc. **48**, 937–950 (1952)
7. S.J. Garcia et al., Unravelling the corrosion inhibition mechanisms of bi-functional inhibitors by EIS and SEM-EDS. Corros. Sci. **69**, 346–358 (2013)
8. J.M. Hu, J.Q. Zhang, C.N. Cao, Determination of water uptake and diffusion of Cl- ion in epoxy primer on aluminum alloys in NaCl solution by electrochemical impedance spectroscopy. Prog. Org. Coat. **46**, 273–279 (2003)
9. V.B. Miskovic-Stankovic et al., Corrosion behaviour of epoxy coatings on modified aluminium surfaces. Bull. Electrochem. **18**(8), 343–348 (2002)
10. V.B. Mišković-stanković, D.M. Dražić, M.J. Teodorović, Electrolyte penetration through epoxy coatings electrodeposited on steel. Corros. Sci. **37**(2), 241–252 (1995)

11. V.B. Mišković-Stanković, M.R. Stanić, D.M. Dražić, Corrosion protection of aluminium by a cataphoretic epoxy coating. Prog. Org. Coat. **36**(1), 53–63 (1999)
12. J.M. McIntyre, H.Q. Pham, Electrochemical impedance spectroscopy; a tool for organic coatings optimizations. Prog. Org. Coat. **27**(1–4), 201–207 (1996)
13. F. Mansfeld, Models for the impedance behavior of protective coatings and cases of localized corrosion. Electrochim. Acta. **38**(14), 1891–1897 (1993)
14. I.M. Zin et al., The mode of action of chromate inhibitor in epoxy primer on galvanized steel. Prog. Org. Coat. **33**(3–4), 203–210 (1998)
15. M. Musiani et al., Constant-phase-element behavior caused by coupled resistivity and permittivity distributions in films. J. Electrochem. Soc. **158**(12), C424–C428 (2011)
16. B. Hirschorn et al., Constant-phase-element behavior caused by resistivity distributions in films: II Applications. J. Electrochem. Soc. **157**(12), C452–C457 (2010)
17. B. Hirschorn et al., Constant-phase-element behavior caused by resistivity distributions in films: I. theory. J. Electrochem. Soc. **157**(12), C458–C463 (2010)
18. S. Amand et al., Constant-phase-element behavior caused by inhomogeneous water uptake in anti-corrosion coatings. Electrochim. Acta. **87**, 693–700 (2013)
19. F. Mansfeld, M.W. Kendig, S. Tsai, Evaluation of corrosion behavior of coated metals with AC impedance measurements. Corrosion **38**(9), 478–485 (1982)
20. M. Tomkiewicz, B. Aurian-Blajeni, Impedance of composite materials. J. Electrochem. Soc. **135**(11), 2743–2747 (1988)
21. A. Perrotta et al., Analysis of nanoporosity in moisture permeation barrier layers by electrochemical impedance spectroscopy. ACS Appl. Mater. Interfaces **7**(29), 15968–15977 (2015)
22. C.H. Hsu, F. Mansfeld, Concernng the conversion of the constant phase element parameter Y 0 into a capacitance. Corrosion **57**(9), 747–748 (2001)
23. N.K. Mehta, M.N. Bogere, Environmental studies of smart/self-healing coating system for steel. Prog. Org. Coat. **64**(4), 419–428 (2009)
24. S.J. Garcia et al., Self-healing anticorrosive organic coating based on an encapsulated water reactive silyl ester: synthesis and proof of concept. Prog. Org. Coat. **70**(2–3), 142–149 (2011)
25. S.J. Garcia, X. Wu, S. Van Der Zwaag, A combined electrochemical impedance spectroscopy and x-ray-computed tomography study of the effect of a silyl ester on delamination and underfilm pit formation in a coated AA7050 sample. Corrosion **70**(5), 475–482 (2014)
26. T. Breugelmans et al., Odd random phase multisine electrochemical impedance spectroscopy to quantify a non-stationary behaviour: theory and validation by calculating an instantaneous impedance value. Electrochim. Acta. **76**, 375–382 (2012)
27. Y. Van Ingelgem et al., Advantages of odd random phase multisine electrochemical impedance measurements. Electroanalysis **21**(6), 730–739 (2009)
28. B.A. Boukamp, Linear Kronig-Kramers transform test for immittance data validation. J. Electrochem. Soc. **142**(6), 1885–1894 (1995)
29. B.A. Boukamp, Practical application of the Kramers-Kronig transformation on impedance measurements in solid state electrochemistry. Solid State Ionics **62**(1–2), 131–141 (1993)
30. J.B. Jorcin et al., Investigation of the self-healing properties of shape memory polyurethane coatings with the 'odd random phase multisine' electrochemical impedance spectroscopy. Electrochim. Acta. **55**(21), 6195–6203 (2010)
31. T. Breugelmans et al., Odd random phase multisine EIS for organic coating analysis. Prog. Org. Coat. **69**(2), 215–218 (2010)
32. J.B. Jorcin et al., Delaminated areas beneath organic coating: a local electrochemical impedance approach. Corros. Sci. **48**(7), 1779–1790 (2006)
33. P.L. Bonora, F. Deflorian, L. Fedrizzi, Electrochemical impedance spectroscopy as a tool for investigating underpaint corrosion. Electrochim. Acta. **41**(7–8), 1073–1082 (1996)
34. M. AbdolahZadeh, S. van der Zwaag, S.J. Garcia, Assessment of healed scratches in intrinsic healing coatings by AC/DC/AC accelerated electrochemical procedure. J. Coat. Sci. Technol. (2015) accepted

35. S.J. García et al., Evaluation of cure temperature effects in cataphoretic automotive primers by electrochemical techniques. Prog. Org. Coat. **60**(4), 303–311 (2007)
36. S.J. Garcia, J. Suay, Application of electrochemical techniques to study the effect on the anticorrosive properties of the addition of ytterbiurn and erbium triflates as catalysts on a powder epoxy network. Prog. Org. Coat. **57**(3), 273–281 (2006)
37. S.J. García, J. Suay, A comparative study between the results of different electrochemical techniques (EIS and AC/DC/AC): application to the optimisation of the cataphoretic and curing parameters of a primer for the automotive industry. Prog. Org. Coat. **59**(3), 251–258 (2007)
38. S.J. García, J. Suay, Optimization of deposition voltage of cataphoretic automotive primers assessed by EIS and AC/DC/AC. Prog. Org. Coat. **66**(3), 306–313 (2009)
39. T.A. Markley, M. Forsyth, A.E. Hughes, Corrosion protection of AA2024-T3 using rare earth diphenyl phosphates. Electrochim. Acta **52**(12), 4024–4031 (2007)
40. D. Ho et al., Cerium dibutylphosphate as a corrosion inhibitor for AA2024-T3 aluminum alloys. J. Electrochem. Soc. **153**(9), B392–B401 (2006)
41. S.J. García et al., The influence of pH on corrosion inhibitor selection for 2024-T3 aluminium alloy assessed by high-throughput multielectrode and potentiodynamic testing. Electrochim. Acta. **55**(7), 2457–2465 (2010)
42. Y.J. Tan, An experimental comparison of three wire beam electrode based methods for determining corrosion rates and patterns. Corros. Sci. **47**(7), 1653–1665 (2005)
43. Y.J. Tan, S. Bailey, B. Kinsella, Mapping non-uniform corrosion using the wire beam electrode method. II. Crevice corrosion and crevice corrosion exemption. Corros. Sci. **43**(10), 1919–1929 (2001)
44. Y.J. Tan et al., Mapping corrosion kinetics using the wire beam electrode in conjunction with electrochemical noise resistance measurements. J. Electrochem. Soc. **147**(2), 530–539 (2000)
45. S.R. Taylor, B.D. Chambers, in *Proceedings of the 4th International Symposium on Aluminium Surface Science and Technology, Beaune, France*, May (2006), Proceedings to be published in: ATB Metallurgie, 45 (1–4) (2006) 418. 2006
46. B.D. Chambers, S.R. Taylor, M.W. Kendig, Rapid discovery of corrosion inhibitors and synergistic combinations using high-throughput screening methods. Corrosion **61**(5), 480–489 (2005)
47. T.H. Muster et al., A rapid screening multi-electrode method for the evaluation of corrosion inhibitors. Electrochim. Acta. **54**(12), 3402–3411 (2009)
48. S.J. García, J.M.C. Mol, T.H. Muster, A.E. Hughes, J. Mardel, T. Miller, T. Markely, H. Terryn, J.H.W. de Wit, in *Green Inhibitors*, ed. by L. Fedrizzi. Advances in the Selection and use of Rare-Earth-Based Inhibitors for Self Healing Organic Coatings, Accepted for publication in Self-Healing Properties of New Surface Treatments. (EFC-Maney Publishing, 2011)
49. T.H. Muster et al., A review of high throughput and combinatorial electrochemistry. Electrochim. Acta. **56**(27), 9679–9699 (2011)
50. S.J. Garcia et al., Validation of a fast scanning technique for corrosion inhibitor selection: influence of cross-contamination on AA2024-T3. Surf. Interface Anal. **42**(4), 205–210 (2010)
51. B.D. Chambers, S.R. Taylor, The high throughput assessment of aluminium alloy corrosion using fluorometric methods. Part II - A combinatorial study of corrosion inhibitors and synergistic combinations. Corros. Sci. **49**(3), 1597–1609 (2007)
52. B.D. Chambers, S.R. Taylor, High-throughput assessment of inhibitor synergies on aluminum alloy 2024-T3 through measurement of surface copper enrichment. Corrosion **63**(3), 268–276 (2007)
53. P.A. White et al., High-throughput channel arrays for inhibitor testing: proof of concept for AA2024-T3. Corros. Sci. **51**(10), 2279–2290 (2009)
54. P.A. White et al., A new high-throughput method for corrosion testing. Corros. Sci. **58**, 327–331 (2012)

55. H. Böhni, T. Suter, F. Assi, Micro-electrochemical techniques for studies of localized processes on metal surfaces in the nanometer range. Surf. Coat. Technol. **130**(1), 80–86 (2000)
56. T. Suter, H. Böhni, Microelectrodes for studies of localized corrosion processes. Electrochim. Acta. **43**(19–20), 2843–2849 (1998)
57. M.M. Lohrengel, A. Moehring, M. Pilaski, Capillary-based droplet cells: limits and new aspects. Electrochim. Acta. **47**(1), 137–141 (2001)
58. T. Suter, R.C. Alkire, Microelectrochemical studies of pit initiation at single inclusions in Al 2024-T3. J. Electrochem. Soc. **148**(1), B36–B42 (2001)
59. F. Andreatta, H. Terryn, J.H.W. De Wit, Corrosion behaviour of different tempers of AA7075 aluminium alloy. Electrochim. Acta. **49**(17–18), 2851–2862 (2004)
60. F. Andreatta, H. Terryn, J.H.W. de Wit, Effect of solution heat treatment on galvanic coupling between intermetallics and matrix in AA7075-T6. Corros. Sci. **45**(8), 1733–1746 (2003)
61. J.O. Park, T. Suter, H. Böhni, Role of manganese sulfide inclusions on pit initiation of super austenitic stainless steels. Corrosion **59**(1), 59–67 (2003)
62. T. Suter, H. Böhni, A new microelectrochemical method to study pit initiation on stainless steels. Electrochim. Acta. **42**(20–22), 3275–3280 (1997)
63. V.S. Rao, H.S. Kwon, Corrosion Studies of Fe3Al – Fe3AlC intermetallics in 0.25 N H2SO4 using microelectrochemical method and SAES analysis. J. Electrochem. Soc. **154**(5), C255–C260 (2007)
64. N. Birbilis et al., Inhibition of AA2024-T3 on a phase-by-phase basis using an environmentally benign inhibitor, cerium dibutyl phosphate. Electrochem. Solid State Lett. **8**(11), C180–C183 (2005)
65. F. Andreatta et al., Localized corrosion inhibition by cerium species on clad AA2024 aluminium alloy investigated by means of electrochemical micro-cell. Corros. Sci. **65**, 376–386 (2012)
66. I. Recloux, Y. Gonzalez-Garcia, M-E. Druart, K. Khelifa, Ph. Dubois, J.M.C. Mol, M-G. Olivier, Active and passive protection of AA2024-T3 by an hybrid inhibitor doped mesoporous sol-gel and top coating system. Prog. Org. Coat. (2015) submitted
67. H.S. Isaacs, The measurement of the galvanic corrosion of soldered copper using the scanning vibrating electrode technique. Corros. Sci. **28**(6), 547–558 (1988)
68. H.S. Isaacs, G. Kissel, Surface preparation and pit propagation in stainless steels. J. Electrochem. Soc. **119**(12), 1628–1632 (1972)
69. F. Zou et al., in *Materials Science Forum*. Application of Scanning Vibrating Electrode Techniques to Study the Degradation of Coil-Coated Steel at Edges (1998), p. 83–92
70. H.S. Isaacs, Use of the scanning vibrating electrode technique for detecting defects in ion vapor-deposited aluminum on steel. Corrosion **43**(10), 594–598 (1987)
71. R.M. Souto et al., Investigating corrosion processes in the micrometric range: a SVET study of the galvanic corrosion of zinc coupled with iron. Corros. Sci. **49**(12), 4568–4580 (2007)
72. M.J. Franklin, D.C. White, H.S. Isaacs, A study of carbon steel corrosion inhibition by phosphate ions and by an organic buffer using a scanning vibrating electrode. Corros. Sci. **33** (2), 251–260 (1992)
73. A.C. Bastos, M.G. Ferreira, A.M. Simões, Corrosion inhibition by chromate and phosphate extracts for iron substrates studied by EIS and SVET. Corros. Sci. **48**(6), 1500–1512 (2006)
74. S. Kallip et al., A multi-electrode cell for high-throughput SVET screening of corrosion inhibitors. Corros. Sci. **52**(9), 3146–3149 (2010)
75. S. Kallip et al., Synergistic corrosion inhibition on galvanically coupled metallic materials. Electrochem. Commun. **20**(1), 101–104 (2012)
76. S.V. Lamaka et al., Nanoporous titania interlayer as reservoir of corrosion inhibitors for coatings with self-healing ability. Prog. Org. Coat. **58**(2–3), 127–135 (2007)
77. D. Raps et al., Electrochemical study of inhibitor-containing organic-inorganic hybrid coatings on AA2024. Corros. Sci. **51**(5), 1012–1021 (2009)

78. M.L. Zheludkevich, J. Tedim, M.G.S. Ferreira, “Smart” coatings for active corrosion protection based on multi-functional micro and nanocontainers. Electrochim. Acta. **82**, 314–323 (2012)
79. M.L. Zheludkevich et al., On the application of electrochemical impedance spectroscopy to study the self-healing properties of protective coatings. Electrochem. Commun. **9**(10), 2622–2628 (2007)
80. M. Marenzana et al., Bone as an ion exchange organ: evidence for instantaneous cell-dependent calcium efflux from bone not due to resorption. Bone **37**(4), 545–554 (2005)
81. J.G. Kunkel et al., in *Plant Electrophysiology: theory and Methods.* Use of Non-Invasive Ion-Selective Microelectrode Techniques for the Study of Plant Development (2006), p. 109–137
82. L. Shabala et al., Non-invasive microelectrode ion flux measurements to study adaptive responses of microorganisms to the environment. FEMS Microbiol. Rev. **30**(3), 472–486 (2006)
83. C.E.M. Berger, B.R. Horrocks, H.K. Datta, Application of ion-selective microelectrodes to the detection of calcium release during bone resorption. Electrochim. Acta. **44**(16), 2677–2683 (1999)
84. S.V. Lamaka et al., Monitoring local spatial distribution of Mg2+, pH and ionic currents. Electrochem. Commun. **10**(2), 259–262 (2008)
85. F. Sundfors et al., Microcavity based solid-contact ion-selective microelectrodes. Electroanalysis **18**(13–14), 1372–1378 (2006)
86. R.E. Gyurcsányi et al., Novel polypyrrole based all-solid-state potassium-selective microelectrodes. Analyst **123**(6), 1339–1344 (1998)
87. G. Gyetvai et al., Solid contact micropipette ion selective electrode for potentiometric SECM. Electroanalysis **19**(10), 1116–1122 (2007)
88. M.G. Taryba, S.V. Lamaka, Plasticizer-free solid-contact pH-selective microelectrode for visualization of local corrosion. J. Electroanal. Chem. **725**, 32–38 (2014)
89. E.A. Zdrachek et al., H+-selective microelectrodes with optimized measuring range for corrosion studies. Sens. Actuat. B: Chem. **207**(PB), 967–975 (2015)
90. P. Bühlmann, E. Pretsch, E. Bakker, Carrier-based ion-selective electrodes and bulk optodes. 2. Ionophores for potentiometric and optical sensors. Chem. Rev. **98**(4), 1593–1687 (1998)
91. J. Bobacka, A. Ivaska, A. Lewenstam, Potentiometric ion sensors. Chem. Rev. **108**(2), 329–351 (2008)
92. A.C. Bastos et al., Localised measurements of pH and dissolved oxygen as complements to SVET in the investigation of corrosion at defects in coated aluminum alloy. Electroanalysis **22**(17–18), 2009–2016 (2010)
93. M. Taryba et al., The combined use of scanning vibrating electrode technique and micro-potentiometry to assess the self-repair processes in defects on “smart” coatings applied to galvanized steel. Electrochim. Acta. **56**(12), 4475–4488 (2011)
94. O.V. Karavai et al., Localized electrochemical study of corrosion inhibition in microdefects on coated AZ31 magnesium alloy. Electrochim. Acta. **55**(19), 5401–5406 (2010)
95. A.J. Bard et al., Scanning electrochemical microscopy. Introduction and principles. Anal. Chem. **61**(2), 132–138 (1989)
96. A.J. Bard, M.V. Mirkin (ed.), *Scanning Electron Microscopy* (Marcel-Dekker, New York, 2001)
97. L. Niu et al., Application of scanning electrochemical microscope in the study of corrosion of metals. J. Mater. Sci. **44**(17), 4511–4521 (2009)
98. K. Mansikkamäki et al., Adsorption of benzotriazole on the surface of copper alloys studied by SECM and XPS. J. Electrochem. Soc. **153**(8), B311–B318 (2006)
99. J. Izquierdo et al., Scanning microelectrochemical characterization of the anti-corrosion performance of inhibitor films formed by 2-mercaptobenzimidazole on copper. Prog. Org. Coat. **74**(3), 526–533 (2012)

100. A.C. Bastos et al., Imaging concentration profiles of redox-active species in open-circuit corrosion processes with the scanning electrochemical microscope. Electrochem. Commun. **6** (11), 1212–1215 (2004)
101. A.M. Simoes et al., SVET and SECM imaging of cathodic protection of aluminium by a Mg-rich coating. Corros. Sci. **49**(10), 3838–3849 (2007)
102. A.C. Bastos et al., Application of the scanning electrochemical microscope to the examination of organic coatings on metallic substrates. Prog. Org. Coat. **53**(3), 177–182 (2005)
103. R.M. Souto et al., On the use of mercury-coated tips in scanning electrochemical microscopy to investigate galvanic corrosion processes involving zinc and iron. Corros. Sci. **55**, 401–406 (2012)
104. Y. González-García et al., SECM study of defect repair in self-healing polymer coatings on metals. Electrochem. Commun. **13**(2), 169–173 (2011)
105. Y. Gonzalez-Garcia et al., A combined redox-competition and negative-feedback SECM study of self-healing anticorrosive coatings. Electrochem. Commun. **13**(10), 1094–1097 (2011)
106. A. Pilbáth et al., SECM study of steel corrosion under scratched microencapsulated epoxy resin. Prog. Org. Coat. **75**(4), 480–485 (2012)
107. L.C. Abodi et al., Modeling localized aluminum alloy corrosion in chloride solutions under non-equilibrium conditions: steps toward understanding pitting initiation. Electrochim. Acta. **63**, 169–178 (2012)
108. M.G. Taryba et al., Novel use of a micro-optode in overcoming the negative influence of the amperometric micro-probe on localized corrosion measurements. Corros. Sci. **95**, 1–5 (2015)
109. R.S. Lillard, P.J. Moran, H.S. Isaacs, Novel method for generating quantitative local electrochemical impedance spectroscopy. J. Electrochem. Soc. **139**(4), 1007–1012 (1992)
110. V.M. Huang et al., Local electrochemical impedance spectroscopy: a review and some recent developments. Electrochim. Acta. **56**(23), 8048–8057 (2011)
111. R.M. Souto et al., in *ECS Transactions. Local Electrochemical Impedance Spectroscopy Investigation of Corrosion Inhibitor Films on Copper* (2012)
112. D. Snihirova, S.V. Lamaka, M.F. Montemor, "SMART" protective ability of water based epoxy coatings loaded with CaCO 3 microbeads impregnated with corrosion inhibitors applied on AA2024 substrates. Electrochim. Acta. **83**, 439–447 (2012)
113. D. Snihirova, S.V. Lamaka, M. Taryba, A.M. Salak, S. Kallip, M.L. Zheludkevich, M.G.S. Ferreira, M.F. Montemor, Hydroxyapatite microparticles as feedback-active reservoirs of corrosion inhibitors. ACS Appl. Mater. Interfaces **2**(11), 3011–3022 (2010)

Chapter 10
Physico-Chemical Characterisation of Protective Coatings and Self Healing Processes

Anthony E. Hughes, Sam Yang, Berkem Oezkaya, Ozlem Ozcan and Guido Grundmeier

10.1 Introduction

Achieving high adhesion strength and corrosion resistance at polymer/metal interfaces, even in hostile environments, is one of the most important challenges of interface engineering and is relevant to many technical applications. In most cases, the materials susceptible to corrosive attack are protected by means of polymeric films such as coatings, paints or adhesives [1–8]. All organic coatings applied to metals such as steel, zinc-coated steel, aluminum, and magnesium require strong interfacial adhesion forces that are able to withstand severe forming operations and long-time exposure to corrosive environments.

In certain cases, polymeric films such as coatings or adhesives are directly applied on the metal surface, which is covered by its native oxide. However, nowadays it is quite common that the metal is processed via chemical, electrochemical, physical, or mechanical pre-treatments to improve the interfacial stability [4, 9–20]. It is well known today that, the long-term stability of the polymer/metal interface in corrosive environments determines the performance of the composite interface and the necessary initial adhesion is rather easily achieved.

To understand the interaction mechanisms between polymers and metal/metal oxide surfaces, as well as to clarify how such interfaces degrade when exposed to corrosive environments, various in situ techniques have been developed in electrochemistry, spectroscopy and microscopy. Such in situ techniques have emerged as advantageous alternatives to the classical ex situ methods and enabled the

A.E. Hughes · S. Yang
CSIRO, Gate 5, Normanby Road, Clayton, VIC 3168, Australia

B. Oezkaya · O. Ozcan · G. Grundmeier (✉)
Technical and Macromolecular Chemistry, University of Paderborn, Warburgerstr. 100, 33098 Paderborn, Germany
e-mail: g.grundmeier@tc.uni-paderborn.de

A.E. Hughes et al. (eds.), *Active Protective Coatings*,
Springer Series in Materials Science 233, DOI 10.1007/978-94-017-7540-3_10

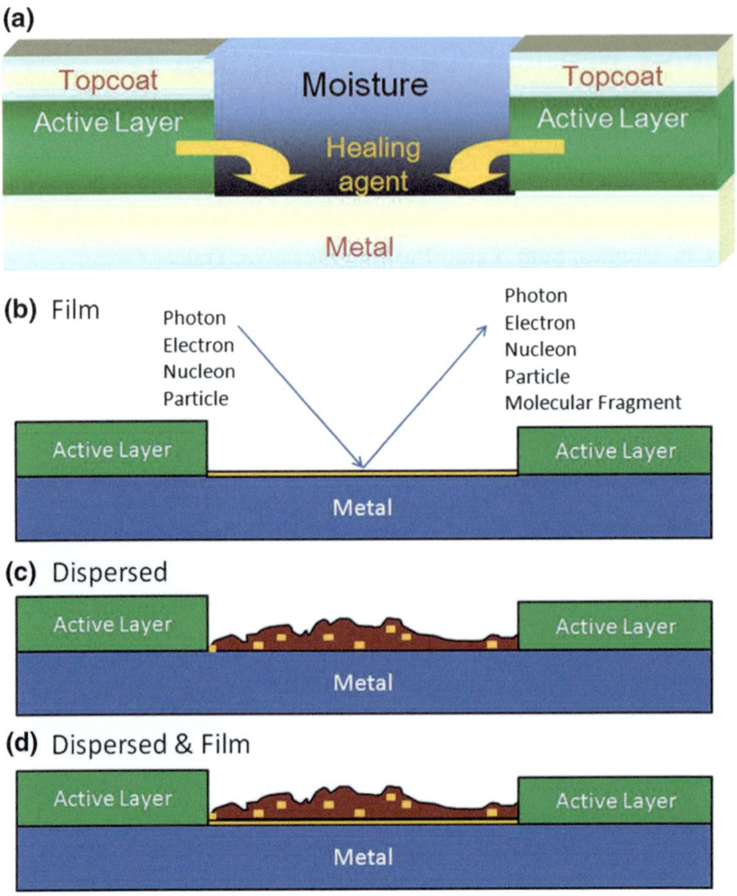

Fig. 10.1 Schematic for the detection of Self Healing at a Defect. **a** The defect is filled with moisture that acts as the trigger for the healing agent which, in this case is likely to be an inhibitor. **b** There are numerous techniques that can be used to characterise the reaction in the defect but the choice will depend on the nature of the healing process. For film formation, self healing can be detected by surface sensitive techniques. **c** If a thick product is formed, but the healing agent is dispersed, then a technique that has a good sensitivity for the healing agent is desirable. **d** If a thin film is formed, but the healing agent is also dispersed in an overlying layer then a profiling technique is desirable

investigation of formation of interfaces, as well as their degradation as the processes are taking place in the relevant environment [3, 15, 21–35].

Self healing coatings expand the challenges regarding interface design and characterisation, since for the development of self-healing coatings not only is the coating and interface analysis of importance but also the characterisation of the formation and healing of defects. To determine whether self healing has occurred the experiment demonstrating self healing needs to be designed with the characterisation technique in mind. With respect to characterisation in a general context, consider Fig. 10.1a displaying a self healing action in a defect. In this instance the defect is filled with moisture that acts as the trigger for the healing agent which, in this case, is likely to be either an inhibitor [36–38] or a film forming agent [39–43]. There are numerous techniques that can be used to study the reaction in the defect with some of the exciting radiations listed in Fig. 10.1b including photons (infrared, ultraviolet, X-ray, optical), electrons, nucleons (protons and neutrons) and particles/ions (α-particles, ion beam sources (see below under SIMS)). Emitted radiation can include photons, electrons, nucleons, particles and molecular fragments. Characterisation of the reaction in the defect will depend on the nature of the healing process. For film formation, self healing can be detected by surface sensitive techniques such as XPS, Static SIMS, TOF-SIMS, vibrational spectroscopies (FTIR, Raman) and nuclear techniques that use either protons or α-particles as exciting radiation such Rutherford Backscattering Spectroscopy (RBS) or Particle Induced X-ray Emission (PIXE) all have good sensitivity for thin films. However, if a thick product is formed and the healing agent is dispersed within the product at low levels Fig. 10.1c; this may require a combination of depth profiling and a technique that is sensitive to the healing agent. If a thin film is formed, but the healing agent is also dispersed in an overlying layer then again a profiling technique is desirable with high sensitivity to the healing agent. While the approach outlined here sounds straight forward there is not always a technique or combination of techniques, which will be able to identify the healing agent. One final point to make about characterization of self healing is that it needs to be performed in combination with a technique that demonstrates recovery (see Chap. 9).

In the following sections, suitable methods for the analysis of the protective coatings and self healing processes will be summarized with a special focus on spectroscopic, microscopic and tomographic methods.

10.2 Tomography and Microscopy

10.2.1 Tomographical Methods

The word tomography is derived from the Greek for 'cut' and thus tomographic methods allow us to characterize sample in three dimensions by sectioning the object either physically in the case of destructive methods such as micro-toming and focused ion beam milling, or virtually by using non-destructive techniques such as X-ray tomography.

The study of three dimensional structures within materials and changes in those structures is of growing importance in modern materials science. The key materials science challenges in this area are (i) probing multiple and smaller length scales, (ii) revealing chemistry and (iii) revealing changes to structures as a result of processing or other external influences such as degradation mechanisms.

In the area of active protective coatings the key areas are alloy microstructures and related corrosion processes, and the protective coating, including the distribution of inhibitor and other inorganic phases and changes to it in response to damage that invokes a self healing response. The study of these effects requires information on a range of different length and time scales, generally meaning that a range of probing techniques might be required to give a complete overview of the structure and processes within these types of systems.

As we will see later in this section there are many probes that can be used for computed tomography (CT) studies of materials; some are non-destructive methods for characterization whereas others are destructive. X-ray techniques based on absorption or phase contrast are certainly one class of emerging techniques in materials science for characterization, particularly with three dimensional imaging of the internal structures of materials, but electron beam approaches are also being developed. Much of this work is based on developments in three dimensional scanning of biological systems, particularly with a focus on medical applications. In materials science, the focus has been on improving the resolution of the techniques as well as on obtaining chemical information.

10.2.1.1 Non-destructive Tomographical Methods

Non-destructive approaches to tomography all rely on imaging the sample in transmission using penetrating radiation such as X-rays, electrons or light, though X-ray computed tomography is by far most widely used method. Tomographic data collection consists of acquiring a series of transmission images of the sample as it rotates through a range of angles. In X-ray tomography the angular range is typically 180 or 360° on a (usually vertical) rotation axis.[1] In electron tomography in a TEM it is typically 140° (−70° to +70°) depending on the tilt range of the TEM sample holder. Computational methods such as filtered back projection, or the Feldkamp, Davis, and Kress (FDK) [44] algorithm are then used to transform these views into a three-dimensional reconstruction of the sample. This can then be digitally sectioned and segmented to probe the internal structure of the sample.

X-rays are the probe of choice when it comes to the application of tomography to materials. This is largely because the X-ray spectrum covers a broad range of high energy photons with good penetration in many materials. Standard absorption contrast X-ray CT relies on the attenuation of the X-rays propagating through a material sample to form an image. The X-rays are attenuated on passing through a sample according to Beer's law,

$$I/I_0 = e^{-\mu t}$$

where I/I_0 is the ratio of transmitted to incident intensity for an X-ray passing through a thickness t of material with X-ray linear attenuation coefficient μ. The tomographic reconstruction of absorption-contrast CT data produces a three-dimensional

[1]This is the case for non-medical CT, however medical and small animal CT scanners rotate the X-ray source and detector around the stationary person or animal to achieve the same result.

microscopic map of the linear attenuation coefficient in the sample, however this is often insufficient to determine compositional distributions within heterogeneous materials as different materials may have similar attenuation coefficients.

Phase-contrast imaging utilizes the refraction of X-rays. It is based on the principle that different materials produce different shifts in the phase of the X-rays passing through them. It is more sensitive to edge effects and is thus suitable for the determination of structure gradients, particularly where there is a well-defined boundary between different regions in a material. X-rays passing through an object of varying refractive index will have their propagation direction modified by the sample. Unlike visible light this effect is small and subtle in the case of X-rays. Nonetheless, at boundaries in the sample there will be sharp changes in the propagation direction of adjacent X-rays passing through either side of the boundary. With a micro-focus or synchrotron X-ray source this refractive effect can easily be transformed into intensity changes by leaving a gap between the sample and detector. The propagation of these deflected X-rays from sample to detector results in interference and produces characteristic fringes in the X-ray image, enhancing the visibility of high gradient regions such as edges, voids, cracks and fine features of the sample. This form of X-ray imaging is known as in-line phase-contrast and is the simplest X-ray imaging mode which can make use of X-ray refraction.

It should be noted that, in X-ray CT, both amplitude attenuation and phase contrast effects exist at the same time. It is often the case that careful considerations are made during projection image acquisitions and CT reconstruction to reduce the effect of one or the other.

Sample requirements

In order to successfully carry out X-ray tomography on a sample the thickest or most strongly absorbing part of the sample should not completely attenuate the X-rays passing through it. This means that a very absorbing sample may require harder, more penetrating X-rays to image it, or else a small subsection of the sample will be required. Conversely the X-rays should have sufficiently strong interaction with the sample to form good contrast in the resulting images. In the case of absorption contrast this means that the sample should absorb a significant fraction of the X-rays passing through it. For phase-contrast imaging absorption is not required but the X-ray refraction should be sufficient to result in good fringes in the image corresponding to surfaces and boundaries within the sample. This will depend on the X-ray energy and the imaging geometry and detector resolution. Softer X-rays will generate stronger fringes, and the detector resolution and magnification of the image should be sufficient that the fringes are well resolved by the detector.

Types of X-ray sources

Two main classes of X-ray sources are widely used for X-ray tomography. Lab-based sources and synchrotron sources. Lab-based sources generate X-rays by focusing a multi-keV electron beam onto a target, usually a solid metal such as tungsten or molybdenum, though some new systems use a liquid metal jet as the target [45]. The X-rays produced by lab sources comprise broad spectrum bremsstrahlung radiation with some sharp spectral peaks corresponding to characteristic X-ray emission lines of the target material. The maximum photon energy of X-rays generated has an upper limit set by the electron energy which is in turn determined by the accelerating voltage setting. The peak energy in the X-ray spectrum will typically be around half this value. Accelerating voltages for a typical lab source go up to around 200 kV and are used to produce X-rays in the 10–100 keV range, though in some special industrial CT equipment, the accelerating voltage can be as high as multiple MV.

Lab sources used for micro-CT are typically micro focus sources, in which the electron beam is focused into a tight spot of a few microns in size on the target to produce a quasi-point source of X-rays (Fig. 10.2). This enables images to be acquired in a magnified geometry with the sample relatively closer to the source and further from the detector without resulting in a blurred image. This magnifies the image of the sample on the detector so that the imaging is not limited by the detector resolution but is instead limited to the source size. A microfocus source is also suitable for inline phase-contrast imaging if the sample-detector distance is sufficient. A typical microfocus source system will be capable of tomographic scans with a resolution of ~5 μm, and more specialised systems can reach submicron resolution.

The majority of micro-CT characterisation is done using lab sources, however they do have a couple of drawbacks. Using microfocus sources and achieving higher resolutions results in longer data collection times of up to several hours. Secondly the polychromatic spectrum can result in beam-hardening where the softer

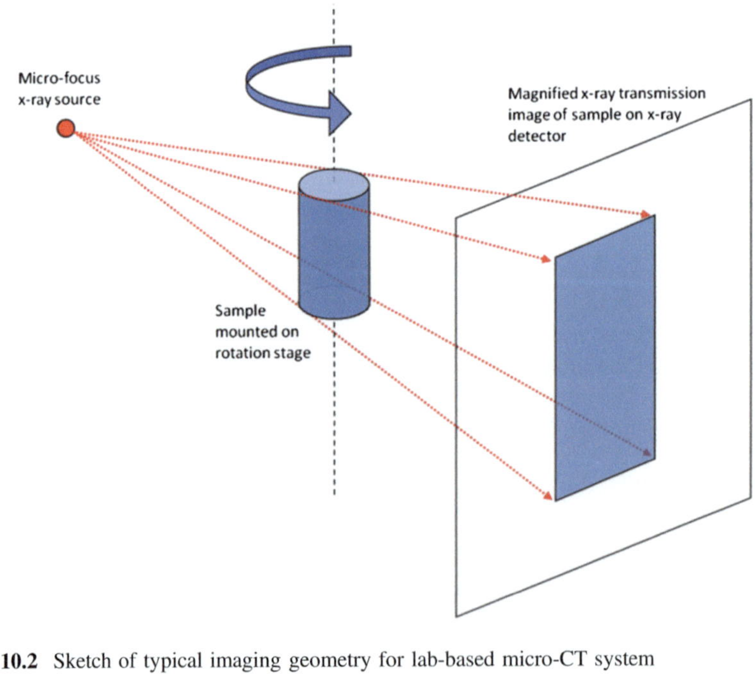

Fig. 10.2 Sketch of typical imaging geometry for lab-based micro-CT system

components of the spectrum are much more strongly attenuated than the hard (high energy) components. This results in reconstruction artefacts but these can be ameliorated by using filters to remove the softest X-rays from the spectrum, or by doing a beam hardening correction after data collection. In some special cases with soft X-ray systems it is also possible to get quasi-monochromatic spectra from a lab source using K emission lines [46].

Synchrotron sources are highly specialised X-ray sources available at synchrotron facilities rather than in the lab. However, the special qualities of synchrotron radiation, particularly the very high-power monochromatic X-ray beam, means that synchrotron micro-CT is ideal for certain types of experiments that cannot be performed in the lab.

The synchrotron is based on a particle accelerator which accelerates electrons to the GeV energy range and then injects them into a circular storage ring. The electrons travelling round the storage ring have their trajectories guided and modified by bending magnets and insertion devices. As the electrons are accelerated into curved trajectories by these devices they emit a powerful, highly focused beam of radiation in the forward direction. The resulting X-ray beams come off at a tangent to the main ring and are collected in a series of beamlines set around the storage ring, where they are used for a variety of experiments including X-ray imaging and tomography.

Thus X-rays have a broad polychromatic spectrum with a far higher intensity than a laboratory source. The beam is usually filtered using a monochromator to produce a monochromatic X-ray beam with sufficient intensity for rapid image acquisition. The photon energy of the X-ray beam can be adjusted conveniently in a wide energy range, typically in 8–60 keV. A monochromatic beam avoids the beam-hardening artefacts that can occur in tomography with polychromatic X-rays. Tomographic scans can also be performed in the same sample at different energies which can give more information on sample composition by exploiting the variation in attenuation coefficients with energy, which vary for different materials, particularly near absorption edges of their constituent elements. The high intensity of the X-ray beam also makes synchrotron sources very suitable for time-resolved, or '4D' tomography, enabling the observation of dynamic processes in three dimensions. Typical tomographic scan times at a synchrotron are of the order of a few minutes, but sub-second speed have been achieved for high-speed studies.

At a synchrotron the X-ray beam is approximately parallel so there is no magnification of the image as seen in a typical lab-system. Consequently the resolution is determined by the detector with a best resolution of around half a micron. Some very specialised beamlines using X-ray optics and softer X-rays are capable of resolution down to tens of nm, however, these are limited to very small samples and low density materials.

Practicalities of data collection

X-ray tomographic data is typically collected using a CCD or CMOS based detector as the imaging device, optically coupled to an X-ray sensitive phosphor or scintillator via optic fibres or lenses and mirrors. The sample is mounted on a high-precision rotation stage, which normally rotates around a vertical axis. A sequence of projection images are acquired with an equal angle increment for a total sample rotation of 180 or 360°. For adequate reconstruction of the 3D map of the sample, the maximum linear displacement at any part of the sample should not be greater than the detector size. Consequently, for a sample with a maximum horizontal projection size of n pixels, the number of required projections should be as high as

$$N = \frac{\pi}{2} n$$

In practice, acceptable reconstructions can be achieved with

$$N = \frac{\pi}{6} n$$

projections.

In addition to acquiring sample projection images, *flat-field* and *dark-field* images are acquired to correct for detector and illumination artifacts. Flat-field images are acquired with the X-ray source on and the sample out of the field-of-view. They are used to compensate for detector and detector-ray illumination inhomogeneity. The latter is particularly important with synchrotron sources where flat-field images are often acquired at intervals during the image projection, by moving the sample out of the field of view. This is required when there are time-varying illumination or detector inhomogeneities during the data collection. Dark-field images are acquired with no X-ray illumination and are used to compensate for the 'dark signal' that accumulates in the detector in the absence of illumination. Some detectors, particularly those with optic-fibre coupling between the detector and X-ray scintillator/phosphor additionally require a geometric correction to account for distortions in the image introduced by the optical coupling.

Data Processing

After correction for the various imaging artefacts the projection data is transformed into a three-dimensional digital reconstruction of the sample using one of a number of tomographic reconstruction algorithms. Filtered-back-projection is a typical algorithm for parallel-beam (e.g. synchrotron) geometry data, and the FDK algorithm for a cone-beam, magnifying geometry as found on a lab system, though there are many alternatives [47]. Tomographic reconstruction is computationally intensive but continuing developments in computer hardware and parallelised algorithms have enabled reconstruction of tomographic data on standalone multi-core computers in reasonable time-frames of a few minutes to an hour depending on the dataset size. Phase-contrast data can be processed in much the same way as conventional absorption contrast data with the addition of a phase-retrieval step [48]. Phase retrieval is a specialised low-pass filter that precisely compensates for the edge-enhancement of phase-contrast images (Fig. 10.3). This can be carried out prior or during tomographic reconstruction depending on the type of data and algorithm

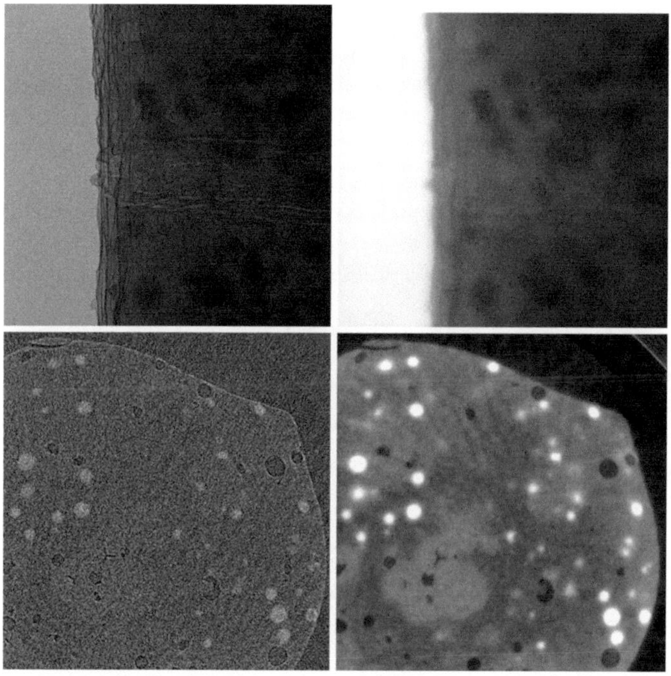

Fig. 10.3 *Top row*—high magnification projection image of a self-healing polymer showing phase-contrast edge-enhancement (*left*) and the same image after phase-retrieval (*right*). *Bottom row*—reconstructed slices from a noisy, lower-magnification dataset of the same sample, without phase retrieval (*left*) and with phase-retrieval (*right*)

used. This process leads to a tomographic reconstruction that resembles that of a conventional absorption CT scan, but with much improved signal-to-noise, particularly for low density samples which weakly absorb X-rays.

10.2.1.2 Transmission Electron Microscopy (TEM)

Of course, tomographic reconstruction can be performed with any dataset where a series of views (section or transmission) can be collected and features within the dataset can be uniquely identified. Thus electron transmission in a TEM as a function of rotation can also be used to collect datasets for tomographic reconstruction. As with the X-ray approaches electron transmission requires that the densest part of the sample be transparent. The electron transparency will be dependent on beam energy and sample thickness and density. With the TEM-based approach the resolution can be much higher that of X-ray tomography, but the field of view and thickness of the sample that can be imaged is also much smaller.

10.2.1.3 Destructive Tomographical Methods—Microtomy

While X-ray tomography in its various forms provides a non-destructive approach to the study of the internal structure of materials, there are many instances where it cannot be used for a variety of reasons. In some materials the absorption path length may be much shorter than the microstructure of interest, in other instances it may be that the features of interest are small compared to the volume of material that is required to be examined, i.e. there may not be enough sensitivity. Sample sensitivity to X-ray doses may also be a problem. Consequently there may be a need for other types of tomography to be used.

Microtomy has long been used in biology and materials science to study sections of materials for the investigation of internal structure. More recently electron microscopes have become available where multiple slices can be made using a diamond knife whilst collecting images in between each slice. This technique is called Serial Block Face Scanning Electron Microscopy (SBFSEM). The key issue here is to find features in the image that have a unique signature. In corrosion and coating studies, the features of interest are the distribution of microstructural features of the underlying metal and well as corrosion attack of these features, distribution of inhibitor and other phases within the protective paint system and transport of inhibitors to attack sites and the formation of protective layers or compounds.

To investigate some of these targets for the application of tomography, aluminium alloy 2024-T3 (AA2024-T3) is used here by way of example. AA2024-T3 is a high strength aluminium alloy used in aircraft manufacture. It has alloying additions which form hardening precipitates at the submicron level. However, the alloying additions also form larger particles (typically a few microns to tens of microns) with themselves and residual impurities [49–51]. These IM particles are one source of corrosion in these alloys so their study, particularly their spatial relationship to corrosion sites is of interest. The IM particles have several different compositions, but for the purposes of demonstration here they can be defined as AlCuFeMnMgSi particles, S-phase (Al_2CuMg) and θ-Phase (Al_2Cu). (Within the category that contains AlCuFeMnMgSi particles are a number of distinct compositions that will not be discussed further here).

Figure 10.4 shows a phase map obtained using electron microprobe analysis as described elsewhere [49]. The three phases, AlCuFeMnMgSi, S-phase and θ-Phase, are clearly identified in red, blue and green respectively. These phases have different levels of Cu with the lowest for AlCuFeMnMgSi particles, then at higher Cu levels there are the S-phase particles and θ-phase has the highest levels. This means that in backscatter mode they have different contrast levels which can be used to uniquely identify each particle type. This is demonstrated in Fig. 10.5. For Fig. 10.5a, the brightest greyscales are those for the θ-phase (see the red rectangle in the inset) which has the highest levels of Cu and therefore the largest backscatter electron signal. Figure 10.5b includes the S-phase in addition to the θ-phase. S-phase has less Cu than θ-phase and thus a lower backscatter signal on the greyscale which is reflected Fig. 10.5b. Figure 10.5c includes the AlCuFeMnMgSi particles which have lower Cu levels than either s-phase or θ-phase. Finally, Fig. 10.5d shows the addition

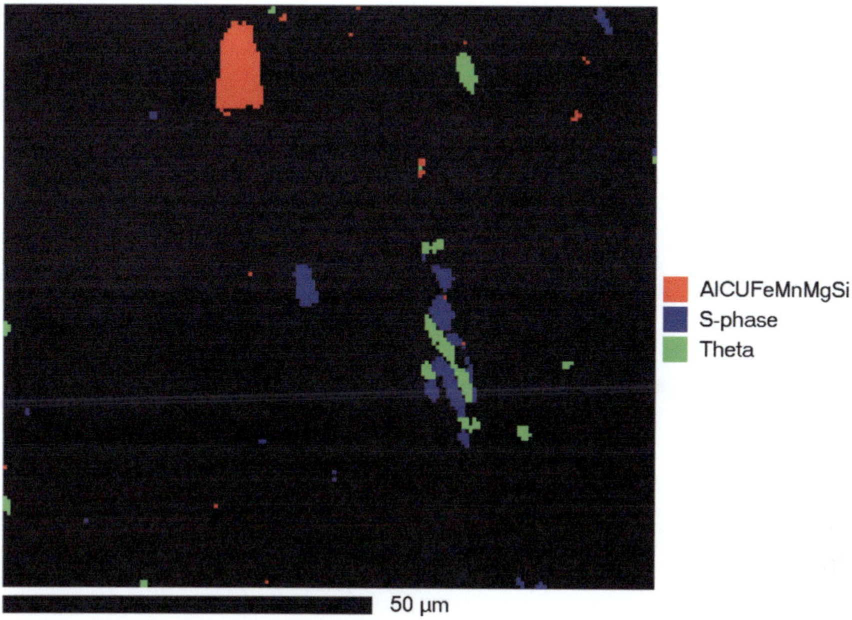

Fig. 10.4 Phase composition map obtained from microprobe analysis of a region of AA2024-T3 showing the distribution of three phases. Details are described in the text

of all the other greyscale pixels which represent the matrix. So, windows can be set for each intermetallic particle type and a three dimensional picture can be constructed based on the difference in greyscale.

10.2.1.4 Tomographical Reconstructions

As mentioned above tomographic reconstructions can be performed from many different types of datasets including from elemental maps collected in an SEM. Reports are now emerging on studies of pitting corrosion and intergrnaular attack [52–55]. Figure 10.6 shows a reconstruction from Al elemental maps of a region of attack beneath filiform corrosion on AA2024-T3 that has been coated with a clear polyurethane topcoat. An inset of a typical Al elemental map is included in Fig. 10.6 for example. The thick surface band represents the corrosion product within the filament but on the surface of the alloy. The intergranular attack is a region where the intergranular network has been preferentially attacked and is a common form of corrosion in high strength aluminium alloys. In this instance the intergranular attack region is highlighted because at the time of examination the filiform was dehydrated and any electrolyte solution in the grain boundary attack region had formed a solid product which "decorated" the attacked region. For the intermetallic particles, they have a lower concentration to the alloy matrix and therefore a lower intensity in the Al map.

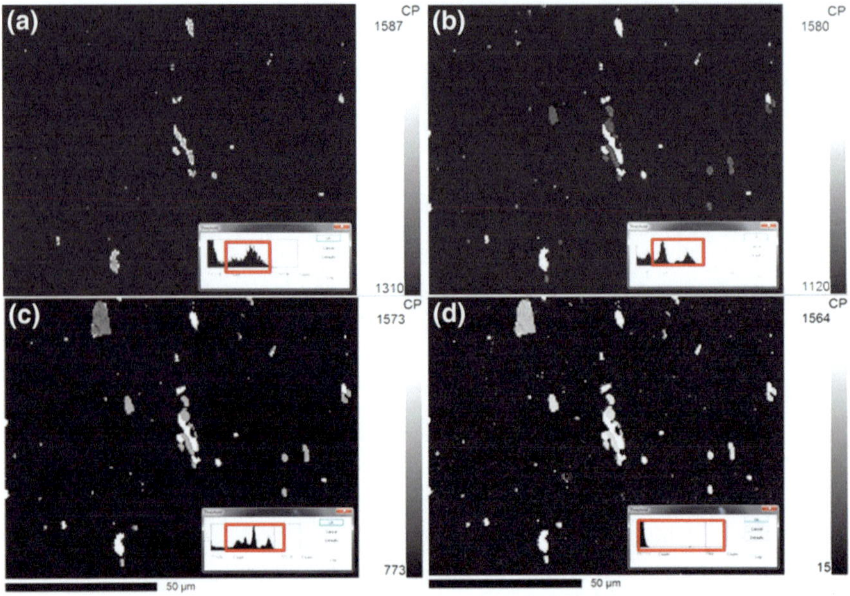

Fig. 10.5 Backscatter images of the surface of polished AA2024-T3 showing various intermetallic particles. Tomographic reconstructions of intermetallic phases can be made by taking a series of images and using diamond microtomy in between each image. The "chemical" information is contained in the intensity of the backscatter image which, in this case reflects the amount of Cu in the intermetallic particles (see text). **a** θ-Phase. **b** θ and S-phase. **c** AlCuFeMnMgSi and θ and S-phase. **d** All backscattered intensity

Fig. 10.6 Reconstruction from Al elemental maps of an attack site in AA2024-T3 beneath filiform corrosion (A. M. Glenn and A.E. Hughes, unpublished)

10.2.1.5 4-D Tomography

The term 4D tomography refers to the use of tomographic techniques to study time dependant changes to materials. It comes from the ballistics studies where there is a need to obtain tomographic data at the μ-second level for the study of impact on armour. This is leading to new developments with experimental configurations and high intensity and energy sources [56, 57].

Of course high speed tomography is not required for the types of issues related to coating applications where the fastest events that are likely to be probed using tomographic techniques are a million to a hundred million times slower, i.e. seconds to minutes. Nevertheless, even at these longer time scales there are still issues with obtaining information on appropriate timescales. The major areas where tomography is perhaps the only type of techniques that provides a complete insight into physical and chemical changes with time is with three dimensional structures, particularly with the *relationship* between structures such as the connectivity. These structures include the distribution of intermetallic compounds and other microstructure within alloys, corrosion, particularly the extent intergranular attack and particles within coatings such as inhibitor particles.

As an example of 4-D tomography, we look at the dissolution of $SrCrO_4$ inhibitor particles in an epoxy-based primer. X-ray tomography of these types of systems can be fraught with problems of sample stability, particularly if dehydration occurs during tomography where sample movement makes it difficult to align images after the data has been collected.

Nevertheless some very interesting progress has been made in studies on $SrCrO_4$—inhibited epoxy. For example, tomographically reconstructed sections of a slither of an epoxy-based primer containing $SrCrO_4$ inhibitor particles are presented in Fig. 10.7. The purpose of this study was to use tomography to determine the degree of connectivity between these inhibitor particles and to see how that changed upon leaching of the inhibitor from the primer.

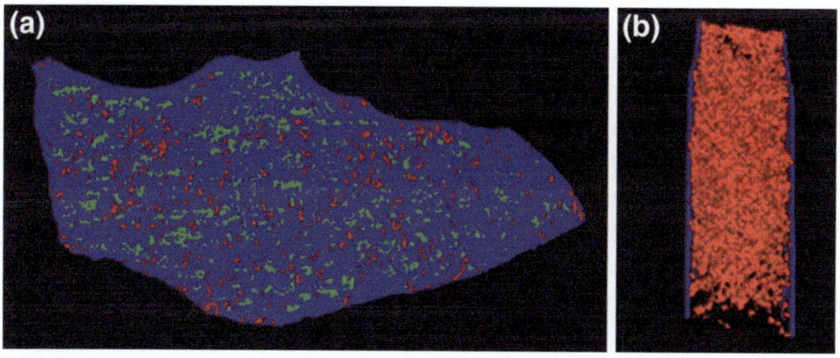

Fig. 10.7 Reconstructed sections from X-ray tomography of a $SrCrO_4$ inhibited epoxy-based primer. **a** reconstructed end slice showing the $SrCrO_4$ particles (*red*), epoxy matrix (*blue*) and a low density part of the epoxy (*green*). **b** "Sandwich" showing the distribution of only the $SrCrO_4$ particles between two end slices which have all components highlighted

Table 10.1 Statistics for clustered $SrCrO_4$ particles prior to and after leaching

	Before	After
Slab		
Number of clusters	2,725	3,365
Largest	185,393	21,438
Smallest*	3	3
Mean (± SD)	223 ± 724	190 ± 419
Smaller volume		
Number of clusters	634	804
Largest	55,525	10,835
Mean (± SD)	187 ± 397	120 ± 203

The density and distribution of the $SrCrO_4$ inhibitor particles (red) in the full reconstructed section is similar to that observed in SEM studies of section of the same primer system [58]. The full slice also has the epoxy matrix in blue as well as another feature thought to be lower density epoxy in green. Figure 10.7a is one end of a sandwich section of the inhibited primer displayed in Fig. 10.7b which shows a dense concentration of inhibitor particles between the two end slices of the section. In this view the inhibitor particles appear to densely populate the epoxy even though they only constitute 10 wt% of the sample. However, this image provides no information on the level of connectivity between particles or the nature of void formation which might result from inhibitor dissolution. This later point (void formation) is where 4D tomography becomes important.

There are 2,725 clusters in the sandwich with the largest cluster having over 185,000 $SrCrO_4$ voxels in it (Table 10.1). These clusters include all the inhibitor particles depicted in red in Fig. 10.7b. It should be pointed out that the clusters here represent both the voxels and the inhibitor particles since the clusters represent all the voxels that contain $SrCrO_4$. The presence of separate clusters in this material may explain why there is a considerable variation in the depth of the depletion front observed in SEM studies of this system and reported above for RBS studies of inhibitor depletion from the cut edge of a slot. The presence of isolated clusters suggests the depletion depth may only reflect the size of individual and isolated clusters of $SrCrO_4$ particles with the largest clusters representing the deepest penetration into the film and the smallest the shallowest depletion depth.

This sample was then immersed in 0.1 M NaCl solution for 20 min at room temperature, dried and re-examined. It was obvious from the absorption contrast that some of the strontium chromate has been dissolved. When the clustering was performed on the same slice and it was evident that the largest clusters were much smaller than those observed prior to leaching. The statistics for the clusters are presented in Table 10.1.[2] The number of clusters increased as a result of dissolution

[2]The data presented here is from the same study as reported elsewhere, however, the slice thickness is narrower than the previous report. For this reason the largest cluster size reported here is much smaller as a result of the largest cluster dimension being larger than the slice dimension.

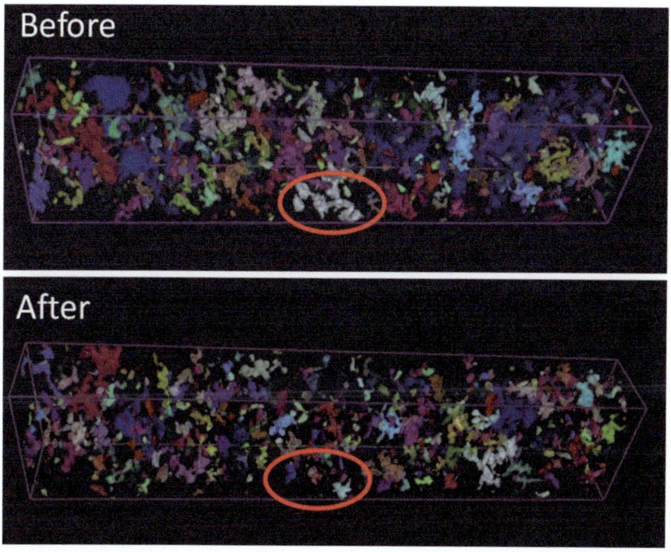

Fig. 10.8 Magnification of a region within the sandwich presented in previous sections. It is evident that the amount of $SrCrO_4$ has been reduced after exposure to 0.1 M NaCl. The region highlighted in a red ellipse shows where one cluster after leaching was been divided into more than three clusters

$SrCrO_4$ particles connecting different parts of the largest cluster size was significantly smaller at around 22,000 voxels compared to prior exposure where the largest cluster contained 185,000 voxels. Figure 10.8 shows a region from within the sandwich which has been magnified. It is evident that the amount of $SrCrO_4$ was reduced after exposure to 0.1 M NaCl. The voids created by the dissolution of the $SrCrO_4$ are not shown. The region, highlighted by a red ellipse, shows where one cluster, after leaching, was divided into more than three clusters as a result of the dissolution of some intervening particles.

10.2.1.6 X-ray K-edge Subtraction Tomography

X-ray K-edge subtraction tomography is a technique which can be used to selectively identify the 3D distribution of a particular element within a sample. This technique requires a monochromatic, tunable source of X-rays and is therefore largely limited to synchrotron sources, or lab systems with highly specialized detectors.

K-edge subtraction makes use of X-ray absorption edges which occur at specific energies characteristic to a given chemical element as shown in Fig. 10.9. X-rays with energies just above the absorption edge are absorbed much more strongly than those below it. Two X-ray tomographic datasets are acquired at closely spaced X-ray energies just above and below the absorption edge for the element of interest.

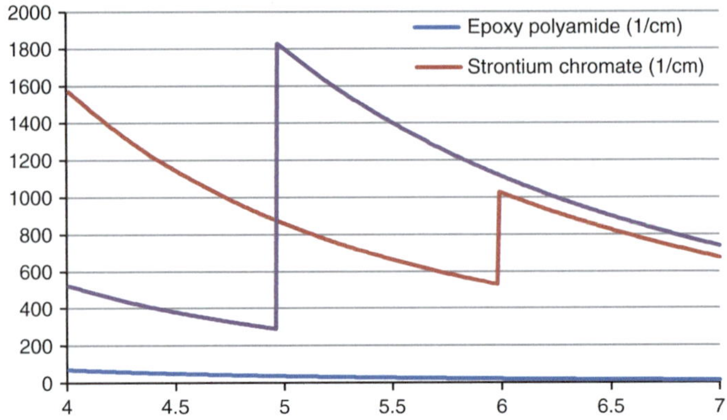

Fig. 10.9 X-ray absorption spectra or the various components of a $SrCrO_4$-inhibited epoxy-based primer containing TiO_2 and $BaSO_4$. By performing tomography experiments at several different energies, it is possible to obtain data that has different combination of absorption features which can be processed to give chemical information

Since the X-ray attenuation of the other elements within the sample will change very little between these two energies, an effective tomographic reconstruction for the targeted element can be obtained by subtracting the reconstruction below the absorption edge from that above.

This method can only detect one chemical element at a time and is restricted to those elements which have an absorption edge at an X-ray energy which is also suitable for X-ray tomographic imaging. Even so, it has been used in a variety of materials characterization applications [59, 60].

10.2.1.7 Data Constrained Tomography

Although single energy X-ray micro-CT is becoming a powerful and widely used method for characterizing 3D microstructures non-sample-destructively [61, 62], there are some limitations. One of the limitations is the inherent resolution issue. With modern X-ray CT equipment the pixel size (on the detector) cannot be smaller than 10^{-4} of the sample size without loss of image signal to noise ratio. For instance, if we need to image a 1 cm sample, the pixel size will need to be larger than 1 μm. For a factor 10 improvement in resolution, the number of projection images will need to be increased by the same factor, and the data volume will be increased by 1000 times. To maintain a similar level of reconstruction accuracy, the exposure dosage will be increased by the same factor. These limitations make it impractical to resolve finer details of the materials by simply increasing the imaging resolution. In post-processing of the CT images to obtain the microscopic distributions of materials in a sample, we also face the issues of non-uniqueness of attributing a particular image gray-scale level to a material composition. Consequently, identification of mixed composition at X-ray CT voxel level is difficult, although it is potentially very useful when attempting to reconstruct a faithful image of the material microstructure distribution.

Considerable effort has been devoted to dual energy (spectrum) approach in characterisation of various materials [63, 64]. These approaches were targeted to resolve the density and atomic number distributions. The dual energy X-ray CT works well in resolving materials with large atomic number difference. However, it is still inadequate in resolving mixing of multiple materials phases and void at the X-ray CT pixel level.

Recently, a data-constrained model (DCM) has been developed [65, 66]. DCM is based on a lattice statistical mechanics model. Multiple sets of X-ray CT data acquired at different X-ray beam energies are used as pervasive boundary conditions to constraint the model. Computationally, the DCM model is defined on a simple-cubic lattice grid with a cubic voxel on each grid point and with the total number of voxels as N. The X-ray absorption coefficients (μ_n) for each voxel are obtained from X-ray CT. For a voxel at n $(n = 1, 2, \ldots, N)$, the computational model minimizes the following objective function:

$$T_n = \sum_{l=1}^{L} \left[\delta\mu_n^{(l)}\right]^{2\zeta} + E_n \tag{10.1}$$

where L is the number of X-ray CT data sets acquired from the same sample using L different beam energies.

For each X-ray spectrum (beam energy) l, the relative CT error is expressed as

$$\delta\mu_n^{(l)} = \frac{1}{\mu_{\max}^{(l)}} \left| \sum_{m=0}^{M} \mu^{(m,l)} v_n^{(m)} - \hat{\mu}_n^{(l)} \right|, \tag{10.2}$$

where $\mu^{(m,l)}$ is the linear absorption coefficient (or linear phase-shift coefficient, or real or imaginary part of the complex refractive coefficient) for material m, $\hat{\mu}_n^{(l)}$ is the CT reconstructed linear absorption coefficient (or linear phase-shift coefficient, or real or imaginary part of the complex refractive coefficient), and $\mu_{\max}^{(l)} = \max\{\mu^{(m,l)}; m = 0, 1, 2, \ldots, M\}$. The optimization is achieved by adjusting the volume fraction variables $v_n^{(m)}$ $(m = 0, 1, 2, \ldots, M)$ for each material composition m, where M is the total number of compositions in the system, subject to the constraints:

$$\sum_{m=0}^{M} v_n^{(m)} = 1 \quad \text{where} \quad 0 \le v_n^{(m)} \le 1 \quad \text{for } m = 0, 1, 2 \ldots, M \tag{10.3}$$

The dimensionless phenomenological interaction energy is expressed as

$$E_n = \sum_{m=0}^{M} v_n^{(m)} S^{(m)} + \sum_{k=0}^{K} \sum_{j=1}^{N^{(k)}} \sum_{m_1=0}^{M} \sum_{m_2=m_1}^{M} \left(v_n^{(m_1)} v_{n+n_j^{(k)}}^{(m_2)} \right)^{\eta} I_k^{(m_1,m_2)} \tag{10.4}$$

where $S^{(m)}$ denotes the self-energy (chemical potential) for composition m, $I_k^{(m_1,m_2)}$ denotes the interfacial energy between neighboring voxel compositions m_1 and m_2 at distance k, where = 0 (k > 0) denotes intra- (inter-) voxel. The maximum interaction range is denoted by K. The number of neighboring voxels with neighboring range k is denoted by $N^{(k)}$ and the jth neighboring vector with a neighboring range k is denoted by $n_j^{(k)}$. The parameter η with a default value of $\eta = 0.5$ is used to approximate the volume to interface relation.

DCM has the same voxel size as the original CT data. It differs from the more conventional image segmentation techniques in that it allows co-existence of void and multiple material compositions in the same pixel (partial volume effect). Such co-existence of multiple materials in the same voxel is related to the structures finer than the X-ray CT resolution.

By ignoring the phenomenological self energy and interaction energy, the linear approximation of the DCM model can be expressed as [67]

$$\begin{cases} 0 \le v_{m,n} \le 1 \quad m = 0, 1, \ldots, M \\ \sum_{m=0}^{M} v_{m,n} = 1 \\ \sum_{m=0}^{M} \mu_m(\lambda_l) v_{m,n} = \bar{\mu}_n(\lambda_l) \quad l = 1, 2, \ldots, L \end{cases} \qquad n = 1, 2, \ldots, N \tag{10.5}$$

The linear DCM implementation is very CPU efficient. Satisfactory results can be obtained under certain circumstances [55, 68, 69].

DCM requires some prior knowledge about the chemical phases within the sample. For example, the calculated absorption coefficients for a chromate inhibited epoxy containing $SrCrO_4$ and TiO_2 are shown in Fig. 10.9. Optimal distinction of compositions is achieved if the beam energies are so chosen that the corresponding absorption coefficients of the material compositions have maximum linear independence. For the chromate inhibitor, the optimal beam energies would be one below about 4.8 keV and another one above around 6.2 keV and one in between. This type of CT image acquisition experiment can be performed in a synchrotron where a selection of incident X-ray energies can be selected or can be performed in a X-ray ultramicroscope (XuM) which is an SEM that has been adapted to use and electron beam focused onto a target which will emit characteristic X-ray that can be used as source X-rays. Pursuing the XuM approach the target materials for the X-ray source are Fe (6.4 keV) and Ti (4.5 keV). The corresponding rendered tomographs are shown in Fig. 10.10a for the Fe source and in Fig. 10.10b for the Ti source, where the X-ray linear absorption coefficients are represented as image intensities. The rendered volume in these two images contains both TiO_2 and $SrCrO_4$ since there is only absorption contrast in these images. Assuming that the X-ray attenuation by a voxel is small, a linear constraint relationship between CT reconstructed linear absorption coefficients and the volume fractions of the compositions can be established [46, 68, 70, 71].

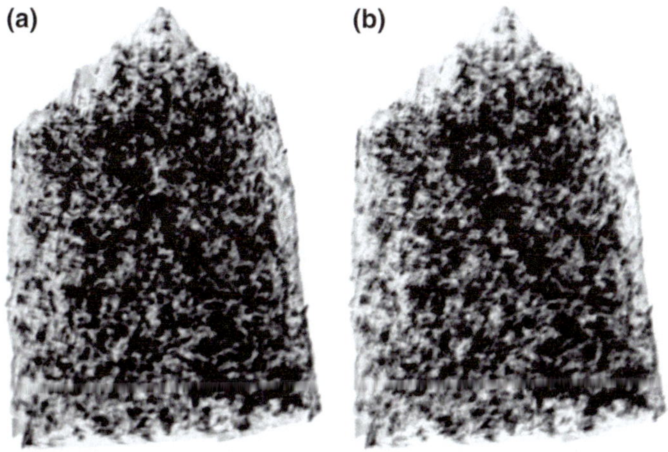

Fig. 10.10 Reconstruction of the combined distribution of TiO_2 and $SrCrO_4$ in a section of an epoxy-based primer. **a** Use a Ti target X-ray source. **b** Use a Fe target X-ray source

Fig. 10.11 DCM chemical phase reconstruction showing $SrCrO_4$ in *green* and the TiO_2 in *red*

Figure 10.11 is a 3D image of the materials components reconstructed using DCM where the actual $SrCrO_4$ is represented as green colour and TiO_2 as red. The validity of the DCM approach to chemical phase reconstruction has been demonstrated using scanning electron microscopy where it has been shown that there is good correspondence between the two, providing confidence that the technique

reproduces the distribution of phases faithfully. The beauty of the DCM technique is that it can be used to provide more information than either the X-ray maps alone or absorption contrast by itself. DCM has also been used in characterisation of Zn corrosion product [55] and nano-structured gold particles in Al [72].

10.2.2 AFM Based Analysis of Surface Topography, Adhesive Properties, and Mechanics

Atomic force microscope (AFM) is one of the widely used tools for surface characterization. It was developed to overcome the shortcomings of scanning tunneling microscope (STM), to enable measurements on non-conductive surfaces. Possibility to perform imaging and spectroscopy in liquid environments or under UHV conditions makes AFM a powerful characterization tool to perform in situ measurements. AFM, which started as a method for the investigation of substrate topography, can provide today, thanks to the developments in imaging modes, chemical and mechanical information at a higher spatial resolution, which made it a valuable asset for surface science studies.

In a standard AFM mode, a very sharp tip—with a typical radius of curvature on the order of a few nanometers—at the end of a cantilever scans the surface. In the close proximity of the surface of interest, the forces between the tip and the surface lead to a deflection of the cantilever. This deflection is precisely measured with a beam of laser reflected from the back of the cantilever to a position sensitive detector (PSD). After the spring constant calibration, measured deflection values can be converted to force with accuracy down to a few piconewtons (pN). Conventionally, a feedback mechanism is used to maintain a constant force between the tip and the sample. Piezoelectric elements are used to control the tip-sample separation (z direction) and to perform the scanning motion on the sample surface (x and y directions).

In addition to high resolution imaging with AFM, force distance spectroscopy, or so called chemical force microscopy (CFM) is an AFM based technique to obtain quantitative information on the interaction forces between tip and the sample. Figure 10.12 demonstrates a typical force-distance curve, where so-called pull-off force is recorded to obtain the mean adhesion force between the tip and the surface. The tip approaches to the surface and attractive forces dominate in the close vicinity of the surface. At this point, so-called jump-to-contact is observed. A predetermined load is applied to the surface before the cantilever is retracted from the surface. The separation (indicated between the points E and F) is the force value to detach the tip from the surface, which is recorded as the pull-off force value.

In the study of Tanem and coworkers, AFM force-distance measurements have been used to measure the adhesion forces between functionalized AFM tips and aluminum alloys before and after application of different conversion coatings [73]. Moreover, the results were compared with a model aluminum surface. Conventionally, Si or Si_3N_4 tips are coated with a thin Au layer and thiol-Au

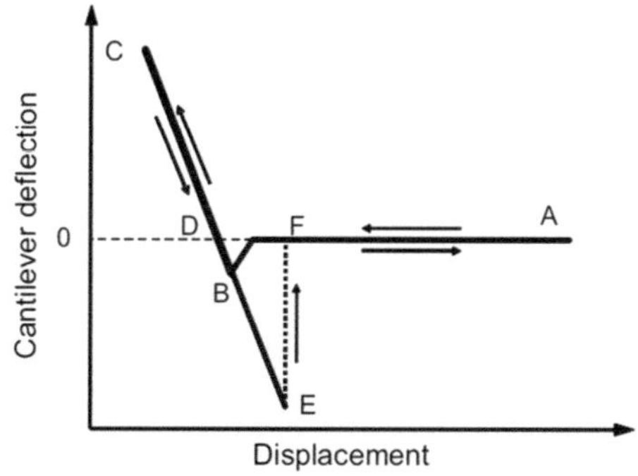

Fig. 10.12 Schematic *curve* of cantilever deflection versus displacement (distance to surface) in AFM adhesion force measurements. Reprinted from [73], Copyright (2009), with permission from Elsevier

chemistry is used to deposit self-assembled monolayers with desired end-groups. Figure 10.13 compares the adhesion forces between a—COOH functionalized AFM tip and aluminum substrates. It has been reported that the adhesion force between the chromate conversion-coated (CCC) aluminum alloy and the tip was significantly increased in comparison to uncoated surface likely due to the increase of hydrogen bonding. However, a decrease of adhesion force was observed after 24 h due to aging of the CCC. Moreover, it has been reported that chromate free Ti-Zr based treatment also increased the adhesion although the measured forces varied significantly due to the non-uniform nature of the technical alloy. The measured AFM adhesion forces were qualitatively correlated with macroscopic

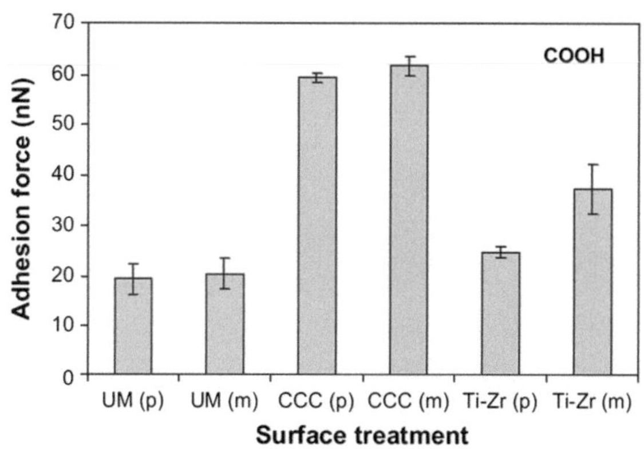

Fig. 10.13 Average adhesion force measured on the α-Al (Fe,Mn)Si particles (p) and aluminum matrix (m) of the uncoated ultra-microtomed (UM), chromated (CCC) and Ti–Zr-treated surfaces. Measurements were performed within 2 h after surface treatment, using a COOH functionalized tip. Reprinted from [73], Copyright (2009), with permission from Elsevier

adhesion tests on various conversion coatings. Overall, AFM based adhesion measurements provide useful information on the adhesion behavior of heterogeneous conversion-coated aluminum surfaces.

Several AFM modes allow not only collecting topographical information but also mechanical [74], electrical [75, 76], magnetic [77] and chemical [78] properties of the surface. Typically, nanoindentation, or AFM based force volume imaging is employed to obtain quantitative mechanical or adhesive properties. However, the conventional phase and force modulation methods fail to provide the resolution comparable to the topographical images. In addition to the poor lateral resolution, the time required to obtain a complete surface map is the drawback of these techniques.

HarmoniX, is a recently introduced mode to map the elastic modulus of a sample with high spatial resolution as fast as conventional AFM imaging. The elastic modulus is derived from force-indentation curves based on Derjaguin–Muller–Toporov (DMT) model [79]. HarmoniX typically work with small loads (<5 nN) and indentation depths are as small as 1 nm [80].

HarmoniX mode uses the so called torsional harmonic cantilever (THC) which has a tip that is offset from the long axis of the cantilever, as can be seen Fig. 10.14 [81]. When the cantilever is oscillated in the tapping mode, a torque is generated due to the tip-sample interaction (Fig. 10.14b). This torsional motion along the long axis of the cantilever moves the reflected laser spot horizontally on the PSD. Similar to the tapping mode, the conventional vertical oscillations are used to follow topography while the torsional vibrations are used to calculate tip-sample interactions in sub microsecond resolution.

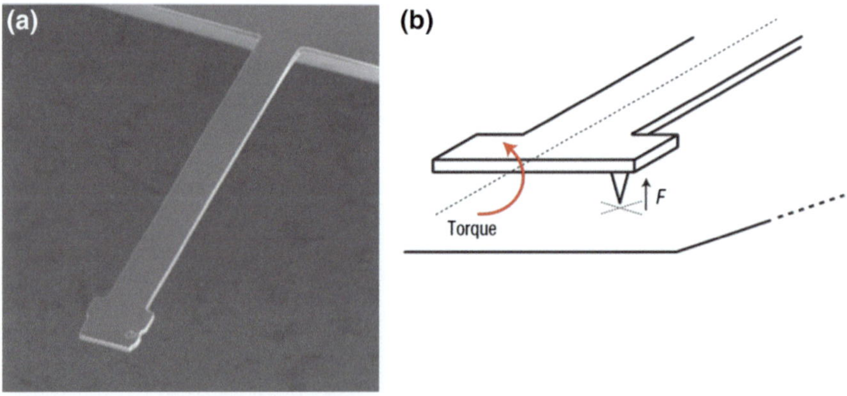

Fig. 10.14 Design of the torsional harmonic cantilever with an off-axis tip. **a** a scanning electron micrograph image of a torsional harmonic cantilever. The cantilever is 300 mm long, 3 mm thick and nominally 30 mm wide (50 mm near the free end). The tip is offset 15 mm from the centre of the cantilever. **b** An illustration of the THC interacting with the surface. The offset position of the tip results in a torque around the axis of the cantilever. Reprinted by permission from Macmillan Publishers Ltd: Nature Nanotechnology [81], copyright (2007)

One of the prominent examples of surface characterization with AFM topography and mechanical measurements is the work by Yoon and coworkers [82]. They have performed AFM imaging to monitor the self-healing of induced cracks on polymeric films and characterized the mechanical properties to assess the degree of healing. Defects of various sizes were induced with tapping or contact mode on reversibly cross-linked star polymer gel films and the evolution of the appearance of the cut region was monitored with the continuous tapping mode AFM imaging. Figure 10.15 indicates that the size of the damaged area decreased within minutes and the depth of the cut decreased to 5–10 nm.

Measurements on the center and edge position revealed that the healing response is driven by the bond re-formation initiated from the bottom of the cut, which indicates that the cut surfaces are accessible to each other for the healing process to start. Figure 10.16 demonstrates the case where a penetrating cut with 300 nm width remains unrepaired even after 3 h. It is concluded that the stiffness of the polymer film prevents the flow and consequently, the contact of two damaged surfaces does not take place. In order to accelerate this process, the damaged area was gently stroked under contact mode scanning perpendicular to the cut direction. Figure 10.16b and c show that the areas stroked with contact mode became as smooth as the original film (arrows). It is argued that the AFM tip in contact with the surface causes the cut surfaces to be accessible to each other and thus, enhance the healing response. Figure 10.16d demonstrates the modulus map of the cut and the healed areas acquired with HarmoniX mode. The arrows indicate that the damaged areas are indistinguishable from the intact area, indicating a successful healing process.

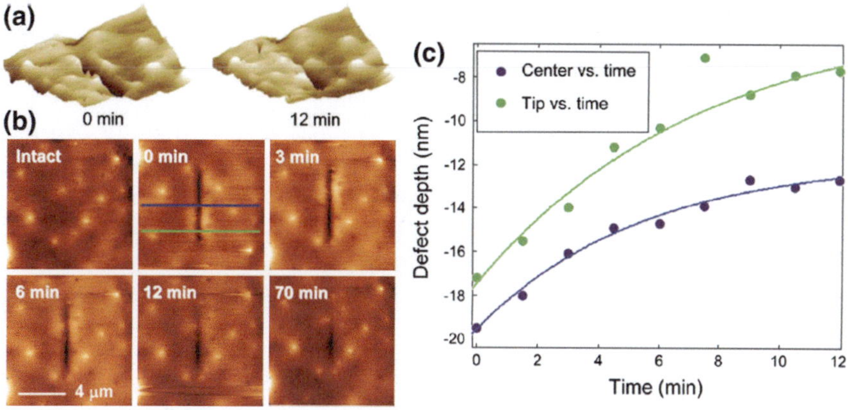

Fig. 10.15 Time-dependent change of height mode AFM image for the cut on the surface of X3 film: **a** 3D images, **b** 2D height mode images, and **c** evolution of damage depth for 12 min measured at two positions. The surface of film is referenced to zero for the Y-axis. The "center" position describes the center of the damage and the "tip" position, 2 μm away from the "center". Reprinted with permission from [82]. Copyright (2012) American Chemical Society

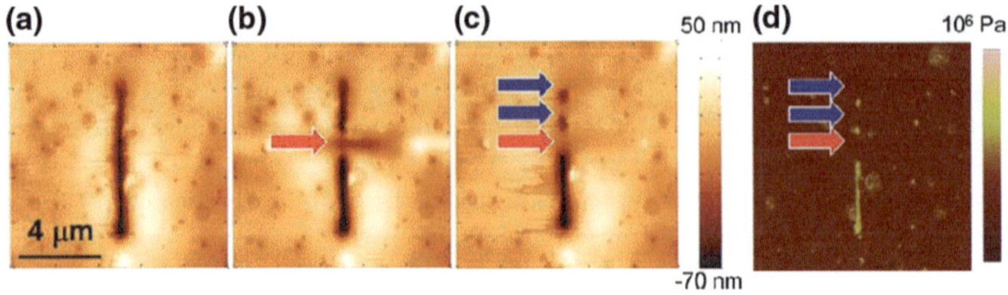

Fig. 10.16 Healing response of X3 film accelerated by AFM tip stroke with contact mode scanning: **a** original cut, **b** after the first series of strokes at the position marked by a *red arrow* (+0.5 V), **c** after the second series of strokes at two positions marked by of *blue arrows* (0 V), and **d** HarmoniX mode modulus map in a logarithmic scale. *Brighter color* represents higher modulus. The *brightest regions* in the damaged area correspond to high Young's modulus of silicon wafer exposed at the *bottom* of the deep cut. Reprinted with permission from [82]. Copyright (2012) American Chemical Society

10.3 Electron and Nuclear Spectroscopy

10.3.1 Electron Spectroscopy

There are two major electron spectroscopies of interest for characterisation which include photoelectron spectroscopy and Auger electron spectroscopy (AES). Photoelectron spectroscopy uses photons to ionise electrons from a surface and is therefore based on the photoelectric effect. AES most commonly uses an electron beam as the exciting source and examines the Auger electron emission. In photoelectron spectroscopy, the exciting photon can be low energy such as ultraviolet (UV) used in UV photoelectron spectroscopy (UPS) or can be higher energy such as X-ray energies (soft X-rays) that are more typically used for characterisation of materials of interest for coatings using X-ray photoelectron spectroscopy (XPS). UPS is used to study orbitals involved in bonding, and laboratory UPS equipment generally use a He source which can typically probe from 0 to 40 eV, whereas X-rays are used to probe core level electron orbitals. Auger lines can be measured in XPS experiments. Binding energies of core photoelectron lines can be plotted against X-ray generated Auger lines in two dimensional chemical state plots which can be useful for distinguishing oxidation states.

10.3.1.1 X-Ray Photoelectron Spectroscopy (XPS)

Laboratory XPS equipment generally comes with a dual anode X-ray source comprising a Mg (1253.6 eV) and Al (1486.6 eV) electrode from which the characteristic Kα lines are excited. In XPS parlance, Kα means a superposition of the $K\alpha_1$ and $K\alpha_2$ X-ray lines. These lines are not monochromated hence the emitted electrons also display broad peaks which make spectroscopic analysis more difficult. These sources also include other target X-ray lines which can interfere with the determination of low levels of material. Modern XPS instruments often have monochromators in which only the Kα_1 line is the source after monochromation. Accessibility to synchrotrons means that monochromatic energies can be tuned to suite the experiment if desired.

One of the most important outputs from an XPS experiment is the core level binding energy which reflects the local chemistry. However, because the sample is continually emitting electrons due to the photoelectric effect, it develops a positive charge that must be taken into account before any useful information can be extracted from the position of the photoelectron peak. In many modern instruments the charge is neutralised using electron flood techniques where very low energy electrons flood the surface of the sample and neutralise the positive charge that develops from the emission of the photoelectrons. For other instruments (and also to check the charge neutralisation of the flood guns) an internal reference peak should be used which can be the C 1s peak which falls at 284.6 eV for inorganic materials or 285.0 eV for organic materials although this is an area of debate [83, 84]. All other peaks are then adjusted accordingly. Another standard is evaporated Au with a peak Au 4f binding energy at 84.0 eV, but care must be taken that the evaporated gold particles are large enough to have the bulk gold binding energy (very smaller particles do not have the same peak binding energy as large particles) [85, 86]. Implanted Ar at with an Ar 2p BE of 243.3 eV has also been used for binding energy calibration [87, 88].

While XPS is widely used to determine surface composition, the interpretation of XPS spectra however, is not always so straight forward as the literature might suggest. Consideration should be given to the analysis volume of XPS compared to the microstructure of the sample since the analysis depth typically has a maximum of 10 nm. If the sample under examination is nanoparticulate with a size of 5 nm, then XPS is effectively a bulk analysis technique. On the other hand if the particles are microns in size or a rough film forms on a surface then the surface that is actually examined may be very complicated.

An advantage of XPS is that it can give chemical state information based on peak positions. Often there are multiple oxidation states as well as multiple species. For example, in the O 1s spectrum there may be oxygen anions as well as hydroxyl ions, the positions of these peaks varies with the type of oxide so in a mixed oxide system with surface hydroxyl ions there may be several peaks contributing to the O 1s envelope. A further issue related to the study of the corrosion protection of the transition metals, is the complicated nature of the transition metal 2p lines, since these have quantum mechanical features (multiplet lines) and loss features that make interpretation difficult [89–95]. This is most extreme for the Fe 2p spectra but while there are good examples of fitting of the Fe 2p spectra for Fe-based corrosion product [96, 97], the multiplet structure is often ignored, perhaps leading to erroneous interpretation [98]. As an example Fig. 10.17a shows the Fe 2p spectrum for the corrosion product generated during atmospheric corrosion. It consists of Fe metal, Fe^{2+} and Fe^{3+}. The top spectrum shows the breakdown of the Fe^{2+} and Fe^{3+} components into all of the multiplet states revealing a complex shape for both oxidation states. Figure 10.17b shows the three oxidation state components where all the multiplets are summed for each oxidation state. In addition to the multiplets there are a number of loss lines as well as shake-up peaks leading to a very complicated spectrum. If a single oxidation state is present, then interpretation of the Fe 2p spectra is relatively straight forward since details of the multiplet splitting are not important (i.e. only the total area under the peak is important [99].

The data in Fig. 10.17 has been generated using a Shirley background subtraction where the background at any particular point is defined as the difference between the background intensity above (at higher binding energy) a photoelectron peak to that below the peak multiplied by the fraction of the integral intensity below the peak divided by the total integral intensity. The fraction is determined by the integral amount of intensity below the peak to the total integral intensity for the peak. This type of background is one of the most common backgrounds used in XPS. The full Fe 2p spectrum with a Shirley background is displayed in Fig. 10.18a

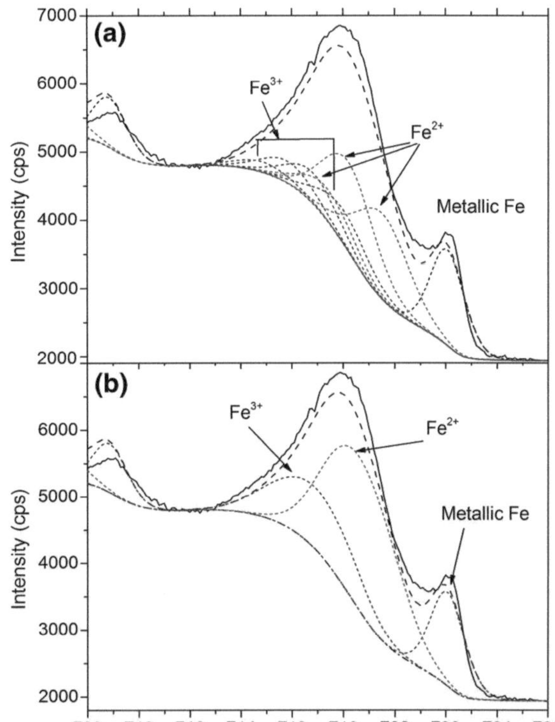

Fig. 10.17 XPS Fe 2p spectrum of mixed oxidation state Fe corrosion product containing metal as well as Fe^{2+} and Fe^{3+}

[100–102]. Linear backgrounds are also commonly used in XPS, although not so commonly for photoelectron peaks where there is a large increase in the background intensity such as for the Fe 2p spectrum. Figure 10.18b shows a linear background. A third type of background is the Tougaard background [103, 104] which is based on the fundamental physics of electron scattering [103–105]; it is shown in Fig. 10.18c.

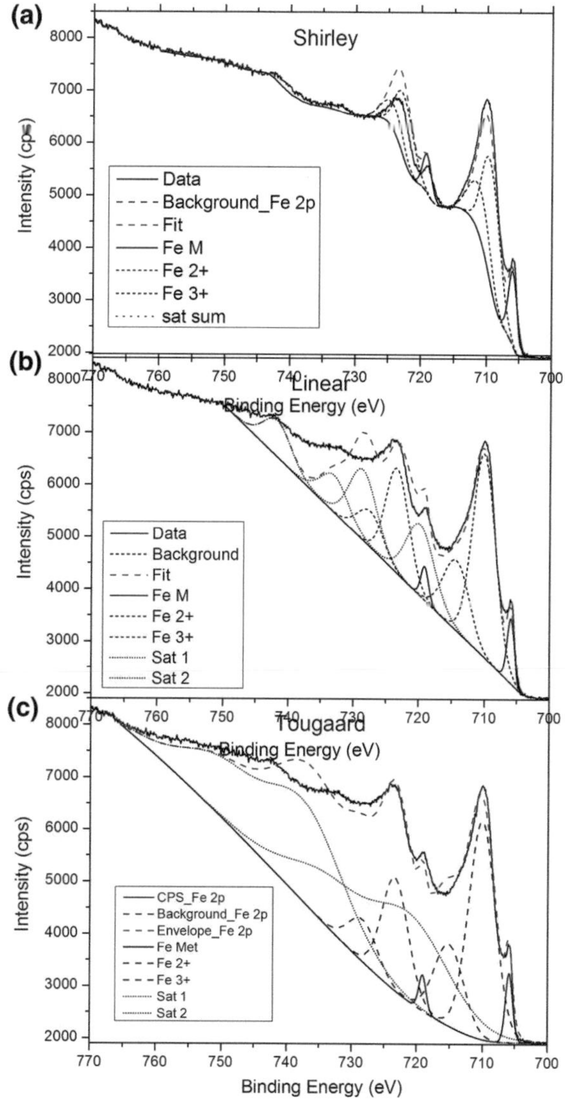

Fig. 10.18 Various methods of fitting of Fe 2p3 core level peaks. **a** Shirley, **b**, linear and **c** Tougaard backgrounds. The *solid black line* is the data, the *dashed blue line* is the fit and *dashed black lines* are the individual component for the different Fe oxidation states and the *dotted lines* are the shakeup satellites

It can be seen that there is a huge increase in the intensity of inelastically scattered electrons going from Shirley to linear to Tougaard backgrounds (area between the spectrum and the background line). This increase has an enormous impact on the fitting of the multiplets for the oxidation state components as well as the satellite peaks. The percentages for each of Fe metal, Fe^{2+} and Fe^{3+} for the Fe 2p3/2 section of the Fe 2p peak using the different backgrounds are shown in Table 10.2. The Fe 2p3/2 of the Fe 2p part of the spectrum has been chosen since the quality of the fit with all three backgrounds is much better in this region of the Fe 2p spectrum. It can be seen that while there is rough agreement between the percentages for each component using the different backgrounds, the results do not agree in detail and cannot be considered as quantitative. There is currently no generally accepted approach to fitting the Fe 2p spectra although there are some very good studies examining this problem [94].

Other first row transition metals also have complex spectra. Copper spectroscopy has intense shakeup satellite peaks, but the main photoelectron lines are not complicated by the presence of multiplet splitting lines [89, 106]. However, the Cu metal and Cu (I) 2p photoelectron lines have similar binding energies which make it nearly impossible to breakdown the measured envelope into its component peaks. Fortunately the Auger KLL lines (generated by X-rays in this case) can be used to distinguish the different components. These two peaks can also be used together to identify different Cu compounds by plotting the Cu 2p3/2 binding energy against the Cu KLL kinetic energy in a plot that is called a 2D chemical state plot. These types of plot are very useful for identifying different compounds.

The spectroscopy of the light metals aluminium and magnesium is straight forward, partly because there are only two oxidation states for each metal (Al metal and Al^{3+} and Mg metal and Mg^{2+}) and the strongest photoelectron lines are from relatively simple core shells with the 2p line for Al and the 1s line for Mg. As with Cu, the 2D chemical state plots can be useful for identifying different oxidation states as well as different classes of compounds. The other light metal worth mentioning here is Si since silanes are studied extensively for coatings and for self healing agents (see Chaps. 5 and 8), Both the Al and Si 2D chemical state plots can be used to provide rich information about surface species and phases.

The surface sensitivity of both XPS and AES means that profiling techniques are required to obtain information at depths greater than around 10 nm. Traditionally this has been achieved using ion milling, particularly with Ar^+ ions. Another, newer, ion source is C_{60} which is also discussed in the SIMS sections. In principal, any source can be used which provides even sputtering of all species.

Table 10.2 Percentage of Fe metal, Fe^{2+} and Fe^{3+} from the Fe 2p3/2 peak for three different background fits as displayed in Fig. 10.18

Background type	Fe metal	Fe^{2+}	Fe^{3+}
Shirley	12.0	60.5	27.5
Linear	6.0	57.2	36.8
Tougaard	5.8	67.2	30.0

Physical sectioning is another technique that can be used to obtain depth information. This can be achieved by a variety of methods including mechanical milling, polishing at an angle to the normal or cutting at an angle to the normal. In this latter process the objective is to "spread" out the layer in the section to suit the lateral resolution of the technique, so the thinner the layer the closer the angle has to be to the surface parallel. Hinder et al. demonstrated this process for XPS [107] and also for Tof-SIMS approach [108].

The application of XPS to self-healing coatings is limited and generally used to support other techniques. It can be used to study components of coating systems, such as the polymer resin, inhibitors and other additives as well as the effects of self healing such as inhibitor or healing agent reactions on a surface. For example, double layer hydroxides have been used extensively as carriers for inhibitors [36–38, 109–113]. XPS has been used for characterisation of the inhibitor in these types of double layer hydroxides structures such as hydrotalcites [114] and Fe-based structures [115], the latter occurring in green rust [96, 97]. In self healing studies XPS can be very useful in studying well prepared surfaces exposed to extracts from new inhibitor pigments. For example, Granizo et al. [99, 116] examined steel surfaces after exposure to extracts from various vanadates exchanged hydrotalcite pigments and found the surfaces after exposure had films incorporating the inhibitor species. Similarly XPS has been used to examined Ce release and interaction with AA2024-T3 [114, 117] and Mg alloys [118].

For novel inhibitors, it has been applied to determine the layer thickness by monitoring the Si photoelectron lines of an organo-silane coating on latex particles [119]. It has also been used to study the Mo 3d lines during molybate inhibition [120]. The rare earths and in particular, Ce (Ce 3d) which have been used extensively in aluminium inhibition [121–125]. Ho et al. [126] used XPS to determine oxidation states on AA2024-T3 surfaces and to examine Ce diphenyl phosphate interaction with steel surfaces [126]. In this case the self healing properties of these films were determined using EIS.

In the case of coatings, there is a wealth of information to be obtained from the spectroscopy of C 1 s and N1 s peaks which have been used extensively to study conducting polymer coatings [125, 127, 128]. Trabelsi et al. [121] used XPS to determine the cerium oxidation state in silane derived sol-gel films. On a variation on the self healing theme Jakab and Scully used XPS to confirm the segregation of Co and Ce for forming a protective, self healing coating on an Al-Co-Ce alloy [129].

10.3.1.2 Spatially Resolved XPS

Spatially resolved XPS is now a common technique in many XPS instruments with spatial resolution down to 1 μm, although the fine resolution comes with a loss of signal to noise [130]. Some of the early applications of XPS included failure analysis in adhesively bonded structures where the large area analysis of early instruments was not an impediment to studying different area of adhesive failure [7, 131]. While it has been used to study Pb pigments in paintings [132] and coating interfacial chemistry, particularly cohesive or adhesive failure [133], its use in self healing applications is not widespread. Mapping has also been used in corrosion studies [130, 134]. The approach used in these applications can be logically extended to self healing, but only if there is a unique "marker" in the healing agent that can be used to reveal the spatial distribution of the healing agent. Thus, in self healing systems where there is encapsulated monomer which is released and polymerises, then XPS is not likely to be able to detect the healing processes.

10.3.1.3 Auger Electron Spectroscopy (AES)

AES as an electron beam characterisation technique, has the potential for excellent spatial resolution, however, much self healing work is on non-conductors which means that charging is a significant problem. Non-conductors include polymers, hybrid (organic/inorganic composites) and inorganic coatings as well as failure sites where non conducting corrosion product occurs.

Despite these difficulties AES has been used to study inhibition in a defect through a coating on galvanised steel by Montemor and Ferreira [123]. In their case they used a variation of the techniques described in Fig. 10.1 by taking lines scans across and along the length of a scribe. They detected that cerium was present in the defect and had leached from an adjacent sol-gel coating during immersion.

10.3.2 Nuclear and Ion Spectroscopy

There are two nuclear techniques that are worth mentioning for characterisation of self healing systems. The techniques are Rutherford Backscattering Spectroscopy (RBS) and Particle induced X-ray Emission (PIXE) spectroscopy. Both techniques use either protons or α-particles (2 protons and 2 neutrons—i.e. a He nucleus) which are accelerated in an accelerator up to energies in the vicinity of MeV. These probes then interact with the system under study; in the case of RBS the quasi elastically scattered particles are measured (the measurement angle is typically 170°), but in PIXE the X-ray emission spectrum is measured. Both PIXE and RBS have spatial resolution that can resolve features down to the micron scale and are therefore ideal to measure interactions in a defect that reflect a healing process. The disadvantage of these techniques is that they require an accelerator which typically is the size of a small building.

The sensitivity and data analysis requirements of the two techniques are quite different because the emitted radiation is so different (backscattered particles versus X-ray emission). As an example Fig. 10.19 shows an optical image and

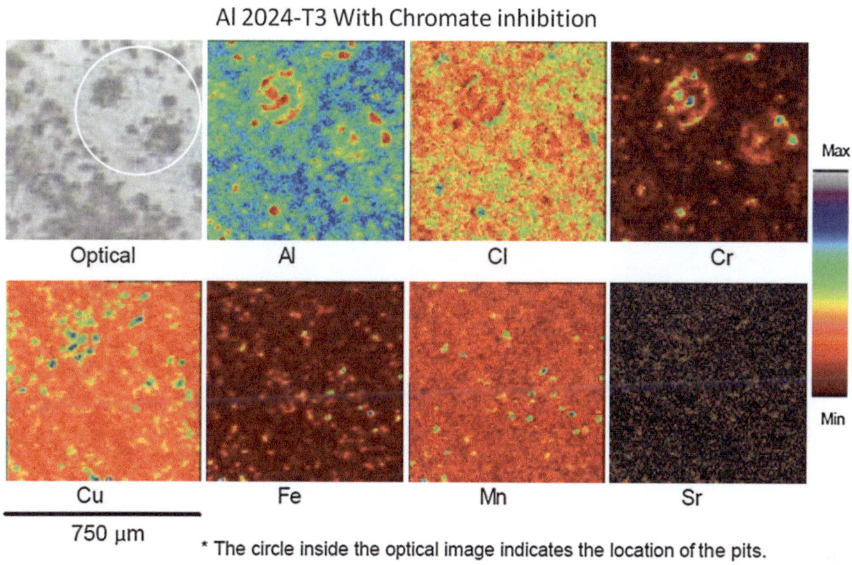

Fig. 10.19 PIXE images of the surface of AA2024-T3 that has been exposed to 0.5 M NaCl solution for 30 min followed by the introduction of 4.5×10^{-4} M potassium dichromate (unpublished). NB: The colour code is an *inverse* thermal code

corresponding elemental maps for a sample of aluminium alloy 2024-T3 (AA2024-T3) which has been exposed to 0.5 M NaCl solution for 30 min prior to the introduction of 4.5×10^{-4} M $K_2Cr_2O_7$. The Al, Cl and Cr maps show the typical ring patterns of corrosion product that has been reported previously as being the sites for stable pits [135, 136]. The Fe, Cu and Mn maps show that the attack sites are associated with clusters of intermetallic particles which are typical of the heterogeneous microstructure of this alloy [49, 50, 137]. In this figure it can be seen that chromate has reacted both with the aluminium oxide corrosion product as well as the many of the intermetallic particles, particularly the Cu-containing ones.

In the context of this chapter it can be seen that 2D mapping using PIXE provides good quality elemental imaging, but relating features of the maps to the sample is limited by optical microscopy. Some advantages of mapping with PIXE are similar to electron microprobe work in that the sample is scanned beneath the particle beam, so the beam-sample geometry remains constant and there is no distortion at the perimeter of the field of view, as can occur at low magnification using a standard scanning electron microscope. Thus large sample areas can be examined. Quantitative analysis from the X-ray emission is also a standard output from PIXE and many facilities which offer PIXE will collect hyperspectral datasets (each pixel in a map has an associated X-ray spectrum).

In the case of RBS the ideal type of application is for heavy elements forming a thin film over light elements. Figure 10.20 shows an RBS spectrum for samples of AA2024-T3 treated for 1 h at 95 °C in 10 mM of either $CeCl_3$ or $CrCl_3$ [138].

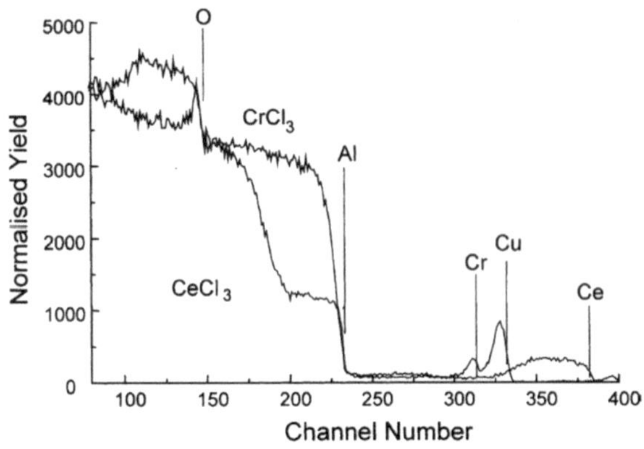

Fig. 10.20 RBS spectra for samples of AA2024-T3 treated for 1 h at 95 °C in 10 mM of either $CeCl_3$ or $CrCl_3$. In the case of $CeCl_3$ there is a buildup of a pseudoboehmite coating which contains Ce. In the case of $CrCl_3$ only a thin Cr-containing oxide was developed. Reprinted from [138], Copyright (2003), with permission from Elsevier

In RBS the position of any element at the surface is known from the backscattered energy position (horizontal scale) and a number of element surface positions are shown in Fig. 10.20. If there is intensity to the left of this position then this means that the element is beneath the surface. It can be seen that the transition elements and Ce fall in a high signal to noise region of their respective spectra making RBS an ideal technique for detecting small quantities of heavier elements in a lighter matrix. In this case the RBS spectra can be used to distinguish between the thick pseudoboehmite coating developed during treatment in $CeCl_3$ solution and the very thin passivating layer during treatment in $CrCl_3$. In addition the cerium profile indicates that Ce is distributed throughout the film since the Ce peak is broad, and for the chromium treated sample there is Cu enrichment at the surface.

As an example of mapping using RBS, Fig. 10.21 shows a Cr map in plan view obtained from the cut edge of a primer that has been exposed to neutral salt spray test (NSS). Figure 10.21a shows the schematic of the sample configuration. The sample was manufactured by cutting a defect through the protective paint system to the underlying metal (using a milling machine). This is a classic configuration for measuring self repair since recovery from damage can be monitored. The paint system consisted of a primer and topcoat. The sample was then put into a neutral salt fog chamber for 16 days and chromate was released from the cut edge with some going to runoff and some reacting with the surface of the slot. Figure 10.21b shows the Cr map collected from the interfacial region between the coating and the milled slot. There is a depletion zone that extends from the cut edge back into the primer. Using RBS it was possible to follow the development of the depletion zone with time. The depletion zone developed rapidly in the early stages but it eventually appeared to reach a maximum depletion depth into the primer. It can be seen in Fig. 10.21 that there appear to be filaments extending into the primer beyond the depletion front. Both the limit to the depletion depth and the appearance of

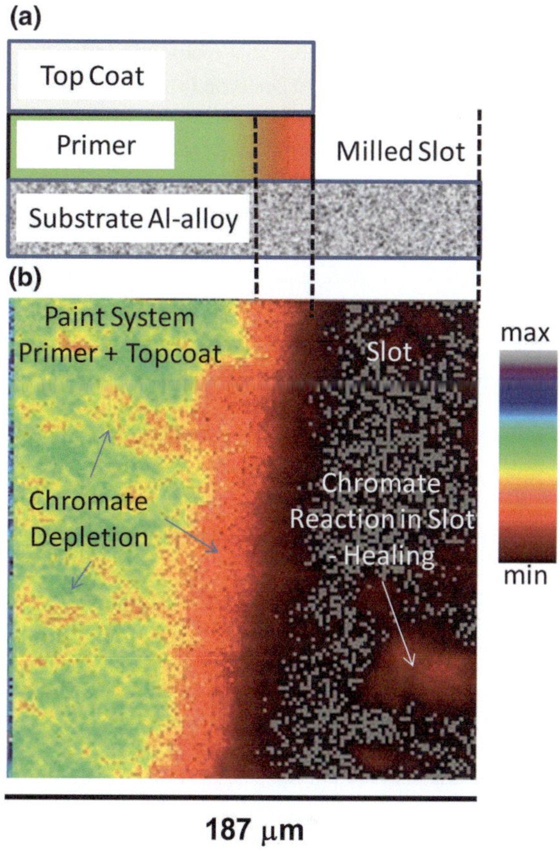

Fig. 10.21 Cr map obtained using Rutherford Backscattering Spectroscopy. **a** Schematic showing the configuration of the sample in section. The sample was manufactured by cutting a defect through the protective paint system the underlying metal (using a milling machine). The paint system consisted of a primer and topcoat. The sample was then put into a neutral salt fog chamber for 16 days. **b** Plan view of the slot (defect, *right*) and the paint system (*left*) and the depletion zone at the cut edge

filaments are consistent with some recently published work that identified clusters of connected $SrCrO_4$ particles within these types of primers [58, 139]. It was observed that only the clusters connected to the surface were the ones that were leached out leading to a general depletion depth for those clusters connected to the surface but also some deeper penetration due to the dissolution of larger clusters. This model is extremely important for the design of paint systems more generally, but also for strategies for self healing where the internal structures of coatings can be used for delivery systems for various agents. It is also discussed in more detail in the tomography section.

The area of Secondary Ion (or Neutral) Mass Spectroscopy (SIMS or SNMS) is an area where there has been considerable development in a range of techniques which are covered in a number of good reviews in the literature [140–142]. Considerable effort has been put into the development of the techniques for soft material analysis, particularly for biological applications. The developments have focused on ionisation techniques that deliver large molecules into the gas phase. These include matrix assisted laser desorption/ionisation (MALDI) and electrospray ionisation [140]. Two dimensional information can be obtained by scanning the laser across the surface but this technique is limited to at best 1 μm and generally higher [140]. On the other hand an ion beam can be focused down to 100 nm. There are two different modes of operation for SIMS instruments including time-of-flight (Tof)—SIMS and dynamic SIMS. In Tof-SIMS the ion beam is pulsed onto the surface and the total mass spectrum is analysed, whereas in dynamic SIMS the ion beam etches the surface and preselected mass to ion ratios are monitored.

The sensitivity of SIMS techniques depends strongly on the ion source. There are a number of different sources that are used including gallium (Ga^+) [142], cesium (Cs^+) [142, 143], O [142], bismuth (Bi^{3+}) [143–145], gold (Au^{3+}) [108, 146, 147] and fullerene (C_{60}) [143, 144, 148, 149]. Secondary ion (SI) yields are highly dependent on the source and the target material as discussed in the literature [142, 143, 150, 151]. The metal ion cluster sources and fullerene are able to produce higher yields of larger mass fragments improving both the scope and sensitivity of the technique [143, 152]. Giannuzzi and Utlaut reported that different ion sources made a difference of a couple of orders of magnitude in SI yields [142].

In the context of self healing in coating applications, SIMS could be applied for characterisation of surfaces generally and SIMS with lateral resolution capability could be used to monitor repair process as a function of time such as repair in a defect.

In practice SIMS and related techniques have not seen widespread use for self healing applications, but are increasingly being used as characterisation techniques in coatings and corrosion work. For example, XPS and SIMS were used to study a range of exposure and curing effects on the adhesion of cathodic e-coats on galvanised steel [8]. It has also been used to study plasma coatings deposited onto water soluble inhibitor particles to reduce the release rate [153] as well as for studying coating defects for a range of coatings [154] and also for interfacial chemistries [133]. It has widespread application in sol-gel technology as well [155, 156].

10.4 Optical Spectroscopy

10.4.1 Fourier Transform Infrared Reflection-Absorption Spectroscopy (FT-IRRAS)

Fourier Transform Infrared reflection-absorption spectroscopy (FT-IRRAS) is generally used for characterization of thin films or monolayers on metal substrates. FT-IRRAS is also an ideal method for the investigation of inhibitor loading in self-healing coatings. The advantages are the high surface sensitivity and the surface selection rule. FT-IRRAS with polarization modulation (PM-IRRAS) enables the differentiation between the surrounding bulk medium and adsorbates on the metal surface. Consequently, the interfering effect of water vapor and carbon oxide

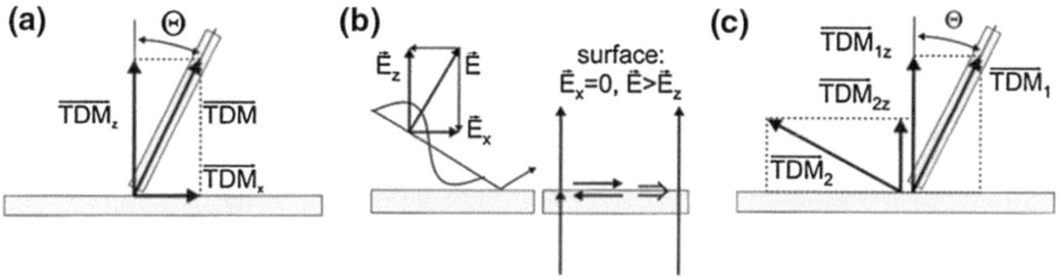

Fig. 10.22 Orientation of the transition dipole moments (TDM) of molecular vibrations (**a**, **c**) and the electric field (**b**) at the surface. Reprinted with permission from [157]. Copyright (2001) American Chemical Society

can be eliminated. IRRAS is dependent on the optical constants of the thin film and substrate, and the angle of incidence, as well as the polarization of the incident IR radiation can be varied to obtain the optimum information. The surface selection rule enables the investigation of the molecular orientation of thin films or monolayers deposited on metal or dielectric substrates.

The methods for the evaluation of the molecular orientation is recently summarized by Arnold et al. [157]. The transition dipole moment (TDM) of a molecular vibration can be decomposed into two components, one oriented normal (TDM_z) and the second one oriented parallel (TDM_x) to the surface (see Fig. 10.22a). The tilt angle Θ can be calculated from the equation:

$$\tan \Theta = \frac{TDM_x}{TDM_z}$$

At a metal surface the parallel component of the electrical field E_x is almost completely screened by the electrons of the metal substrate (Fig. 10.22b) [157, 158]. As a consequence, TDM_x does not contribute and for a single vibrational mode, the tilt angle can only be calculated if the absolute magnitude of the transition dipole moment, |TDM|, is known.

If two vibrational modes with different orientations of their transition dipole moments can be measured (Fig. 10.22c) the ratio $(|TDM_1|/|TDM_2|)^2$ can be calculated from the ratio of their intensities ($I_{1\ bulk}/I_{2\ bulk}$) of the bulk spectrum. Similarly, the ratio $(TDM_{1,z}/TDM_{2,z})^2$ is equal to the ratio ($I_{1\ SAM}/I_{2\ SAM}$). If these ratios can be evaluated the tilt angle can be calculated using [157]:

$$(\tan \Theta)^2 = \frac{I_1^{bulk} I_2^{SAM}}{I_2^{bulk} I_1^{SAM}}$$

The details on the relative method was described by Debe and the interested reader is referred to this article [159]. Parikh and Allara have proposed in 1992 an alternative method, based on the determination of the absolute excitation probability, which corresponds to |TDM| of a given mode in the bulk [157, 160]. Considering the geometrical approximation by Tillman et al. [161], Bram et al. [162] calculated the

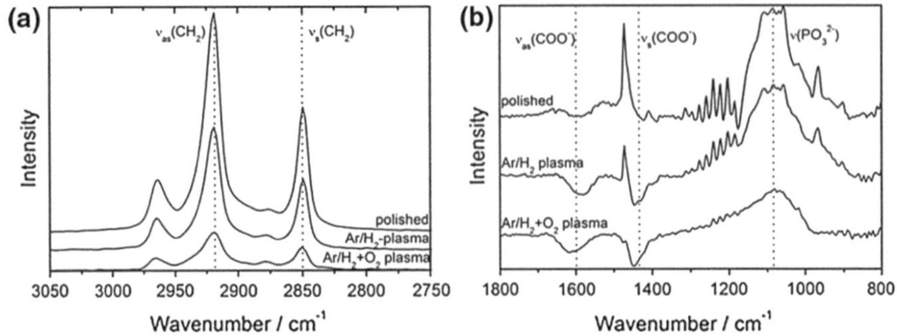

Fig. 10.23 A and B: PM-IRRAS of Zn–Mg–Al alloy polished and after plasma modifications immersed in 10^{-3} mol/l ODPA. All spectra are referenced to the surface spectra taken before ODPA adsorption. Reprinted from [10], Copyright (2013), with permission from Elsevier

tilt angle of self-assembled n-alkanoic phosphonic acid molecules on aluminium surfaces using the following formula:

$$\frac{I_s \nu_{CH_2}}{I_s \nu_{CH_3}} = \frac{nCH_2 \cdot 2(H\,per\,CH_2) \cdot \cos^2(90^\circ - \Theta)}{nCH_3 \cdot 3(H\,per\,CH_3) \cdot \cos^2(35^\circ - \Theta)}$$

A recent applications of the same formalism were reported by Maxisch et al. and Pohl et al. for the investigation of self-assembled monolayer formation on oxyhydroxide-covered aluminium and plasma treated Zn-Mg–Al alloy surfaces, respectively [10, 12].

In the region of 2800–3000 cm^{-1} of PM-IRRA spectra acquired on plasma treated Zn-Mg–Al alloy surfaces after octadecylphosphonic acid (ODPA) adsorption (see Fig. 10.23a), the C-H stretching modes of CH_2 and CH_3 vibrations belonging to the aliphatic chains of ODPA are visible. The spectra shown in Fig. 10.23b indicate the absence of the peaks belonging to the free acid group P-OH and P = O (1230 cm^{-1}). The absence of the free acid groups was interpreted by the authors such that the adsorbed phosphonic acid functionality is deprotonated, which is supported by the presence broad PO_3^{2-} stretching mode. Thus, the interaction of the ODPA molecule and the oxide layer was explained by the presence of coordinative binding modes. Moreover, the tilt angles of the monolayers to the surface normal were calculated as 44°, 39° and 28° for just polished, reductive plasma treated and oxidative plasma treated Zn–Mg–Al surfaces, respectively. The results indicated that the highest molecular ordering of the ODPA self-assembled monolayer was obtained on the oxygen plasma treated Zn-Mg–Al alloy surface [10].

With the high sensitivity for changes in the molecular structure, FT-IRRAS is also an ideal method for the investigation of self-healing processes. In a recent study, Garcia et al. applied FT-IRRAS spectroscopy for the investigation of a reactive silyl ester as a new organic reactive healing agent [39]. The aim of the work was to develop a coating with encapsulated silylester, which in contact with moisture should undergo hydrolysis and forms silanol and oleic acid. These two

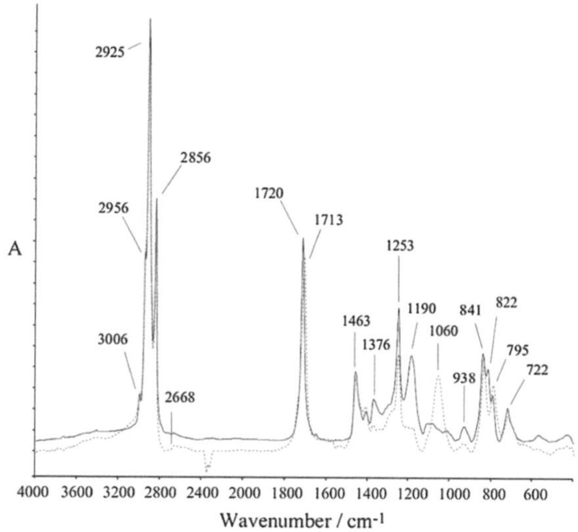

Fig. 10.24 FTIR spectra. Un-reacted silyl ester (*black top line*) and reacted silyl ester on AA2024-T3 exposed to aqueous solution at pH4 (*dashed red bottom line*). Reprinted from [39], Copyright (2011), with permission from Elsevier

components then adhere to the metallic surface creating a polymeric hydrophobic barrier layer and undergo further crosslinking upon prolonged exposure to corrosive environment. As seen in Fig. 10.24, the silyl ester undergoes chemical changes upon exposure to an aqueous solution at pH 4. As seen in Fig. 10.24, the disappearance of the Si–O–CO–R peak (1190 cm^{-1}) together with a the formation of a peak at 1060 cm^{-1} (corresponding to Si–OH, Si–O–Si or Si–O–Al bonds) and shift in the ester carbonyl peak from 1720 cm^{-1} to 1713 cm^{-1} along with a decrease in the intensity of the ester peak 1376 cm^{-1} indicates that the silyl ester bond has broken by hydrolysis and new bond has been formed [39].

Moreover, FTIR and Raman spectroscopy was reported for the characterisation of molybdates incorporated into HTs [163]. FTIR spectroscopy was used to analyse Zn and Fe benzoates (insoluble forms) [164, 165]. FTIR spectroscopy has also been applied to confirm encapsulation of oils for alkyd self healing [40] and epoxies [41, 166]. It has been used to study a variety of inhibitors including Ce-salicylate [167], Ce-dibutyl phosphate [168] and Cu-Al DLH with either carbonates or sulphonates [169, 170].

Raman has been used to study epoxy healing in polymers [171] as well as chromate depletion zones as a result of leaching from scribes [172]. Raman is used extensively to study organic molecules and therefore has seen application to the identification of organic inhibitors such as organophosphates [71, 173–175]. It has less sensitivity for keys lines in inorganic compounds, but there are reports on CeO_2 thin films [176], single crystals [177] and mixed lanthanide oxides [178] as well as Ce-based conversion coatings [179, 180], aluminium oxides and hydroxides [181] and SiO_2-based sol-gel studies [182].

10.4.2 Attenuated Total Reflection Infrared Spectroscopy (ATR-FTIR) and Surface Enhanced ATR-Spectroscopy

ATR-FTIR is based on the phenomenon of total internal reflection and the generation of evanescent fields at an internal reflection element (IRE)/sample interface. An electromagnetic radiation travelling from a medium with a higher refractive index (n_1) to a medium with a lower refractive index (n_2) is found to totally interlay reflect in the incident medium if the incidence angle of the beam exceeds a certain value, known as the critical angle (θ_c) ZnSe, silicon, germanium and diamond are used typically as IRE due to their high refractive index. The penetration depth of the evanescent wave depends on the wavelength of the incident radiation, angle of incidence and the refractive index of the IRE and the sample. However, the spectral information obtained is the average over the wave that is decaying exponentially with the distance from the surface. The evanescent field penetration depth is usually in the low nanometer or the micrometer range.

Non-enhanced attenuated total reflection (ATR) spectroscopy provides information from the interphase region within depths between a few hundred nanometers up to few micrometers, where the information depth is dependant on on the angle of incidence and the refractive index of the ATR crystal [183]. However, the deposition of nanoparticles of coinage metals on surfaces of ATR crystals enable much more localized and enhanced fields at the interface between the analyzed medium and the optical crystal. For more information on the evolution of the SEIRAS technique, the readers are referred to the review article by Grundmeier and Stratmann and the references therein [5].

Attenuated total reflection infrared spectroscopy [184] has been applied for the determination of water ingress into polymer films. Due to the high sensitivity to the local environment of the adsorbates, ATR spectroscopy allows for an identification of different interaction modes of water at the interface, such as the formation of clusters or molecular water layers, hydrogen bonding between the polymer and water and surface hydroxylation of the metal oxide [29, 30, 33–35, 185].

Öhman and coworkers recently reported the application of ATR spectroscopy to the investigation of the burried zinc—conversion layer—polyester and aluminium —silane—epoxy interfaces [31, 32, 186]. Figure 10.25a shows the ATR-FTIR absorption spectra when the thin zinc film with a conversion layer was exposed to the NaSCN electrolyte. Figure 10.25b and c show the corresponding Bode impedance and phase angle plots.

Figure 10.25a, the broad band at about 3400 cm^{-1} was assigned to hydroxyl stretching vibrations of water and zinc hydroxide. The narrow peaks at 1650 and 2075 cm^{-1} were assigned to bending vibrations of water and C-N stretching vibrations of thiocyanate, respectively. The authors report that water-induced reactions during the exposure to the electrolyte result in the appearance of a band between 1150 and 850 cm^{-1} with a maximum at about 1030 cm^{-1} which is in agreement with the bending vibrations of $Zn(OH)_2$ at 1080 and 1025 cm^{-1} indicating a less ordered zinc hydroxide

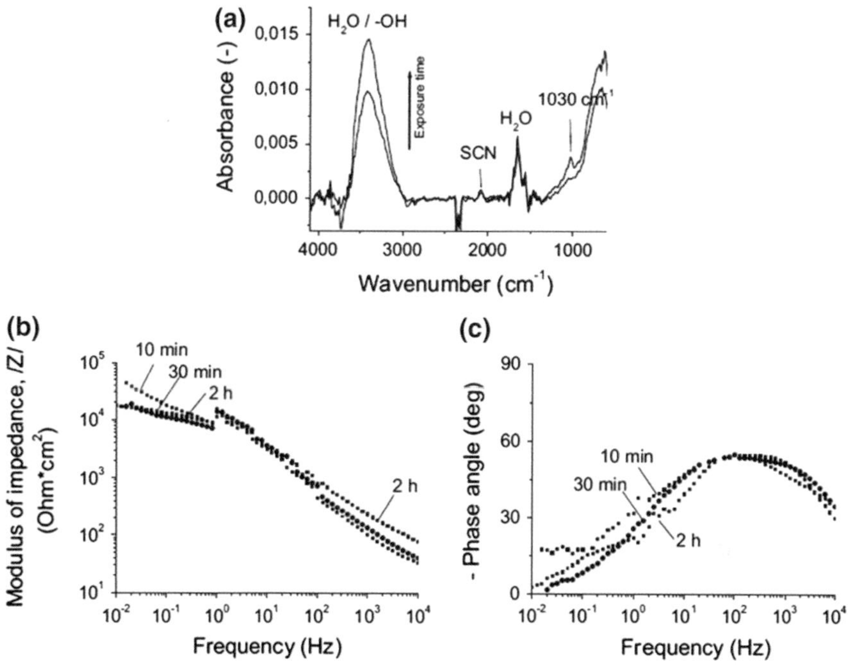

Fig. 10.25 **a** ATR-FTIR spectra of the zinc-conversion layer samples after 15 min and 2 h of exposure to a 1 M NaSCN solution and the corresponding **b** Bode impedance plot and **c** Bode phase angle plot after 10 min, 30 min and 2 h exposure. Reprinted from [32], Copyright (2011), with permission from Elsevier

film. The simultaneous increase of the band at 3400 cm^{-1} with exposure time and the water peak at 1650 cm^{-1} remaining constant further supports the hydroxide formation. Moreover, the formation of e.g. $Zn(OH)_2$ also contributes to the lower region of the hydroxyl stretching band, with a maximum at around 3260–3240 cm^{-1}. The bands between 1550 and 1350 cm^{-1} were assigned by the authors to carbonate containing reformation/oxidation products of the conversion layer. The corrosive activity at the substrate surface was also detected by the decrease of the modulus of impedance at low frequencies, indicating active processes at the metal surface [32].

In Fig. 10.26a and b, ATR-FTIR absorption spectra obtained at different times and the time dependence of the absorption peaks at 3400 and 1030 cm^{-1} are presented. In agreement with the results obtained on conversion films, the band between 3700 and 3000 cm^{-1} was present in the spectra obtained on polyester coated conversion films due to hydroxyl vibrations of water and of metal hydroxides within the buried conversion layer. The occurrence of the peak at about 2100 cm^{-1} indicates the presence of thiocyanate species at the interface, assuring that the electrolyte solution can ingress through the polymer film and reaches the interface. It is reported that, during the first 3 h of exposure, water diffuses to the interface through the polymer layer resulting in an increase of the absorbance at

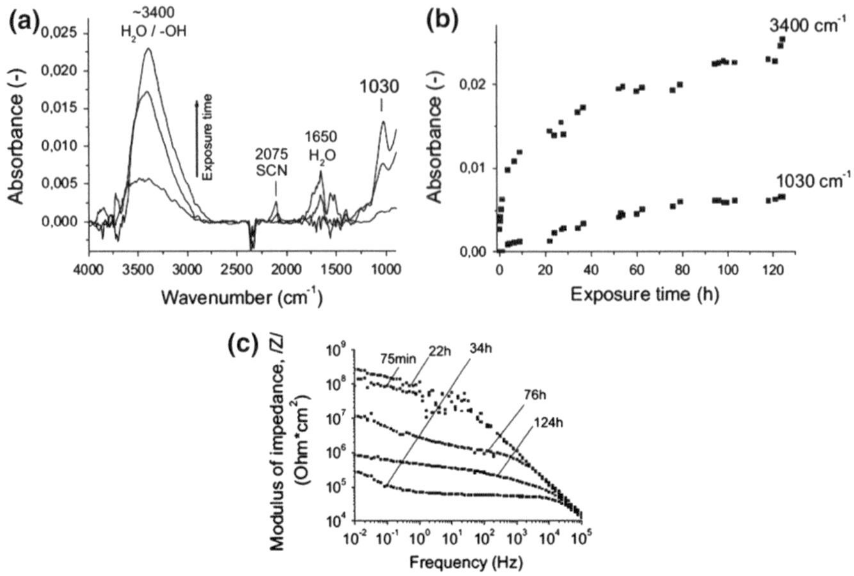

Fig. 10.26 **a** ATR-FTIR spectra of the zinc-conversion layer—polyester samples after 75 min, 22 and 120 h exposure to a 1 M NaSCN solution, **b** the time dependence of the peak maxima at 3400 and 1030 cm^{-1}, and **c** the Bode impedance plots after 75 min, 22, 34, 76 and 120 h exposure. Reprinted from [32], Copyright (2011), with permission from Elsevier

3400 cm^{-1} (see Fig. 10.26b). With the prolonged exposure, the absorbance of the band at 3400 cm^{-1} follows a similar trend as that of the band at 1030 cm^{-1} and does not reach a steady state, which is attributed by the authors to the formation of zinc corrosion products within the surface film. The authors support this conclusion with the increase in the modulus of impedance at low to intermediate frequencies within the first 22 h of exposure (see Fig. 10.26c) indicating that the formation of the corrosion products at the interface result in the sealing of conductive pathways and thus, prevent the transport of electrolyte to the metal surface [32].

In a recent study Öhman and coworkers have investigated the aluminium/silane/epoxy interface by means of ATR spectroscopy and electrochemical impedance spectroscopy (EIS) [186]. As seen in Fig. 10.27a, within 1 h of exposure a broad and strong FTIR band is detected between 3700 and 3000 cm^{-1} with the maximum at ~3430 cm^{-1}, which is assigned to symmetric and asymmetric —OH stretching vibrations of water molecules and of hydroxides possibly formed at the aluminum surface. The narrower peak observed at ~1650 cm^{-1} was assigned by the authors to the H–O–H bending vibrations of the water molecule. Öhman et al. observed that while—OH vibrations were detected within 1 h of exposure and the bending vibrations of water were not detected until after ~7 h of exposure. The negative bands observed at 1510 and 830 cm^{-1} were assigned to a decreased density of the aromatic structure of the epoxy backbone. After 10 h of exposure, the broad band observed below 1000 cm^{-1} with a maximum at 950 cm^{-1} was attributed

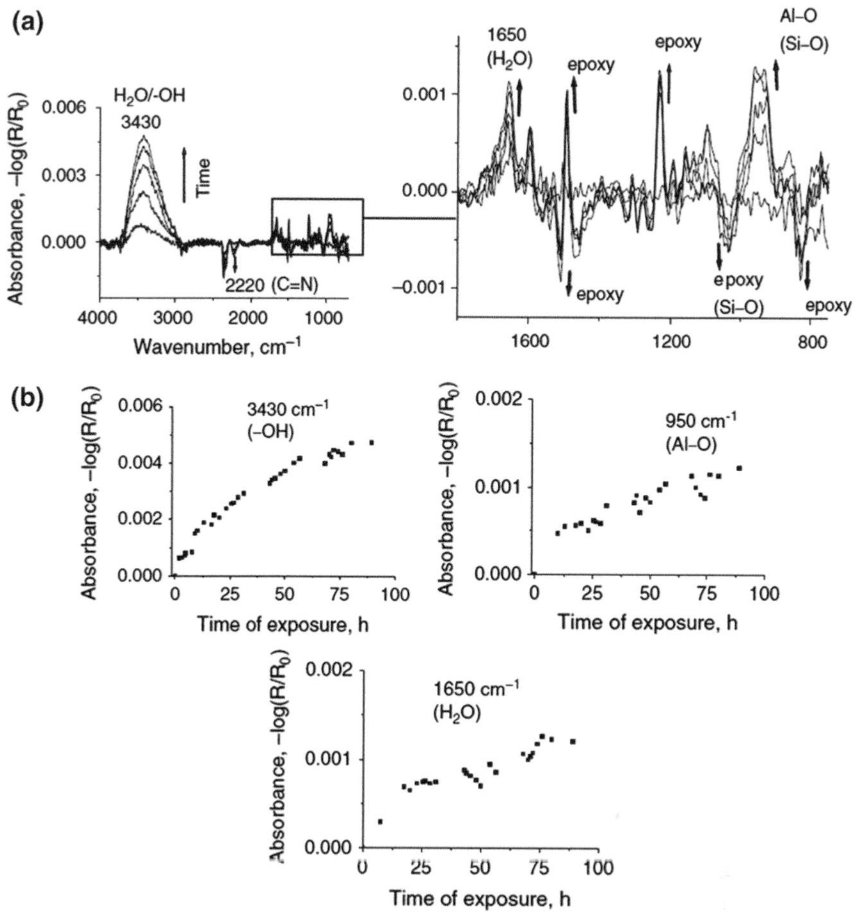

Fig. 10.27 **a** ATR-FTIR Kretschmann spectra for the aluminium film coated with silane and epoxy after 8, 24, 45, 70, 89 h exposure to 0.1 M NaCl and **b** the absorption intensity of the maxima at 3430, 950 and 1650 cm^{-1} as a function of the exposure time. Reprinted with permission from [186]. Copyright (2011) John Wiley & Sons, Ltd.

to water-induced alterations of the aluminum surface. According to the authors, this observation confirmed ongoing surface processes, which require at least a partial delamination of the silane film. The broad negative band observed between 1050 and 1000 cm^{-1} was assigned to more loosely bonded Si–O–Si, keeping in mind that changes in this spectral region could also stem from possible structural alterations of the epoxy network. The results from supporting EIS measurements (not presented here, for further information please refer to Ref. [186]) indicated that after 45 h of exposure, a resistive behavior can also be verified from the Bode impedance and Nyquist plots where the lowest resistance was registered after about 70 h of exposure. The authors discuss the possibility of an active surface process with electrolytic pathways opened and physically sealed by reaction products [186].

In 1980, Hartstein et al. reported the enhancement effect of IR-absorption on thin gold or silver films in the so-called Kretschmann-ATR configuration [187]. After this pioneering study, many reports were published addressing the nature of the enhancement effect [188, 189]. Osawa and coworkers broadened the application of this technique to equilibrium and kinetic studies at electrochemical interfaces [190–192]. Wandlowski et al. investigated the structure and the orientation of water at gold electrodes in contact with aqueous sulfuric acid by means of ATR-SEIRAS [193]. Hue et al. recently reported the application of ATR-SEIRAS for the investigation the electrosorption of a corrosion inhibitor benzotriazole (BTAH) on Fe electrode, expanding the applicability of the method for corrosion studies of iron. The measurements were performed as a function of the applied potential, providing valuable spectral information at positive potentials indicating the formation of a polymer-like surface complex [194].

The AFM images for the Au underfilm chemically deposited on a Si wafer without and with the Fe overfilm are presented in Fig. 10.28. Regular and uniform nanoparticles with a mean size of ca. 60–80 nm in diameter are present in the Au underfilm, whereas the Fe overfilm exhibits irregular nano-islands with significantly larger particle size.

The transmission spectra of the bulk BTAH powder and potential dependent SEIRA spectra for the adsorption of BTA on the Fe electrode are presented in Fig. 10.29. At potentials negative of −0.6 V, no characteristic modes corresponding to the BTAH/BTA molecule were observed in the SEIRA spectra. The intensity of the bands at 1487, 1447, 1390, 1288, and 1187 cm^{-1} increased with increasing potential, interpreted by the authors with structural changes occurring at the interface at positive potentials. The deprotonation of BTAH to form BTA- on the Fe surface was deduced from the comparison of the characteristic bands such as

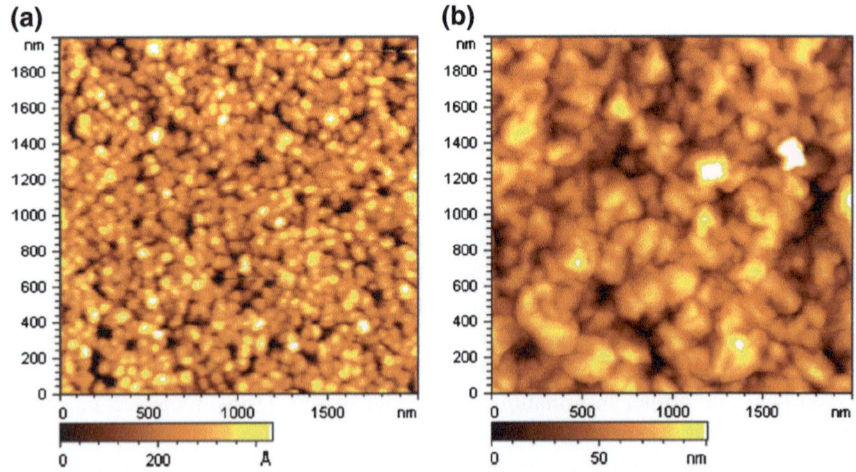

Fig. 10.28 AFM images for **a** Au underfilm chemically deposited on a Si wafer and **b** Fe overfilm electrodeposited on the Au underfilm. Reprinted with permission from [194]. Copyright (2010) American Chemical Society

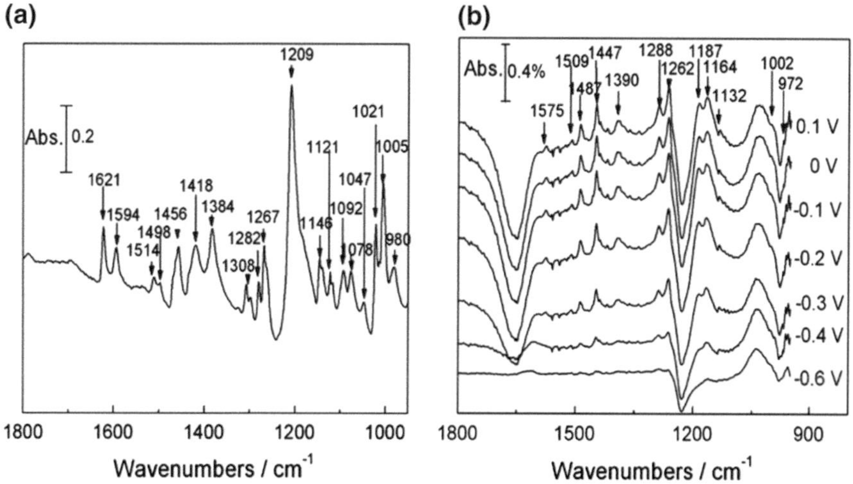

Fig. 10.29 **a** IR spectrum of BTAH powder in transmission mode. **b** Potential dependent SEIRA spectra for the adsorption of BTA on a Fe film electrode in 0.1 M $KClO_4$ + 10 mM BTAH; the potential was indicated as the spectra, and the reference spectrum was collected at −1.0 V. Reprinted with permission from [194]. Copyright (2010) American Chemical Society

in-plane N-H bending at 1092 cm^{-1} and the asymmetric triazole ring stretching at 1418 cm^{-1} corresponding to BTAH (see Fig. 10.29a), which disappeared upon adsorption of BTAH on the Fe surface (see Fig. 10.29b). Moreover, a red-shift from 1209 to 1187 cm^{-1} was observed with the triazole ring breathing mode accompanied with a decreased intensity, which was interpreted by the authors as the deprontonation of BTAH upon its adsorption on Fe electrode. This example illustrates the applicability of the ATR-SEIRAS technique for the study of the inhibition mechanism of BTAH on Fe, in particular, at relative high potentials [194].

A recent development for the application of ATR-SEIRAS technique was reported by Xue et al. by sandwiching an ultrathin water interlayer between a hemicylindrical ZnSe prism and a Si wafer as an integrated window [195] (Fig. 10.30). The hemicylindrical prism ZnSe crystal was used as the primary ATR element and the Pt-coated Si wafer was pressed with its uncoated side against the water-wetted reflecting plane of the ZnSe prism. The novel combined ATR element offers a wider spectral range in comparison to standard Si prisms, which suffer from strong IR absorption at frequencies lower than 1000 cm^{-1}. Important spectral fingerprints in this region containing crucial information regarding the molecular structure could be resolved by the combined ATR element approach which revealed that adsorbed formate acts as the intermediate moiety in the electrooxidation of methanol. Moreover, the results reported by Xue et al. clarified the orientational configuration of p-nitrobenzoate species on Pt electrode surfaces in acid solutions [195].

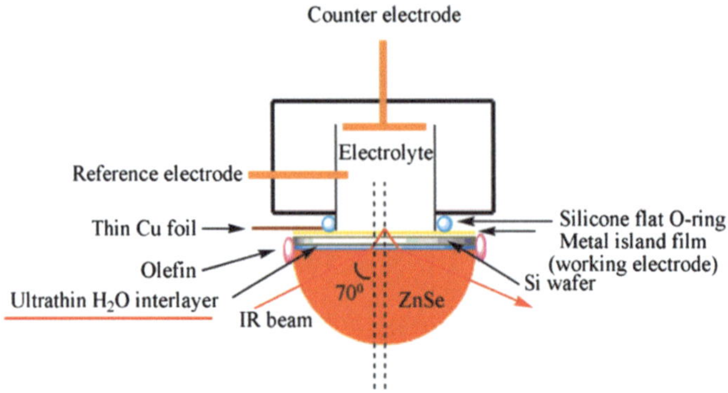

Fig. 10.30 Schematic diagram of the practically modified ATR-SEIRAS configuration containing an ultrathin H_2O interlayer in between two solid surfaces of a ZnSe hemicylindrical prism and a Si wafer. Reprinted with permission from [195]. Copyright (2008) American Chemical Society

10.4.3 Surface Enhanced Raman Spectroscopy (SERS)

SERS is currently one of the most sensitive techniques available to surface and materials science. Its ability of delivering specific chemical identification and the ease of combination with other experimental methods such as electrochemistry makes SERS an excellent tool for the in situ investigation of adsorption and corrosion processes, as well as self-healing mechanisms.

The enhancement of the Raman signal on coinage metals is explained by two mechanisms. The first one is the classical electromagnetic (EM) enhancement mechanism, where the creation of a localized surface plasmons (LSP) on the substrate surface transfer energy through an electric field to the target molecules allowing otherwise inaccessible vibrational transitions to be determined. The second mechanism is the charge transfer (CT) or chemical enhancement mechanism, where the incident radiation striking the roughened metallic surface results in a photon being excited within the metal to higher energy levels and promotes a charge transfer process to a vibrational level of the same energy. Vibrations in the energetic states take place, which result in the transfer of a photon of different frequency being passed back to the metallic energy levels and returned to the ground state of the metal [5, 183].

SERS has been successfully utilized for the investigation of the geometry, conformation, and chemical reaction of corrosion inhibitors on metal surfaces. The most prominent systems studied by SERS are pyridine and its derivatives adsorbed on silver electrodes, the surface reaction of thiols adsorbed on silver surfaces the orientation of organic molecules adsorbed on metal surfaces and protein adhesion [196–209].

In a recent paper, Maciel et al. presented the application of SERS technique to the investigation of the nature of the protective benzotriazole (BTAH) film formed on the surface of the 90/10 CuNi alloy [210]. The spectra presented in Fig. 10.31

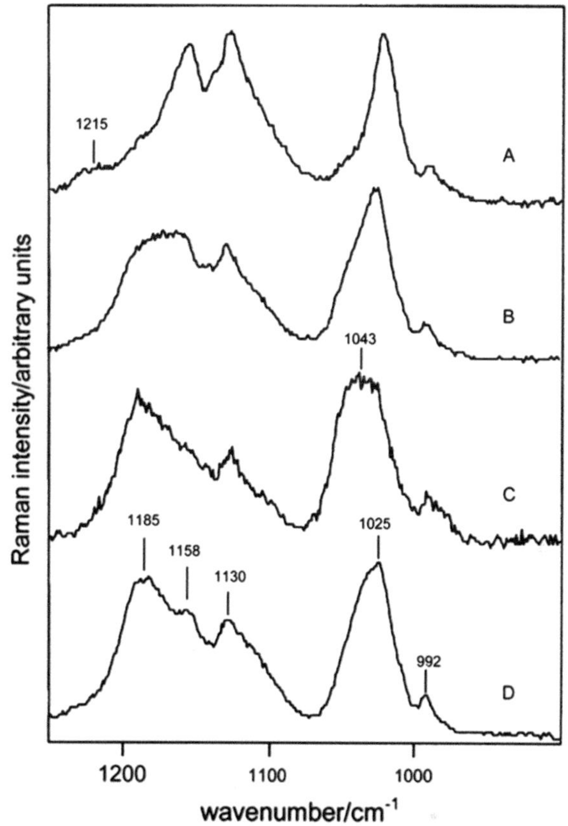

Fig. 10.31 In situ Raman spectra for the interface of the 90/10 CuNi alloy in 0.5 mol L^{-1} H_2SO_4 containing 3.0×10^{-2} mol L^{-1} BTAH, polarized at 0.4 V (*curve A*), 0.1 V (*curve B*), 0.0 V (*curve C*) and 0.1 V (*curve D*) versus Ag/AgCl reference electrode (numbers quoted in *curves* are peak position, in cm^{-1}). Reprinted from [210], Copyright (2008), with permission from Elsevier

were acquired in situ on a 90/10 CuNi alloy substrate during a polarisation experiment in the presence of the BTAH molecule. A very thorough spectral analysis was performed by Maciel et al., enabling the characterization of the film formed on the alloy surface as the cuprous benzotriazolate complex. A similar nickel complex was not detected in the SERS analysis, indicating a selective reaction mechanism. As seen in this example, the application of SERS can elucidate the protection mechanisms of corrosion inhibitors on metals and alloys, which a mechanistic understanding.

10.5 Concluding Remarks

A variety of spectroscopic and microscopic techniques are available for a detailed understanding of repair processes in materials and coatings. The healing process has to be addressed microscopically in most cases as the size of defects strongly influences the repair process. As in many cases the defect formation occurs at buried interfaces, the spectroscopic and microscopic methods have to be adapted to the studied system. Moreover in situ techniques which can monitor the defect formation and self-healing under ambient and corrosive conditions will become more and more important.

References

1. H. Leth-Olsen, K. Nisancioglu, Filiform corrosion of aluminium sheet. I. Corrosion behaviour of painted material. Corros. Sci. **40**(7), 1179–1194 (1998). doi:10.1016/S0010-938X(98)00026-2
2. R. Posner, T. Titz, K. Wapner, M. Stratmann, G. Grundmeier, Transport processes of hydrated ions at polymer/oxide/metal interfaces Part 2. Transport on oxide covered iron and zinc surfaces. Electrochim. Acta. **54**(3), 900–908 (2009). doi:10.1016/j.electacta.2008.07.011
3. J. Wielant, R. Posner, R. Hausbrand, G. Grundmeier, H. Terryn, Cathodic delamination of polyurethane films on oxide covered steel - Combined adhesion and interface electrochemical studies. Corros. Sci. **51**(8), 1664–1670 (2009). doi:10.1016/j.corsci.2009.04.014
4. F. Deflorian, S. Rossi, L. Fedrizzi, Silane pre-treatments on copper and aluminium. Electrochim. Acta **51**(27), 6097–6103 (2006). doi:10.1016/j.electacta.2006.02.042
5. G. Grundmeier, M. Stratmann, Adhesion and De-adhesion mechanisms at polymer/metal interfaces: mechanistic understanding based on in situ studies of buried interfaces. Annu. Rev. Mater. Res. **35**, 571–615 (2005). doi:10.1146/annurev.matsci.34.012703.105111
6. P. Bonora, F. Deflorian, L. Fedrizzi, Electrochemical impedance spectroscopy as a tool for investigating underpaint corrosion. Electrochim. Acta. **41**(7–8), 1073–1082 (1996). doi:10.1016/0013-4686(95)00440-8
7. L.P. Haack, M.A. DeBolt, S.L. Kaberline, J.E. Devries, R.A. Dickie, XPS mapping technique for the identification of adhesion loss mechanisms. Surf. Interface Anal. **20**(2), 115–120 (1993). doi:10.1002/sia.740200205
8. W.J. van Ooij, A. Sabata, A.D. Appelhans, Application of surface analysis techniques to the study of paint/metal interfaces related to adhesion and corrosion performance. Surf. Interface Anal. **17**(7), 403–420 (1991). doi:10.1002/sia.740170703
9. O. Ozcan, K. Pohl, B. Ozkaya, G. Grundmeier, Molecular studies of adhesion and De-Adhesion on ZnO nanorod film-covered metals. J. Adhes. **89**(2), 128–139 (2013). doi:10.1080/00218464.2012.731928
10. K. Pohl, J. Otte, P. Thissen, M. Giza, M. Maxisch, B. Schuhmacher, G. Grundmeier, Adsorption and stability of self-assembled organophosphonic acid monolayers on plasma modified Zn-Mg-Al alloy surfaces. Surf. Coat. Technol. **218**, 99–107 (2013). doi:10.1016/j.surfcoat.2012.12.035
11. G. Fonder, I. Minet, C. Volcke, S. Devillers, J. Delhalle, Z. Mekhalif, Anchoring of alkylphosphonic derivatives molecules on copper oxide surfaces. Appl. Surf. Sci. **257**(14), 6300–6307 (2011). doi:10.1016/j.apsusc.2011.02.071
12. M. Maxisch, P. Thissen, M. Giza, G. Grundmeier, Interface chemistry and molecular interactions of phosphonic acid self-assembled monolayers on oxyhydroxide-covered aluminum in humid environments. Langmuir **27**(10), 6042–6048 (2011). doi:10.1021/la200445x
13. M. Santa, R. Posner, G. Grundmeier, Wet- and corrosive De-Adhesion processes of water-borne epoxy film coated steel. J. Electrochem. Soc. **158**(3), C36–C41 (2011). doi:10.1149/1.3525240

14. K. Wapner, M. Stratmann, G. Grundmeier, Structure and stability of adhesion promoting aminopropyl phosphonate layers at polymer/aluminium oxide interfaces. Int. J. Adhes. Adhes. **28**(1–2), 59–70 (2008). doi:10.1016/j.ijadhadh.2007.05.001
15. G. Klimow, N. Fink, G. Grundmeier, Electrochemical studies of the inhibition of the cathodic delamination of organically coated galvanised steel by thin conversion films. Electrochim. Acta **53**(3), 1290–1299 (2007). doi:10.1016/j.electacta.2007.05.045
16. G. Grundmeier, B. Rossenbeck, K. Roschmann, P. Ebbinghaus, M. Stratmann, Corrosion protection of Zn-phosphate containing water borne dispersion coatings - Part 2: investigations of the corrosive de-adhesion of model latex coatings on iron. Corros. Sci. **48**(11), 3716–3730 (2006). doi:10.1016/j.corsci.2006.01.007
17. T. Narayanan, Surface pretreatment by phosphate conversion coatings - a review. Rev. Adv. Mater. Sci. **9**(2), 130–177 (2005)
18. G. Grundmeier, C. Reinartz, M. Rohwerder, M. Stratmann, Corrosion properties of chemically modified metal surfaces. Electrochim. Acta **43**(1–2), 165–174 (1998). doi:10.1016/S0013-4686(97)00221-1
19. I. Maege, E. Jaehne, A. Henke, H. Adler, C. Bram, C. Jung, M. Stratmann, Self-assembling adhesion promoters for corrosion resistant metal polymer interfaces. Prog. Org. Coat. **34**(1–4), 1–12 (1998)
20. I. Maege, E. Jaehne, A. Henke, H. Adler, C. Bram, C. Jung, M. Stratmann, Ultrathin organic layers for corrosion protection. Macromol. Symp. **126**, 7–24 (1998). doi:10.1002/masy.19981260104
21. M. Wiesener, R. Regenspurger, M. Pilz, D. Shchukin, A. Latnikova, J. Yang, G. Grundmeier, In-situ contact angle studies of the release of water displacing agents from capsule filled organic coatings. Surf. Coat. Technol. **206**(21), 4481–4487 (2012). doi:10.1016/j.surfcoat.2012.05.021
22. M. Giza, G. Grundmeier, Combination of FTIR reflection absorption spectroscopy and work function measurements for in situ studies of plasma modified passive films on MgZn2. Plasma Processes Polym. **8**(7), 607–616 (2011). doi:10.1002/ppap.201000198
23. R. Posner, M. Santa, G. Grundmeier, Wet- and corrosive De-Adhesion processes of water-borne epoxy film coated steel. J. Electrochem. Soc. **158**(3), C29–C35 (2011). doi:10.1149/1.3525239
24. R. Posner, K. Wapner, S. Amthor, K. Roschmann, G. Grundmeier, Electrochemical investigation of the coating/substrate interface stability for styrene/acrylate copolymer films applied on iron. Corros. Sci. **52**(1), 37–44 (2010). doi:10.1016/j.corsci.2009.08.038
25. T. Titz, F. Horzenberger, K. Van den Bergh, G. Grundmeier, Correlation of interfacial electrode potential and corrosion resistance of plasma polymer coated galvanized steel. Part 1: ultra-thin plasma polymer films of varying thickness. Corros. Sci. **52**(2), 369–377 (2010). doi:10.1016/j.corsci.2009.09.024
26. T. Titz, F. Horzenberger, K. Van den Bergh, G. Grundmeier, Correlation of interfacial electrode potential and corrosion resistance of plasma polymer coated galvanized steel. Part 2: influence of forming induced defects. Corros. Sci. **52**(2), 378–386 (2010). doi:10.1016/j.corsci.2009.09.027
27. M. Geisler, D. Horinek, T. Hugel, Single molecule adhesion mechanics on rough surfaces. Macromolecules **42**(23), 9338–9343 (2009). doi:10.1021/ma9017372
28. R. Posner, G. Giza, R. Vlasak, G. Grundmeier, In situ electrochemical scanning Kelvin probe blister-test studies of the de-adhesion kinetics at polymer/zinc oxide/zinc interfaces. Electrochim. Acta **54**(21), 4837–4843 (2009). doi:10.1016/j.electacta.2009.03.089
29. K. Wapner, M. Stratmann, G. Grundmeier, In situ infrared spectroscopic and scanning Kelvin probe measurements of water and ion transport at polymer/metal interfaces. Electrochim. Acta. **51**(16), 3303–3315 (2006). doi:10.1016/j.electacta.2005.09.024
30. I. Linossier, F. Gaillard, M. Romand, T. Nguyen, A spectroscopic technique for studies of water transport along the interface and hydrolytic stability of polymer/substrate systems. J. Adhes. **70**(3–4), 221–239 (1999). doi:10.1080/00218469908009557

31. M. Ohman, D. Persson, ATR-FTIR Kretschmann spectroscopy for interfacial studies of a hidden aluminum surface coated with a silane film and epoxy I. Characterization by IRRAS and ATR-FTIR. Surf. Interface Anal. **44**(2), 133–143 (2012). doi:10.1002/sia.3779
32. M. Ohman, D. Persson, D. Jacobsson, In situ studies of conversion coated zinc/polymer surfaces during exposure to corrosive conditions. Prog. Org. Coat. **70**(1), 16–22 (2011). doi:10.1016/j.porgcoat.2010.09.012
33. M. Ohman, D. Persson, C. Leygraf, A spectroelectrochemical study of metal/polymer interfaces by simultaneous in situ ATR-FTIR and EIS. Electrochem. Solid State Lett. **10**(4), C27–C30 (2007). doi:10.1149/1.2436647
34. M. Ohman, D. Persson, An integrated in situ ATR-FTIR and EIS set-up to study buried metal-polymer interfaces exposed to an electrolyte solution. Electrochim. Acta. **52**(16), 5159–5171 (2007). doi:10.1016/j.electacta.2007.02.007
35. M. Ohman, D. Persson, C. Leygraf, In situ ATR-FTIR studies of the aluminium/polymer interface upon exposure to water and electrolyte. Prog. Org. Coat. **57**(1), 78–88 (2006). doi:10.1016/j.porgcoat.2006.07.002
36. M. Kendig, M. Hon, A hydrotalcite-like pigment containing an organic anion corrosion inhibitor. Electrochem. Solid State Lett. **8**(3), B10–B11 (2005). doi:10.1149/1.1857743
37. H.N. McMurray, G. Williams, Inhibition of filiform corrosion on organic-coated aluminum alloy by hydrotalcite-like anion-exchange pigments. Corrosion **60**(3), 219–228 (2004)
38. G. Williams, H.N. McMurray, Anion-exchange inhibition of filiform corrosion on organic coated AA2024-T3 aluminum alloy by hydrotalcite-like pigments. Electrochem. Solid State Lett. **6**(3), B9–B11 (2003). doi:10.1149/1.1539771
39. S.J. Garcia, H.R. Fischer, P.A. White, J. Mardel, Y. Gonzalez-Garcia, J.M.C. Mol, A.E. Hughes, Self-healing anticorrosive organic coating based on an encapsulated water reactive silyl ester: synthesis and proof of concept. Prog. Org. Coat. **70**(2–3), 142–149 (2011). doi:10.1016/j.porgcoat.2010.11.021
40. C. Suryanarayana, K.C. Rao, D. Kumar, Preparation and characterization of microcapsules containing linseed oil and its use in self-healing coatings. Prog. Org. Coat. **63**(1), 72–78 (2008). doi:10.1016/j.porgcoat.2008.04.008
41. T. Yin, M.Z. Rong, M.Q. Zhang, G.C. Yang, Self-healing epoxy composites - Preparation and effect of the healant consisting of microencapsulated epoxy and latent curing agent. Compos. Sci. Technol. **67**(2), 201–212 (2007). doi:10.1016/j.compscitech.2006.07.028
42. A. Kumar, L.D. Stephenson, J.N. Murray, Self-healing coatings for steel. Prog. Org. Coat. **55** (3), 244–253 (2006). doi:10.1016/j.porgcoat.2005.11.010
43. E.N. Brown, S.R. White, N.R. Sottos, Microcapsule induced toughening in a self-healing polymer composite. J. Mater. Sci. **39**(5), 1703–1710 (2004). doi:10.1023/B:JMSC.0000016173.73733.dc
44. L.A. Feldkamp, L.C. Davis, J.W. Kress, PRACTICAL CONE-BEAM ALGORITHM. J. Opt. Soc. Am. a-Optics Image Sci. Vis. **1**(6), 612–619 (1984). doi:10.1364/josaa.1.000612
45. O. Hemberg, M. Otendal, H.M. Hertz, Liquid-metal-jet anode electron-impact X-ray source. Appl. Phys. Lett. **83**(7), 1483–1485 (2003). doi:10.1063/1.1602157
46. S.C. Mayo, A.M. Tulloh, A. Trinchi, S.Y.S. Yang, Data-constrained microstructure characterization with multispectrum X-ray micro-ct. Microsc. Microanal. **18**(3), 524–530 (2012)
47. X.C. Pan, E.Y. Sidky, M. Vannier, Why do commercial CT scanners still employ traditional, filtered back-projection for image reconstruction? Inverse Prob. **25**(12) (2009). doi:10.1088/0266-5611/25/12/123009
48. D. Paganin, S.C. Mayo, T.E. Gureyev, P.R. Miller, S.W. Wilkins, Simultaneous phase and amplitude extraction from a single defocused image of a homogeneous object. J Microsc. - Oxford **206**, 33–40 (2002)
49. A.E. Hughes, A.M. Glenn, N. Wilson, A. Moffatt, A.J. Morton, R.G. Buchheit, A consistent description of intermetallic particle composition: an analysis of ten batches of AA2024-T3. Surf Interface Anal, n/a-n/a (2013). doi:10.1002/sia.5207

50. R.G. Buchheit, R.P. Grant, P.F. Hlava, B. McKenzie, G.L. Zender, Local dissolution phenomena associated with S phase (Al2CuMg) particles in aluminum alloy 2024-T3. J. Electrochem. Soc. **144**(8), 2621–2628 (1997)
51. A.E. Hughes, N. Birbilis, J.M.C. Mol, S.J. Garcia, X. Zhou, G.E. Thompson, High strength Al-Alloys: microstructure, corrosion and principles of protection, in *Recent Trends in Processing and Degradation of Aluminium Alloys*, ed. by Z. Ahmad (Intech Publishing, Rijeka, 2011)
52. X. Zhou, C. Luo, T. Hashimoto, A.E. Hughes, G.E. Thompson, Study of localized corrosion in AA2024 aluminium alloy using electron tomography. Corros. Sci. **58**, 299–306 (2012). doi:10.1016/j.corsci.2012.02.001
53. S.M. Ghahari, A.J. Davenport, T. Rayment, T. Suter, J.P. Tinnes, C. Padovani, J.A. Hammons, M. Stampanoni, F. Marone, R. Mokso, In situ synchrotron X-ray micro-tomography study of pitting corrosion in stainless steel. Corros. Sci. **53**(9), 2684–2687 (2011)
54. S.P. Knight, M. Salagaras, A.M. Wythe, F. De Carlo, A.J. Davenport, A.R. Trueman, In situ X-ray tomography of intergranular corrosion of 2024 and 7050 aluminium alloys. Corros. Sci. **52**(12), 3855–3860 (2010). doi:10.1016/j.corsci.2010.08.026
55. Y. Wang, Y. Yang, T. Xiao, K. Liu, B. Clennell, G. Zhang, H.Wang, Synchrotron-based data-constrained modeling analysis of microscopic mineral distributions in limestone. Int. J. Geosci. **4**(2), 344–351 (2013). doi:10.4236/ijg.2013.42032
56. S. Miller, B. Singh, S. Cool, G. Entine, L. Campbell, R. Bishel, R. Rushing, V.V. Nagarkar, Ultrahigh-speed X-ray imaging of hypervelocity projectiles. Nucl. Instrum. Methods Phys. Res., Sect. A-Accelerators Spectrometers Detectors and Associated Equipment **648**, S293–S296 (2011). doi:10.1016/j.nima.2010.11.048
57. G.A. Johansen, U. Hampel, B.T. Hjertaker, Flow imaging by high speed transmission tomography. Appl. Radiat. Isot. **68**(4–5), 518–524 (2010). doi:10.1016/j.apradiso.2009.09.004
58. A.E. Hughes, A. Trinchi, F.F. Chen, Y.S. Yang, I.S. Cole, S. Sellaiyan, J. Carr, P.D. Lee, G. E. Thompson, T.Q. Xiao, Revelation of intertwining organic and inorganic fractal structures in polymer coatings. Adv. Mater. n/a-n/a (2014). doi:10.1002/adma.201400561
59. D.M.L. Cooper, L.D. Chapman, Y. Carter, Y. Wu, A. Panahifar, H.M. Britz, B. Bewer, W. Zhouping, M.J.M. Duke, M. Doschak, Three dimensional mapping of strontium in bone by dual energy K-edge subtraction imaging. Phys. Med. Biol. **57**(18), 5777–5786 (2012). doi:10.1088/0031-9155/57/18/5777
60. H. Toda, K. Shimizu, K. Uesugi, Y. Suzuki, M. Kobayashi, Application of dual-energy K-Edge subtraction imaging to assessment of heat treatments in Al-Cu alloys. Mater. Trans. **51**(11), 2045–2048 (2010). doi:10.2320/matertrans.L-M2010819
61. C.H. Arns, F. Bauget, A. Ghous, A. SakellarioU, T.J. Senden, A.P. Sheppard, R.M. Sok, W. V. Pinczewski, J.C. Kelly, M.A. Knackstedt, Digital core laboratory: petrophysical analysis from 3D imaging of reservoir core fragments. Petrophysics **46**(4), 260–277 (2005)
62. M. Van Geet, R. Swennen, M. Wevers, Towards 3-D petrography: application of microfocus computer tomography in geological science. Comput. Geosci. **27**(9), 1091–1099 (2001)
63. Z.R. Ying, R. Naidu, C.R. Crawford, Dual energy computed tomography for explosive detection. J X-Ray Sci. Technol. **14**(4), 235–256 (2006)
64. S. Neethirajan, D.S. Jayas, C. Karunakaran, Dual energy X-ray image analysis for classifying vitreousness in durum wheat. Postharvest Biol. Technol. **45**(3), 381–384 (2007). doi:10.1016/j.postharvbio.2007.03.009
65. Y.S. Yang, A data-constrained non-linear optimisation approach to a data-constrained model for compositional microstructure prediction. Lecture Notes Inform. Technol. **15**, 198–205 (2012)
66. H.P. Wang, Y.S. Yang, Y.D. Wang, J.L. Yang, J. Jia, Y.H. Nie, Data-constrained modelling of an anthracite coal physical structure with multi-spectrum synchrotron X-ray CT. Fuel **106**, 219–225 (2013)

67. Y.S. Yang, A. Tulloh, I. Cole, S. Furman, A. Hughes, A data-constrained computational model for morphology structures. J. Aust. Ceram. Soc. **43**(2), 159–164 (2007)
68. S. Yang, T.E. Gureyev, M.B. Tulloh, M.B. Clennell, M. Pervukhina, Feasibility of a data constrained prediction of hydrocarbon reservoir sandstone microstructures. Meas. Sci. Technol. **21** (2010). doi:10.1088/0957-0233/21/4/047001
69. Y.S. Yang, K.Y. Liu, S. Mayo, A. Tulloh, M.B. Clennell, T.Q. Xiao, A data-constrained modelling approach to sandstone microstructure characterisation. J. Petrol. Sci. Eng. **105**, 76–83 (2013)
70. S. Yang, D.C. Gao, T. Muster, A. Tulloh, S. Furman, S. Mayo, A. Trinchi, Microstructure of a paint primer - a data-constrained modeling analysis. In: J.F. Nie, A. Morton (eds.) Pricm 7, Pts 1–3, vol 654–656. Materials Science Forum, pp. 1686–1689 (2010)
71. A.E. Hughes, S. Mayo, Y.S. Yang, T. Markley, S.V. Smith, S. Sellaiyan, A. Uedono, S.G. Hardin, T.H. Muster, Using X-ray tomography, PALS and Raman spectroscopy for characterization of inhibitors in epoxy coatings. Prog. Org. Coat. **74**(4), 726–733 (2012)
72. A. Trinchi, Y.S. Yang, J.Z. Huang, P. Falcaro, D. Buso, L.Q. Cao, Study of 3D composition in a nanoscale sample using data-constrained modelling and multi-energy X-ray CT. Model. Simul. Mater. Sci. Eng. **20**(1) (2012)
73. B.S. Tanem, O. Lunder, A. Borg, J. Mardalen, AFM adhesion force measurements on conversion-coated EN AW-6082-T6 aluminium. Int. J. Adhes. Adhes. **29**(5), 471–477 (2009). doi:10.1016/j.ijadhadh.2008.09.005
74. B. Bhushan, V.N. Koinkar, Nanoindentation hardness measurements using atomic-force microscopy. Appl. Phys. Lett. **64**(13), 1653–1655 (1994). doi:10.1063/1.111949
75. F. Hane, B. Moores, M. Amrein, Z. Leonenko, Effect of SP-C on surface potential distribution in pulmonary surfactant: atomic force microscopy and Kelvin probe force microscopy study. Ultramicroscopy **109**(8), 968–973 (2009). doi:10.1016/j.ultramic.2009.03.046
76. E. Finot, Y. Leonenko, B. Moores, L. Eng, M. Amrein, Z. Leonenko, Effect of cholesterol on electrostatics in lipid-protein films of a pulmonary surfactant. Langmuir **26**(3), 1929–1935 (2010). doi:10.1021/La904335m
77. A. Wadas, P. Grutter, H.J. Guntherodt, Analysis of magnetic bit pattern by magnetic force microscopy. J. Vac. Sci. Technol. A **8**(1), 416–420 (1990). doi:10.1116/1.576410
78. I. Sokolov, Q.K. Ong, H. Shodiev, N. Chechik, D. James, M. Oliver, AFM study of forces between silica, silicon nitride and polyurethane pads. J. Colloid. Interf. Sci. **300**(2), 475–481 (2006). doi:10.1016/j.jcis.2006.04.023
79. B.V. Derjaguin, V.M. Muller, Y.P. Toporov, Effect of contact deformations on the adhesion of particles. Prog. Surf. Sci. **45**(1–4), 131–143 (1994). doi:10.1016/0079-6816(94)90044-2
80. M.E. Dokukin, I. Sokolov, Quantitative mapping of the elastic modulus of soft materials with HarmoniX and peak force QNM AFM modes. Langmuir **28**(46), 16060–16071 (2012). doi:10.1021/La302706b
81. O. Sahin, S. Magonov, C. Su, C.F. Quate, O. Solgaard, An atomic force microscope tip designed to measure time-varying nanomechanical forces. Nat. Nanotechnol. **2**(8), 507–514 (2007). doi:10.1038/nnano.2007.226
82. J.A. Yoon, J. Kamada, K. Koynov, J. Mohin, R. Nicolay, Y.Z. Zhang, A.C. Balazs, T. Kowalewski, K. Matyjaszewski, Self-Healing polymer films based on thiol-disulfide exchange reactions and self-healing kinetics measured using atomic force microscopy. Macromolecules **45**(1), 142–149 (2012). doi:10.1021/Ma2015134
83. P. Swift, Adventitious carbon - the panacea for energy referencing. Surf. Interface Anal. **4**(2), 47–51 (1982). doi:10.1002/sia.740040204
84. C.D. Wagner, L.H. Gale, R.H. Raymond, 2-dimensional chemical-state plots - standardized data set for use in identifying chemical-states by X-ray photoelectron-spectroscopy. Anal. Chem. **51**(4), 466–482 (1979). doi:10.1021/ac50040a005
85. Y. Uwamino, T. Ishizuka, H. Yamatera, Charge correction by gold deposition onto non-conducting samples in X-ray photoelectron-spectroscopy. J. Electron Spectrosc. Relat. Phenom. **23**(1), 55–62 (1981). doi:10.1016/0368-2048(81)85036-0

86. S. Kohiki, K. Oki, An appraisal of evaporated gold as an energy reference in X-ray photoelectron-spectroscopy. J. Electron Spectrosc. Relat. Phenom. **36**(1), 105–110 (1985). doi:10.1016/0368-2048(85)80011-6
87. A.E. Hughes, B.A. Sexton, The use of implanted ar as a binding-energy reference - comment. J. Electron Spectrosc. Relat. Phenom. **50**(3), C15–C18 (1990)
88. S. Kohiki, T. Ohmura, K. Kusao, A new charge-correction method in X-ray photoelectron-spectroscopy. J. Electron Spectrosc. Relat. Phenom. **28**(4), 229–237 (1983). doi:10.1016/0368-2048(83)80001-2
89. C.S. Fadley, D.A. Shirley, Multiplet splitting of metal-atom electron binding energies. Phys. Rev. A **2**(4), 1109–1120 (1970). doi:10.1103/PhysRevA.2.1109
90. D. Briggs, V.A. Gibson, Direct observation of multiplet splitting in 2p photoelectron peaks of cobalt complexes. Chem. Phys. Lett. **25**(4), 493–496 (1974). doi:10.1016/0009-2614(74)85350-9
91. H. Basch, Multiplet splitting of core level binding-energies in paramagnetic species and unpaired orbital spin density distribution. Chem. Phys. Lett. **20**(3), 233–237 (1973). doi:10.1016/0009-2614(73)85165-6
92. R.P. Gupta, S.K. Sen, Calculation of multiplet structure of core p-vacancy levels .Phys. Rev. B **12**(1), 15–19 (1975). doi:10.1103/PhysRevB.12.15
93. R.P. Gupta, S.K. Sen, Calculation of multiplet structure of core para-vacancy levels. Phys. Rev. B **10**(1), 71–77 (1974). doi:10.1103/PhysRevB.10.71
94. A.P. Grosvenor, B.A. Kobe, M.C. Biesinger, N.S. McIntyre, Investigation of multiplet splitting of Fe 2p XPS spectra and bonding in iron compounds. Surf. Interface Anal. **36**(12), 1564–1574 (2004). doi:10.1002/sia.1984
95. M. Aronniemi, J. Sainio, J. Lahtinen, Chemical state quantification of iron and chromium oxides using XPS: the effect of the background subtraction method. Surf. Sci. **578**(1–3), 108–123 (2005). doi:10.1016/j.susc.2005.01.019
96. M. Mullet, V. Khare, C. Ruby, XPS study of Fe(II)-Fe(III) (oxy)hydroxycarbonate green rust compounds. Surf. Interface Anal. **40**(3–4), 323–328 (2008). doi:10.1002/sia.2758
97. M. Mullet, Y. Guillemin, C. Ruby, Oxidation and deprotonation of synthetic Fe-II-Fe-III (oxy)hydroxycarbonate green rust: an X-ray photoelectron study. J. Solid State Chem. **181**(1), 81–89 (2008). doi:10.1016/j.jssc.2007.10.026
98. Y.S. Hao, F.C. Liu, E.H. Han, Protection of epoxy coatings containing polyaniline modified ultra-short glass fibers. Prog. Org. Coat. **76**(4), 571–580 (2013). doi:10.1016/j.porgcoat.2012.11.012
99. N. Granizo, J.M. Vega, D. de la Fuente, B. Chico, M. Morcillo, Ion-exchange pigments in primer paints for anticorrosive protection of steel in atmospheric service: anion-exchange pigments. Prog. Org. Coat. **76**(2–3), 411–424 (2013). doi:10.1016/j.porgcoat.2012.10.009
100. D.A. Shirley, High-resolution X-ray photoemission spectrum of valence bands of gold. Phys. Rev. B **5**(12), 4709–4714 (1972). doi:10.1103/PhysRevB.5.4709
101. J. Vegh, The analytical form of the shirley-type background. J. Electron Spectrosc. Relat. Phenom. **46**(4), 411–417 (1988). doi:10.1016/0368-2048(88)85038-2
102. A. Proctor, P.M.A. Sherwood, Data-analysis techniques in X-ray photo-electron spectroscopy. Anal. Chem. **54**(1), 13–19 (1982). doi:10.1021/ac00238a008
103. S. Tougaard, B. Jorgensen, Inelastic background intensities in XPS spectra. Surf. Sci. **143**(2–3), 482–494 (1984). doi:10.1016/0039-6028(84)90554-5
104. S. Tougaard, P. Sigmund, Influence of elastic and inelastic-scattering on energy-spectra of electrons emitted from solids. Phys. Rev. B **25**(7), 4452–4466 (1982). doi:10.1103/PhysRevB.25.4452
105. S. Tougaard, Inelastic background correction and quantitative surface-analysis. J. Electron Spectrosc. Relat. Phenom. **52**, 243–271 (1990). doi:10.1016/0368-2048(90)85022-2
106. C.S. Fadley, D.A. Shirley, X-ray photoelectron spectroscopic study of iron cobalt nickel copper and platinum. Phys. Rev. Lett. **21**(14), 980–983 (1968). doi:10.1103/PhysRevLett.21.980

107. S.J. Hinder, J.F. Watts, C. Lowe, Interface analysis and compositional depth profiling by XPS of polymer coatings prepared using ultra-low-angle microtomy. Surf. Interface Anal. **36** (8), 1032–1036 (2004). doi:10.1002/sia.1830
108. S.J. Hinder, C. Lowe, J.T. Maxted, J.F. Watts, A ToF-SIMS investigation of a buried polymer/polymer interface exposed by ultra-low-angle microtomy. Surf. Interface Anal. **36** (12), 1575–1581 (2004). doi:10.1002/sia.1985
109. G. Williams, H.N. McMurray, Inhibition of filiform corrosion on polymer coated AA2024-T3 by hydrotalcite-like pigments incorporating organic anions. Electrochem. Solid State Lett. **7**(5), B13–B15 (2004). doi:10.1149/1.1691529
110. S. Bohm, H.N. McMurray, S.M. Powell, D.A. Worsley, Novel environment friendly corrosion inhibitor pigments based on naturally occurring clay minerals. Materials Corrosion-Werkstoffe Und Korrosion **52**(12), 896–903 (2001)
111. A.N. Salak, J. Tedim, A.I. Kuznetsova, M.L. Zheludkevich, M.G.S. Ferreira, Anion exchange in Zn-Al layered double hydroxides: in situ X-ray diffraction study. Chem. Phys. Lett. **495**(1–3), 73–76 (2010). doi:10.1016/j.cplett.2010.06.041
112. M.L. Zheludkevich, S.K. Poznyak, L.M. Rodrigues, D. Raps, T. Hack, L.F. Dick, T. Nunes, M.G.S. Ferreira, Active protection coatings with layered double hydroxide nanocontainers of corrosion inhibitor. Corros. Sci. **52**(2), 602–611 (2010). doi:10.1016/j.corsci.2009.10.020
113. S.P.V. Mahajanarn, R.G. Buchheit, Characterization of inhibitor release from Zn-Al-V10O28 (6-) hydrotalcite pigments and corrosion protection from hydrotalcite-pigmented epoxy coatings. Corrosion **64**(3), 230–240 (2008)
114. R.G. Buchheit, S.B. Mamidipally, P. Schmutz, H. Guan, Active corrosion protection in Ce-modified hydrotalcite conversion coatings. Corrosion **58**(1), 3–14 (2002)
115. C. Ruby, M. Usman, S. Naille, K. Hanna, C. Carteret, M. Mullet, M. Francois, M. Abdelmoula, Synthesis and transformation of iron-based layered double hydroxides. Appl. Clay Sci. **48**(1–2), 195–202 (2010). doi:10.1016/j.clay.2009.11.017
116. N. Granizo, J.M. Vega, D. de la Fuente, J. Simancas, M. Morcillo, Ion-exchange pigments in primer paints for anticorrosive protection of steel in atmospheric service: cation-exchange pigments. Prog. Org. Coat. **75**(3), 147–161 (2012). doi:10.1016/j.porgcoat.2012.04.013
117. E.A. Matter, S. Kozhukharov, M. Machkova, V. Kozhukharov, Comparison between the inhibition efficiencies of Ce(III) and Ce(IV) ammonium nitrates against corrosion of AA2024 aluminum alloy in solutions of low chloride concentration. Corros. Sci. **62**, 22–33 (2012). doi:10.1016/j.corsci.2012.03.039
118. H. Ardelean, C. Fiaud, P. Marcus, Enhanced corrosion resistance of magnesium and its alloys through the formation of cerium (and aluminium) oxide surface films. Materials Corrosion-Werkstoffe Und Korrosion **52**(12), 889–895 (2001). doi:10.1002/1521-4176(200112)52:12<889:aid-maco889>3.0.co;2-0
119. B.L. Liu, B.T. Zhang, S.S. Cao, X.B. Deng, X.H. Hou, H.L. Chen, Preparation of the stable core-shell latex particles containing organic-siloxane in the shell. Prog. Org. Coat. **61**(1), 21–27 (2008). doi:10.1016/j.porgcoat.2007.08.008
120. A.S. Hamdy, A clean low cost anti-corrosion molybdate based nano-particles coating for aluminum alloys. Prog. Org. Coat. **56**(2–3), 146–150 (2006). doi:10.1016/j.porgcoat.2006.03.002
121. W. Trabelsi, P. Cecilio, M.G.S. Ferreira, M.F. Montemor, Electrochemical assessment of the self-healing properties of Ce-doped silane solutions for the pre-treatment of galvanised steel substrates. Prog. Org. Coat. **54**(4), 276–284 (2005). doi:10.1016/j.porgcoat.2005.07.006
122. A.S. Hamdy, Advanced nano-particles anti-corrosion ceria based sol gel coatings for aluminum alloys. Mater. Lett. **60**(21–22), 2633–2637 (2006). doi:10.1016/j.matlet.2006.01.049
123. M.F. Montemor, M.G.S. Ferreira, Analytical characterization of silane films modified with cerium activated nanoparticles and its relation with the corrosion protection of galvanised steel substrates. Prog. Org. Coat. **63**(3), 330–337 (2008). doi:10.1016/j.porgcoat.2007.11.008

124. M.L. Zheludkevich, R. Serra, M.F. Montemor, K.A. Yasakau, I.M.M. Salvado, M.G.S. Ferreira, Nanostructured sol-gel coatings doped with cerium nitrate as pre-treatments for AA2024-T3 - Corrosion protection performance. Electrochim. Acta. **51**(2), 208–217 (2005). doi:10.1016/j.electacta.2005.04.021
125. L.S. Kasten, J.T. Grant, N. Grebasch, N. Voevodin, F.E. Arnold, M.S. Donley, An XPS study of cerium dopants in sol–gel coatings for aluminum 2024-T3. Surf. Coat. Technol. **140** (1), 11–15 (2001). doi:10.1016/s0257-8972(01)01004-0
126. D. Ho, N. Brack, J. Scully, T. Markley, M. Forsyth, B. Hinton, Cerium dibutylphosphate as a corrosion inhibitor for AA2024-T3 aluminum alloys. J. Electrochem. Soc. **153**(9), B392–B401 (2006). doi:10.1149/1.2217260
127. N. Sheng, T. Ohtsuka, Preparation of conducting poly-pyrrole layer on zinc coated Mg alloy of AZ91D for corrosion protection. Prog. Org. Coat. **75**(1–2), 59–64 (2012). doi:10.1016/j.porgcoat.2012.03.008
128. Y. Hao, F. Liu, E.-H. Han, Protection of epoxy coatings containing polyaniline modified ultra-short glass fibers. Prog. Org. Coat. (0). doi:http://dx.doi.org/10.1016/j.porgcoat.2012.11.012
129. M.A. Jakab, J.R. Scully, On-demand release of corrosion-inhibiting ions from amorphous Al-Co-Ce alloys. Nat. Mater. **4**(9), 667–670 (2005). doi:10.1038/nmat1451
130. N.S. McIntyre, R.D. Davidson, G. Kim, J.T. Francis, New frontiers in X-ray photoelectron spectroscopy. Vacuum **69**(1–3), 63–71 (2002). doi:10.1016/s0042-207x(02)00308-1
131. D.J. Bland, A.J. Kinloch, V. Stolojan, J.F. Watts, Failure mechanisms in adhesively bonded aluminium: an XPS and PEELS study. Surf. Interface Anal. **40**(3–4), 128–131 (2008). doi:10.1002/sia.2651
132. B.J. James, R. Cameron, C. Baskcomb, Selected area XPS analysis for identification of pigment compounds in microscopic paint flakes. Adv. Mater. Sci. Eng. **2008** (2008)
133. J.E. Devries, L.P. Haack, T.J. Prater, S.L. Kaberline, J.L. Gerlock, J.W. Holubka, R.A. Dickie, J. Chakel, Characterization of interfacial chemistries associated with polymer systems by spatially-resolved surface analytical methodologies. Prog. Org. Coat. **25**(1), 95–108 (1994). doi:10.1016/0300-9440(94)00505-2
134. J.T. Francis, N.S. McIntyre, R.D. Davidson, S. Ramamurthy, A.M. Brennenstuhl, A. McBride, A. Roberts, Mechanisms for pitting corrosion in alloy N04400 as revealed by imaging XPS, ToF-SIMS and low-voltage SEM. Surf. Interface Anal. **33**(1), 29–34 (2002). doi:10.1002/sia.1157
135. A.E. Hughes, A. Boag, A.M. Glenn, D. McCulloch, T.H. Muster, C. Ryan, C. Luo, X. Zhou, G.E. Thompson, Corrosion of AA2024-T3 Part II Co-operative corrosion. Corros. Sci. **53**(1), 27–39 (2011). doi:10.1016/j.corsci.2010.09.030
136. A. Boag, R.J. Taylor, T.H. Muster, N. Goodman, D. McCulloch, C. Ryan, B. Rout, D. Jamieson, A.E. Hughes, Stable pit formation on AA2024-T3 in a NaCl environment. Corros. Sci. **52**(1), 90–103 (2010). doi:10.1016/j.corsci.2009.08.043
137. A.E. Hughes, C. MacRae, N. Wilson, A. Torpy, T.H. Muster, A.M. Glenn, Sheet AA2024-T3: a new investigation of microstructure and composition. Surf. Interface Anal. **42** (4), 334–338 (2010). doi:10.1002/sia.3163
138. J.D. Gorman, A.E. Hughes, D. Jamieson, P.J.K. Paterson, Oxide formation on aluminium alloys in boiling deionised water and NaCl, CeCl3 and CrCl3 solutions. Corros. Sci. **45**(6), 1103–1124 (2003)
139. S. Sellaiyan, A.E. Hughes, S.V. Smith, A. Uedono, J. Sullivan, S. Buckman, Leaching properties of chromate-containing epoxy films using radiotracers, PALS and SEM. Prog. Org. Coat. **77**(1), 257–267 (2014)
140. S.G. Boxer, M.L. Kraft, P.K. Weber, Advances in imaging secondary ion mass spectrometry for biological samples. In: Ann. Rev. Biophys. **38**, 53–74 (2009)
141. N.P. Lockyer, J.C. Vickerman, Progress in cellular analysis using ToF-SIMS. Appl. Surf. Sci. **231**, 377–384 (2004). doi:10.1016/j.apsusc.2004.03.103
142. L.A. Giannuzzi, M. Utlaut, A review of Ga + FIB/SIMS. Surf. Interface Anal. **43**(1–2), 475–478 (2011). doi:10.1002/sia.3454

143. K.M. Ng, Y.T.R. Lau, C.M. Chan, L.T. Weng, J.S. Wu, Surface studies of halloysite nanotubes by XPS and ToF-SIMS. Surf. Interface Anal. **43**(4), 795–802 (2011). doi:10.1002/sia.3627
144. S.J. Hinder, C. Lowe, J.F. Watts, ToF-SIMS depth profiling of a complex polymeric coating employing a C-60 sputter source. Surf. Interface Anal. **39**(6), 467–475 (2007). doi:10.1002/sia.2546
145. S.J. Hinder, C. Lowe, J.F. Watts, An XPS and ToF-SIMS investigation of the outermost nanometres of a poly(vinylidene difluoride) coating. Prog. Org. Coat. **60**(3), 255–261 (2007). doi:10.1016/j.porgcoat.2007.07.016
146. S.J. Hinder, J.F. Watts, C. Lowe, Surface and interface analysis of complex polymeric paint formulations. Surf. Interface Anal. **38**(4), 557–560 (2006). doi:10.1002/sia.2325
147. N. Davies, D.E. Weibel, P. Blenkinsopp, N. Lockyer, R. Hill, J.C. Vickerman, Development and experimental application of a gold liquid metal ion source. Appl. Surf. Sci. **203**, 223–227 (2003). doi:10.1016/s0169-4332(02)00631-1
148. S.J. Hinder, J.F. Watts, G.C. Simmons, C. Lowe, An investigation of the distribution of minor components in complex polymeric paint formulations using ToF-SIMS depth profiling. Surf. Interface Anal. **40**(3–4), 436–440 (2008). doi:10.1002/sia.2712
149. S.C.C. Wong, R. Hill, P. Blenkinsopp, N.P. Lockyer, D.E. Weibel, J.C. Vickerman, Development of a C-60(+) ion gun for static SIMS and chemical imaging. Appl. Surf. Sci. **203**, 219–222 (2003). doi:10.1016/s0169-4332(02)00629-3
150. A. King, T. Henkel, D. Rost, I.C. Lyon, Determination of relative sensitivity factors during secondary ion sputtering of silicate glasses by Au + , Au-2(+) and Au-3(+) ions. Rapid Commun. Mass Spectrom. **24**(1), 15–20 (2010). doi:10.1002/rcm.4351
151. T. Henkel, D. Rost, I.C. Lyon, Improvements in quantification accuracy of inorganic time-of-flight secondary ion mass spectrometric analysis of silicate materials by using C-60 primary ions. Rapid Commun. Mass Spectrom. **23**(21), 3355–3360 (2009). doi:10.1002/rcm.4257
152. N. Winograd, The magic of cluster SIMS. Anal. Chem. **77**(7), 142A–149A (2005). doi:10.1021/ac053355f
153. H. Yang, W.J. van Ooij, Plasma deposition of polymeric thin films on organic corrosion-inhibiting paint pigments: a novel method to achieve slow release. Plasmas Polym. **8**(4), 297–323 (2003). doi:10.1023/a:1026389311431
154. M. Brenda, R. Doring, U. Schernau, Investigation of organic coatings and coating defects with the help of time-of-flight-secondary ion mass spectrometry (TOF-SIMS). Prog. Org. Coat. **35**(1–4), 183–189 (1999). doi:10.1016/s0300-9440(99)00035-1
155. J. Fang, B.J. Flinn, Y.L. Leung, P.C. Wong, K.A.R. Mitchell, T. Foster, A characterization of the gamma-glycidoxypropyltrimethoxysilane and aluminium interface by SIMS and XPS. J Mater. Sci. Lett. **16**(20), 1675–1676 (1997). doi:10.1023/a:1018530231931
156. M.L. Abel, R.P. Digby, I.W. Fletcher, J.F. Watts, Evidence of specific interaction between gamma-glycidoxypropyltrimethoxysilane and oxidized aluminium using high-mass resolution ToF-SIMS. Surf. Interface Anal. **29**(2), 115–125 (2000). doi:10.1002/(sici)1096-9918(200002)29:2<115:aid-sia702>3.3.co;2-y
157. R. Arnold, A. Terfort, C. Woll, Determination of molecular orientation in self-assembled monolayers using IR absorption intensities: the importance of grinding effects. Langmuir **17**(16), 4980–4989 (2001). doi:10.1021/la010202o
158. M. Born, E. Wolf, Principles of Optics, 7th edn. Cambridge University Press, Cambridge (1999)
159. M. Debe, Extracting physical structure information from thin organic films with reflection absorption infrared-spectroscopy. J. Appl. Phys. **55**(9), 3354–3366 (1984). doi:10.1063/1.333374
160. A. Parikh, D. Allara, Quantitative-determination of molecular-structure in multilayered thin-films of biaxial and lower symmetry from photon spectroscopies.1. reflection infrared vibrational spectroscopy. J. Chem. Phys. **96**(2), 927–945 (1992). doi:10.1063/1.462847

161. N. Tillman, A. Ulman, J. Schildkraut, T. Penner, Incorporation of phenoxy groups in self-assembled monolayers of trichlorosilane derivatives - effects on film thickness, wettability, and molecular-orientation. J. Am. Chem. Soc. **110**(18), 6136–6144 (1988). doi:10.1021/ja00226a031
162. C. Bram, C. Jung, M. Stratmann, Self assembled molecular monolayers on oxidized inhomogeneous aluminum surfaces. Fresenius J. Anal. Chem. **358**(1–2), 108–111 (1997). doi:10.1007/s002160050357
163. X. Yu, J. Wang, M. Zhang, P. Yang, L. Yang, D. Cao, J. Li, One-step synthesis of lamellar molybdate pillared hydrotalcite and its application for AZ31 Mg alloy protection. Solid State Sci. **11**(2), 376–381 (2009). doi:10.1016/j.solidstatesciences.2008.08.003
164. G. Blustein, R. Romagnoli, J.A. Jaén, A.R. Di Sarli, B. del Amo, Zinc basic benzoate as eco-friendly steel corrosion inhibitor pigment for anticorrosive epoxy-coatings. Colloids Surf., A **290**(1–3), 7–18 (2006). doi:10.1016/j.colsurfa.2006.04.043
165. G. Blustein, A.R. Di Sarli, J.A. Jaén, R. Romagnoli, B. Del Amo, Study of iron benzoate as a novel steel corrosion inhibitor pigment for protective paint films. Corros. Sci. **49**(11), 4202–4231 (2007). doi:10.1016/j.corsci.2007.05.008
166. L. Mascia, L. Prezzi, G.D. Wilcox, M. Lavorgna, Molybdate doping of networks in epoxy-silica hybrids: domain structuring and corrosion inhibition. Prog. Org. Coat. **56**(1), 13–22 (2006). doi:10.1016/j.porgcoat.2006.01.013
167. M. Forsyth, C.M. Forsyth, K. Wilson, T. Behrsing, G.B. Deacon, ATR characterisation of synergistic corrosion inhibition of mild steel surfaces by cerium salicylate. Corros. Sci. **44**, 2651–2656 (2002)
168. J. Mardel, S.J. Garcia, P.A. Corrigan, T. Markley, A.E. Hughes, T.H. Muster, D. Lau, T.G. Harvey, A.M. Glenn, P.A. White, S.G. Hardin, C. Luo, X. Zhou, G.E. Thompson, J.M.C. Mol, The characterisation and performance of Ce(dbp)3-inhibited epoxy coatings. Prog. Org. Coat. **70**(2–3), 91–101 (2011). doi:10.1016/j.porgcoat.2010.10.009
169. R. Trujillano, M.J. Holgado, V. Rives, Obtention of low oxidation states of copper from Cu2 + -Al3 + layered double hydroxides containing organic sulfonates in the interlayer. Solid State Sci. **11**(3), 688–693 (2009). doi:10.1016/j.solidstatesciences.2008.10.015
170. R. Trujillano, M.J. Holgado, F. Pigazo, V. Rives, Preparation, physicochemical characterisation and magnetic properties of Cu-Al layered double hydroxides with CO32- and anionic surfactants with different alkyl chains in the interlayer. Physica B-Condens. Matter **373**(2), 267–273 (2006). doi:10.1016/j.physb.2005.11.154
171. Y.C. Yuan, M.Z. Rong, M.Q. Zhang, J. Chen, G.C. Yang, X.M. Li, Self-healing polymeric materials using epoxy/mercaptan as the healant. Macromolecules **41**(14), 5197–5202 (2008). doi:10.1021/ma800028d
172. F.H. Scholes, S.A. Furman, A.E. Hughes, T. Nikpoura, N. Wrighta, P.R. Curtis, C.M. Macrae, S. Intem, A.J. Hill, Chromate leaching from inhibited primers - Part I. Characterisation of leaching. Prog. Org. Coat. **56**(1), 23–32 (2006). doi:10.1016/j.porgcoat.2006.01.015
173. F.H. Scholes, A.E. Hughes, D. Jamieson, K. Inoue, S.A. Furman, T.H. Muster, S.G. Hardin, D. Lau, T.G. Harvey, P. Corrigan, M. Glenn, P.A. White, J. Mardel, M. Forsyth, Interaction of Ce(dbp)(3) with surface of aluminium alloy 2024-T3 using macroscopic models of intermetallic phases. Corros. Eng. Sci. Technol. **44**(6), 416–424 (2009). doi:10.1179/147842208x320333
174. T.A. Markley, A.E. Hughes, T.C. Ang, G.B. Deacon, P. Junk, M. Forsyth, Influence of praseodymium - Synergistic corrosion inhibition in mixed rare-earth diphenyl phosphate systems. Electrochem. Solid State Lett. **10**(12), C72–C75 (2007). doi:10.1149/1.2790724
175. B.R.W. Hinton, N. Dubrule, A.E. Hughes, M. Forsyth, T. Markley, D. Ho, F.H. Scholes, S. A. Furman, Raman, EDS and SEM studies of the interaction of corrosion inhibitor $Ce(dbp)_3$ with AA2024-T3. In: Paper presented at the 4th International Symposium on Aluminium Surface Science and Technology, Beaune, France

176. S.Y. Wang, W. Wang, J. Zuo, Y.T. Qian, Study of the Raman spectrum of CeO2 nanometer thin films. Mater. Chem. Phys. **68**(1–3), 246–248 (2001). doi:10.1016/s0254-0584(00)00357-6
177. W.H. Weber, K.C. Hass, J.R. McBride, Raman-study of ceo2 - 2nd-order scattering, lattice-dynamics and particle-size effects. Phys. Rev. B **48**(1), 178–185 (1993). doi:10.1103/PhysRevB.48.178
178. J.R. McBride, K.C. Hass, B.D. Poindexter, W.H. Weber, Raman and X-ray studies of CE1-XREXO2-Y, where RE = LA, PR, ND, EU, GD and TB. J. Appl. Phys. **76**(4), 2435–2441 (1994). doi:10.1063/1.357593
179. A.E. Hughes, F.H. Scholes, A.M. Glenn, D. Lau, T.H. Muster, S.G. Hardin, Factors influencing the deposition of Ce-based conversion coatings, part I: the role of Al3 + ions. Surf. Coat. Technol. **203**(19), 2927–2936 (2009). doi:10.1016/j.surfcoat.2009.03.022
180. F.H. Scholes, C. Soste, A.E. Hughes, S.G. Hardin, P.R. Curtis, The role of hydrogen peroxide in the deposition of cerium-based conversion coatings. Appl. Surf. Sci. **253**(4), 1770–1780 (2006). doi:10.1016/j.apsusc.2006.03.010
181. H.D. Ruan, R.L. Frost, J.T. Kloprogge, D.G. Schulze, L. Duong, FT-Raman spectroscopy and SEM of gibbsite, bayerite, boehmite and diaspore in relation to the characterization of bauxite. 2001 - a Clay Odyssey, (2003)
182. A. Bertoluzza, C. Fagnano, M.A. Morelli, M. Guglielmi, G. Scarinci, N. Maliavski, Raman-spectra of sio2 gel glasses prepared from alkoxide, colloidal and amine silicate solutions. J. Raman Spectrosc. **19**(4), 297–300 (1988). doi:10.1002/jrs.1250190414
183. P. Bohn, Localized optical phenomena and the characterization of materials interfaces. Annu. Rev. Mater. Sci. **27**, 469–498 (1997). doi:10.1146/annurev.matsci.27.1.469
184. G. Fieldson, T. Barbari, The use of FTIR-ATR spectroscopy to characterize penetrant diffusion in polymers. Polymer **34**(6), 1146–1153 (1993). doi:10.1016/0032-3861(93)90765-3
185. R. Vlasak, I. Klueppel, G. Grundmeier, Combined EIS and FTIR-ATR study of water uptake and diffusion in polymer films on semiconducting electrodes. Electrochim. Acta **52**(28), 8075–8080 (2007). doi:10.1016/j.electacta.2007.07.003
186. M. Ohman, D. Persson, ATR-FTIR Kretschmann spectroscopy for interfacial studies of a hidden aluminum surface coated with a silane film and epoxy II. Analysis by integrated ATR-FTIR and EIS during exposure to electrolyte with complementary studies by in situ ATR-FTIR and in situ IRRAS. Surf. Interface Anal. **44**(1), 105–113 (2012). doi:10.1002/sia.3780
187. A. Hartstein, J. Kirtley, J. Tsang, Enhancement of the infrared-absorption from molecular monolayers with thin metal overlayers. Phys. Rev. Lett. **45**(3), 201–204 (1980). doi:10.1103/PhysRevLett.45.201
188. A. Hatta, Y. Suzuki, W. Suetaka, Infrared-absorption enhancement of monolayer species on thin evaporated ag films by use of a kretschmann configuration - evidence for 2 types of enhanced surface electric-fields. Appl. Phys. Mater. Sci. Process. **35**(3), 135–140 (1984). doi:10.1007/BF00616965
189. M. Osawa, K. Ataka, K. Yoshii, Y. Nishikawa, Surface-enhanced infrared-spectroscopy - the origin of the absorption enhancement and band selection rule in the infrared-spectra of molecules adsorbed on fine metal particles. Appl. Spectrosc. **47**(9), 1497–1502 (1993). doi:10.1366/0003702934067478
190. M. Osawa, K. Ataka, K. Yoshii, T. Yotsuyanagi, Surface-enhanced infrared atr spectroscopy for in-situ studies of electrode-electrolyte interfaces. J. Electron Spectrosc. Relat. Phenom. **64–5**, 371–379 (1993). doi:10.1016/0368-2048(93)80099-8
191. Y. Yan, Q. Li, S. Huo, M. Ma, W. Cai, M. Osawa, Ubiquitous strategy for probing ATR surface-enhanced infrared absorption at platinum group metal-electrolyte interfaces. J.Phys. Chem. B **109**(16), 7900–7906 (2005). doi:10.1021/jp044085s

192. K. Ataka, G. Nishina, W. Cai, S. Sun, M. Osawa, Dynamics of the dissolution of an underpotentially deposited Cu layer on Au(111): a combined time-resolved surface-enhanced infrared and chronoamperometric study. Electrochem. Commun. **2**(6), 417–421 (2000). doi:10.1016/S1388-2481(00)00053-9
193. T. Wandlowski, K. Ataka, S. Pronkin, D. Diesing, Surface enhanced infrared spectroscopy - Au(1 1 1-20 nm)/sulphuric acid - new aspects and challenges. Electrochim. Acta. **49**(8), 1233–1247 (2004). doi:10.1016/j.electacta.2003.06.002
194. S. Huo, J. Wang, J. Yao, W. Cai, Exploring electrosorption at iron electrode with in situ surface-enhanced infrared absorption spectroscopy. Anal. Chem. **82**(12), 5117–5124 (2010). doi:10.1021/ac1002323
195. X. Xue, J. Wang, Q. Li, Y. Yan, J. Liu, W. Cai, Practically modified attenuated total reflection surface-enhanced IR absorption spectroscopy for high-quality frequency-extended detection of surface species at electrodes. Anal. Chem. **80**(1), 166–171 (2008). doi:10.1021/ac7017487
196. Q. Huang, X. Lin, Z. Yang, J. Hu, Z. Tian, An investigation of the adsorption of pyrazine and pyridine on nickel electrodes by in situ surface-enhanced Raman spectroscopy. J. Electroanal. Chem. **563**(1), 121–131 (2004). doi:10.1016/j.jelechem.2003.08.015
197. J. Chowdhury, M. Ghosh, T. Misra, pH-Dependent surface-enhanced Raman scattering of 8-hydroxy quinoline adsorbed on silver hydrosol. J. Colloid Interface Sci. **228**(2), 372–378 (2000). doi:10.1006/jcis.2000.6977
198. Y. Xie, X. Wang, X. Han, X. Xue, W. Ji, Z. Qi, J. Liu, B. Zhao, Y. Ozaki, Sensing of polycyclic aromatic hydrocarbons with cyclodextrin inclusion complexes on silver nanoparticles by surface-enhanced Raman scattering. Analyst **135**(6), 1389–1394 (2010). doi:10.1039/c0an00076k
199. Y. Nishikawa, T. Nagasawa, K. Fujiwara, M. Osawa, Silver island films for surface-enhanced infrared-absorption spectroscopy - effect of island morphology on the absorption enhancement. Vib. Spectrosc. **6**(1), 43–53 (1993). doi:10.1016/0924-2031(93)87021-K
200. E. Cortes, P. Etchegoin, E. Le Ru, A. Fainstein, M. Vela, R. Salvarezza, Strong correlation between molecular configurations and charge-transfer processes probed at the single-molecule level by surface-enhanced raman scattering. J. Am. Chem. Soc. **135**(7), 2809–2815 (2013). doi:10.1021/ja312236y
201. A. Kudelski, Structures of monolayers formed from different HS-(CH2)(2)-X thiols on gold, silver and copper: comparitive studies by surface-enhanced Raman scattering. J. Raman Spectrosc. **34**(11), 853–862 (2003). doi:10.1002/jrs.1062
202. Y. Chen, G. Carter, S. Tripathy, Study of langmuir-blodgett polydiacetylene polymer-films by surface enhanced raman-scattering. Solid State Commun. **54**(1), 19–22 (1985). doi:10.1016/0038-1098(85)91024-5
203. C. Constantino, J. Duff, R. Aroca, Surface enhanced resonance Raman scattering imaging of Langmuir-Blodgett monolayers of bis (benzimidazo) thioperylene. Spectrochimica Acta. Part a-Molecular and Biomolecular Spectrosc. **57**(6), 1249–1259 (2001). doi:10.1016/S1386-1425(00)00470-4
204. D. Jeanmaire, R. Vanduyne, Surface raman spectroelectrochemistry.1. heterocyclic, aromatic, and aliphatic-amines adsorbed on anodized silver electrode. J. Electroanal. Chem. **84**(1), 1–20 (1977). doi:10.1016/S0022-0728(77)80224-6
205. J. Neng, M. Harpster, W. Wilson, P. Johnson, Surface-enhanced Raman scattering (SERS) detection of multiple viral antigens using magnetic capture of SERS-active nanoparticles. Biosens. Bioelectron. **41**, 316–321 (2013). doi:10.1016/j.bios.2012.08.048
206. K. Kneipp, H. Kneipp, I. Itzkan, R. Dasari, M. Feld, Surface-enhanced Raman scattering and biophysics. J. Phys. Condens. Matter **14**(18), R597–R624 (2002). doi:10.1088/0953-8984/14/18/202
207. H. Long, Z. Zhen, L. Tang, J. Jiang, Surface-enhanced raman scattering based biosensor for histone acetylation detection. Acta. Chim. Sinica **71**(5), 739–742 (2013). doi:10.6023/A13020171

208. K. Kim, D. Shin, H. Lee, K. Shin, Surface-enhanced Raman scattering of 4-aminobenzenethiol on gold: the concept of threshold energy in charge transfer enhancement. Chem. Commun. **47**(7), 2020–2022 (2011). doi:10.1039/c0cc04925e
209. R. Aroca, C. Constantino, Surface-enhanced Raman scattering: imaging and mapping of Langmuir-Blodgett monolayers physically adsorbed onto silver island films. Langmuir **16** (12), 5425–5429 (2000). doi:10.1021/la991478n
210. J. Maciel, R. Jaimes, P. Corio, J. Rubim, P. Volpe, A. Agostinho, S. Agostinho, The characterisation of the protective film formed by benzotriazole on the 90/10 copper-nickel alloy surface in H2SO4 media. Corros. Sci. **50**(3), 879–886 (2008). doi:10.1016/j.corsci.2007.10.011

Chapter 11
Transport in Protective Coatings

Niteen Jadhav, Joseph Byrom, Abhijit Suryawanshi and Victoria Gelling

11.1 Introduction

Corrosion is a thermodynamically favorable electrochemical process in which metal is transformed to its original oxide form. Coatings are the effective means of protecting metals and their alloys from the corrosion. Inorganic and organic coatings are applied to metal substrates for corrosion protection. Water/moisture, oxygen, corrosive ions are the components which induce corrosion. The process of corrosion in the case of metals coated with coatings is dependent on following factors [1–3].

1. Water, oxygen, and ionic species diffusion through the coating
2. Delamination, loss of adhesion at coating metal interface
3. Changes in composition and continuity of coating
4. Charge transfer process
5. Surface treatment/modification of metal substrate

Water and moisture permeation through the polymeric coating is factually different as the kinetic energy of the water molecules in the vapor form (ca. 450 m/s) is much higher, than the water molecules resulting in the different behavior of the permeation by water and water vapors. Accumulation of the diffused water at the coating/substrate interface results in delamination of the coating leading to loss in adhesion and finally leads to the increased rate of corrosion. The permeation of water/moisture, oxygen, corrosive ions leads to changes in the composition of the coating binder and induces the inhomogeneity in the coating matrix resulting in the increased corrosion rate. The active nature of the corrosive inhibiting pigments in

N. Jadhav · J. Byrom · A. Suryawanshi · V. Gelling (✉)
Department of Coatings & Polymeric Materials, North Dakota State University,
1735 NDSU Research Park Drive, Fargo, ND, USA
e-mail: v.j.gelling@ndsu.edu; v.j.gelling@gmail.com

A.E. Hughes et al. (eds.), *Active Protective Coatings*,
Springer Series in Materials Science 233, DOI 10.1007/978-94-017-7540-3_11

the coating for corrosion protection results in the charge transfer process at metal substrate interface. Proper surface treatment is important for the adhesion of the coating on the substrate as good adhesion results in better corrosion protection.

11.2 Water and Oxygen Transport in Polymers and Coatings

Water is the main means of transportation of aggressive ions as well as water soluble gases through the coating [4]. Water has high hydrogen bonding capacity which leads to changes in the coatings resulting in loss in mechanical, electrical and adhesive properties. Paint film properties such as water uptake, permeability, permselectivity, and ion exchange capacity could be analyzed for studying the transport properties. Furthermore, water transport is also the function of binder composition, molecular weight, concentration of additives, solvents, coating curing conditions, pigment volume concentration, polymer crystallinity, orientation, crosslinking and functionality [5]. Water diffusion through the coatings is important phenomenon which has tremendous impact on the protective properties of the coating.

Diffusion in the steady state is given by Fick's first law of diffusion as demonstrated in Eq. (11.1)

$$J = -D(\partial c/\partial x) \tag{11.1}$$

where, J is the flux in the flow direction, c, *is the concentration*, $\partial c/\partial x$ is the concentration gradient in the direction x and D is the diffusion coefficient. Non-steady state diffusion is given by Fick's second law of diffusion as demonstratated in Eq. (11.2). It states the effect of time on the changes in the concentration of the permeating species.

$$\partial c/\partial t = D(\partial c^2/\partial x^2) \tag{11.2}$$

Diffusion processes occurring in the coating have been characterized by several techniques such as Fourier transform infrared spectroscopy (FTIR), attenuated total reflection Fourier transform infrared spectroscopy (ATR-FTIR), solid state nuclear magnetic resonance (NMR), positron annihilation lifetime spectroscopy (PALS), gravimetry (Gravimetric cup method, Manometric cell method), neutron reflectivity technique, capacitance, molecular simulations, dielectric relaxation measurements, ultraviolet (UV) reflection, and radiotracer technology [6–8].

Changes in the capacitance values of the coating can be determined using EIS at various time intervals. The capacitance values of organic coatings vary with permeation of water through the coating, thus the volume fraction of penetrated water can be calculated. Water uptake, in terms of volume fraction of water absorbed (V_t), can be determined using Brasher/Kingsbury expression (as shown in Eq. 11.3) at capacitance values C_0 and C_t after 0 and t hours respectively, where ε is the dielectric constant of water [9].

$$V_t = \frac{100\log(C_t/C_0)}{\log\varepsilon} \quad (11.3)$$

Furthermore, if saturated coatings of thickness L have capacitance of C_∞, diffusion coefficient of water can be determined using the following Eq. (11.4)

$$D = \frac{L^2\pi}{4t}\left[\frac{\log(C_t/C_0)}{\log(C_\infty/C_0)}\right]^2 \quad (11.4)$$

Diffusion kinetics can also be monitored using FTIR-ATR, increase in absorption intensity at approximately 3000–3700 cm^{-1} corresponding to characteristic OH absorption, in FTIR spectrum can be observed with water permeation through coating [6]. A combination of FTIR-ATR and EIS has also been studied recently; simultaneously two sets of data were obtained from a single set-up, however complex data was obtained due to consideration of space charge distribution of the semiconducting electrodes used in the study [10].

Water vapor permeability can be measured using the gravimetric cup method (ASTM D96), wherein, the weight of the cup containing distilled water or anhydrous calcium chloride, attached with the film of interest is measured for the movement of water vapor through the film. The test is conducted in a controlled atmosphere. Infrared spectroscopy can also be utilized to measure water vapor permeability. Similar to gravimetric cup method oxygen permeability can be determined using a chamber which is separated in two parts by the film of interest. Oxygen free environment is created in one chamber, while oxygen is introduced in the other chamber. The permeation of oxygen to the former chamber is measured by a coulorimetric detector which produces electric current proportional to permeated oxygen per unit time. The gaseous permeation can also be determined using manometric cell method which relates to the drop/increase in pressure because of permeation of gas from one chamber to the other through the film that separates the two chambers [11].

Scanning Kelvin probe (SKP) and scanning acoustic microscopy (SAM) are two techniques which are often used to study transport phenomenon at the interlayer of coating and metal substrate [12]. Some of the advanced quantitative techniques for studying water transport in the protective coatings are summarized in Table 11.1.

Permeability can be explained with solution-diffusion theory according to which first sorption of penetrating species occurs followed by diffusion of penetrating species from high concentration to low concentration, ending with desorption of penetrating species from other side of film. The solution-diffusion theory can be expressed by Eq. (11.5).

Table 11.1 Advanced techniques for water transport study in protective coatings

Technique	Measurements	Processes	References
Electrochemical impedance spectroscopy	Capacitance measurement	Diffusion/Absorption	[13–15]
FTIR-ATR	Peak intensities corresponding to OH absorptions	Diffusion	[16, 17]
Scanning Kelvin probe	Measures work function difference between a sample and a vibrating probe	Diffusion in the coating-substrate interlayer	[18, 19]
Scanning acoustic microscopy	Sound pulses are transmitted to the sample which are reflected by various interfaces of the sample	Diffusion in the coating-substrate interlayer	[20, 21]

$$P = D \times S \tag{11.5}$$

where, P is the coefficient of permeability, D is the coefficient of diffusivity, and S is the coefficient of solubility of the penetrating species. For the penetrating molecules exhibiting simple dissolution behavior and molecules exhibiting pre-cavity existence before dissolution were demonstrated by the modifications given by Henry's law and Langmuir isotherm respectively. Free volume theory model was employed which demonstrated permanent diffusion of polymer segments from one hole to another hole.

For the O_2 permeation, the coefficients of permeability, diffusivity, and solubility are almost constant as the polymer does not undergo rearrangement's or strains as its interaction with O_2 is minimal. However depending on the level of polymer unsaturation, crosslinking, intermolecular forces such as hydrogen bonding, orientation, crystallinity, polarity, glass transition temperature, and symmetry of polymer chains differences in the permeation rates are observed. Polymer unsaturation results in greater mobility around the polymer chains leading to increases in the diffusion of O_2. Increase in the saturation due hydrogenation leads to the restriction for the diffusion and permeability of O_2 [22]. Increased crosslinking density results in chain entanglement thereby reducing coefficients of oxygen solubility, diffusion, and permeation [23]. Hydrogen bonding results in increased inter-chain interactions thereby reducing the rate of O_2 permeation [24]. Crystallinity represents the order in the polymers. A decrease in the O_2 occurs as the percent crystallinity in the polymer increases [25]. Orientation results in better packing and reduction in O_2 permeation rate [26]. Polar groups on the polymer backbone decreases O_2 permeability. Again steric hindrance posed by bulky side groups increased permeability. O_2 permeability decreases as the symmetry of the side groups on the polymer backbone increases [26].

O_2 permeation measurement is generally performed by monitoring increases in pressure at constant volume or by monitoring increases in volume at constant pressure by ISO 2782 and ISO 1399 respectively [26]. Luminescence quenching for measuring diffusion and permeability of O_2 was also employed [27, 28]. A six port measurement device, with hexagonal dissolved O_2 sensors consisting electrochemical system, was employed for the measurement of O_2 permeability in six different polymers at the same time [29]. A spectrophotochemical method was also demonstratated for the measurement of O_2 permeability through the polymer film [30]. The singlet oxygen technique was also employed for the measurement of O_2 permeability of films made up of polymers [31, 32]. Apart from the coatings industry, O_2 permeation is important in waste water treatment, food and beverage packaging, and artificial lungs development [33].

11.3 Effect of Coating Constituents on the Transport Properties

Paint is a mixture of several components including binders, pigments, solvents, and various additives. By definition, paint is also an inhomogeneous mixture of these components in the dispersed form. Once applied to the substrate, paint transforms into a film by the process of curing (air drying, radiation curing, heat curing, crosslinking etc.). The process of drying may also introduce porosity in the coating.

11.3.1 Effect of Pigmentation

In order to decrease permeability, the pigments should be impermeable and should also exhibit maximum interaction between themselves and binder. Proper pigment dispersion also leads to reduced permeability. If proper dispersion of fillers is not achieved due to the incompatibility between filler particles and binder, free spaces might be created leading to the increased porosity for the permeation of the corrosive species. Use of flaky pigments and extenders (aluminum flakes, mica, micaceous iron oxide, glass flakes, talc, and steel and nickel flakes) can reduce the permeability by lengthening the path for species such as water, O_2, ions as shown in Fig. 11.1.

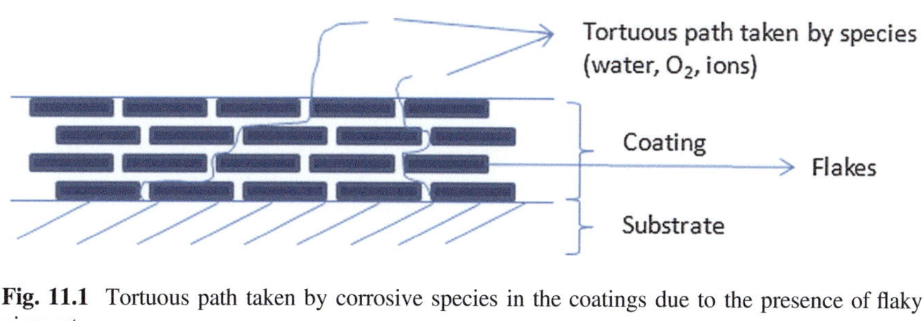

Fig. 11.1 Tortuous path taken by corrosive species in the coatings due to the presence of flaky pigments

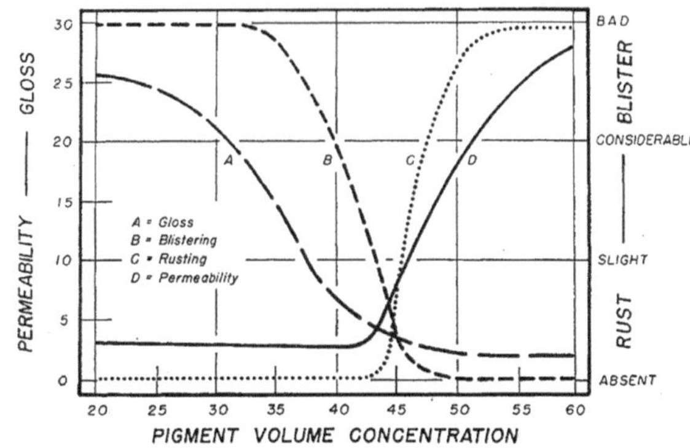

Fig. 11.2 Effect of CPVC on paint characteristics. Adapted with permission from Ref. [34] Copyright (1949) American Chemical Society

Protective properties of the coatings dramatically change above critical pigment volume concentration (CPVC). Above the CPVC there is not enough binder present in the coating to hold pigment and filler particles together which induces porosity in the coating [34]. As shown in Fig. 11.2, permeability and rusting increases above the CPVC. Above the CPVC the tendency to blistering also increases. This is valid for barrier coatings whereas in case of metal rich coatings, where electrical contact between two particles is necessary, the coatings are formulated at or near CPVC. This ensures mutual contact of metal particles as well as their contact with underlying substrates (steel substrate in case of zinc-rich primers and aluminum substrate in case of magnesium-rich primers) [35]. Pigment and filler addition also affects the adhesion of the paint film to the substrate. There is also the possibility of adsorption of moisture onto the pigment particles during storage. The adsorbed moisture might result in incompatibility in dispersing the pigment in non-aqueous media resulting in poor dispersion and compromising water and O_2 barrier properties.

11.3.2 Effect of Binder

Most of the information regarding the effect of the binder on transport properties has been explained in the earlier section (Water and oxygen transport in polymers and coatings) of this chapter. It has been found that water necessary for the corrosion of unpainted steel (0.003–0.06 mg cm^{-2} d^{-1}) is an order of magnitude less than the actual permeation rate of water through most of the binder systems. Water vapor permeability on a 100 μm film thickness for alkyd resin was observed to be 2.3 mg cm^{-2} d^{-1} and chlorinated rubber was observed to be 1 mg cm^{-2} d^{-1} [11]. Oxygen necessary for corrosion of unpainted steel was observed to be 0.008–0.150 mg cm^{-2} d^{-1} whereas the O_2 permeability for at 100 μm film thickness for alkyd resin was observed to be 0.0103 mg cm^{-2} d^{-1} and chlorinated rubber was observed to be 0.0022 mg cm^{-2} d^{-1} [11]. It is still a matter of debate whether O_2 permeation through the coating is the rate determining step for the corrosion occurring on the metal substrate.

Solvent borne coatings usually cure by a crosslinking mechanism in which small molecular weight precursors get transformed into the high molecular weight coating. Increased crosslinking density leads to a lower water and O_2 permeation. Water based coatings generally cure by coalescence of the latex particles. Finally a polyhydric structure is formed but there are still doubts about the homogeneity of such films which might compromise the protective nature of the coating towards permeation of water and O_2. Interfaces of latex particles might serve as an avenue for water vapor transportation [36].

11.3.3 Effect of Additives

Coatings contains several additives such as wetting and dispersing agents, rheology modifiers, flow and leveling agents, defoamers, biocides, colorants, high boiling solvents, etc. Some of the additives have affinity for the water and so there is a tendency for them to leach out. Leached additive might leave pores behind in coating leading to increased permeation of water and O_2 in the coating. Surfactants employed for the latex synthesis for water-based coating are water sensitive. They increase the water transmission rate. Anionic surfactants have been found to show lower permeation (due to ability to form denser films) as compared to cationic surfactant [37]. Plasticized films due to addition of plasticizers show increased segmental mobility thereby increasing O_2 permeability.

11.4 Transport in Rebar Reinforced Concrete

Corrosion underneath porous materials is also a very important phenomenon with rebar reinforced concrete being one of the most widely used building materials. Rebar located inside a concrete matrix typically does not corrode due to the passive layer formed on the iron. Penetration of corrosive species is problematic, however, due to the porous structure of the concrete surrounding the rebar. With sufficient carbon dioxide penetration, the pH surrounding the rebar can drop low enough to cause dissolution of the passive layer and facilitate corrosion processes to occur [38]. Corrosion byproducts of rebar have much lower densities and as such, can induce stresses that can cause mechanical failures in concrete which can end up being costly to repair, or life threatening.

Permeation of these corrosive species can be facilitated through three different mechanisms: diffusion, absorption, and permeation [39]. Permeation of the invasive materials occurs slightly differently than described in the previous section. The permeation of corrosive species in porous materials can best be described by Darcy's law shown in Eq. (11.6).

$$\upsilon = Q/A = -(k/\mu)(\partial P/\partial L) \tag{11.6}$$

where υ is the velocity of flow, Q is flow rate, A is cross sectional area, P is the pressure $\partial P/\partial L$ is the pressure gradient in the direction L, μ is viscosity of the permeating liquid, and k is the intrinsic permeability of the medium. With permeability being an important parameter in determining the lifetime of concrete, ASTM standards have been developed to effectively characterize concrete samples. ASTM 1202 is a rapid permeability test where concrete samples are vacuum saturated for 24 h followed by exposure of a DC potential of 60 V for 6 h. The resulting charge of the concrete samples determines the chloride ion permeability of the samples.

Absorption of corrosive species is another important pathway for the introduction of corrosive species. Absorption takes place due to the capillary forces that exist due to the porous channels that appear in concrete [40]. This phenomenon can be understood by Eq. (11.7)

$$S = V_w/\left(A\sqrt{t}\right) \tag{11.7}$$

where S is the sorptivity of the material, V_w is the volume of water absorbed, A is the cross sectional area, and t is time. Water absorption tests include the Autoclam sorptivity test, and a variety of ASTM tests for different grades of concrete.

11.5 Measurement and Monitoring of Transport of Corrosion Inhibitors in Coatings

Measurement and monitoring of transport of corrosion inhibitors in coatings can be done directly using an electrochemical quartz microbalance (ECQM) or indirectly using cyclic voltammetry, scanning vibrating electrode technique (SVET), and DC polarization techniques.

Barrier coatings fail to perform once a defect is created. Corrosion inhibitors are added to coatings so that they can be released on stimulus (creation of defect or change in pH). Inhibitors form, or cause to form, a passivating layer on the metal substrate thereby restricting further corrosion. Anodic inhibitors inhibit anodic reaction whereas cathodic inhibitors inhibit cathodic reactions. Some of the inhibitors are of mixed types which inhibit both anodic and cathodic reactions. Performance of these inhibitors can be studied by electrochemical characterization techniques such as DC polarization to study the effect of the inhibitor on either the anodic or cathodic arm. Results can be further quantified.

Leaching of inhibitive species from the coating to the substrate is a critical process for the anticorrosion performance. This is primary mode of action for soluble inhibitors. Chromate is one of the examples of such types. However

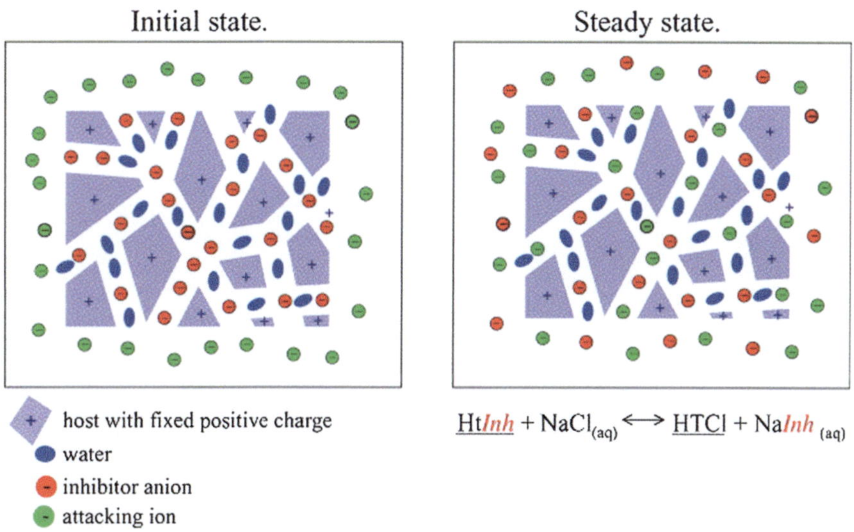

Fig. 11.3 Anion exchange in hydrotalcite clays, Reprinted from Ref. [41] with permission from Elsevier

chromates are toxic and carcinogenic, and when leached into environment cause serious problems.

Al–Zn hydroxide-based hydrotalcite (HT) with decavanadate as corrosion-inhibiting, charge-balancing anions were dispersed as additive in the epoxy coating. This coating was applied to an aluminum 2024-T3 substrate. Decavanadate gets released under appropriate conditions of the corrosion and chlorides and sulfates are adsorbed on HT for charge compensation as evidenced in Fig. 11.3. In Fig. 11.3, Ht*Inh* refers to inhibitor and HTCl refers to chloride in the gallery of HT. This exchange resulted in a change in crystal structure as evidenced in X-ray diffraction (XRD) studies. Changes in the diffraction pattern of the coating were monitored which can serve as a tool for observing corrosion inhibition in the coating [41].

In smart coatings, dopants are released on demand. The efficiency of the Cr(VI) and Ce^{3+} ions was studied by rotating disc electrode experiments. It was found that the current density (Fig. 11.4) was less in the case of inhibitor as against the one without inhibitor [42].

Conducting polymers such as polyaniline (PAni) and polypyrrole (PPy) can be doped with corrosive inhibiting dopants which can be released once the polymer is reduced in the corrosive environment (Fig. 11.5). Incorporation of dopants also improves PPy properties such as solubility and electrical conductivity. The released dopant then can combine with metal cations to form an impervious layer due to passivation. This passivation results in reduced current and shift in the positive direction in DC polarization experiments. In electrochemical impedance spectroscopy (EIS) experiments, passivation leads to increased low frequency impedance suggesting better corrosion protection.

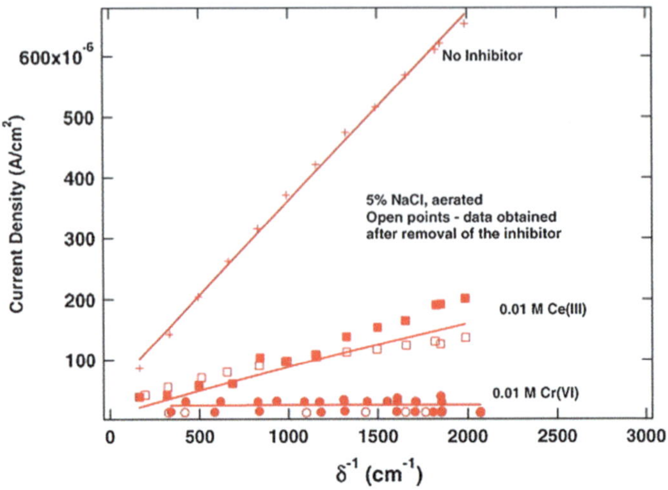

Fig. 11.4 Plot of current density versus inverse diffusion length for a Cu RDE in aerated 5 % NaCl with and without Cr(VI) and Ce(III) inhibitors. Reprinted from Ref. [42] with permission from Elsevier

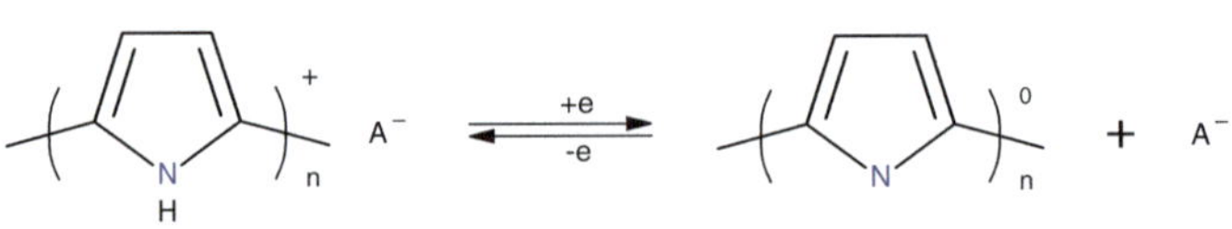

Fig. 11.5 Anion (A−) release by PPy in redox reaction

PAni doped with molybdate anions showed reduced current in the defect as observed by SVET experiments after 24 h of immersion in 1 % NaCl (Fig. 11.6). The reduction in current was attributed to the self healing ability of PANi due to the

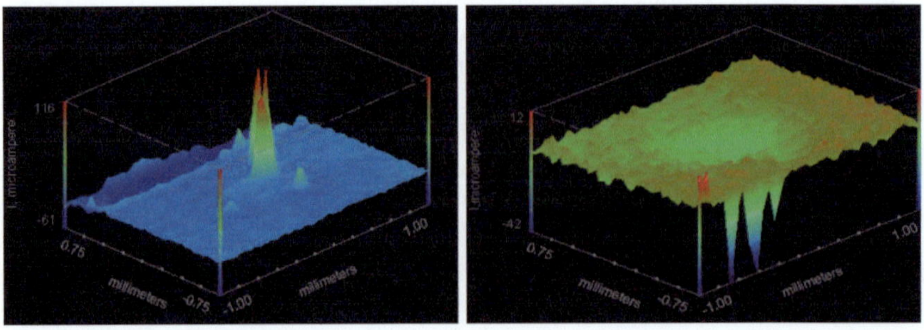

Fig. 11.6 Current distribution map for PAni-molybdate on steel immersed in 1 % NaCl (*Left image* after 15–20 min, *Right image* after 24 h). Reprinted from Ref. [43] with permission from Elsevier

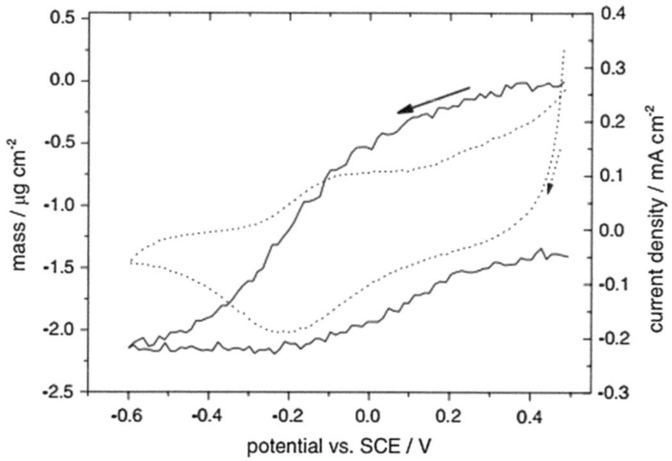

Fig. 11.7 Cyclic voltammogram (…) and the EQCM mass change (—) versus potential of Ppy (MoO4) in 0.1 M tetrabutylammoniumbromide (N(Bu)4Br) solution. Reprinted from Ref. [45] with permission from Springer

release of molybdate dopant and its combination with iron to form an iron-molybdate complex [43, 44].

In another study, mass change and mobility study of the molybdate dopant was performed by employing EQCM in combination with cyclic voltammetry [45]. From the mass-potential curve a mass decrease was observed as reduction of PPy occurred suggesting the release of dopant molybdate (Fig. 11.7). A gain an increase in the mass was observed when oxidation occurred suggesting bromide ion countercharge.

11.6 Summary

As discussed in this chapter, physical processes such as diffusion, permeation, and absorption play important roles in the transport of water, O_2 and other corrosive species through the coatings. The role of coating ingredients (pigments, binders, and additives) significantly affect the protective properties of the coatings. Several techniques including specialized electrochemical techniques can be employed for studying and quantifying the transport of corrosive species as well as inhibiting dopants in the protective coatings.

References

1. K. Doblhofer, I. Eiselt, The adhesion and inhibitor properties of organic coatings investigated with very thin polymer films on noble metal electrodes. Corros. Sci. **27**(9), 947–956 (1987). doi:10.1016/0010-938X(87)90061-8
2. P. Carbonini, T. Monetta, L. Nicodemo, P. Mastronardi, B. Scatteia, F. Bellucci, Electrochemical characterisation of multilayer organic coatings. Prog. Org. Coat. **29**(1–4), 13–20 (1996). doi:10.1016/S0300-9440(96)00628-5
3. V.B. Mišković-Stanković, M.D. Maksimović, Z. Kačarević-Popović, J.B. Zotović, The sorption characteristics and thermal stability of epoxy coatings electrodeposited on steel and steel electrochemically modified by Fe–P alloys. Prog. Org. Coat. **33**(1), 68–75 (1998). doi:10.1016/S0300-9440(98)00011-3
4. A. Barbucci, M. Delucchi, G. Cerisola, Organic coatings for concrete protection: liquid water and water vapour permeabilities. Prog. Org. Coat. **30**(4), 293–297 (1997). doi:10.1016/S0300-9440(97)00007-6
5. D.Y. Perera, P. Selier, Water transport in organic coatings. Prog. Org. Coat. **2**(1), 57–80 (1973). doi:10.1016/0300-9440(73)80016-5
6. L. Li, Y. Chen, S. Li, Water diffusion behavior in epoxy resins with various fluorine contents. Appl. Spectrosc. **60**(4), 392–397 (2006)
7. T. Nguyen, D.P. Bentz, W.E. Byrd, Method for measuring water diffusion in a coating applied to a substrate. J. Coat. Technol. **67**(844), 37–46 (1995)
8. Z. Németh, L. Erdei, A. Kolics, A new approach to studying ion transport in corrosion protective coatings using an in situ radiotracer method. Corros. Sci. **37**(7), 1163–1166 (1995). doi:10.1016/0010-938X(95)00024-E
9. Z. Kolek, Characterization of water penetration inside organic coatings by capacitance measurements. Prog. Org. Coat. **30**(4), 287–292 (1997). doi:10.1016/S0300-9440(97)00006-4
10. R. Vlasak, I. Klueppel, G. Grundmeier, Combined EIS and FTIR–ATR study of water uptake and diffusion in polymer films on semiconducting electrodes. Electrochim. Acta **52**(28), 8075–8080 (2007). doi:10.1016/j.electacta.2007.07.003
11. N.L. Thomas, The barrier properties of paint coatings. Prog. Org. Coat. **19**(2), 101–121 (1991). doi:10.1016/0033-0655(91)80001-Y
12. H. Bi, J. Sykes, Cathodic disbonding of an unpigmented epoxy coating on mild steel under semi- and full-immersion conditions. Corros. Sci. **53**(10), 3416–3425 (2011). doi:10.1016/j.corsci.2011.06.021
13. F. Deflorian, L. Fedrizzi, S. Rossi, P.L. Bonora, Organic coating capacitance measurement by EIS: ideal and actual trends. Electrochim. Acta **44**(24), 4243–4249 (1999). doi:10.1016/S0013-4686(99)00139-5
14. J.-T. Zhang, J.-M. Hu, J.-Q. Zhang, C.-N. Cao, Studies of water transport behavior and impedance models of epoxy-coated metals in NaCl solution by EIS. Prog. Org. Coat. **51**(2), 145–151 (2004). doi:10.1016/j.porgcoat.2004.08.001
15. J.T. Zhang, J.M. Hu, J.Q. Zhang, C.N. Cao, Studies of impedance models and water transport behaviors of polypropylene coated metals in NaCl solution. Prog. Org. Coat. **49**(4), 293–301 (2004). doi:10.1016/S0300-9440(03)00115-2
16. L. Philippe, C. Sammon, S.B. Lyon, J. Yarwood, An FTIR/ATR in situ study of sorption and transport in corrosion protective organic coatings: 1. Water sorption and the role of inhibitor anions. Prog. Org. Coat. **49**(4), 302–314 (2004). doi:10.1016/j.porgcoat.2003.07.002
17. L. Philippe, C. Sammon, S.B. Lyon, J. Yarwood, An FTIR/ATR in situ study of sorption and transport in corrosion protective organic coatings: paper 2. The effects of temperature and isotopic dilution. Prog. Org. Coat. **49**(4), 315–323 (2004). doi:10.1016/j.porgcoat.2003.10.007
18. M. Stratmann, A. Leng, W. Fürbeth, H. Streckel, H. Gehmecker, K.H. Groβe-Brinkhaus, The scanning Kelvin probe; a new technique for the in situ analysis of the delamination of organic coatings. Prog. Org. Coat. **27**(1–4), 261–267 (1996). doi:10.1016/0300-9440(94)00542-7

19. K. Wapner, M. Stratmann, G. Grundmeier, In situ infrared spectroscopic and scanning Kelvin probe measurements of water and ion transport at polymer/metal interfaces. Electrochim. Acta. **51**(16), 3303–3315 (2006). doi:10.1016/j.electacta.2005.09.024
20. B. Reddy, J.M. Sykes, Degradation of organic coatings in a corrosive environment: a study by scanning Kelvin probe and scanning acoustic microscope. Prog. Org. Coat. **52**(4), 280–287 (2005). doi:10.1016/j.porgcoat.2004.04.004
21. M. Doherty, J.M. Sykes, Micro-cells beneath organic lacquers: a study using scanning Kelvin probe and scanning acoustic microscopy. Corros. Sci. **46**(5), 1265–1289 (2004). doi:10.1016/j.corsci.2003.09.016
22. I. Auerbach, W.R. Miller, W.C. Kuryla, S.D. Gehman, A diffusivity approach for studying polymer structure. J. Polym. Sci. **28**(116), 129–150 (1958). doi:10.1002/pol.1958.1202811612
23. S.-A. Chen, H.-L. Ju, Permeation of oxygen through polyurethane–polyepoxide interpenetrating polymer networks. J. Appl. Polym. Sci. **25**(6), 1105–1116 (1980). doi:10.1002/app.1980.070250613
24. M. Salame, Prediction of gas barrier properties of high polymers. Polym. Eng. Sci. **26**(22), 1543–1546 (1986). doi:10.1002/pen.760262203
25. K.S. Miller, J.M. Krochta, Oxygen and aroma barrier properties of edible films: a review. Trends Food Sci. Technol. **8**(7), 228–237 (1997). doi:10.1016/S0924-2244(97)01051-0
26. N.S. Sangaj, V.C. Malshe, Permeability of polymers in protective organic coatings. Prog. Org. Coat. **50**(1), 28–39 (2004). doi:10.1016/j.porgcoat.2003.09.015
27. Y. Rharbi, A. Yekta, M.A. Winnik, A method for measuring oxygen diffusion and oxygen permeation in polymer films based on fluorescence quenching. Anal. Chem. **71**(22), 5045–5053 (1999). doi:10.1021/ac990193c
28. W.J. Bowyer, W. Xu, J.N. Demas, Determining oxygen diffusion coefficients in polymer films by lifetimes of luminescent complexes measured in the frequency domain. Anal. Chem. **76**(15), 4374–4378 (2004). doi:10.1021/ac0497961
29. T.-J. Kim, T.-H. Jurng, U.-H. Chung, S.-I. Hong, Simultaneous determination of oxygen transport characteristics of six membranes by hexagonal dissolved oxygen sensor system. Sens Actuators B: Chem. **72**(1), 11–20 (2001). doi:10.1016/S0925-4005(00)00628-6
30. K. Petrak, Permeability of oxygen through polymers. I. A novel spectrophotochemical method. J. Appl. Polym. Sci. **23**(8), 2365–2371 (1979). doi:10.1002/app.1979.070230814
31. R.V. Holland, M.L. Rooney, R.A. Santangelo, Measuring oxygen permeabilities of polymer films by a new singlet oxygen technique. Die Angewandte Makromolekulare Chemie **88**(1), 209–221 (1980). doi:10.1002/apmc.1980.050880116
32. N.J. Turro, M.-F. Chow, M. Blaustein, Generation, diffusivity, and quenching of singlet oxygen in polymer matrixes investigated via chemiluminescence methods. J. Phys. Chem. **85**(20), 3014–3018 (1981). doi:10.1021/j150620a034
33. S.C. George, S. Thomas, Transport phenomena through polymeric systems. Prog. Polym. Sci. **26**(6), 985–1017 (2001). doi:10.1016/S0079-6700(00)00036-8
34. W.K. Asbeck, M.V. Loo, Critical pigment volume relationships. Ind. Eng. Chem. **41**(7), 1470–1475 (1949). doi:10.1021/ie50475a042
35. M. Nanna, G. Bierwagen, Mg-rich coatings: a new paradigm for Cr-free corrosion protection of Al aerospace alloys. J. Coat. Technol. Res. **1**(2), 69–80 (2004). doi:10.1007/s11998-004-0001-7
36. J. Richard, Structure–property relationships of latex films: a special emphasis on viscoelastic behavior and water vapor permeability. Polym. Adv. Technol. **6**(5), 270–275 (1995). doi:10.1002/pat.1995.220060503
37. B.J. Roulstone, M.C. Wilkinson, J. Hearn, Studies on polymer latex films: II. Effect of surfactants on the water vapour permeability of polymer latex films. Polym. Int. **27**(1), 43–50 (1992). doi:10.1002/pi.4990270107
38. B. Elsener, Macrocell corrosion of steel in concrete – implications for corrosion monitoring. Cement Concr. Compos. **24**(1), 65–72 (2002). doi:10.1016/S0958-9465(01)00027-0

39. L. Basheer, J. Kropp, D.J. Cleland, Assessment of the durability of concrete from its permeation properties: a review. Constr. Build. Mater. **15**(2–3), 93–103 (2001). doi:10.1016/S0950-0618(00)00058-1
40. M.H.F. Medeiros, P. Helene, Surface treatment of reinforced concrete in marine environment: influence on chloride diffusion coefficient and capillary water absorption. Constr. Build. Mater. **23**(3), 1476–1484 (2009). doi:10.1016/j.conbuildmat.2008.06.013
41. R.G. Buchheit, H. Guan, S. Mahajanam, F. Wong, Active corrosion protection and corrosion sensing in chromate-free organic coatings. Prog. Org. Coat. **47**(3–4), 174–182 (2003). doi:10.1016/j.porgcoat.2003.08.003
42. M. Kendig, M. Hon, L. Warren, 'Smart' corrosion inhibiting coatings. Prog. Org. Coat. **47**(3–4), 183–189 (2003). doi:10.1016/S0300-9440(03)00137-1
43. V. Karpakam, K. Kamaraj, S. Sathiyanarayanan, G. Venkatachari, S. Ramu, Electrosynthesis of polyaniline–molybdate coating on steel and its corrosion protection performance. Electrochim. Acta **56**(5), 2165–2173 (2011). doi:10.1016/j.electacta.2010.11.099
44. P.J. Kinlen, Y. Ding, D.C. Silverman, Corrosion protection of mild steel using sulfonic and phosphonic acid-doped polyanilines. Corrosion **58**(6), 490–497 (2002). doi:10.5006/1.3277639
45. U. Rammelt, L. Duc, W. Plieth, Improvement of protection performance of polypyrrole by dopant anions. J. Appl. Electrochem. **35**(12), 1225–1230 (2005). doi:10.1007/s10800-005-9033-7

Part IV
Applications

Chapter 12
Aerospace Coatings

Peter Visser, Herman Terryn and Johannes M.C. Mol

12.1 Introduction

Since the first flight of Orville and Wilbur Wright on Dec. 17, 1903, the aviation industry evolved from pioneers that wanted to fly, into a global high-tech engineering industry that enables transport and travel around the world, with an impressive safety record. Aviation has become an integral part of society; it brings people closer together, contributes to economic growth, and plays an important role in today's global economy. As air travel continues to grow and new business models expand, the evolution of the industry continues and provides new aircraft designs and construction materials. Over the years, aircraft design evolved from wooden structures covered with fabric into complex multi-material models optimized in terms of strength to weight ratio, fatigue, aerodynamics and operational performance.

Aerospace coatings play an important role in aircraft design. They have a decorative role, but an active protective function as well. Coatings for aircraft must cope with extreme conditions. In daily service they have to withstand rapid temperature changes when taking off for example from the desert at +50 °C, and climbing to a cruising altitude of 10 km with temperatures as low as −55 °C and high levels of UV irradiation. On the other hand, these coatings need to withstand repeated dry and wet cycles due to condensation, erosion and impact effects, the vibrations of the engines and structural flexing due to turbulence, pressurizing

P. Visser (✉)
AkzoNobel Performance Coatings, Sassenheim, The Netherlands
e-mail: peter.visser@akzonobel.com

H. Terryn · J.M.C. Mol
Delft University of Technology, Delft, The Netherlands

H. Terryn
Vrije Universiteit Brussel, Brussel, Belgium

A.E. Hughes et al. (eds.), *Active Protective Coatings*,
Springer Series in Materials Science 233, DOI 10.1007/978-94-017-7540-3_12

effects and wing movements. In addition to this, the coating systems are from time to time exposed to various fluids and chemicals, such as de-icing fluids, aviation fuel, cleaning chemicals, and hydraulic fluids. Last but not least, other environmental stresses such as accumulated dirt and potentially corrosive salts from coastal and maritime environments, and acids formed by industrial air pollution and volcano eruptions have a detrimental effect on the coating system integrity [1].

Each aircraft is exposed to unique conditions during its operational life. For example, long haul with low cycles or regional service with many cycles and the operational route (transatlantic or continental, northern hemisphere or southern hemisphere) and geographical areas (Europe or South-East Asia). Whereas commercial aircraft may fly up to 4000 h per year, defence aircraft or business jets can have only limited flight time. Some aircraft are even parked in the desert for a significant period of time. Despite all these factors and impacts originating from the in-service environment the aircraft needs to be colourful and glossy on the outside and the coating systems should protect the structure against corrosion for the entire service life. Aerospace coatings have been designed to address these specific needs of this highly regulated and demanding industry.

12.2 Aerospace Market

The Aerospace coatings market can be divided in four segments: commercial passenger aircraft, defence aircraft, general aviation and space. The first two markets account for most of the sales and volume of the market. The commercial segment is a growing segment whereas the military and defence markets are expected to decline over the coming years [2]. The general aviation market produces private and business jets and depends strongly on the economical climate and is slowly recovering after production dropped from 4277 units in 2007 to about 2100 per year in 2012 [3].

According to the long-term outlook of large commercial aircraft manufacturers, air travel will keep growing together with the expansion of the world's aviation network and connectivity and more capacity between major populations centres. Based on this, they produce between 27,000 and 34,000 commercial aircraft in the coming 20 years [4, 5]. The in-service fleet of 20,341 commercial aircraft with a capacity of more than 100 passengers is expected to grow an average of 3.6 % per year to double its size to more than 41,000 aircraft by 2032. Growth of passenger travel in Asia and Middle East and the need for more fuel-efficient aircraft are driving this growth. Over a 23-year period the 7-year moving average of aircraft deliveries grew from 500 to more than 1000 in 2013 and is expected to grow even further [2].

The aerospace coatings business has evolved to a global business. All segments have their Original Equipment Manufacturers (OEM's) and Maintenance, Repaint and Overhaul (MRO) facilities. In addition to their final assembly lines, OEM's have an extensive network of contractors and subcontractors all over the world that supply a variety of parts and assemblies. These suppliers must apply coatings specified by the OEM's. Most airlines repaint their aircraft every 5–7 years of service during maintenance intervals or for rebranding purposes. Considering the average service life of an aircraft being more than 20 years, the volume of paint for this MRO business is significant [6].

Historically, the final assembly lines of the OEM's were concentrated in Europe and North America. Now, manufacturing and maintenance operations are moving to Asia and South America to take advantage of the lower labour costs. In particular the aerospace industry of China is growing rapidly in terms of airline operations, maintenance and aircraft production [7].

12.3 Corrosion Control in the Aerospace Industry

Air transport plays an important role in the world's economic activities. Safety is a key element next to efficient and environmentally sustainable operations on regional and global level. Due to many safety management programs and systems that are based on risk management principles, the safety record of the industry dramatically improved, with 2013, as the safest year ever in terms of fatalities for scheduled international air transport operations [8]. Despite this achievement corrosion control remains a very important aspect of the aerospace industry. The cost of corrosion is a significant percentage of the total aircraft maintenance costs [9]. In addition to this, corrosion impacts safety and aircraft availability. Corrosion compromises the structural integrity of aircraft structures and in the event of corrosion consequences can be dramatic. In the design phase of an aircraft many requirements such as weight, aerodynamics, fatigue and fuel properties are determined. An aircraft design comprises many dissimilar metals, complex geometrical structures, joining techniques and many difficult-to-inspect areas. Corrosion is a very important but difficult parameter to take into account because the prediction of corrosion initiation and extent of structural damage is hard to determine. Despite advances in corrosion knowledge, computational design capabilities, maintenance data analysis and simulations, there is no reliable model for predicting corrosion [10]. Aircraft structures are exposed to a number of stresses and conditions depending on the operating environment of the aircraft. The aircraft is designed to resist various types of corrosion through selection of the proper materials, alloys, tempers, pre-treatments, coating systems (defined as the corrosion protective scheme), and the use of drainage points, sealants and corrosion inhibitors. These designs are based on knowledge and experience of what causes corrosion and the types of corrosion that occur in airplane structures. In addition to this design, a corrosion control program is needed throughout the service life of the aircraft. This is essential to control corrosion to a predictable, manageable level that does not degrade structure or jeopardize the ability of the airplane to carry its intended design loads. The use of the current corrosion schemes and corrosion management programs has made it possible for aircraft to operate beyond their designed life without compromising safety performance.

12.3.1 Introduction to the Corrosion Protection Scheme

The corrosion protection scheme is a key element of the aircraft design. Each part of the structure has several protecting layers to ensure the structural integrity during the service life. In general it can be stated that the 'state of the art' corrosion protection scheme is applied on top of the substrate, aerospace aluminium alloys, and consists of 2 or 3 layers, which typically is a pre-treatment and a coating system. The pre-treatment can be a chemical conversion coating, wash-primer or an anodic film (anodising layer). Whereas, the second layer is an organic coating, which can be a primer or a primer-topcoat system. The main function of such a corrosion protection scheme is to act as a barrier between the alloy and the operating environment (passive protection). In addition, the corrosion protection scheme needs to provide active corrosion protection, through the ability to release an inhibitor (leaching mechanism) to provide corrosion protection in case this barrier is damaged. Today, this leaching mechanism is a key feature in the active corrosion protection mechanism of the "state of the art" corrosion protective scheme. This means, when the corrosion protective scheme is damaged and the substrate is exposed to the environment, the inhibitor dissolves from the coating due to moisture ingress and will be transported to the exposed or damaged area and forms a new barrier layer and protects the alloy from the aggressive environment and corrosion.

For decades, the aerospace industry has been, and still is relying on hexavalent chromium-based (Cr(VI)) compounds (chromates) as active corrosion inhibiting material for the corrosion protection of the structures. These hexavalent chromium compounds are used in many processes such as: deoxidizers, anodising processes, conversion coatings, sealants and organic coatings. The corrosion inhibition mechanism of these compounds has been studied extensively [11]. Their solubility combined with the ability to act as both an anodic or cathodic inhibitor and the complex oxidation/reduction chemistry, provide them with unique properties to inhibit corrosion on ferrous and non-ferrous metals at low concentrations and over a wide pH range. Chromates have a proven track record for many years. They are not perfect, but over the years, the industry has based all the designs, specifications, processes, and corrosion management programs around the solid performance of this class of inhibitors. Unfortunately, these hexavalent chromium compounds are toxic and proven to be carcinogenic [12]. Hence, elimination of these compounds has been an important topic for more than 20 years [13].

12.3.1.1 Aluminium Alloys Used in Aerospace

Aluminium is the most dominant substrate in aircraft design, because of its lightweight properties and high strength-to-weight ratio. The mechanical, physical and chemical properties of aluminium alloys depend on the composition and microstructure. The addition of selected alloying elements in combination with the mechanical and thermal treatments (tempering) provide the desired properties for the intended application of the aluminium alloy [14]. Wrought alloys such as AA2024 and AA7075 are most commonly used for the legacy aircraft. The first digit of the alloy indicates the major alloying element. This is copper for AA2024 and zinc for AA7075. The second digit indicates modifications from the original alloy and the last two digits identify the specific alloy. The alloy designation is followed by a temper designation, for example T3 or T6. This designates the heat treatment used. T3 is solution heat treated, cold worked, and naturally aged. The T6 temper is solution heat treated and artificially aged [15–17]. AA2024 is mainly used for the fuselage skin and frame along with other structural parts, AA7075 is often used for a wide variety of structural parts.

The addition of alloying elements such as copper and zinc to aluminium has significant benefits for the strength-to-weight ratio but it also reduces the corrosion resistance. The heterogeneous microstructure of these aluminium alloys, crystallography, grain size and shape, residual stresses and intermetallic particles, plays an important role in the reduced corrosion resistance of these alloys. In recent years, understanding of the microstructure and the influence on corrosion of high strength aluminium alloys advanced significantly [18].

Weight reduction is one of the most important driving forces in new aircraft design [19]. Lower weight reduces fuel consumption, increases payload and range. New construction materials should also provide improved and optimised mechanical properties to increase maintenance intervals and reduced repair costs. Reduction of the material density is the most effective way to reduce the weight of the aircraft. Hence, composites and fibre metal laminates (GLARE) have been introduced and new alloys are being developed. Although, the use of composite in new aircraft models reduced the role of high strength aluminium alloys to some extend, the aluminium alloys remain important in the aircraft design due to their low cost, lightweight properties and easy low cost workability. Besides the 2000 and 7000 series aluminium alloys, the 3rd generation aluminium-lithium (Al–Li) alloys are finding their application in new aircraft designs. This new generation of alloys provides weight savings, but also higher strength and improved fracture toughness. Al–Li 2199 is used for the fuselage and lower wing applications. Al–Li 2099 is used for extrusions for the internal structure (Table 12.1).

12.3.1.2 Surface Treatment of Aluminium

The surface treatment is the first step of the corrosion protective scheme after the selection of the alloy. These treatments are the first defensive layer to protect the

Table 12.1 Chemical composition of some aerospace aluminium alloys [19]

	Cu	Zn	Li	Mg	Mn	Fe	Si	Cr	Zr	Ti	Al
2024	4.4	–	–	1.5	0.6	≤0.5	≤0.5	0.1		0.15	Balance
7075	1.2–2.0	5.1–6.1	–	2.1–2.9	0.3	0.5	0.4	0.18–0.28	–	0.2	Balance
2099	2.4–3.0	0.4–1.0	1.6–2.0	0.1–0.5	0.1–0.5	0.07	0.05	0.1–0.5	0.05–0.12	0.1	Balance
2199	2.0–2.9	0.2–0.9	1.4–1.8	0.05–0.4	0.1–0.5	0.07	0.05	–	0.05–0.12	0.1	Balance

substrate from the environment and are the foundation for the second step, the application of the organic coating. Typically, three types of treatments are being used in the aerospace coatings industry: chemical conversion coatings, anodising and wash primers. Each of them has its specific characteristics and application properties. Their use has been described in specifications and application manuals. These treatments are mostly used in combination with an organic coating but provide stand-alone corrosion protection as well.

Chromated chemical conversion coatings have been an important part of the corrosion protective scheme for commercial and military aircraft for many decades. These coatings provide a thin protective layer on the substrate generated by a chemical reaction of the chemical conversion solution and the metal. The layer is a very good base for paint adhesion and it provides good stand-alone corrosion protection. Chemical conversion coatings can be applied by dipping, spraying and no-rinse processes.

The chromated chemical conversion process has been studied extensively to resolve the mechanism [20]. When the chemical conversion solution is applied on the aluminium, a redox process forms the chemical conversion film. In this process aluminium is oxidized and hexavalent chromium (Cr(VI)) is reduced to trivalent chromium (Cr(III)) and the coating formation evolves by hydrolysis, polymerization, and condensation of Cr(III), triggered by the pH increase near the surface of the aluminium alloy [21, 22]. Fluoride species are added to the conversion solutions to activate the aluminium and to prevent re-passivation of the surface. Higher concentrations of fluoride (F^-) activate the surface faster and results in a thicker conversion coating in the same time [23]. Studies demonstrated the inclusion of Cr(VI) in the chemical conversion coating [22, 24]. This inclusion is thought to be the reason for the excellent corrosion protective properties and the "self-healing" nature of these chromated chemical conversion coatings. The model for self-healing activity assumes the leaching or diffusion of Cr^{6+} species from the chemical conversion film to cracks or other damage whereas it forms a chromate corrosion product consisting of an Al^{3+} and Cr^{3+} mixed oxide which repassivates the damaged area [25]. This storage, release, diffusion and inhibition are essential to the excellent corrosion protective properties of these chromated chemical conversion coatings.

Anodising of aluminium alloys is another common treatment in the aerospace industry [26]. This electrochemical process converts the surface of the aluminium surface to a strong and stable aluminium oxide with a porous morphology. These porous aluminium oxide films provide very good corrosion protection and wear properties. Also, it serves as a good foundation for adhesive bonding and adhesion of organic paint films.

Such an anodic layer is typically 2–5 μm thick and consists of millions of pores per cm^2. The film has a thin non-porous barrier layer of 10–50 nm thick. On top of this layer there is the porous structure with a close packed hexagonal (honeycomb) cell arrangement, each cell consists of an amorphous anodic alumina with a central pore. The pore and cell dimensions can be influenced by selection of the anodising electrolyte and the anodising parameters, such as pH, time, temperature, current density, and voltage etc. [27]. The desired morphology and anodic film characteristics depend on the properties needed for the application, for example adhesion bonding, painting or stand-alone corrosion protection.

There are many types of anodising processes. In the aerospace industry, chromic acid and sulphuric acid electrolytes are widely used for the protection of aircraft parts. Chromic Acid Anodising (CAA) provides excellent corrosion protection with a relatively thin coating and without significant decrease in the fatigue strength of the material [28]. Sulphuric Acid Anodising (SAA) and Phosphoric Acid Anodising (PAA) are frequently used as well. Boric sulphuric acid electrolytes have been introduced to overcome the fatigue issues, which are the main drawback of the traditional sulphuric acid anodising processes. Tartaric Sulphuric acid (TSA) and Boric Sulphuric acid anodising (BSAA) have been introduced as a replacement of the hexavalent chromium CAA process.

In some cases, anodised parts are subjected to a sealing procedure. Sealing in hot water or hot dilute dichromate solution will convert the amorphous aluminium oxide to a monohydrate. This process partially or fully closes the pores and enhances the corrosion resistance. However, sealed anodised layers are not very suitable for adhesive bonding or coating adhesion. If painting or bonding is required, the sealing step must be eliminated or carefully controlled to ensure good adhesion. Anodising is much more expensive than a chemical conversion process and not suitable for all types of parts. Critical structures are often anodized for maximum corrosion protection.

Wash primers, also known as etch primers, have been used as metal conditioner and alternative to the "wet" chemical conversion process. Polyvinyl butyral based wash primers have been used extensively on military and commercial aircraft. The reactions of a curing chromated wash primer are complex. They are available in two component and one-pack systems. A typical two-component wash primer consists of a base component and an activator. The base component contains: polyvinylbutyral, zinc tetraoxychromate, talc, and solvents such as isopropanol and butanol. The activator is a solution of water isopropanol and phosphoric acid [29]. Upon mixing of the base with the activator many different reactions take place that are unique for this chemistry and make it very difficult to develop alternative chromate free formulations. Wash primers were mainly used in repair schemes and in the maintenance area for repainting of commercial aircraft using chromated or chromate-free epoxy or urethane intermediate primer over it. Various wash primer systems are qualified according to maintenance specifications. However, the use of wash primers has declined over recent years due to the introduction of new pre-treatments and coating solutions.

12.3.1.3 Organic Coatings

The organic coatings are the last part of the corrosion protection scheme. They act as the first barrier against the environmental conditions and need to fulfil a wide range of functions. Secondly, they have to be able to actively protect the substrate when a damage, crack or delamination occurs. Typically, the organic coating is applied in one (primer only) or two (primer/topcoat) layers. The primer is designed to maintain adhesion on the various treatments and contains corrosion inhibitors for active corrosion protection. Their composition has been carefully formulated to ensure the performance throughout the service life of the aircraft. In other cases, a second layer (top coat) is applied to the corrosion protective scheme. This topcoat can have a decorative or an additional protective function. Together with the pre-treatment, the coating system protects the substrate or alloy against environmental effects. There are many different coating systems specified in an aircraft design. For example, a fuel tank coating system has a different build-up and requirements differ compared to exterior decorative systems. The specific requirements of each coating system is defined in specifications defined by the OEM's.

12.3.2 Types of Corrosion and Failures

Corrosion occurs in many forms on the aircraft structure. These complex phenomena, although mechanistically different, are similar in terms of their potential effects in reducing the structural integrity. The corrosion processes of the aluminium in the aerospace industry are electrochemically driven. The local electrochemical potential differences at the alloy surface create the driving force for corrosion initiation [30–32]. Active spots become anodes, whereas the more noble areas act as a cathode. The form of corrosion is influenced by factors such as alloy and temper used, environmental conditions, corrosive agents present, geometry, design, and mechanical loading. The various types of corrosion are briefly discussed below.

Filiform corrosion is a very common type of corrosion in aircraft structures. This worm-like type of corrosion occurs between the metal and the polymeric coating. It initiates at a crack or rupture around a rivet head or lap joint under acidic, chlorine-rich and humid conditions. Once initiated, the propagation is determined by anodic undermining [33]. In addition to the environmental effects, filiform corrosion is also affected by the microstructure of the aluminium alloys and the properties of the applied coating system [34].

Pitting is another type of corrosion that occurs frequently in aerospace structures. It is different from general corrosion because of the absence of measurable material thinning. This is an extremely localized form of corrosion that results in holes and pits that penetrate in the structure and leads to degradation of the structural capabilities. The typical 2000 and 7000 series aluminium alloys used in the aerospace industry are susceptible to pitting corrosion and therefore have been studied extensively [35–41]. These studies have led to improved knowledge on the nucleation of pits on high strength alloys such as AA2024-T3. However, the establishment and propagation of stable pits is still being investigated.

Intergranular corrosion and to a greater extent *exfoliation* corrosion typically occurs on 2000 and 7000 series high strength aluminium alloys [42–44]. Especially, the AA7075-T6 alloy is susceptible to intergranular and exfoliation corrosion, this alloy contains both copper and zinc. Due to processing and heat treatments copper and/or zinc accumulate at the grain boundaries. These grain boundaries preferentially corrode because of the galvanic couple between the grain bodies and boundaries and the anodic behaviour of aluminium towards copper. Intergranular corrosion can start at a pit, second phase intermetallic particle or at grain boundaries at the surface and it penetrates much faster into the alloy compared to pitting corrosion. Exfoliation corrosion has a very characteristic appearance of leafing aluminium layers due to corrosion of a very elongated and flattened grain structure parallel to the alloy surface due to the rolling. It originates from intergranular corrosion at the edges of riveted and bolted aircraft components. The susceptibility to exfoliation corrosion can be reduced by the modification of the annealing process to obtain a more favourable distribution of precipitates [45, 46].

Stress corrosion cracking (SCC) [47–49] or environmental assisted stress corrosion can be one of the most destructive types of corrosion. This form of corrosion causes a loss of load-carrying capability from the combined effect of tensile stress and corrosion. There are two forms of SCC: intergranular and transgranular SCC. Aluminium alloys typically show intergranular SCC. Under corrosive conditions (humid environments with chloride) stress corrosion is initiated rapidly and follows the grain boundaries in aluminium alloys. There are two mechanisms proposed for SCC: anodic dissolution and hydrogen embrittlement. However, the exact mechanisms have not been resolved yet, despite extensive work in modelling of SCC and elucidating microstructural features along grain boundaries that influence stress assisted corrosion. The most effective way to control this kind of corrosion is to use materials that are not susceptible to SCC at design stress levels and protection from corrosion initiation by coatings and excluding electrolytes.

Crevice corrosion [47] or concentration cell corrosion is the most common type of corrosion in aircraft. It occurs when moisture is trapped between surfaces. This can be at fasteners, delaminated paint, and unsealed joints or within a delaminated bond line. This type of corrosion is aggressive and can initiate pitting and exfoliation corrosion depending on the alloy, temper and form of the material that is affected. It is driven by a differential concentration mechanism [50]. The most effective way to prevent crevice corrosion is to avoid water penetration by the application of sealants and water-displacing corrosion inhibiting compounds [51].

The aircraft design does not only contain aluminium alloys but also many dissimilar materials like: titanium, magnesium, stainless steel and composites. Without proper protection the combination of these dissimilar materials can lead to *galvanic corrosion*. This occurs when two dissimilar materials are connected in the presence of an electrolyte. Recent aircraft designs use more and more carbon-reinforced plastics. These composite materials can induce galvanic potential differences due to the very conductive carbon fibres in the material. When joined to the aluminium structure, composite materials can act as a cathode and accelerate the electrochemical dissolution of the aluminium. The rate of corrosion depends on the surface ratio between the cathode and the anode design geometry. A large cathode surface area with respect to the anode surface area can lead to fast and severe corrosion. The only way to prevent galvanic corrosion is to prevent direct contact of the dissimilar metals and the ingress of moisture (electrolyte). This can be done with the use of a physical barrier in the form of a coating, sealant and/or corrosion inhibiting compounds (CIC's).

12.3.3 Corrosion Protection and Design

Corrosion can be initiated by many sources and affect the aircraft structure. These sources can be identified at the manufacturer and operator level [50]. The basic design combined with manufacturing and processing are the most important factors in corrosion protection at the manufacturing level. At the operator level there are different factors that potentially could initiate corrosion of the structure, among them improper maintenance, deterioration of the coating system, operational environment, accidental contaminations, and the environmental conditions in the aircraft.

Corrosion can initiate when the following three parameters are present:

- A cathode and an anode.
- A metallic connector between the anode and cathode.
- An electrolyte (water).

Elimination of these three conditions is limited by feasibility, practicality, and functionality. Each part of the aircraft is designed to withstand the stresses and operating environment for the service life of the aircraft. Therefore, six key elements need to be considered when designing for corrosion control:

- Selection of materials
- Coating system
- Sealants
- Corrosion inhibiting compounds
- Avoidance of contact between dissimilar metals

- Drainage
- Access for maintenance

For the *material selection*, the design considers many parameters such as, alloy, shape, weight to strength ratio strength, durability and fatigue properties. Whereas the high-strength aluminium alloys are used most widely in the aircraft design, coupling to other materials like Titanium (6-AL-4 V), Magnesium, Stainless steel, low alloy, high strength carbon steel, and increasing amounts of Carbon Fibre-Reinforced Plastics (CFRP) can set up galvanic corrosion cells which need to be eliminated in the design stage.

The *coating system* is a key element to protect the selected material from the operating environment. This barrier typically consists of multiple layers which may include anodised layer or chemical conversion coatings that are coated with corrosion inhibiting primers. Exterior surfaces and some structural areas are coated with chemical resistant topcoats. Titanium alloys are anodized or coated and stainless steel is most commonly cadmium plated before being primed. The active protective primers are a key element in the protection scheme and will be discussed in more detail.

It is not always possible to avoid dissimilar metal contact, crevices and the use of specific alloys because of weight, cost, and functional issues. However, the potential for corrosion can be minimized by using polysulfide *sealants* that are typically applied to faying surfaces of the joints to prevent moisture ingress leading lap joint or crevice corrosion or to prevent dissimilar metal contact. They are applied at many connections like stringer to stringer and skin-to-shear tie joints, skin splices and skin doublers and pressure the bulkhead. In addition to this, exterior fasteners and fasteners that penetrate pressurized areas are installed with a sealant.

Moisture cannot be eliminated during operations. Therefore, it needs to be controlled by *drainage* and *corrosion-inhibiting compounds* (CIC's). Drainage of the aircraft structure by the use of drain paths, drain holes is important to prevent entrapment of moisture and corrosive fluids in crevices and lower fuselage area. The design is optimized to direct the fluids towards the drain holes. CIC's are used to provide additional corrosion protection. These CIC's are petroleum based water-displacing compounds with corrosion inhibitors and are sprayed on the structure to penetrated faying surfaces. The CIC's are applied in areas that are prone to corrosion (lower fuselage). Low viscosity CIC's are able to penetrate crevices and displace water, whereas viscous CIC's will act as a coating.

The features discussed above must ensure a safe and economical service life. However, the aircraft design must ensure easy *access for maintenance* and corrosion inspections as part of a corrosion management/maintenance program. Corrosion cannot be eliminated but, timely inspections will help to detect corrosion, trapped moisture, plugged drain-holes, and chipped or missing paint at an early stage and combined with sealant and CIC's applications much corrosion can be prevented.

12.4 Introduction into Aerospace Coatings

Aerospace coatings serve as decorative and protective layers for all types of passenger, freight and military aircraft. The systems evolved over the years from very simple formulations into complicated materials with specific functions and an important role in the corrosion protective scheme of today's aircraft industry. Three types of coating systems can be recognized. The exterior decorative system, which provides image and recognisability. Structural coatings that are an integral part of the corrosion protection scheme of the aircraft structure, and finally the special purpose systems. Each of these coating systems has their own requirements and must be able to withstand a wide range of stresses during their operational life.

12.4.1 History of Aerospace Coatings

In the early days, 1910, the first aircraft were painted with colourless single component cellulose lacquers to provide tension to the linen (fabric). Around 1915, pigmented or coloured coatings were introduced to provide camouflage for military aircraft during World War I. Up to 1935, aircraft were mainly constructed from wood, linen and steel. Cellulose lacquers and alkyd enamels were mainly used as coatings for the protection of these aircraft. Around 1935, the first metal aircraft were designed. New coatings were developed to protect these aircraft from corrosion, such as zinc chromate alkyd primers and chromated chemical conversion coatings (CCC), wash or etch primers, acrylic lacquers and alkyd enamels. The products were used on commercial as well as on military aircraft. In 1960, the first generation of jet engine powered aircraft, McDonnell-Douglas DC-8 and the Boeing B-707, were introduced. New highly chemical resistant coatings were required to withstand the hydraulic fluids that were introduced with these new aircraft. These developments are the basis of the coating technologies used today. Two component paint systems were developed such as epoxy -amine or -amide primer technology and polyurethane (polyester/isocyanate) topcoats. In the mid 1980s clear coat technologies were introduced because of atmospheric acidification due to volcanic eruptions (Mt Pinatubo) caused severe damage to the paint systems. At the same time environmental awareness started to influence the aerospace coatings industry. Toxic materials such as lead and cadmium were removed from the coatings and the quest for the replacement of carcinogenic hexavalent chromium was started. Solvent emissions needed to be reduced. This led to the introduction of high solid primers and topcoats and water-based coatings. Benzyl alcohol based paint removers replaced methylene chloride and phenol containing paint removers. These can be neutral, acid or hydro peroxide activated. This initiated the introduction of selective strippable paint systems. In early 2000, developments were focused on efficiency improvements like super durable topcoats, weight reduction, less layers, easy application fast curing systems. This led to the introduction of base coat/clear coat systems for the aerospace industry. Over the years aerospace coatings evolved into high performance coatings that are able to withstand aggressive environments. Currently the aerospace coatings industry is facing new challenges with the introduction of all composite fuselage designs and the replacement of hexavalent chromium for structural applications.

12.4.2 Aerospace Coating Systems and Their Requirements

The main reasons for painting aircraft are corrosion protection, airline identification, appearance and survivability. The paint system on the aircraft protects the metal and composite structures from the environment. Many aircraft components are designed for structural and fatigue performance and are not very resistant to the atmospheric exposure. Aerospace coating systems have to cope with extreme in-service conditions. Because of this, the systems evolved over the years in terms of chemical and fluid resistance, corrosion resistance and exterior durability. Many different systems were developed to protect the various areas of the aircraft. In general, aerospace coating systems can be categorized into 3 main groups: exterior systems, structural systems, and special purpose systems.

12.4.2.1 Exterior Systems

Exterior coating systems are applied on the fuselage and other exterior areas of the aircraft. These systems reflect the identity and image of the airline. Historically aircraft were mainly white. Today, aircraft liveries have to be glossy and often feature vibrant colours which are used as a marketing tool (Fig. 12.1). The exterior coating systems have to be able to withstand a wide range of environmental stresses like temperature changes, high levels of UV irradiation, humidity exposure, erosion and exposure to aggressive media including fuel, de-icing fluids, hydraulic fluids, potentially corrosive salts due to maritime areas and acidic aerosols in the atmosphere due to volcano eruptions and industrial pollution. These stresses combined with the vibration effects of the powerful engines and structural flexing due to pressurizing of the fuselage and turbulence put the coatings under severe stress.

A typical exterior coating system is a multi-layered system and consists of a pre-treatment layer (anodised layer or chemical conversion coating), and a primer layer of 15–25 μm, which is covered with a pigmented decorative topcoat of 60–120 μm (Fig. 12.2). The airline livery is applied with the decoration colours on top of this basic topcoat layer.

Recently, the base coat/clear coat topcoat philosophy has been introduced into the aerospace market. In this philosophy the decorative topcoat is replaced by a fast-drying, highly pigmented coloured base coat making it possible to reduce the process time of painting an aircraft, particularly complex liveries. Once all the

Fig. 12.1 Exterior coating systems are used for identity and marketing

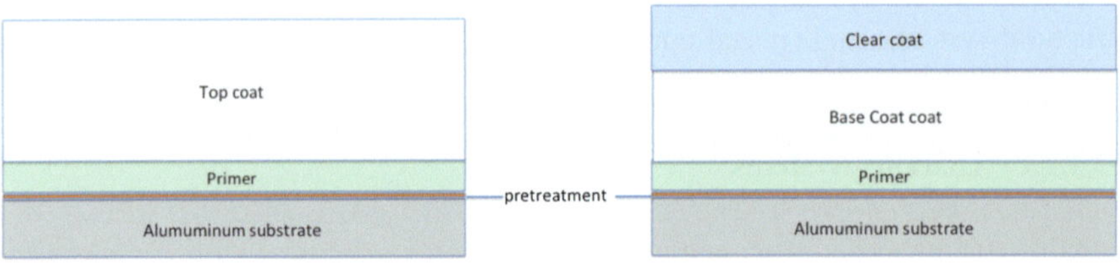

Fig. 12.2 Generic system build-up of exterior coating systems

colours are applied, the clear coat is applied providing an improved gloss and colour retention, thus longer durability.

Typical failures of external systems during aircraft operation are gloss reduction, loss of adhesion (rivet rash), cracking, and filiform corrosion.

12.4.2.2 Structural Coating Systems

Structural coating systems are designed to protect the structural parts of the aircraft. These systems are applied on small parts and sub-assemblies before the final assembly (Fig. 12.3).

Usually the system consists of a pre-treatment layer, which can be a chemical conversion coating or an anodic film, coated with 15–20 μm of structural primer. In some areas of the aircraft, for example in lower areas (bulge area) and cargo areas, an additional 20 μm epoxy or polyurethane topcoat is applied for additional protection (Fig. 12.4). Structural coating systems are usually not stripped at regular maintenance intervals and are applied in areas that cannot be inspected. The system

Fig. 12.3 Impression of the aircraft structure with the structural coating system

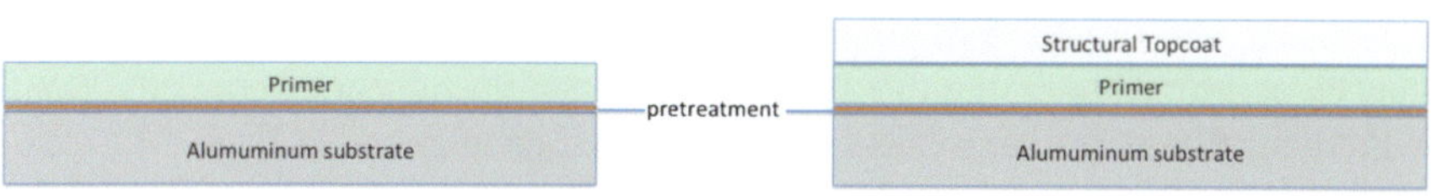

Fig. 12.4 General build-up of a structural coating system

needs to provide passive and active corrosion protection for the entire service-life of the aircraft. Therefore these systems are excellent barriers and provide long-term corrosion protection and excellent adhesion to a wide variety of substrates. During the operational life of aircraft these coatings are exposed to many different stresses and wet and dry cycles [52].

12.4.2.3 Special Purpose Coating Systems

In addition to the exterior and structural coating systems, specific area's of the aircraft need specific coatings. Examples are: composite areas, the fuel tank and wings. But as well, there are many different types of coatings with specific purposes. Among them, coatings for non-slip, conductivity, selective strip ability, heat resistance, abrasion resistance, solar heat-reflection, fuel vapour barrier, and camouflage (Infra-red absorption or reflection).

12.4.2.4 Coating System Requirements and Specifications

The requirements for coating systems used in the aerospace industry are controlled by a large number of specifications. Various specifications can be considered: MIL-Specifications issued by the US Government for military aircraft, OEM specifications of commercial aircraft manufacturers and maintenance specifications defined and controlled by the SAE group. Specifications describe the general properties and performance requirements of a coating system for a specific purpose like: structural primer, fuel tank primer, exterior paint system, paint system for composites or a wing coating. The requirements are based on critical properties that are needed for a coating to demonstrate its intended function. Commercial aircraft manufacturers, governmental research laboratories or certified private testing institutes perform tests on submitted coating products to determine whether it complies with the requirements of a particular specification. Once the product meets the specification requirements, it will be added to the Qualified Product List (QPL) of the specification. Specifications of commercial aircraft manufacturers are usually not public documents, whereas military specifications that are managed by governments are publicly available.

Specifications are designed for each specific coating. They describe a variety of tests that ensure that application, appearance, and performance characteristics of the coating match the intended use. The tests listed in these specifications often follow ISO, or ASTM standards. Specifications describe qualification and quality control tests that coatings must pass in order to be approved for use in the industry. The specifications contain a list of requirements for wet paint properties, application, appearance, and the dry film properties. Quality control tests describe which tests must be performed to ensure that the batch has been manufactured correctly. Some of the typical tests for aerospace coatings are described below [53].

Wet paint properties:

Properties such as *viscosity, pot life* and *dry-time* are typical application related properties. The paint viscosity influences the atomization, substrate wetting, flow, levelling, and sagging resistance. Usually, the paint viscosity is measured with a flow cup. Coatings with thixotropic and pseudo-plastic behaviour need more sophisticated techniques such a Brookfield viscometer. Most aerospace paints are two-component chemically curing systems. These systems have a so-called “pot life”. This important property can be defined as the period of time in which the product can be applied with the required finish quality (appearance) and final dry film is performing according to the specification. Viscosity increase is an important measure for this property. In addition to this, dry-time is an important factor in the aerospace industry. The curing profile is an important characteristic when developing an aerospace formulation. Various dry times can be identified as relevant for the aerospace business. Set to touch time, dust free time, tack free time, dry to stack-time, dry to tape time, dry hard time and dry to recoat.

Appearance and durability

Gloss and *colour retention* are the most important appearance related properties for exterior coatings. Gloss can measured at various angles (20°, 60° and 85°) with a gloss meter. Colour retention and colour accuracy is very important to the aerospace business. Colour is usually measured with a spectrophotometer using the CIELAB colour system [54] and compared with colour standards. Exterior systems need to retain their gloss and colour during their service life. Aerospace coatings are exposed to high levels of ultraviolet light, which has a detrimental effect on the binder system and colour retention of the pigments. Thermic cycling and moisture from humidity, condensation and rain accelerate this degradation. Artificial weathering tests are used to simulate these degradation phenomena. UVCON and Xenon Weather-O-Meter are the regularly used methods for accelerated weathering. Unfortunately, neither are representative of real life conditions. Alternatively, south Florida exposure is used as another reference exposure for durability.

Dry Film properties

There are many properties that should be considered when evaluating the performance aerospace coating systems. *Adhesion* is one of the prime properties and can be tested by scribing the coating with a knife, followed by the application, and removal of a pressure sensitive tape. Pass or fail is specified by the amount of coating that was removed by this test. The method, scribe pattern and tape are specified in the specification. Adhesion tests are normally executed before and after water or *fluid immersion*. During operation of the aircraft, the coating will be exposed to several kinds of fluids like hydraulic fluids, kerosene, lubricating oil, toilet fluids, and de-icing fluids. Both primers and topcoats are designed to withstand these fluids and their resistance is tested by immersion in these fluids for a specific duration at a specific temperature. After immersion the system is tested for adhesion, softening, and blistering.

Flexibility and impact resistance

Aerospace systems are exposed to a range of fast and slow deformation tests. Therefore, the systems are subjected to various flexibility tests with conical and cylindrical mandrels at ambient and low temperature (−55 °C). Impact tests are typical fast deformation using a specified force and or shape. Impact resistance is reported in different ways varying from pounds per inch, elongation or degree of cracking with fixed deformation and mass. After impact the coating is evaluated for cracks and adhesion. During the service-life of an aircraft, the exterior paint system will be removed and re-applied a number of times. Reasons for such strip and re-paint exercises include periodical maintenance, inspection and overhaul schedules but as well changing aircraft owners and new liveries (re-branding). In order to ensure adequate and efficient strip ability process, the system needs to be tested for *strippability* using aerospace qualified chemical paint removers.

Corrosion Resistance.

In order to assess the corrosion resistance of aerospace coatings, a number of corrosion testing methods are used. *Salt spray* and *Filiform* corrosion testing are the most commonly applied corrosion tests. In these tests, coated panels, primer only or primer—topcoat, are scribed through the coating into the bare metal and subjected to the corrosion test. In neutral salt spray, panels are positioned in a slight inclined vertical position and exposed to a 5 % sodium chloride solution up to 3000 h. After this exposure the panels are evaluated for corrosion extended along the scribe and blistering. In addition to this regular neutral salt spray, some specifications require acidified salt spray. Whereas in salt spray, corrosion develops under a continuous salt fog, filiform develops under high temperature and humid conditions in the presence of chloride ions. Therefore, the scribed paint systems are exposed first to the fumes of concentrated hydrochloric acid for one hour and then immediately placed in elevated temperature (30–42 °C) and high-humidity (85–95 %) environment for 1000 h. Although, these accelerated corrosion tests do not represent the in-service conditions, these are currently the only way to assess the corrosion protective properties of the active corrosion protective mechanism of the coatings before implementation.

12.4.3 Coating Chemistry

Several reviews on protective coatings for aerospace applications can be found in the literature [55, 56]. The organic coatings that are applied in the aerospace industry are complex materials. These heterogeneous materials are the basis for the corrosion protection scheme of aircraft structures. The complex resin matrix contains pigments, fillers, solvents, and additives. Each type of coating is carefully designed to meet the industry standards. In the 1970's, the fundament of today's generation of coatings and the protection of aerospace structures was set with the development of hexavalent chromium containing, hydraulic fluid resistant epoxy/amine primers, together with the development of flexible but chemical resistant polyol/isocyanate polyurethane topcoat technology.

This section discusses the different coating technologies, primer and topcoat chemistries in the marketplace and the active protective mechanism used to provide long term corrosion protection.

12.4.3.1 Coating Technologies

There are many different types of coatings in the aerospace industry. Over the years, many coating technologies have been developed due to pressure to reduce solvent emissions of coatings. The regulations for Volatile Organic Compounds (VOC) vary all over the world but in general the current limits for VOC are 350 g/L for primers and 420 g/L for topcoats. Currently, there are 4 different types of coatings in the market place:

- Conventional coatings
- VOC exempt coatings
- High Solid coatings
- Water Based coatings

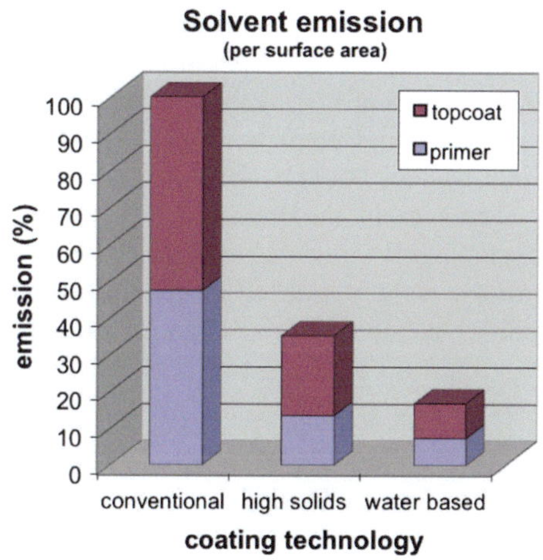

Fig. 12.5 Relative solvent emission reduction by introduction of high solids and water based coating technology

Together with this pressure to reduce VOC in the coatings many different resin technologies were developed to maintain the application properties and performance of the aerospace coating systems. Whereas in conventional coatings solid resins were used, high solid coatings are based on liquid resins. Whereas, in some instances the solvent reductions in grams per litre seems limited, Fig. 12.5 shows the real solvent emission reduction per applied surface area when replacing conventional coatings by low VOC coating technologies.

Conventional aerospace coatings contain typically 550–650 grams of solvent per mixed litre (g/L) paint. The coatings are often formulated with high molecular weight resins. In addition to their physical properties and the ability to comply with the aerospace applications, these systems have very favourable application properties such as: long pot life, fast physical drying properties, good film thickness control and a smooth appearance.

VOC exempt coatings are applied as a low VOC (340 g/L) alternative in primer technology. Some VOC's have a negligible atmospheric photochemical reactivity and are considered as exempt solvents and therefore these solvents are not considered a VOC. Examples of such solvents are acetone and parachlorobenzotrifluoride. The use of exempt solvents offers many advantages to the coating formulator to formulate low VOC coatings with similar resin technologies as conventional coatings but maintain low viscosity and the application characteristics.

Today, new technologies are often based on high solid coating technology is another common approach to formulate VOC compliant primers and topcoats. Respectively they have a VOC of 350 and 420 g/L. High solid coating technology is based on the crosslinking of low viscosity resins with low viscosity curing agents. These resins are much lower in molecular weight compared to the conventional resins and have a higher reactivity. The drying time depends on how fast a high molecular weight network requires to become tack-free. However, a fast tack–free time is in contradiction to a long pot life. Therefore, formulators have to find the right balance between these properties. Additionally, low molecular resins and curing agents provide very highly cross-linked systems. This is very beneficial for a good chemical resistance but can be detrimental to the flexibility. Usually high solids coating formulations are blends of various resins and curing agents to find the right balance between the dry film and application properties.

Progress in the resin industry has helped it to develop water-based coating technologies for aerospace applications. In the late 1980's, the first water reducible primers were introduced for military aircraft. Today, there are multiple water based primer technologies available for structural applications. Water-reducible coatings are based on high solid resins made in solvents that form a dispersion of resin in water when reduced with water. Modern water-based systems are based on dispersions and emulsions of resins that easily mix and can be used without the addition of additional water. These systems provide a further reduction of solvent emissions (250 g/L excluding water) and application properties of conventional coating systems such as fast cure and good film thickness control.

12.4.3.2 Aerospace Primer Coating Technologies

Primers are used for exterior and structural protection of the aircraft. They provide active corrosion protection towards the substrate and provide adhesion to the topcoats. These primers have to comply with many criteria. These include:

- Room temperature cure
- Excellent adhesion to multiple substrates
- High chemical resistance
- Sufficient flexibility
- Long term corrosion resistance
- Good compatibility with the topcoat
- Applicable under various climatic conditions

Epoxy primer technology

The majority of the primers in the aerospace coatings market are 2 component epoxy/amine primers. The base component comprises a mixture of epoxy resins, pigments, corrosion inhibitors, extenders, additives, and solvents. The second component, the hardener or curing solution, contains curing agents (amine or poly amide) and solvents. Corrosion inhibitors such as barium chromate and strontium chromate have been and are still most commonly used as corrosion inhibitor. Epoxy primer technology is used because of its very versatile properties. It has good adhesion to many substrates, good corrosion resistance, and can be formulated to find a balance between the chemical resistance, flexibility, impact resistance and hardness requirements of the aerospace industry. On the other hand, epoxy amine primers are difficult to strip, have relative long drying times and limited exterior durability.

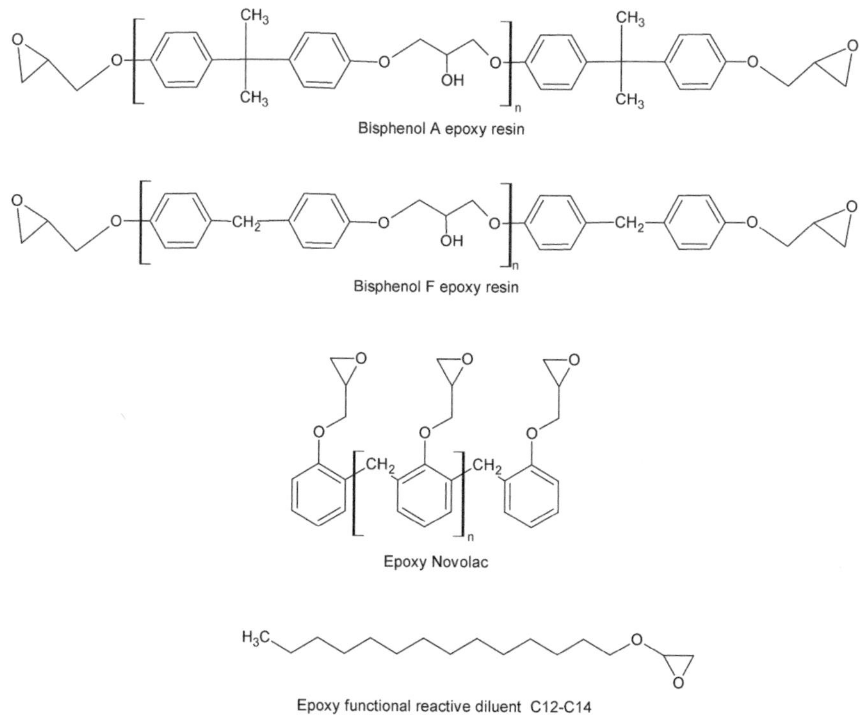

Fig. 12.6 Types of epoxy resins

Epoxy resins

There is a wide range of epoxy resins available. The majority of the epoxy resins are derived from the reaction between epichlorohydrin and diphenylol propane (Bisphenol A) [57]. There are mainly 4 groups of epoxy resins (Fig. 12.6):

- Bisphenol A epoxy resins
- Bisphenol F epoxy resins
- Epoxy Novolacs
- Epoxy functional diluents and modifiers

Each type of resin has a range of variations in molecular weight and functionality. The selection of an epoxy resin depends on for which specification or application the coating needs to be developed. Many products contain mixtures of epoxy resins to obtain the specific coating properties needed for aerospace applications. High functional resins are blended with flexible epoxies for increased elongation properties whereas solid high functional resins are blended for fast physical drying and chemical resistance. Bisphenol A and F resins are blended to adapt the cure speed and epoxy functional reactive diluents are added to reduce viscosity and increase flexibility.

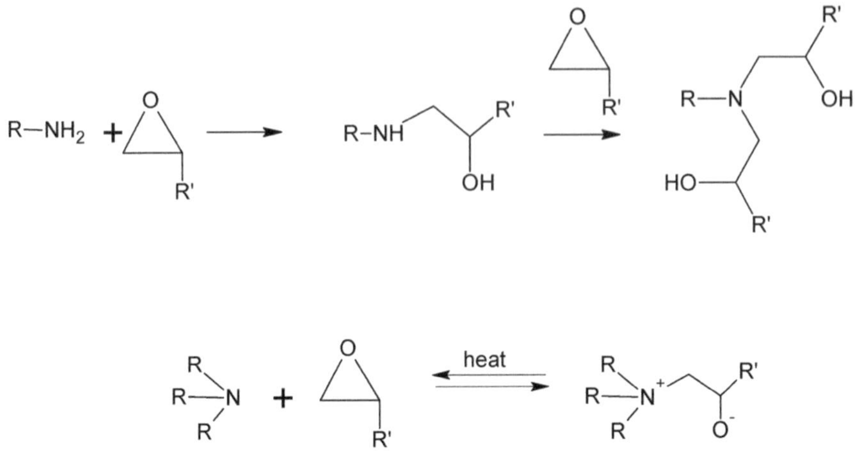

Fig. 12.7 Epoxy curing reactions

Epoxy Curing agents: There are many possibilities to cross-link epoxy resins (amines, thiols, carboxylic acids) via an addition reaction. In aerospace coatings, most epoxy primers are formulated with amine functional curing agents (Fig. 12.7). Epoxy resins react under ambient conditions with the primary and secondary amines. The reaction rates depend on the epoxy and amine structures.

Terminal epoxy groups are more reactive than internal and more sterically hindered epoxy groups. Reactivity of amines tends to increase with base strength and decreases with steric hindrance. Cycloaliphatic amines have reduced reactivity; the second reaction of such an amine is particularly slow. Aliphatic amines are less basic than aromatic amines and therefore more reactive. The epoxy amine reaction is catalysed by water, alcohols, tertiary amines, and weak acids (e.g. phenols), which promote ring opening by proton complexation with the epoxide oxygen.

The selection of epoxy resins and curing agents depends on the desired coating properties. Most commonly the formulation will consist of a mixture various epoxy resins reactive diluents of aliphatic or cyclo-aliphatic amines blended with polyamides. This all to adjust the various coating properties for adhesion, flexibility, chemical resistance, pot life, and cure speed needed for its specific application and specification.

12.4.3.3 Aerospace Topcoat Technologies

Aerospace topcoats are applied both in structural and exterior applications. Although, both with a different purpose. Epoxy topcoats are applied for structural applications due to their good adhesion and excellent fluid resistance to ensure the good corrosion protection properties of the coating system. The basic chemistry of epoxy topcoats is similar to the primer technology. Epoxy topcoat technology is

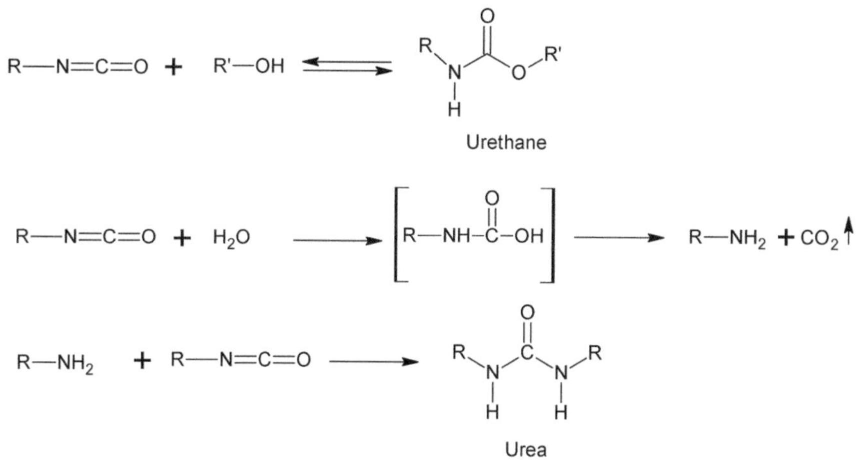

Fig. 12.8 Polyurethane curing reactions

only used for structural parts since they are not durable enough for exterior applications.

Since the late sixties, polyurethane topcoats have been applied in the aerospace industry due to their very good exterior durability combined with excellent flexibility and good fluid resistance. Polyurethane topcoats are based on 2 K formulations of which one part contains a polyol (polyester or acrylate resin) that is cured with the other part that contains aliphatic isocyanates. Sometimes, a third part that contains a catalyst is added to adjust the cure rate for different application conditions.

The polyurethane coating technology is based on the main reaction between primary and secondary hydroxyl groups of the polyol and isocyanates forming urethane crosslinks [57]. On the other hand, urea crosslinks are formed by the reaction of atmospheric water with isocyanates (Fig. 12.8). This secondary reaction is somewhat slower than the reaction with a secondary hydroxyl group but it is an important aspect of formulating chemical resistant aerospace coatings with polyurethane coating technology.

The properties of polyurethane coatings depend on the polyol structure, type of isocyanate and the stoichiometric ratio of the isocyanate and hydroxyl groups (NCO/OH ratio). The chemical or fluid resistance of the coating depends on the crosslink density. Polyols with high hydroxyl functionality will increase crosslink density and provide more chemical resistant but less flexible films. Aerospace finishes are often formulated with a higher NCO/OH ratio to increase fluid resistance especially hydraulic fluids. Over-indexing with isocyanate minimizes the amount of unreacted hydroxyl groups and the excess isocyanate groups react with water to form the urea groups resulting in a more resistant coating with a higher T_g. Hydroxy-terminated polyester and hydroxy-substituted acrylic resins are most commonly used polyol resins in aerospace exterior topcoats. Polyesters permit higher solids and give films with greater solvent resistance. Acrylics provide faster dry, higher gloss and better exterior durability, resulting from superior hydrolytic and photochemical stability. Therefore, polysesters are used for solid colour topcoats and acrylics are used for clear coat applications.

Aliphatic isocyanates such as hexamethylene diisocyanate and isophorone diisocyanate are commonly used in the aerospace topcoats. Because of their toxic nature, these diisocyanate monomers are converted into high molecular weight trimers (triisocyanurates) with a very low concentration of monomeric isocyanate. These high molecular isocyanates are available in liquid but 100 % solid form.

Aerospace coatings are designed for cure at room temperature conditions. However, the reaction between primary or secondary hydroxyl groups and the aliphatic isocyanate are slow at room temperature conditions and therefore needs to be catalysed. Tertiary amines (diazabi-cyclo [2.2.2] octane (DABCO)) and organometallic compounds, such as dibutyltin dilaurate (DBTDL), or carboxylic acid salts, are the most commonly used catalysts for these 2 component polyurethane coatings.

The type of the catalyst and the amount used in the formulation is critical to the balance of pot life, application properties and the drying time of the coating. Extended pot life in combination with rapid cure can be obtained using organotin catalyst (DBTL) in combination with 2, 4-pentanedione (acetyl acetone) or volatile carboxylic acids (formic acid or acetic acid).

The majority of the polyurethane technology is used for exterior applications. Therefore, exterior durability is a key element for these formulations. Exterior durability is referred to the resistance to withstand degradation processes when exposed to environmental conditions and the ability to maintain the aesthetic and functional properties of the coating. Chemical processes such as hydrolysis and photo-oxidation resulting from exposure to the environmental conditions cause coating degradation. UV irradiation from sunlight can initiate photo-oxidation processes in the resin system. In addition to this, environmental conditions such as acid rain due to air pollution or atmospheric degradants (e.g. ozone) can initiate hydrolytic degradation processes in the coating. Temperature and adhesion also can affect these processes. These degradation processes may lead to loss of gloss, discoloration or even to cracking, delamination that can provide an access point for the environment to reach the substrate and cause corrosion. The resins for exterior applications are optimized for hydrolytic stability and are stable against photo-oxidation. Another critical factor for exterior durability is the selection of pigments. The quality and after-treatment of pigments can have a significant effect on the colour stability and gloss of the coating. Photo stabilization additives, such as UV absorbers, and radical scavengers hindered amine light stabilizers (HALS) and anti-oxidants, provide an additional protection against photo initiated oxidative degradation.

12.4.4 Active Corrosion Protection and Leaching

Organic coatings provide protection to the metallic substrates by passive inhibition mechanisms such as barrier function and adhesion to the substrate; this protects

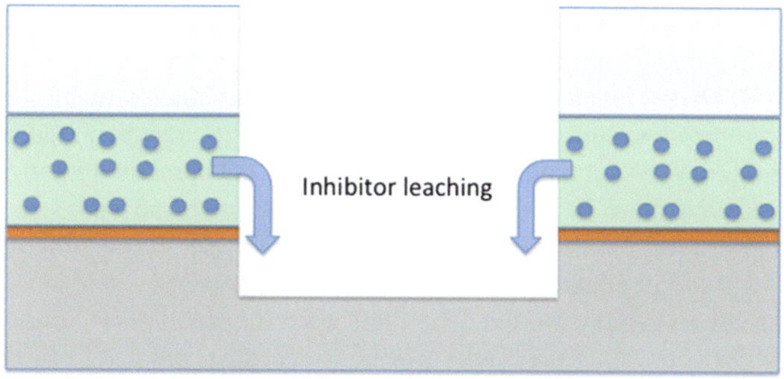

Fig. 12.9 Inhibitor leaching from the coating as active corrosion inhibition mechanism

substrates against corrosive species from the operating environment. However, passive inhibition is not sufficient for aerospace coatings and therefore these coatings require an active protective mechanism as well. This active protection is achieved by the addition of corrosion inhibitors. Hexavalent chromate pigments (i.e. strontium chromate) loaded primers are widely used in the aerospace industry and have demonstrated excellent performance and long-term corrosion resistance. [58, 59]. These coatings provide active corrosion protection by means of a leaching mechanism (Fig. 12.9). This is initiated by the ingress of water into the coating via a crack, void or other damage. This will enable the dissolution of the corrosion inhibitor particles and transportation of these inhibitors to the damaged area that is exposed to the corrosive environment and inhibit the corrosion process. The rate of inhibitor leaching has an influence on the immediate and long-term corrosion protection.

Ideally, coating systems would release inhibitors quickly for fast and effective corrosion protection as well as slow release to ensure long-term protection. The performance of active protective coatings depends on many parameters like the binder system [58] and solubility of the inhibitor, pH, coating composition, pigment volume concentration (PVC) and finally the inhibitor loading. Corrosion inhibiting primers usually are highly loaded coatings with a PVC between 25 and 32 %. The amount of hexavalent chromate pigment depends on the leach rate necessary for the specific application. Although the corrosion protection of hexavalent chromium systems is very good, their leaching properties have been studied only since early 2000. Scholes et al. characterized the leaching properties of chromate from primers [60] and Furman et al. found that the inhibitor release is not controlled by a Fickian diffusion process [61]. However it must be considered that every coating composition is different and leach rates can differ significantly which each different coating. Studies are continuing to improve the understanding of the leaching process Sellaiyan et al. studied the leaching process using radioactive tracers demonstrating that water ingress can cause the complete dissolution of the chromate inhibitor and the generation of small voids which can further enhance the leaching of the inhibitors [62]. Knowledge and understanding of the leaching process is important to understand the leaching mechanisms. Computed tomography techniques are becoming more and more common in corrosion [63, 64] and paint research [65]. These techniques make it possible to study microstructural changes in a paint film. For example, Hughes et al. studied the transport pathways of a chromate inhibitor loaded model primer using Quasi 4D computed tomography and demonstrated that there is no chromate transport through the epoxy resin matrix itself [66]. However, the chromate release occurs via direct dissolution of chromate clusters, which are in contact with the electrolyte at the surface and at a later stage through the micro-capillaries and voids left after the dissolution of the inhibitor pigment. This work defines a model that considers fast initial release due to dissolution and a slower release mechanism of inhibitor dissolution and transport through the fractal network. Clusters that were not connected to the electrolyte were not dissolved and are considered inhibitor reservoirs in case of any further local damage. These inhibitor transport studies and models are important for the development of new hexavalent chromium free systems to tailor the release of alternative inhibitors for immediate and long term protection [67]. All corrosion inhibiting primers in the aerospace industry use this leaching mechanism for the active protection mechanism. The unique solubility and corrosion inhibiting properties of these chromated primers is the basis for the good track record and confidence of the chromated primer technologies for the protection of aircraft structures.

12.4.5 Application

Today, most aerospace coatings are applied with hand-held spray equipment. The large variety of parts that need painting before assembly and the dimension and shape of the aircraft have made automation difficult. Especially, the exterior painting process is still a labour-intensive job due to the pre-treatment steps and the masking and de-masking processes involved prior to and during the painting process (Fig. 12.10). In addition to this, there are several other factors that influence the painting process. Hangar conditions can vary from 23 °C and 55 % Relative Humidity (RH) in climatized hangers to severe conditions with high temperatures and high humidity (35 °C/90 % RH) or low tow temperature conditions with very low humidity conditions. Some hangars are equipped with movable platforms and others have fixed scaffolding. Under these varying conditions, the quality of the finish depends on factors like, teamwork, experience, and skills of the painters. Another important factor is the application equipment.

12.4.5.1 Equipment

There are many different types of equipment for the application of aerospace coatings. The most conventional type is the air-atomized spray. This type of spray guns uses air to atomize the paint in very small droplets that results in a very smooth finish.

Fig. 12.10 Aircraft painting

The equipment is not expensive, easy to clean and is easy to maintain. Although, it has a low transfer efficiency of approximately 20–30 % and is not suitable for large area application. Over the years several alternative types of air atomized spray guns have been developed to improve the transfer efficiency. Examples are HVLP (High Volume Low Pressure) and HR/RP (High Return/Reduced Pressure) spray guns. Air Less spray equipment uses hydraulic pressure instead of air to atomize the paint. The paint is forced through the nozzle with 100–150 bar of pressure to atomize the paint. The velocity of the atomized droplets is less compared to air atomized spray, which results in a higher transfer efficiency. The equipment has a high flow rate providing a faster application speed. Airless air assisted spray equipment combines both of both worlds in air atomized and airless spray. Another way to reduce overspray and increase transfer efficiency is the use of electrostatic spray equipment. The atomized droplets are charged using a high voltage (60–90 kV)/low current generator and attracted to a grounded conductive object. Transfer efficiencies of 70 % are possible. This transfer efficiency depends on the coating formulation. This type of equipment is most frequently in OEM and MRO paint shops.

Plural component spray equipment is becoming more common in the structural coatings market but as well for exterior aircraft painting. The equipment mixes the components (base, hardener and activator) at the desired mixing ratios just before or at the spray gun. This equipment reduces the pot-life factor of 2 component paints and reduces the amount of waste from unused mixed paint.

12.4.5.2 Paint Automation for Exterior and Structural Painting

The use of paint automation in commercial aircraft production is being studied to reduce process cycle times, provide a higher quality paint finish, lower emissions, increased transfer efficiency and increased process consistency [68]. Increasing production rates and the need for increased capacity in the current hangers is driving this need for automation and use of new coating technologies. However, there are differences between aerospace and automotive applications such as paint cure times, number of paint colours, environment control, and part size considerations are some of the issues that make aerospace application of coatings more difficult than automotive applications. Understanding the unique factors involved in the robotic application of commercial aerospace coatings is important for future advancements in application technology, gains in aircraft paint hangar capacity, delivering quality coating finishes, and lowering environmental footprint. In structural painting automation has been introduced. Small parts are applied on conveyer belt systems and robotized paint lines have been introduced on the sub-contracting level e.g. for the wing sections and stringer painting.

12.4.5.3 Application of an Exterior Paint System

Painting aircraft is a labour-intensive, multi-step process that requires a significant amount of handwork and attention to detail. Figure 12.11 illustrates the different steps required for exterior painting. The aircraft is washed and moved into hanger, all the at-risk components (composite parts) and flight controls are covered or removed. Then the chemical stripper is applied to remove the old paint film. After this, the aircraft is cleaned and inspected, flaws are removed, corrosion is treated, and necessary repairs made. Then the aircraft is pre-treated for application using an alkaline soap wash, the aluminium surfaces are etched or abraded prior to the application of the chemical conversion coating, or wash primer. Within the recommended application window the protective primer will be applied followed by the base colour topcoat for the aircraft. With use of decals, the paint scheme will be laid out on the aircraft and the rest of the aircraft is masked and the decoration colours are applied. When the paint scheme is completed, there is the final inspection and the aircraft will be released for service.

12.4.6 Trends in the Aerospace Coatings

The aerospace market [69] is focused to increased efficiency, reduced operational and maintenance costs. There are two important drivers that influence the aspects of innovation in the in the market for aerospace coatings. First, there is the drive for more environment friendly coating systems and second, there is the need for increased efficiency.

Fig. 12.11 Steps required for aircraft painting (images kindly provided by Airbus and Emirates Airlines)

When considering efficiency, there is a drive towards reduced weight, faster application times for coatings, longer service life and increased functionalities. Fuel costs are a significant part of the operational costs, airlines and aircraft manufactures need to consider ways to save money. One way to achieve this is to reduce weight by selecting new materials in design or reduction of the weight of the paint system. The weight of the paint film can be reduced in different ways like, the reduction of the dry film density of the coating, reduction of thickness of the paint system or using fewer layers in the paint film.

The second efficiency driver is the time of the paint process. Airlines merge and rebrand their liveries; therefore there is an increased need for repainting. In addition to this, the market is growing; build rates at the OEM's are increasing to keep up with the demand from the market. Therefore the cycle time of paint process needs to be reduced to get the aircraft back in service quicker to create revenue for the airline and to ensure final assembly line throughput at the OEMs. The livery and interior of the aircraft are part of the airline's brand and image. A clean and glossy appearance with vibrant colours is a key aspect of the branding strategy of an airline. To support this, basecoat/clear coat paint systems have been introduced into the market. These fast curing collared basecoats are covered with a clear coat to ensure long-term durability (gloss and colour). In addition to this, the industry is looking for increased coating functionalities in the coating. Examples of such functionalities are: drag reduction, anti-icing, solar hear reflection/heat dissipation, and self-healing abilities.

The other important aspect is the environmental impact of the coatings. Terms such as eco-efficiency, sustainability, and carbon footprint are an integral part of the development of new products for the aerospace industry. Not only for the coating manufacturers, but for the entire supply chain from raw material to end-user. More importantly, new legislations such as REACH (Registration, Evaluation, Authorization, and Restriction of Chemicals) and regulations from governmental institutions such as the Environmental Protection Agency (EPA) and Occupational Safety and Health Administration (OSHA) are applying enormous pressure to develop more sustainable products with a lower environmental impact. The REACH legislation targets the ban of all chemicals that may be considered hazardous. Over the years, this has resulted in the replacement of many hazardous materials with more acceptable alternatives and solvent emissions have been reduced.

In the field of aerospace coatings there have been many advances [70]. Today, solvent emissions can be reduced significantly. Conventional coatings, with low solids and high solvent content can be replaced by high solids or water-based coatings. Water based cleaners and pre-treatments have been introduced and environmentally friendly paint removers were developed.

The elimination of hexavalent chromium or chromate from the protective coatings and coating processes is a key topic for the aerospace industry, which will be discussed below.

12.5 Developments Towards a Chromate Free Corrosion Protective Scheme

For more than 50 years the aerospace industry has been using hexavalent chromium as corrosion inhibiting compounds for the protection of the aircraft structure in many processes and products. However, the health issues associated with the use of these chromate materials and new regulations and legislations are putting pressure on the industry to find alternative solutions [12, 71, 72]. The European REACH (Registration, Evaluation, Authorization and restriction of Chemicals) regulation (EC N° 1907.2006), hexavalent chromium compounds have been included as a Substances of Very High Concern (SVHC) and are subject to a formal and time limited authorization for use. This means that they cannot be used after the "sunset date" unless there is an authorization. In the USA, the Occupational Safety and

Health Administration (OSHA) reduced the employee permissible exposure limits (PEL) to hexavalent chromium from 52 to 5 $\mu g/m^3$ inducing strict expose controls of these chemicals.

Through significant efforts in the different areas of the industry various chromate free alternatives have been developed and qualified to aerospace standards and specifications, and introduced to the market place [13, 73–79]. Examples of this are chromate free anodising, acid pickling, conversion coatings, sealants and coatings. This part of the chapter provides an overview of the developments in chromate free coating technology from pre-treatments, coatings and some more innovative approaches which should lead eventually towards a fully chromate free corrosion protective scheme for the protection of the aircraft structures.

12.5.1 Chromate-Free Pre-treatments

Various processes can be utilized prior to the application of the organic coating. The specific process used depends on the substrate, and the performance characteristics required for the protective system. Typically, a process will start with a cleaning step to remove organic contaminations. This will be followed by a deoxidation or pickling step and the application of a surface treatment for corrosion protection and paint adhesion [80, 81]. The use of non-chrome processing solutions for cleaning, deoxidation, desmutting and etching has increased over the recent years. In addition to this, several alternative solutions for surface treatment have been developed and introduced to the market place. These hexavalent chromate free surface treatments are the new base for the future hexavalent chromate free corrosion protective scheme.

12.5.1.1 Chemical Conversion Coatings

Significant effort has been spent to develop a suitable alternative to the chromated chemical conversion coatings. The coatings developed from chromate solutions have a robust performance in processing, corrosion protection and adhesion performance. A wide range of technologies has been investigated, demonstrating the array of alternative chemistries available for aluminium. These include molybdates, vanadates [82], permanganate [83–85], phosphates, silanes, sol-gels, self-assembling monolayers, hydrotalcites [86–91], and rare earth (cerium) based chemistries [92]. Only a few chemistries have been able to come close or meet the corrosion protection requirements of aluminium according to the MIL-DTL-81706/5541 specification. Examples of this are conversion coatings based on trivalent chromium (Cr(III)) processes (TCP) and cerium. Both these approaches are commercially available and found their application as part of a coating system [81].

Rare earth based salts are promising options for chromate replacement in chemical conversion coatings for the protection of aluminium alloys. Several rare earths have been identified as cathodic inhibitors [93, 94]. These materials tend to precipitate as metal hydroxide at local areas that are associated with a pH increase due to the oxygen reduction reaction at the cathode [95]. This results in the deposition of a hydrated oxide layer on the aluminium surface [96–98]. Unfortunately, most rare earths are too costly for commercial use. However, cerium is the most active rare and has been studied extensively since the early 1990's [93]. A review by Harvey is the most recent that covers specifically cerium-based conversion coatings [92]. Several cerium-based chemical conversion coating have been developed and commercialised [99–102].

On the other hand, the trivalent chromium (Cr(III)) processes have been developed as alternative to hexavalent chromium pre-treatments. The Naval Air Systems Command (NAVAIR) has developed and patented a range of processes using trivalent chromium and zirconium fluoride [103]. This process has been approved according to MIL-PRF-81706/5541 and has been licensed to various parties. These trivalent chromium conversion coatings have been introduced into the hexavalent chromium free surface finishes of the US Navy [104–106] and slowly finds it way to the civil aviation market. There have been many studies on this Trivalent Chromium Process (TCP). Guo and Frankel studied the conversion layers and the self-healing or active corrosion inhibition properties of TCP [107, 108]. On the coating surface a dense layer of 40–120 nm was found. The protective barrier layer provides corrosion protection by suppression of the oxygen reduction reaction. Moreover, the TCP coating is able to release chromium from the coating proving self-healing or active corrosion inhibition properties. No hexavalent chromium has been detected in these coatings after deposition and corrosion testing [109, 110].

The performance of these new chemical conversion coatings is not at the same level as the hexavalent chromated conversion coatings. Although, these chromate free chemical conversion coatings are finding their place in the aerospace industry and potentially are the new base for the chromate free coatings systems.

12.5.1.2 Anodising

Traditionally, the anti-corrosion performances were achieved by chromic acid anodizing (CAA), followed by painting. However, environmental issues and associated costs for the disposal of chromate wastes, require the development of new approaches for anodizing of aluminium alloys. Phosphoric acid anodising (PAA) and sulphuric acid anodising (SAA) have been available to the industry but were not suitable chromate free alternatives to CAA because of degradation of fatigue life, corrosion resistance, adhesive bonding or paint adhesion. Boric sulphuric acid (BSA) and tartaric sulphuric acid (TSA) electrolyte anodising processes have been developed and introduced as suitable alternatives to CAA for corrosion protection [111–113]. Phosphoric sulphuric acid (PSA) anodizing is used for structural bonding. These processes have been optimized balancing 3 criteria: Adhesion, corrosion resistance and fatigue properties. The new processes provide anodic film with a similar thickness as the conventional chromate containing process and a regular open structure using lower anodizing temperatures, shorter process times and reduced voltages.

Further research is focused to improve and understand the chromate free anodising processes and their performance even further [114, 115]. Reduction of energy costs [116] modification of film morphologies [117] Potentiodynamic anodising is a new tool that allows fundamental and practical investigations of the anodising process. It is suitable to compare different electrolytes and to determine the optimum parameters to obtain porous anodic films. These parameters include current-potential relationship, maximum anodizing potential and current as a function of electrolyte concentration or temperature [118]. Dichromate sealing processes are commonly used after the anodising process to improve the corrosion resistance of the anodic film. Numerous studies were performed to investigate the effect of the addition of inhibitive species to the anodizing electrolyte or as sealing process to improve the anodic film or to replace this dichromate-sealing step. Proposed alternatives are species like cerium nitrate, molybdate, permanganate, phosphates, or hydrophobic carboxylic acids [119–122].

12.5.1.3 Pre-treatments

The need for hexavalent chromate free chemical conversion coatings stimulated the development of several alternative chromate free pre-treatments. An example of such an alternative is a sol-gel based pre-treatment. This coating has a specific formulation of components that provides a sol-gel hybrid network. The formulation is based on a reactive mixture of an organo-functionalized silane with a stabilized zirconium complex and forms a covalently bonded film on the metal surface (Fig. 12.12) and strong and durable bonds with paints adhesives and sealants [123]. The sol-gel pre-treatment does not contain any corrosion inhibitors and has no

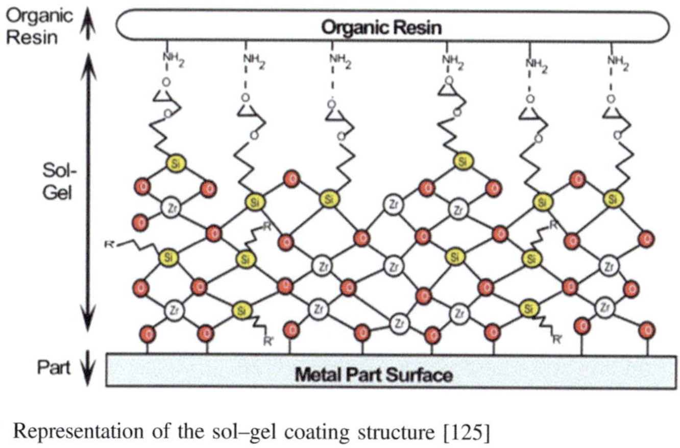

Fig. 12.12 Representation of the sol–gel coating structure [125]

significant "stand-alone" corrosion protection. However, it is compliant to the aerospace specifications when used in combination with appropriate paint systems. In 2008, this pre-treatment was introduced to the OEM end-lines for exterior painting. Since then it was licensed to several parties and is currently used in OEM and the MRO sector as part of chromated and chromate free coating systems. In addition to the environmental and health, and safety benefits, "in-service" flight data demonstrated that the sol-gel pre-treatment reduced the occurrence of "rivet rash" adhesion failures, which has been a serious issue for many years. The material can be applied by spray, brush or flooding. This pre-treatment did not only eliminate the use of hexavalent chromium in the conversion coating step but, it simplified the application process, reduced drying times, water consumption and, the time required to paint an aircraft [81, 124].

In addition to the sol-gel technology, other chromate-free pre-treatment technologies were introduced into the marketplace. These products do not only eliminate hexavalent chromium from the pre-treatment step but as well acid brighteners, solvents and alkaline washes [126]. These particular pre-treatments clean the surface to be painted and deposit a molecular layer that bonds to the metal. This layer improves the adhesion of the coating system to the metal, but do not have any stand-alone corrosion protection. However, they provide an environmentally friendly alternative to the hexavalent chromate containing chemical conversion coatings in combination with aerospace qualified coating systems [73, 77].

12.5.2 Developments in Chromate-Free Organic Coatings

The hexavalent chromate-free surface treatments as discussed above provide the first basis of the corrosion protective scheme. Today, the majority of these environmental friendly surface treatments are coated with chromate containing organic coatings (primers). On the other hand, there is an enormous effort ongoing to find suitable alternatives for hexavalent chromium in coatings for the aerospace industry. Many inhibitors were evaluated over the years, however with limited success [58, 127]. For a long time, the objective was to mimic the activity of chromates which are very effective in low concentrations and in a wide pH range [128]. Many inhibitors were identified in electrochemical studies but, often these were out-performed by the unique properties of hexavalent chromium or they needed the support of chromated chemical conversion coatings, wash primers or anodising. The quest has not yet been resolved but, there has been promising progress in chromate free coatings development and total chromate free coating systems are being introduced into the market place for exterior painting. With this achievement, the industry will be able realign its focus to the elimination of hexavalent chromium from the structure of the aircraft.

12.5.2.1 Corrosion Inhibiting Strategies

Multiple corrosion inhibiting strategies have been developed for chromate free active protective coatings. First, there is the traditional "leaching" process, whereas a corrosion inhibitor is incorporated in a paint film and moisture dissolves the corrosion inhibitor and transports it to the damaged area as discussed in Sect. 12.4.4. Second, there is the mechanism of galvanic inhibition, this inhibiting strategy is based on sacrificial oxidation (cathodic protection) using metallic particles which are electrically connected and more anodic (reactive) than the substrate in the electrochemical series and last, there are the "smart" coating approaches with inhibition through self-healing properties and on-demand inhibitor release. The "leaching" and "galvanic" inhibition strategies have been introduced to the market place [129] and the "smart" coating approaches are still under development.

12.5.2.2 Inhibition Through Leaching

The traditional "leaching" mechanism is the fundament of current aerospace specifications for active protective coatings in the aerospace industry. The performance of hexavalent chromium corrosion inhibiting pigments provided a proven track record and is currently the benchmark for all new chromate free active protective coatings. The leaching inhibitors need to provide active corrosion inhibition when the measures of the corrosion protection scheme are failing (substrate, pre-treatment, sealant, drainage) and the system is damaged by means of a crack, drilling hole or, scratch. At this stage, moisture ingresses into the crack or system, the inhibitor should be dissolved and transported to the damage. Where chromates were unrivalled for a long time, chromate-free alternatives are being introduced into the market place.

Leaching inhibitors have to be versatile and it is the question whether it is possible to find the complete package (like chromates) in a single inhibitor. There are many chromate-free inhibitors that demonstrate activity but do not perform when incorporated into a coating due to various reasons like: solubility, efficacy, or compatibility with the binder system. Therefore the leaching capabilities should be considered from coating/inhibitor perspective and have adequate solubility and leaching characteristics, fast and effective formation of an irreversible protective layer in the damaged area, underneath the coating and, maintain a good coating integrity (adhesion and barrier function). Many inhibitors have been developed and investigated in the academic environment but as well in the industry. Many patents have been filed and published on a range of corrosion inhibiting materials. Several of them have been introduced to the market place but none of them can be really considered as the true replacement and alternative to hexavalent chromium. Some examples are mentioned below.

Most of the inhibitors are based on inorganic materials, among them inhibitors based on rare earths (cerium and praseodymium), molybdates and vanadates [130, 131]. Praseodymium has received considerable attention. Stoffer et al. claim a corrosion inhibiting coating composition based on a rare earth compound (praseodymium oxide) and a neutral to acidic generating extender (Calcium Sulphate) [132]. Praseodymium oxide (Pr_2O_3) provides the best protection across a pH range from 5 to 8. Praseodymium-rich species leach from the coating and precipitate as hydroxy carbonates in the damaged area near cathodic sites [133]. Pipko and Vitner developed and claimed compositions of oxyaminophosphate salts of magnesium as leaching inhibitors. These slow releasing materials were found to form stable magnesia layers on the metal surface [134, 135]. Yu et al. claimed a non-carcinogenic corrosion-inhibiting additive comprises an anodic corrosion inhibitor and/or a cathodic corrosion inhibitor. This leaching inhibitor uses a molybdate salt as anodic inhibitor, cerium as cathodic inhibitor and citrate as a metal-complexing compound [136, 137]. Walters and Schneider claim a coatings composition utilizing nano-sized Magnesium oxide particles as alternative to hexavalent chromium [138]. Visser and Hayes propose lithium salts as corrosion inhibitor for 2024-T3 aluminium alloys [139]. Whereas, organic inhibitors and rare earth species are active on active copper intermetallic (cathodes) [140], leaching of lithium species from organic coatings results in the formation of a protective layer with a characteristic 3 layered morphology on the aluminium in the damaged area [141]. Boocock used a composition with amorphous zirconium vanadium oxide/hydroxide for the protection of high strength aluminium alloys [142], and Becker et al. proposed a coatings composition containing a corrosion inhibitor and at least one quaternary ammonium compound [143].

In addition to the inorganic inhibitors, there are many organic inhibitors that are and have been investigated. Harvey et al. investigated the effect of the inhibitor structure on the corrosion of AA2024-T3 and AA7075-T6 and found good inhibitor efficiencies for most of these well-known “copper” inhibitors, Among them Benzotriazole, 2-mercaptothiazole, dithiocarbamates, dimercapto thiadiazoles [144]. Sinko designed different hybrid organic-inorganic corrosion-inhibiting micro composite pigment grades that contain 2, 5 dimercapto-1, 3, 4-thiadiazole, zinc oxide, zinc phosphate and zinc cyanimide [145–147]. Furthermore, together with Kendig, Sinko proposed a guest–host pigment (hydrotalcite based) that can be loaded with an organic inhibitor [148, 149]. Hayes et al. proposed a pigment comprising a neutralized metal salt of a corrosion inhibiting organic anion, wherein 2.5 % by weight aqueous mixture of the dry salts has a pH above 6 and below 8 [150].

Other researchers have tried to find viable alternatives for hexavalent chromium combining the best of both inorganic and organic inhibitors. Examples are cerium salts of organic inhibitors cerium cinnamate [151], cerium dibutylphosphate [152, 153] and cerium or praseodymium mercapto acetate [154].

12.5.2.3 Galvanic Inhibition

Cathodic protection or galvanic inhibition is a method that is extensively used for the protection of steel structures. The protection is based on a sacrificial mechanism. In the case of steel, zinc (Zn) is used because it is more reactive (anodic) compared to Iron (Fe), which is the main constituent of steel in the electrochemical series. Using this principle metal rich coatings have been developed for the protection of the aluminium structures of aircraft.

There are some basic principles that need to be taken into account to provide cathodic protection. The coatings are composed of organic or inorganic film forming resins and are pigmented with particulate metal particles in either spherical or flake form. The metallic particle selected must be must be more anodic compared to the metal/substrate that needs to be protected. As already mentioned, for steel this is most commonly zinc. The pigment volume concentration (PVC) of the metal particles in the coating needs to be close to or exceed the critical pigment volume concentration (CPVC) enabling mutual contact between the particles and the underlying substrate. The resin matrix of the coating must be able to withstand the alkaline environment created by the oxidation of the metal pigment. Good adhesion to the substrate under corroding conditions is another essential parameter. In order to preserve the inhibitive capacity, the sacrificial primer needs to be covered by a topcoat ensuring good barrier for the structure to the environment and protection in case of damage [155].

In 2004, Nanna and Bierwagen proposed Magnesium-Rich Primer technology (MgRP) as a new paradigm for chromium free corrosion protection for aerospace aluminium alloys [156, 157]. This was an entirely different approach compared to the traditional chromated "leaching" technology. Aluminium is at the bottom of the galvanic series, and magnesium has an even more positive potential [158]. Therefore magnesium can act as a sacrificial metal for the protection of aluminium alloys. The magnesium rich coating has the same features as a zinc rich primer. The magnesium particles are connected to each other and the metal substrate (Fig. 12.13) [159]. Therefore the primer is loaded with magnesium particles to a level close to the

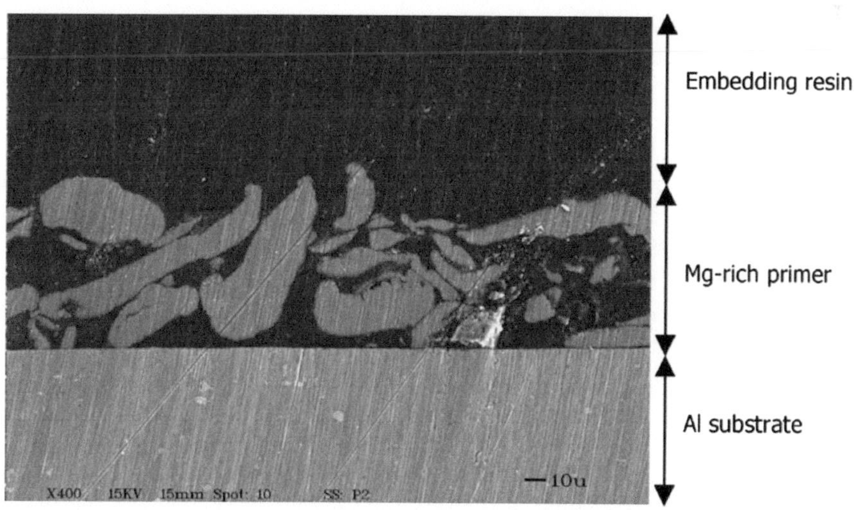

Fig. 12.13 Cross-section micrograph of Mg-rich primer on aluminium alloy substrate [160]

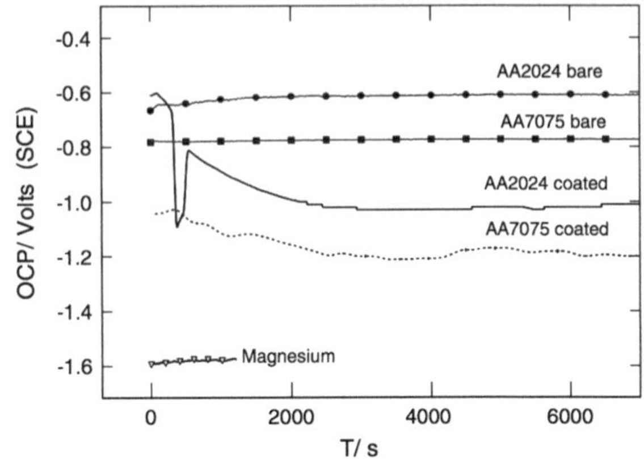

Fig. 12.14 Open circuit potential of the bare substrates, of the magnesium-rich primer coated substrates and of magnesium, in 0.1 % NaCl [160]

critical pigment volume concentration (CPVC). When the MgRP is applied over aluminium alloys, the steady state potentials (OCP) are corresponding to the potential of the galvanic couple of the substrate and magnesium (Fig. 12.14). Magnesium polarizes the system cathodically, shifting the system potential below the pitting corrosion potential. This polarization prevents pit nucleation at exposed or damaged areas or inhibits pit growth for nucleated pits. In addition to this electrochemical protection, a precipitation effect has been postulated due to precipitation of magnesium oxides in damaged areas [160]. The technology was licensed and developed to a mature product that meets aerospace specifications and is currently used in the aerospace industry [161, 162]. Research on this inhibitor continued, addition of metal salts (Li_2CO_3 and $Mg(NO_3)_2$) enabled the reduction of Magnesium metal in the coating while improving the overall corrosion protection [163].

Similar approaches with different alloys have been investigated since the introduction of the magnesium rich technology. Xu et al. [164] evaluated magnesium alloys to understand the effects of particle shape, particle size, particle size distribution, and metal alloy chemical composition whereas, Plageman et al. [165] investigated the effect of spherical zinc and zinc alloys to use the benefits of both zinc and magnesium by alloying. Magnesium would still provide cathodic protection, whereas the zinc would provide a lower reactivity of the particle. In addition, the zinc magnesium particles would lead to more voluminous corrosion products and a densification within the coating and improving the barrier properties of the coating. The alloy with 74 % zinc and 26 % magnesium demonstrated to be effective in accelerated corrosion testing. Thin film zinc rich primers (10–20 μm) were successfully tested on laboratory scale and by flight trials [166].

12.5.2.4 Smart and Self-healing Coating Technologies

Self-healing and smart coatings have become a growing and important concept for coatings for the protection of metals and alloys. One of the strategies is to detect and heal damage in a coating and to recover the initial functionality of the coating system. Since this pioneering work of White et al. [167], self-healing or smart coatings discipline took a flight and resulted in a rapid increase of approaches. The main self-healing strategies are aesthetics and barrier recovery (gap filling) and active corrosion inhibition [168]. The strategies include the release of liquid reactive healing agents from capsules due to a mechanical damage, expansive phases, flow and reflow of material followed by a chemical reaction or restoration of physical bonds. The other approach is the release corrosion inhibitors or surface reactive compounds from microcapsules. Inhibitor release from these capsules can be triggered by mechanical damage, diffusion, ion exchange or change of pH. In 2011 Zheludkevich provided a comprehensive review of these smart and self healing technologies [169]. Since then, new approaches have been proposed. Garcia et al. proposed a self-healing concept based on an encapsulated water reactive silyl ester that is based on wetting and a reaction with ambient humidity and the metallic surface and it finally forms a hydrophobic layer to extend corrosion protection [170]. Other micro-capsule approaches have been proposed including: inhibitor impregnated $CaCO_3$ micro beads [171], TiO_2 nano-containers loaded with 8-hydroxyquinoline [172], Silica nano-containers with a regular shape and diameter between 100 and 150 nm loaded with 2-mercaptobenzothiazole [173], EDTA loaded layered double hydroxides [174], and Cerium and diethyldithiocarbamate (DEDTC) double doped zeolites [175]. Also some patents have been filed to claim the use of such strategies for the protection of aluminium structures for aerospace applications. Gammel et al. propose a coating with embedded micro-capsules containing 2-mercaptobenothiazole [176]. Barbe et al. claim a coating composition that comprises ceramic particles with a releasable active material homogeneously distributed in each particle [177].

Another approach is the use of conducting polymers. Tallman et al. reported corrosion protective properties of conductive polymer coatings [178, 179]. Most of the investigated conductive polymers are based on polyaniline (Pani). The galvanic interaction between the oxidized and conducting form of Pani, the emeraldine salt (ES) and the aluminium substrate leads to positive polarization of the aluminium substrate and the reduction of Pani ES to the leucoemeraldine base (EB) of Pani and the release of a dopant anion [180]. The dopant can play a role in the corrosion protection due to the formation of a metal-dopant interface, Kendig et al. proposed to dope Pani ES with specific anions that could work as cathodic inhibitors and reduce the oxygen reduction reaction at copper-rich intermetallic particles [181, 182]. However the exact protective mechanism for aluminium alloys has not been resolved yet [183, 184]. Concepts of such 'smart' conductive coating have been developed by Kinlen and Kendig et al. [185, 186], and the first field testing studies, as an alternative to chromated chemical conversion coatings, have been reported [187].

Both the self-healing and conductive coating approaches are still under development and do not provide the industry with a drop-in alternative for hexavalent chromium, however the knowledge that is generated from this work can provide future tools for chromate free corrosion protection of aircraft structures.

12.5.2.5 Alternative Coating Techniques

The aerospace industry has been rather conservative in terms of accepting alternative coating technologies. The coating systems have been improved incrementally over the years and no real new coatings concepts have been introduced. However, with the current demands of the industry in terms of productivity and environmental impact, alternative coating techniques can be considered. Examples of such alternative coating techniques are electro-deposition paints, powder coatings and UV-curable coatings.

Electro-deposition primers are widely used in the automotive industry. However, it was not introduced in the aerospace industry because of bath stability issues, coating thickness, high curing temperature and large capital investment. Recent developments make it possible to coat small parts with a chromate-free electro deposition process [188]. In this efficient coating process, an electrical field is applied and water dispersed paint particles will be negatively charged and move to the anode. When the surface is coated, the part will be rinsed and thermally cured. Introduction of this technology will increase transfer efficiency and reduce over spray and solvent emissions, waste, and improved layer thickness control. The technology is based on anaphoretic paint deposition, which cures below 120 °C. An Electro-deposition coating provides anodisation and coating application in the same step. An interfacial oxide layer is formed during the anaphoretic paint deposition [189]. The oxide layer is composed of a dense sub-layer on the substrate with oxide filaments interpenetrating in the coating with a film thickness of 20 μm. The technology is being introduced to the aerospace industry [190].

Since the early 1980's there have been developments on powder coatings for aerospace applications [191]. This technology would eliminate solvent emissions, hazardous air pollutants, and reduce waste. In addition, it can provide improved efficiency due to faster curing, eliminate pot life issues, and increased transfer efficiency. The main issue of powder coatings has been the high temperatures needed to get a full cure of the system. These high curing temperatures, 220 °C and higher, might affect the strength of the aluminium alloy. Research has resulted in low temperature and UV curable approaches, but the technology is not widely used in the industry [192, 193].

Ultraviolet (UV) curing is the other alternative technology that can be used in coatings. These coatings can be formulated as 100 % solids, without the use of VOC and hazardous air pollutants and can be polymerized, cross-linked in a matter of seconds and provide films that are decorative and functional. The UV curing mechanism is a photo-chemical process by which a liquid coating is completely cross-linked to solid polymer through exposure to UV radiation without the need for isocyanate cross-linking agents. There are two classes of UV-curing technology, based on the nature of photo-initiators and resin chemistry: free radical (mostly acrylate) and cationic (epoxy). Acrylate self-priming topcoat approaches were proposed for the protection of aerospace aluminium alloys. Although the approach was over chromated conversion coatings, energy savings and solvent emission reductions were additional benefits to the reduction of chromates in the total paint system [194, 195]. Moreau et al. designed UV-curable hybrid sol-gel formulations based on both organic and inorganic photo polymerization [196]. This process is based on the liberation of photo acids by the UV decomposition of an iodonium salt. This triggers simultaneously the sol-gel process and epoxy cationic polymerization [197]. The one-step UV-curing process hybrid sol-gel coatings can provide high corrosion resistance on metallic substrates and protection of composites.

12.5.3 New Approaches Towards Chromate Free Coatings Technology

Chromate-free coating technology is evolving rapidly. Over the years, many inhibitors have been tested. However, corrosion testing is one of the bottlenecks in the coating development process. There is a need for faster and more accurate screening of new chromate-free candidates. Inhibitors need to be tested for synergistic performance or activity over a wide pH range. In addition to this, chromate-free corrosion inhibiting coating technologies became better and therefore it becomes more difficult to discriminate between the current technologies and possible improvements, resulting in longer testing. Evaluation of corrosion results is still rather old fashioned and difficult to obtain absolute numbers. High throughput experimentation and imaging techniques are two examples of tools that can be used to accelerate the development of new chromate-free inhibiting technology.

12.5.3.1 High-Throughput Experimentation

Corrosion performance has been historically evaluated using electrochemical techniques, accelerated corrosion testing, or even outdoor exposure. These methods require testing times ranging from hours to years. Advances in material science have broadened in the scope for the discovery of new corrosion inhibitors. Recent developments consider the combination of two or more inhibitor technologies (i.e. anodic and cathodic inhibitors) to obtain multifunctional or synergistic performance [198, 199]. There has been a considerable effort to develop high throughput experimentation techniques to increase the rate of new inhibitor screening and to find combinations for synergistic and antagonistic effects. High-throughput combination and electrochemical characterization have provided new tools for the exploration and understanding of new corrosion inhibitors. Chambers et al. investigated the inhibition characteristics of 50 different chemistries within 9 hours using a direct current (DC) polarization between two AA2024-T3 wire electrodes and a multiple—electrode-testing system [200]. Garcia et al. used a multi-metal electrode set-up to identify inhibitor efficacy to different metals simultaneously [201, 202]. White et al. designed a multi-channel array for screening aqueous corrosion inhibitors on aluminium AA2024-T3. This method allows solutions to flow over the surface of the alloy. Inhibitor efficiency can be compared visually and quantitatively via solution analyses [203]. Another approach of White et al. is the use of a multi-well plate that can be filled with a range of solutions and the inhibitor performance is quantified with a computerized optical analysis method [204]. Further high throughput techniques are summarized in a review of Muster et al. [205].

12.5.3.2 Imaging Techniques for Corrosion Evaluation

Quantification of corrosion during and after corrosion testing is another issue in corrosion research. Today, corrosion evaluation of coated panels for specification testing is still a visual process which is supported by a ruler and a camera to capture a value for the maximum corrosion, creep or undercutting from the scribe edge and sometimes corrosion density. Many specifications have only maximum creep from scribe requirements. In addition, corrosion progresses after testing which makes storage of specimen after corrosion testing for future reference more or less useless. Imaging techniques have been proposed to improve the quantification of corrosion during and after corrosion testing [206, 207]. The quantification of corrosion is potentially valuable because it provides researches with absolute numbers, which can be used to discriminate between different inhibitor technologies. Corrosion is dissipating over a certain area and hence using maximum creep values does not provide a realistic value for the degree of corrosion. Imaging has been used for the analysis of coatings with different corrosion inhibitors during the filiform corrosion test, large differences are obvious and can be easily observed. On the other hand, the other systems perform alike according the longest filament evaluation. However, the filiform analysis (Fig. 12.15) method can analyse data over a larger

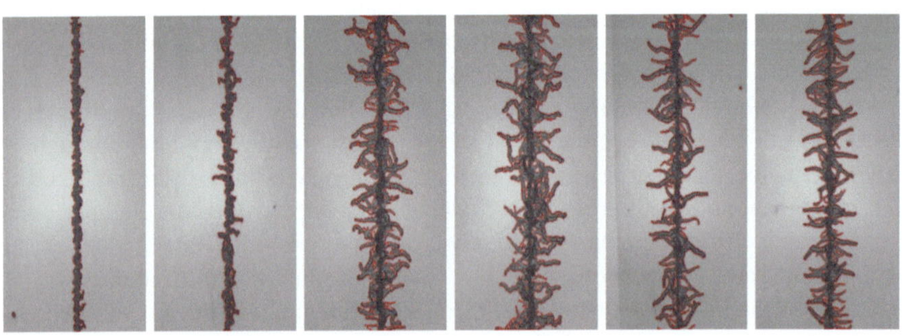

Fig. 12.15 Imaging of filiform corrosion [207]

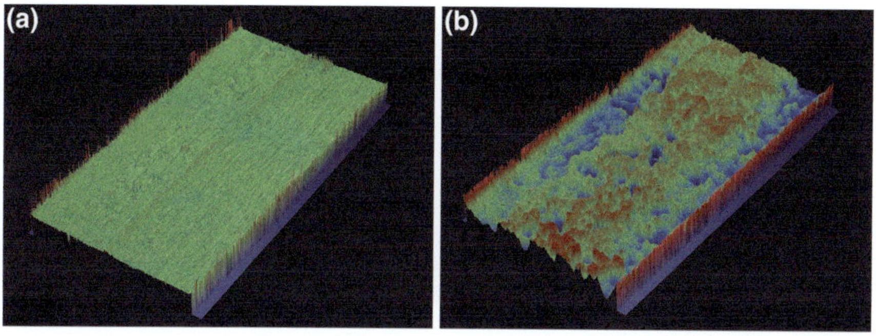

Fig. 12.16 White light interferometry analysis before (**a**) and after (**b**) neutral salt spray (ASTM B117) exposure reveals the change in surface roughness due to pitting corrosion on AA2024-T3 [208]

area, and it provides absolute values that potentially can be used for corrosion rate or corrosion kinetics calculations [207]. Evaluation of the total corrosion area provides a different perspective and makes it easier to discriminate the systems that perform better and have a lower corrosion rate.

Another challenge for the engineers that develop or qualify new corrosion inhibiting systems is the evaluation of panels that have been exposed to salt spray exposure (ASTM B-117). It is very complicated to assign absolute numbers to these panels. Moreover, the time to discriminate between better performing technologies is increasing and limits fast developments. A method has been developed to evaluate the degree of pitting corrosion during and after accelerated corrosion testing. The method involves a surface roughness measurement of the exposed damaged area with the use of white light interferometry (Fig. 12.16). Especially on AA2024-T3, the method demonstrated the ability to monitor the increase of the surface roughness of the damaged area (scribe) due to corrosion phenomena at the early stages of exposure and after longer-term exposure to the corrosive environment [206, 208].

With the use of imaging software it is possible to quantify the surface roughness of the area, increase in surface area, and the amount of areas or pits deeper than a certain threshold value. These values can be related to a degree of pitting [208]. The technique enables the determination and quantification of the performance of corrosion inhibiting technologies before the human eye is able to detect the differences, and support engineers to make decisions based on values rather than opinions. A practical example can be found in Fig. 12.17. This figure compares three chromate-free coatings and a chromated coating. It shows that the surface roughness and degree (%) of pitting corrosion are closely related and that both new chromate free coatings are demonstrating good corrosion protection with a low degree of pitting like the traditional chromated coating.

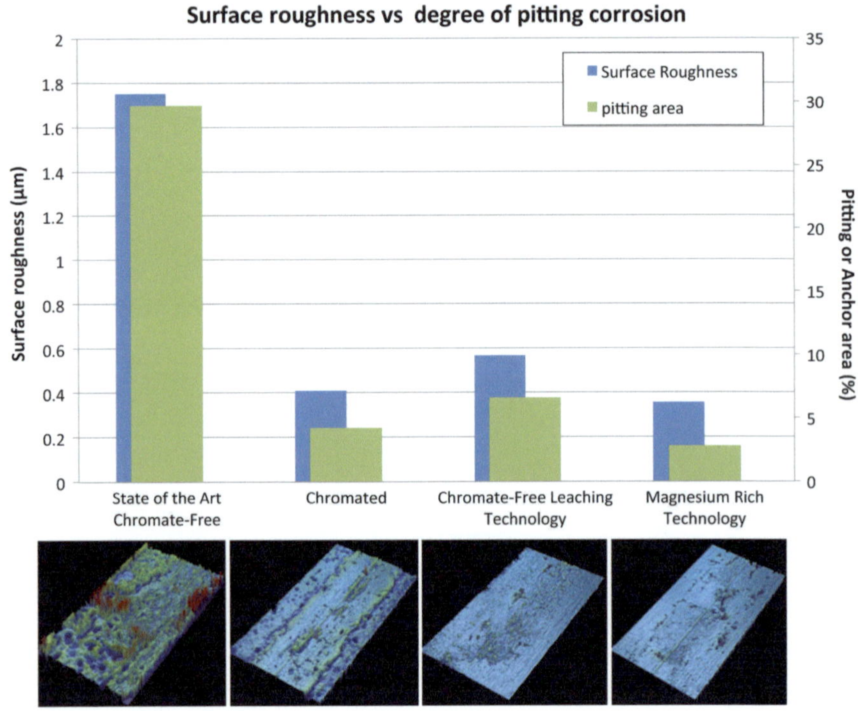

Fig. 12.17 Relation between surface roughness and pitting area for various corrosion inhibiting coating technologies after 3000 h. neutral salt spray (ASTM B-117) exposure [208]

12.6 Remaining Challenges for Chromate Replacement

The replacement and elimination of hexavalent chromium based processes and inhibitors is an enormous challenge for the aerospace industry. The corrosion control requirements in the aerospace industry have been developed and are based on the performance of hexavalent chromium (chromates). All designs, specifications, testing data and in-service experience are based on this versatile and unique inhibitor. The track record of chromates is impressive. However, legislative pressure pushes the industry to look for suitable alternatives for this unique material. A tremendous amount work is ongoing and over recent years, the progress has provided the industry with a number chromate free options for pre-treatments, and coatings. Despite this progress, there are still many challenges for the industry. Processes and mechanisms need to be reviewed or reconsidered and possibly changed to provide the industry with the confidence to eliminate hexavalent chromium. Chromate-free coating technology evolved and chromate-free systems are being qualified at the major aircraft manufacturers for external applications. Nevertheless, it is clear that the current efforts did not result in the perfect replacement and a lot of work is needed to provide confidence to apply these systems.

Multiple chromate free alternative processes have been introduced to the industry [80, 106]. Especially on pre-treatment level, alternative solutions were introduced but still to be used in combination with chromated coating systems Therefore, many challenges have to be overcome before hexavalent chromium and chromates can be eliminated from the aerospace industry. These challenges are:

- Technical performance
- Processes and requirements
- Industry confidence
- Service life prediction

First, the primary challenge is to find suitable alternative inhibitors and coating systems that demonstrate an equal or better technical performance compared to the hexavalent chromium. The superior inhibiting efficiency and versatile behaviour of chromates and hexavalent chromium materials make it very difficult to replace it by a single drop-in material. Chromate inhibition mechanisms have been studied [20], but mimicking chromates has not been very successful and possibly new approaches are necessary to provide a similar protection. Nevertheless, in recent years many inhibitors were identified and tested extensively. Some are promising candidates and some are very close to the performance of chromates in accelerated corrosion tests. However, the performance in service is still unclear.

The second challenge concerns the processes and requirements of the aerospace industry. Hexavalent chromium is the primary corrosion control material. It is embedded in all parts in aerospace industry, from the smallest clip to the largest stringer, fasteners and sealants, coatings and pre-treatments. Chromate containing products processes are specified by material specifications for new aircraft but as well in the maintenance manuals of legacy aircraft currently in service. The replacement of chromates will be a long process with many hurdles.

Industry confidence is the third challenge. First of all, is the performance of the new chromate free technologies and the confidence of the industry sufficient enough to make a switch to total chromate free aircraft when legislation and regulations ban the use of these carcinogenic materials. Today, alternative solutions have been introduced, these solutions passed the material specification requirements but still are not as good or robust as chromated technology. When these individual qualified materials are introduced in a system, it is possible that the total system does not always meet the requirements. The laboratory performance of new systems is known, but the in-service behaviour and performance is unknown. Therefore, new chromate-free systems need to be tested extensively to provide confidence to the aerospace engineers to justify a switch to chromate-free coating technology.

The final challenge is service-life prediction. Chromated systems have a proven track record of more than 50 years within the aerospace industry. The protection of aircraft structure has been ensured for more than 30 years. With the introduction of the chromate-free technology it is a brand new story. Developments demonstrate that it is possible to meet the aerospace specification requirements. However, it remains unclear how the system behaves under in-service conditions. The challenge is how to predict the long-term performance of a new chrome-free coating system so it protects the aircraft structure for more than 30 years. Reduced performance will lead to higher maintenance and depreciation costs.

12.7 Summary and Conclusion

Aviation is an integral part of society of today. Air travel is increasing and the number of aircraft is expected to double over the coming 20 years. Aircraft have very complex designs with a variety of metals and composite materials. With an expected service- life of about 30 years and the harsh operating environment with many flight cycles makes fatigue properties and corrosion protection important concerns.

Aluminium is the predominant construction material for aircraft. Especially, AA2024-T3 and AA7075-T3 are commonly used because of their favourable strength to weight ratio. However, these alloys are susceptible to various types of corrosion. Corrosion control is a key element next to efficient and environmental sustainable operations Coatings and surface treatments is one of the critical factors in the corrosion protection scheme to ensure the long- term protection of the aircraft structure.

Aerospace coatings are used on all types of commercial and military aircraft as protective and decorative finishes. These coatings are designed and formulated to withstand challenging aerospace operating conditions. The requirements and properties for aerospace coatings are controlled by specifications. There are three categories of aerospace coatings: structural systems, exterior systems and special purpose systems. They are usually applied with hand-held spray equipment and cured at ambient conditions.

Current aerospace coating systems are mainly based on a pre-treatment (e.g. chemical conversion coating or anodising) on the aluminium alloy, a corrosion inhibiting epoxy-amine primer, and a protective or decorative polyurethane topcoat. Hexavalent chromium compounds are widely used in these aerospace coatings and pre-treatments, these materials exhibit excellent anticorrosive properties and provided the foundation of the corrosion protection of today's aircraft structures.

The industry has been using hexavalent chromium inhibition since the introduction of the aluminium aircrafts. However, health issues and new legislations and regulations are putting pressure on the industry to find sustainable alternatives to replace hexavalent chromium compounds. Already since the 1980's fundamental and applied research has been searching for alternatives and resolving the inhibition mechanism of hexavalent chromium. The quest for chromate-free coatings systems that provides corrosion protection better than or equal to the chromated technology has resulted in a range of chromate-free pre-treatments and coating systems. Several of them have been qualified and introduced into the market place. The research also provided new inhibition strategies like: cathodic protection, conductive polymers, self-healing approaches, high throughput technologies, and advanced corrosion assessment tools.

Despite the availability of these chromate-free materials and systems, it can be stated that there is no direct replacement for hexavalent chromium available to date. Therefore, research continues to meet the technical performance of chromated technology and to overcome thc challenges associated chromate-free inhibition systems for aerospace coatings and to provide the confidence to the industry to introduce the chromate-free alternatives and eliminate hexavalent chromium compounds from the industry.

References

1. O. Guseva, S. Brunner, P. Richner, Analysis of the environmental parameters for aircraft coatings. Macromol. Symp. **187**, 883–893 (2002). doi:10.1002/1521-3900(200209)187:1<883:AID-MASY883>3.0.CO;2-M
2. Deloitte, 2013 Global aerospace and defense industry outlook (2013), http://www.deloitte.com/aerospace&defense/2013a&doutlook. Accessed 12 March 2014
3. GAMA, 2012 general aviation statistical databook & industry outlook. General aviation manufacturers association (GAMA) (2012), http://www.gama.aero/media-center/industry-facts-and-statistics/. Accessed 10 Oct 2013
4. Airbus, Airbus global market forecast 2012–2031 (2013), http://www.airbus.com/company/market/forecast/?. Accessed Aug 2013
5. Boeing, Current market outlook 2013–2032 (2013), http://www.boeing.com/assets/pdf/commercial/cmo/pdf/Boeing_Current_Market_Outlook_2013.pdf. Accessed Aug 2013
6. H. Jiang, Key Findings on airplane economic life. The boeing company (2013), http://www.boeing.com/boeing/commercial/industry_info.page. Accessed 12 Dec 2013
7. T. Wright, Aerospace coatings. Coat. World **15**(8), 28–31 (2010)
8. Organization ICA, ICAO safety report 2014 edition. International civil aviation organization (2014), http://www.icao.int/safety/Documents/ICAO_2014_Safety_Report_final_02042014_web.pdf. Accessed 13 March 2014
9. G.H. Koch, Cost of corrosion Appendix P aircraft. CC technologies, http://corrosionda.com/transportation/aircraft/index.htm. Accessed 13 March 2014
10. J.A. DeRose, T. Suter, T. Hack, R. Adey, *Aluminium Alloy Corrosion of Aircraft Structures: Modelling and Simulation* (WIT Press, 2013)
11. M. Kendig, S. Jeanjaquet, R. Addison, J. Waldrop, Role of hexavalent chromium in the inhibition of corrosion of aluminum alloys. Surf. Coat. Technol. **140**(1), 58–66 (2001). doi:10.1016/S0257-8972(01)01099-4
12. National Toxicology P NTP, 12th Report on Carcinogens. Report on Carcinogens : Carcinogen profiles/US Dept of Health and Human Services, Public Health Service, National Toxicology Program **12**, 106–109 (2011)
13. D.C.V. Beelen, P. Rouw, R. Boomgaard, K. Zabel, Aviox CF priamer: a new generation primer for aircraft maintenance. Anti-Corros. Methods Mater. **44**(2), 28–33 (1997)
14. J.G. Kaufman, *Introduction to Aluminum Alloys and Tempers* (ASM International, 2000)
15. J.R. Davis, *Corrosion of Aluminum and Aluminum Alloys* (ASM International, 1999)
16. T. Richardson, B. Cottis, D. Scantlebury, R. Lindsay, S. Lyon, M. Graham, *Shreir's Corrosion* (Elsevier, 2010)
17. I.J. Polmear, *Light Alloys* (Elsevier Ltd, 2005)
18. A.E. Hughes, N. Birbilis, J.M.C. Mol, J. Santiago, S.J. Garcia, X. Zhou, G.E Thompson, in *Recent Trends in Processing and Degradation of Aluminium Alloys*, ed by P.Z. Ahmad. Thompson, High Strength Al-Alloys: Microstructure, Corrosion and Principles of Protection. *In Tech.* doi:10.5772/18766

19. T. Dursun, C. Soutis, Recent developments in advanced aircraft aluminium alloys. Mater. Des. **56**, 862–871 (2014). doi:10.1016/j.matdes.2013.12.002
20. G.S. Frankel, Mechanism of Al alloy corrosion and the role of chromate inhibitors final report. STAR 40 (2002)
21. L. Xia, R.L. McCreery, Chemistry of a chromate conversion coating on aluminum alloy AA2024-T3 probed by vibrational spectroscopy. J. Electrochem. Soc. **145**(9), 3083–3089 (1998). doi:10.1149/1.1838768
22. J.H. Osborne, Observations on chromate conversion coatings from a sol-gel perspective. Prog. Org. Coat. **41**(4), 280–286 (2001). doi:10.1016/S0300-9440(01)00143-6
23. P. Campestrini, G. Goeminne, H. Terryn, J. Vereecken, J.H.W. De Wit, Chromate conversion coating on aluminum alloys I. formation mechanism. J. Electrochem. Soc. **151**(2), B59–B70 (2004). doi:10.1149/1.1637355
24. L. Xia, E. Akiyama, G. Frankel, R. McCreery, Storage and release of soluble hexavalent chromium from chromate conversion coatings. Equilibrium aspects of CrVI concentration. J. Electrochem. Soc. **147**(7), 2556–2562 (2000). doi:10.1149/1.1393568
25. J.D. Ramsey, R.L. McCreery, In situ Raman microscopy of chromate effects on corrosion pits in aluminum alloy. J. Electrochem. Soc. **146**(11), 4076–4081 (1999). doi:10.1149/1.1392594
26. G.E. Thompson, H. Habazaki, K. Shimizu, M. Sakairi, P. Skeldon, X. Zhou, G.C. Wood, Anodizing of aluminium alloys. Aircraft Engineering and Aerospace Technology **71**(3), 228–238 (1999). doi:10.1108/00022669910270709
27. G.C. Wood, J.P. O'Sullivan, The anodizing of aluminium in sulphate solutions. Electrochim. Acta **15**(12), 1865–1876 (1970). doi:10.1016/0013-4686(70)85024-1
28. A.D. Juhl, Overview of anodizing in the aerospace industry. Met. Finish. **108**(2), 20–21 (2010). doi:10.1016/S0026-0576(10)00011-5
29. H. Rosenboom, Chemistry of wash primers. Ind. Eng. Chem. **45**(11), 2982 (1953)
30. R.G. Buchheit, A compilation of corrosion potentials reported for intermetallic phases in aluminum-alloys. J. Electrochem. Soc. **142**(11), 3994–3996 (1995). doi:10.1149/1.2048447
31. R.G. Buchheit, R.P. Grant, P.F. Hlava, B. McKenzie, G.L. Zender, Local dissolution phenomena associated with S phase (Al2CuMg) particles in aluminum alloy 2024-T3. J. Electrochem. Soc. **144**(8), 2621–2628 (1997). doi:10.1149/1.1837874
32. N. Birbilis, R.G. Buchheit, Electrochemical characteristics of intermetallic phases in aluminum alloys—an experimental survey and discussion. J. Electrochem. Soc. **152**(4), B140–B151 (2005). doi:10.1149/1.1869984
33. J.H.W. de Wit, New knowledge on localized corrosion obtained from local measuring techniques. Electrochim. Acta **46**(24–25), 3641–3650 (2001). doi:10.1016/S0013-4686(01)00642-9
34. J.M.C. Mol, A.E. Hughes, B.R.W. Hinton, S. van der Zwaag, A morphological study of filiform corrosive attack on chromated and alkaline-cleaned AA2024-T351 aluminium alloy. Corros. Sci. **46**(5), 1201–1224 (2004). doi:10.1016/j.corsci.2003.09.012
35. Z. Szklarska-Smialowska, Pitting corrosion of aluminum. Corros. Sci. **41**(9), 1743–1767 (1999). doi:10.1016/S0010-938X(99)00012-8
36. A. Boag, A.E. Hughes, A.M. Glenn, T.H. Muster, D. McCulloch, Corrosion of AA2024-T3 Part I: localised corrosion of isolated IM particles. Corros. Sci. **53**(1), 17–26 (2011). doi:10.1016/j.corsci.2010.09.009
37. A.E. Hughes, A. Boag, A.M. Glenn, D. McCulloch, T.H. Muster, C. Ryan, C. Luo, X. Zhou, G.E. Thompson, Corrosion of AA2024-T3 Part II: co-operative corrosion. Corros. Sci. **53**(1), 27–39 (2011). doi:10.1016/j.corsci.2010.09.030
38. A.M. Glenn, T.H. Muster, C. Luo, X. Zhou, G.E. Thompson, A. Boag, A.E. Hughes, Corrosion of AA2024-T3 part III: propagation. Corros. Sci. **53**(1), 40–50 (2011). doi:10.1016/j.corsci.2010.09.035
39. A. Boag, R.J. Taylor, T.H. Muster, N. Goodman, D. McCulloch, C. Ryan, B. Rout, D. Jamieson, A.E. Hughes, Stable pit formation on AA2024-T3 in a NaCl environment. Corros. Sci. **52**(1), 90–103 (2010). doi:10.1016/j.corsci.2009.08.043

40. S.T. Pride, J.R. Scully, J.L. Hudson, Metastable pitting of aluminum and criteria for the transition to stable pit growth. J. Electrochem. Soc. **141**(11), 3028–3040 (1994). doi:10.1149/1.2059275
41. M.B. Vukmirovic, N. Dimitrov, K. Sieradski, Dealloying and corrosion of Al alloy 2024-T3. J. Electrochem. Soc. **149**(9), B428–B439 (2002). doi:10.1149/1.1498258
42. C. Luo, X. Zhou, G.E. Thompson, A.E. Hughes, Observations of intergranular corrosion in AA2024-T351: The influence of grain stored energy. Corros. Sci. **61**, 35–44 (2012). doi:10.1016/j.corsci.2012.04.005
43. X. Zhou, C. Luo, Y. Ma, T. Hashimoto, G.E. Thompson, A.E. Hughes, P. Skeldon, Grain-stored energy and the propagation of intergranular corrosion in AA2xxx aluminium alloys. Surf. Interface Anal. **45**(10), 1543–1547 (2013). doi:10.1002/sia.5218
44. F. Andreatta, H. Terryn, J.H.W. De Wit, Corrosion behaviour of different tempers of AA7075 aluminium alloy. Electrochim. Acta **49**(17–18), 2851–2862 (2004). doi:10.1016/j.electacta.2004.01.046
45. T. Marlaud, B. Malki, C. Henon, A. Deschamps, B. Baroux, Relationship between alloy composition, microstructure and exfoliation corrosion in Al-Zn-Mg-Cu alloys. Corros. Sci. **53**(10), 3139–3149 (2011). doi:10.1016/j.corsci.2011.05.057
46. M. Tanaka, C. Hénon, T. Warner, in *Microstructural Evolution of a New Aerospace 7XXX Alloy During Retrogression and Re-ageing Treatment*, vol 519–521 (Vancouver, 2006)
47. R.T. Foley, Localized corrosion of aluminum alloys—a review. Corrosion **42**(5), 277–288 (1986)
48. R. Braun, Environmentally assisted cracking of aluminium alloys. Materialwiss. Werkstofftech. **38**(9), 674–689 (2007). doi:10.1002/mawe.200700204
49. R.C. Newman, in *Stress Corrosion Cracking* (Elsevier, 2010), pp. 864–901
50. D.B. Jam, M. Mohaghegh, Design for corrosion control. Boeing commercial airplanes group (1999), http://www.boeing.com/commercial/aeromagazine/aero_07/corrosn.html—methods. Accessed 22 Dec 2013
51. O.M. Matarredona, K. Mach, M.M. Rieger, E.A. O'Rear, Alteration of wettability and inhibition of corrosion in narrow aluminium 7075 gaps by thin polymer films. Corros. Sci. **45**(11), 2541–2562 (2003). doi:10.1016/s0010-938x(03)00084-2
52. I.S. Cole, W.G. Ganther, P. Corrigan, S. Galea, P. Trathen, B. Hinton, Frequency and duration of wetness periods on surfaces in airframes. Corros. Eng. Sci. Technol. **47**(7), 529–535 (2012). doi:10.1179/1743278212Y.0000000003
53. J.V. Koleske, *Paint and Coating Testing Manual: Fourteenth Edition of the Gardner-Sward Handbook*. (ASTM, 1995)
54. N. Ohta, A.R. Robertson, in *Colorimetry*. CIE Standard Colorimetric System (Wiley, 2006), pp. 63–114. doi:10.1002/0470094745.ch3
55. A.K. Chattopadhyay, M.K.R. Zentner, *Aerospace and Aircraft Coatings*. (Federation of Societies for Coatings Technology, 1990)
56. G.P. Bierwagen, D.E. Tallman, Choice and measurement of crucial aircraft coatings system properties. Prog. Org. Coat. **41**(4), 201–216 (2001). doi:10.1016/S0300-9440(01)00131-X
57. Z.W. Wicks, F.N. Jones, S.P. Pappas, D.A. Wicks, *Organic Coatings: Science and Technology. Organic Coatings: Science and Technology* (Wiley, 2006). doi:10.1002/9780470079072
58. J. Sinko, Challenges of chromate inhibitor pigments replacement in organic coatings. Prog. Org. Coat. **42**(3–4), 267–282 (2001). doi:10.1016/S0300-9440(01)00202-8
59. M.W. Kendig, R.G. Buchheit, Corrosion inhibition of aluminum and aluminum alloys by soluble chromates, chromate coatings, and chromate-free coatings. Corrosion **59**(5), 379–400 (2003). doi:0010-9312/03/000079
60. F.H. Scholes, S.A. Furman, A.E. Hughes, T. Nikpour, N. Wright, P.R. Curtis, C.M. Macrae, S. Intem, A.J. Hill, Chromate leaching from inhibited primers. Part I. characterisation of leaching. Prog. Org. Coat. **56**(1), 23–32 (2006). doi:10.1016/j.porgcoat.2006.01.015

61. S.A. Furman, F.H. Scholes, A.E. Hughes, D. Lau, Chromate leaching from inhibited primers. Part II: modelling of leaching. Prog. Org. Coat. **56**(1), 33–38 (2006). doi:10.1016/j.porgcoat.2006.01.016
62. S. Sellaiyan, A.E. Hughes, S.V. Smith, A. Uedono, J. Sullivan, S. Buckman, Leaching properties of chromate-containing epoxy films using radiotracers, PALS and SEM. Prog. Org. Coat. **77**(1), 257–267 (2014). doi:10.1016/j.porgcoat.2013.09.014
63. X. Zhou, C. Luo, T. Hashimoto, A.E. Hughes, G.E. Thompson, Study of localized corrosion in AA2024 aluminium alloy using electron tomography. Corros. Sci. **58**, 299–306 (2012). doi:10.1016/j.corsci.2012.02.001
64. S.P. Knight, M. Salagaras, A.M. Wythe, F. De Carlo, A.J. Davenport, A.R. Trueman, In situ X-ray tomography of intergranular corrosion of 2024 and 7050 aluminium alloys. Corros. Sci. **52**(12), 3855–3860 (2010). doi:10.1016/j.corsci.2010.08.026
65. A. Trueman, S. Knight, J. Colwell, T. Hashimoto, J. Carr, P. Skeldon, G. Thompson, 3-D tomography by automated in situ block face ultramicrotome imaging using an FEG-SEM to study complex corrosion protective paint coatings. Corros. Sci. **75**, 376–385 (2013). doi:10.1016/j.corsci.2013.06.021
66. A.E. Hughes, A. Trinchi, F.F. Chen, Y.S. Yang, I.S. Cole, S. Sellaiyan, J. Carr, P.D. Lee, G. E. Thompson, T.Q. Xiao, The application of multiscale quasi 4D CT to the study of SrCrO4 distributions and the development of porous networks in epoxy-based primer coatings. Prog. Org. Coat. **77**(11), 1946–1956 (2014). doi:10.1016/j.porgcoat.2014.07.001
67. E. Javierre, S.J. García, J.M.C. Mol, F.J. Vermolen, C. Vuik, S. van der Zwaag, Tailoring the release of encapsulated corrosion inhibitors from damaged coatings: controlled release kinetics by overlapping diffusion fronts. Prog. Org. Coat. **75**(1–2), 20–27 (2012). doi:10.1016/j.porgcoat.2012.03.002
68. P.E. Jennerjohn, Unique aspects involved in the robotic painting of commercial aircraft structures. SAE Tech. Pap. (2011). doi:10.4271/2011-01-2790
69. K. Pianoforte, Aerospace coatings market. coatings world (2013), http://www.coatingsworld.com/issues/2013-05/view_features/aerospace-coatings-market-316334/. Accessed 08 July 2013
70. G. Bierwagen, Next generation of aircraft coatings systems. J. Coat. Technol. **73**(915), 45–52 (2001)
71. M. Costa, C.B. Klein, Toxicity and carcinogenicity of chromium compounds in humans. Crit. Rev. Toxicol. **36**(2), 155–163 (2006). doi:10.1080/10408440500534032
72. U.P.H. Service, Toxicological profile for chromium (1989). Report No ATSDR/TP 88/10
73. W. Robison, Pantheon's preKote surface pretreatment qualifies to AMS 3095A with three coatings systems. Pantheon enterprises (2013). http://www.reuters.com/article/2013/04/30/az-pantheon-enterprises-idUSnBw306717a+100+BSW20130430
74. A. Davis, Boeing delivers KLM 777-300ER in SkyTeam livery (2009), http://www.prnewswire.com/news-releases/boeing-delivers-klm-777-300er-in-skyteam-livery-62024167.html
75. T. Pennington, PPG aerospace chromate-free primer first to be qualified by Boeing. Prod. Finish. **75**(9), 12 (2011)
76. Anonymous, Army approves hex-chrome-free coating. Prod. Finish. **72**(9), 12 (2008)
77. T. Pennington, PPG industries, AkzoNobel launch new aerospace coating products. Prod. Finish. (Cincinnati) **76**(8), 32–33 (2012)
78. T. Pennington, Aerospace drives more calls for non-chrome passivate finishes. Prod. Finish. **76**(8), 16–21 (2012)
79. L. Roberts, A. Galanis, Eliminating hexavalent chrome. Ind. Paint Powder **82**(5), 13–16 (2006)
80. E. Eichinger, J.H. Osborne, Hexavalent chromium elimination: an aerospace industry progress report. Metal Finish.(USA) **95**(3), 36–41 (1997)
81. J.H. Osborne, Development of non-chromium aerospace surface treatments. Paper presented at the Surfair, Biarritz, 5–6 June 2008

82. A.S. Hamdy, I. Doench, H. Möhwald, Intelligent self-healing corrosion resistant vanadia coating for AA2024. Thin Solid Films **520**(5), 1668–1678 (2011). doi:10.1016/j.tsf.2011.05.080
83. A.E. Hughes, J.D. Gorman, T.G. Harvey, A. Galassi, G. McAdam, Development of permanganate-based coatings on aluminum alloy 2024-T3. Corrosion **62**(9), 773–780 (2006)
84. S.A. Kulinich, A.S. Akhtar, P.C. Wong, K.C. Wong, K.A.R. Mitchell, Growth of permanganate conversion coating on 2024-Al alloy. Thin Solid Films **515**(23), 8386–8392 (2007). doi:10.1016/j.tsf2007.04.164
85. G. Yoganandan, J.N. Balaraju, V.K.W. Grips, The surface and electrochemical analysis of permanganate based conversion coating on alclad and unclad 2024 alloy. Appl. Surf. Sci. **258**(22), 8880–8888 (2012). doi:10.1016/j.apsusc.2012.05.108
86. R.G. Buchheit, M.D. Bode, G.E. Stoner, Corrosion-resistant, chromate free Talc coatings for aluminum. Corrosion **50**(3), 205–214 (1994)
87. C.A. Drewien, M.O. Eatough, D.R. Tallant, C.R. Hills, R.G. Buchheit, Lithium-aluminum-carbonate-hydroxide hydrate coatings on aluminum: composition, structure and processing bath chemistry. J. Mater. Res. **11**(6), 1507–1513 (1996)
88. R.G. Buchheit, G.E. Stoner, Formation of protective film on the surfaces of aluminium@ alloys contg. lithium@—by immersion in inexpensive, non-toxic alkaline salt soln. which forms salt film with the lithium, affording improved corrosion resistance. US5266356-A; US35576-E (1993)
89. M.R.S. Castro, J.C. Nogueira, G.P. Thim, M.A.S. Oliveira, Adhesion and corrosion studies of a lithium based conversion coating film on the 2024 aluminum alloy. Thin Solid Films **457**(2), 307–312 (2004). doi:10.1016/j.tsf.2003.10.024
90. C.M. Rangel, M.A. Travassos, Li-based conversion coatings on aluminium: an electrochemical study of coating formation and growth. Surf. Coat. Technol. **200**, 5823–5828 (2006)
91. C.M. Rangel, M.A. Travassos, The passivation of aluminium in lithium carbonate/bicarbonate solutions. Corros. Sci. **33**(3), 327–343 (1992). doi:10.1016/0010-938X(92)90064-A
92. T.G. Harvey, Cerium-based conversion coatings on aluminium alloys: a process review. Corros. Eng. Sci. Technol. **48**(4), 248–269 (2013). doi:10.1179/1743278213Y.0000000089
93. M. Bethencourt, F.J. Botana, J.J. Calvino, M. Marcos, M.A. Rodriguez-Chacon, Lanthanide compounds as environmentally-friendly corrosion inhibitors of aluminium alloys: a review. Corros. Sci. **40**(11), 1803–1819 (1998). doi:10.1016/s0010-938x(98)00077-8
94. A.J. Aldykewicz Jr, H.S. Isaacs, A.J. Davenport, Investigation of cerium as a cathodic inhibitor for aluminum-copper alloys. J. Electrochem. Soc. **142**(10), 3342–3350 (1995). doi:10.1149/1.2049985
95. S.A. Hayes, P. Yu, T.J. O'Keefe, M.J. O'Keefe, J.O. Stoffer, The phase stability of cerium species in aqueous systems. I. E-pH diagram for the Ce-$HClO_4$-H_2O system. J. Electrochem. Soc. **149**(12), C623–C630 (2002). doi:10.1149/1.1516775
96. P. Campestrini, H. Terryn, A. Hovestad, J.H.W. de Wit, Formation of a cerium-based conversion coating on AA2024: relationship with the microstructure. Surf. Coat. Technol. **176**(3), 365–381 (2004). doi:10.1016/s0257-8972(03)00743-6
97. A.E. Hughes, F.H. Scholes, A.M. Glenn, D. Lau, T.H. Muster, S.G. Hardin, Factors influencing the deposition of Ce-based conversion coatings, part I: the role of Al3 + ions. Surf. Coat. Technol. **203**(19), 2927–2936 (2009). doi:10.1016/j.surfcoat.2009.03.022
98. D. Lau, A.M. Glenn, A.E. Hughes, F.H. Scholes, T.H. Muster, S.G. Hardin, Factors influencing the deposition of Ce-based conversion coatings, part II: the role of localised reactions. Surf. Coat. Technol. **203**(19), 2937–2945 (2009). doi:10.1016/j.surfcoat.2009.03.016
99. M.J. O'Keefe, S. Geng, S. Joshi, Cerium-based conversion coatings as alternatives to hex chrome. Rare-earth compounds provide resistance against corrosion for aluminum alloys in military applications. Met. Finish. **105**(5), 25–28 (2007). doi:10.1016/S0026-0576(07)80547-2
100. S. Geng, P. Yu, M.J. O'Keefe, W.G. Fahrenholtz, T.J. O'Keefe, Screening study of spray solution parameters for depositing cerium-based conversion coatings on Al alloy 2024-T3. J. Appl. Electrochem. **40**(3), 551–559 (2010). doi:10.1007/s10800-009-0027-8

101. T.S. Schmidt-Hansberg, A comparative study of innovative aluminium pretreatments. ATB Metallurgie **43**, 9–14 (2003)
102. E. Morris, Implementation of rare earth primers and pretreatments for military applications. Paper presented at the ASETSDefense 2012, Sustainable Surface Engineering for Aerospace and Defense Workshop, San Diego, CA, 27–30 Aug 2012
103. C. Matzdorf, M. Kane, J. Green, Preparation of coating aluminum substrates for protection against corrosion, involves treating substrates with acidic aqueous solution containing trivalent chromium salt, hexafluorozirconate thickener, and surfactant. WO200255758-A; EP1404894-A; US6375726-B1 (2002)
104. H. Bhatt, A. Manavbasi, D. Rosenquist, Trivalent chromium for enhanced corrosion protection on aluminum surfaces. Met. Finish. **107**(7–8), 31–37 (2009). doi:10.1016/S0026-0576(09)80203-1
105. C. Matzdorf, Naval aviation progress in implementing non-hexavalent chromium surface finishes. Paper presented at the ASETS Defense 2012, Sustainable Surface Engineering for Aerospace and Defense Workshop, San Diego, Ca, USA, 27–30 Aug 2012
106. K. Legg, in *DoD Metal Finishing Workshop—Chromate Alternatives for Metal Treatment and Sealing*, Report DoD Metal Finishing Workshop—Chromate Alternatives for Metal Treatment and Sealing, Layton, Utah, USA, 15–17 May 2007
107. Y. Guo, G.S. Frankel, Active corrosion inhibition of AA2024-T3 by trivalent chrome process treatment. Corrosion **68**(4) (2012). doi:10.5006/0010-9312-68-4-3
108. Y. Guo, G.S. Frankel, Characterization of trivalent chromium process coating on AA2024-T3. Surf. Coat. Technol. **206**(19–20), 3895–3902 (2012). doi:10.1016/j.surfcoat.2012.03.046
109. A. Iyer, W. Willis, S. Frueh, W. Nickerson, A. Fowler, J. Barnes, L. Hagos, J. Escarsega, J. La Scala, S.L. Suib, Characterization of NAVAIR trivalent chromium process (TCP) coatings and solutions. Plat. Surf. Finish. **97**(4), 32–41 (2010)
110. S.L. Suib, J. La Scala, W. Nickerson, A. Fowler, N. Zaki, Determination of hexavalent chromium in NAVAIR trivalent chromium process (TCP) coatings and process solutions. Metal Finish. **107**(2), 28–31, 33–34 (2009). doi:10.1016/S0026-0576(09)80028-7
111. F.T. Museux FT, Introducing more eco-efficient chemical treatments for aircraft structure. Flight Airworthiness Support Technology. Airbus Technical Magazine (2009)
112. J.H. Osborne, Nonchromate conversion coatings in use at Boeing. Paper presented at the DoD metal finishing workshop—chromate alternatives for metal treatment and sealing, Layton, Utah, 16 & 17 May 2007
113. Company TB preferred materials and processes. http://www.boeingsuppliers.com/environmental/chromic.html. Accessed 19 Oct 2013
114. L. Zhang, G.E. Thompson, M. Curioni, P. Skeldon, Anodizing of aluminum in sulfuric acid/boric acid mixed electrolyte. J. Electrochem. Soc. **160**(4), C179–C184 (2013). doi:10.1149/2.032306jes
115. M. Curioni, P. Skeldon, E. Koroleva, G.E. Thompson, J. Ferguson, Role of tartaric acid on the anodizing and corrosion behavior of AA 2024 T3 aluminum alloy. J. Electrochem. Soc. **156**(4), C147–C153 (2009). doi:10.1149/1.3077602
116. M. Curioni, P. Skeldon, J. Ferguson, G.E. Thompson, Reducing the energy cost of protective anodizing. J. Appl. Electrochem. **41**(7), 773–785 (2011). doi:10.1007/s10800-011-0295-y
117. M. Curioni, P. Skeleton, G.E. Thompson, J. Ferguson, Graded anodic film morphologies for sustainable exploitation of aluminium alloys in aerospace. vol 38 (2008)
118. M. Curioni, T. Gionfini, A. Vicenzo, P. Skeldon, G.E. Thompson, Optimization of anodizing cycles for enhanced performance. Surf. Interface Anal. **45**(10), 1485–1489 (2013). doi:10.1002/sia.5222
119. M. Mohammadi, A. Yazdani, M.E. Bahrololoom, A. Alfantazi, Corrosion behavior of 2024 aluminum alloy anodized in presence of permanganate and phosphate ions. J. Coat. Technol. Res. **10**(2), 219–229 (2013). doi:10.1007/s11998-012-9459-x

120. F. Mansfeld, C. Chen, C.B. Breslin, D. Dull, Sealing of anodized aluminum alloys with rare earth metal salt solutions. J. Electrochem. Soc. **145**(8), 2792–2798 (1998). doi:10.1149/1.1838716
121. M. Garcia-Rubio, M.P. de Lara, P. Ocon, S. Diekhoff, M. Beneke, A. Lavia, I. Garcia, Effect of postreatment on the corrosion behaviour of tartaric-sulphuric anodic films. Electrochim. Acta **54**(21), 4789–4800 (2009). doi:10.1016/j.electacta.2009.03.083
122. G. Boisier, A. Lamure, N. Pebere, N. Portail, M. Villatte, Corrosion protection of AA2024 sealed anodic layers using the hydrophobic properties of carboxylic acids. Surf. Coat. Technol. **203**(22), 3420–3426 (2009). doi:10.1016/j.surfcoat.2009.05.008
123. K.A. Krienke, K.Y. Blohowiak, J.H. Osborne, Sol for improving corrosion resistance and adhesion of titanium@ or aluminium@ alloys, to organic matrix resin or adhesive—comprises alkoxy:zirconium, organo:silane coupling agent, organic acid catalyst and stabiliser, and e.g. ammonia for disrupting zirconium = acetate complex. US5814137-A
124. K.Y. Blohowiak, J.H. Osborne, J.E. Seebergh, Development and implementation of sol-gel coatings for aerospace applications. SAE Tech. Papers (2009). doi:10.4271/2009-01-3208
125. J.H. Osborne, K.Y. Blohowiak, S.R. Taylor, C. Hunter, G. Bierwagon, B. Carlson, D. Bernard, M.S. Donley, Testing and evaluation of nonchromated coating systems for aerospace applications. Prog. Org. Coat. **41**(4), 217–225 (2001). doi:10.1016/S0300-9440(01)00132-1
126. L. Roberts, A. Galanis, Proven non-chromated technology. Paint Coat. Ind. **19**(10), 86 (2003)
127. R.L. Twite, G.P. Bierwagen, Review of alternatives to chromate for corrosion protection of aluminum aerospace alloys. Prog. Org. Coat. **33**(2), 91–100 (1998). doi:10.1016/S0300-9440(98)00015-0
128. G.S. Frankel, *Mechanism of Al Alloy Corrosion and the Role of Chromate Inhibitors* (AFOSR Multidisciplinary University Research Initiative, 2001)
129. P. Visser, Chrome free corrosion inhibiting coating technology for the protection of high strength aluminum alloys. In, Chicago, IL, 2013. National Association for Surface Finishing Annual Conference and Trade Show, SUR/FIN 2013. pp. 514–526
130. M. Iannuzzi, G.S. Frankel, Mechanisms of corrosion inhibition of AA2024-T3 by vanadates. Corros. Sci. **49**(5), 2371–2391 (2007). doi:10.1016/j.corsci.2006.10.027
131. M. Iannuzzi, T. Young, G.S. Frankel, Aluminum alloy corrosion inhibition by vanadates. J. Electrochem. Soc. **153**(12), B533–B541 (2006). doi:10.1149/1.2358843
132. J.O. Stoffer, T.J. O'Keefe, E. Morris, P. Yu, S. Hayes, J. Stoffer, T. Okeefe, S.A. Hayes, E.L. Morris, T.J. Okeefe, Corrosion inhibiting composition used in coating substrate used for e.g. aerospace comprises a rare earth compound, and a binder. WO2004065497-A2; US2004249043-A1 (2004)
133. M. O'Keefe, B. Fahrenholtz, E. Morris, R. Albers, Corrosion resistant polymer coatings containing rare earth compounds. In, Chicago, IL, 2013. National Association for Surface Finishing Annual Conference and Trade Show, SUR/FIN 2013. pp. 501–513
134. G. Pipko, Anti-corrosive compositions for use in paints for water systems and steam generators—containing an oxy-amino-phosphate pigment comprising oxy-nitrite compounds, oxy-metal-phosphate compounds, oxy-forate compounds or oxy-silicate compounds. WO9840440-A; EP988347-A (1998)
135. G. Pipko, A. Vitner, A. Vinter, Anti-corrosive pigment comprises metal salt comprising metal cation and anion with specific molar ratio of total metal to anion. WO200299002-A (2002)
136. X. Yu, W. Zhang, T. Garosshen, P. Bhatia, M. Jaworowski, F. Lamm, X. Tang, A. Besing, M.A. Kryzman, X. Chang, J. Yu, C. Xiaoyuan, B. Amy, Z. Weilong, Y.J. Xiaomei, B. Promila, L. Foster, T. Xia, G.T. Mar, Non-carcinogenic corrosion inhibiting additive for use in, e.g. adhesives, comprises anodic corrosion inhibitor and/or cathodic corrosion inhibitor, and metal complexing agent for increasing solubility of the inhibitor(s). US2004262580-A1 (2004)

137. J.W. Putnam, M. Jaworowski, Non-carcinogenic corrosion inhibiting additive for adhesives, paints and primers, sealants and epoxies, comprises anodic corrosion inhibitor and cathodic corrosion inhibitor. EP1191073-A2 (2003)
138. D.N. Walters, J.R. Schneider, Composition, useful e.g. in primer, aluminum substrate and architectural coating, comprises thermosetting film-forming resin and corrosion resisting particles comprising magnesium oxide particles. CN101287784-B; US2008090069-A1; US8231970-B2 (2007)
139. P. Visser, S.A. Hayes, Low-temperature-curable coating composition useful as anticorrosive primer coating for non-ferrous metal substrates, particularly aluminum or aluminum alloy, comprises film-forming resin, curing agent, and lithium salt. WO2010112605-A1 (2010)
140. S.J. Garcia, T.A. Markley, J.M.C. Mol, A.E. Hughes, Unravelling the corrosion inhibition mechanisms of bi-functional inhibitors by EIS and SEM-EDS. Corros. Sci. **69**, 346–358 (2013). doi:10.1016/j.corsci.2012.12.018
141. P. Visser, Y. Lui, X. Zhou, T. Hashimoto, G. Thompson, S. Lyon, L.G.J. van der Ven, A.J. M.C. Mol, H. Terryn, The corrosion protection of AA 2024-T3 aluminium alloy by leaching of lithium-containing salts from organic coatings. Faraday Discuss. (2015). doi:10.1039/C4FD00237G
142. S. Boocock, Amorphous zirconium vanadium oxide/hydroxide composition useful as an anti-corrosive pigment in zinc or zinc-aluminum alloy coil stock or aerospace grade aluminum alloy. WO2008118950-A2; US2009092839-A1
143. D. Becker, M. Geistbeck, C. Gerlach, M. Kaune, J. Kuever, A. Rabenstein, S. Schroeder, Y. Wilke, C. Gerlach, T. Pietzker, O. Schramm, S. Schroder, J. Kuver, D. Becq, O. Skram, T. Pertztek, Y.A. Kyufer, A. Rubenstein, I. Wilke, M. Gaystbek, C. Gerlah, Composition, useful e.g. as a corrosion protection agent for metals or metal alloys, preferably aluminum or aluminum alloys, comprises a binder, hardener, corrosion inhibitor and a quaternary ammonium compound. WO2009094975-A1 (2009)
144. T.G. Harvey, S.G. Hardin, A.E. Hughes, T.H. Muster, P.A. White, T.A. Markley, P.A. Corrigan, J. Mardel, S.J. Garcia, J.M.C. Mol, A.M. Glenn, The effect of inhibitor structure on the corrosion of AA2024 and AA7075. Corros. Sci. **53**(6), 2184–2190 (2011). doi:10.1016/j.corsci.2011.02.040
145. J. Sinko, Synergistic corrosion inhibiting composition for application to metals such as steel, copper or aluminium. US5948147-A (1997)
146. J. Sinko, Corrosion inhibitor for metals comprising a hybrid organic-inorganic microcomposite. WO9918044-A (1999)
147. J. Sinko, Protection against atmospheric corrosion, for aluminum substrates, steel, involves coating substrate with material having limited solubility in water such as di-mercapto, poly-mercapto compounds or their derivatives. WO200292880-A; US2002197468-A1 (2002)
148. J. Sinko, M.W. Kendig, Pigment grade corrosion inhibitor for paint formulations and organic coatings, contains host species comprising inorganic compound having layered structure and guest species comprising anion of weak acid. WO2005003408-A2; US2005022693-A1 (2008)
149. M. Kendig, M. Hon, A hydrotalcite-like pigment containing an organic anion corrosion inhibitor. Electrochem. Solid State Lett. **8**(3), B10–B11 (2005). doi:10.1149/1.1857743
150. S. Hayes, P.J. Kinlen, P. Kinlen, Corrosion-responsive agent, useful for reducing or preventing the corrosion of a metal which is a copper-containing aluminum alloy, comprises a neutralized metal salt of a corrosion inhibiting organic anion. US2008216705-A1; WO2008140648-A2 (2008)
151. H. Shi, E.H. Han, S.V. Lamaka, M.L. Zheludkevich, F. Liu, M.G.S. Ferreira, Cerium cinnamate as an environmentally benign inhibitor pigment for epoxy coatings on AA 2024-T3. Prog. Org. Coat. (2014). doi:10.1016/j.porgcoat.2014.01.003
152. M. Van Soestbergen, V. Baukh, S.J.F. Erich, H.P. Huinink, O.C.G. Adan, Release of cerium dibutylphosphate corrosion inhibitors from highly filled epoxy coating systems. Prog. Org. Coat. **77**(10), 1562–1568 (2014). doi:10.1016/j.porgcoat.2013.12.018

153. D. Ho, N. Brack, J. Scully, T. Markley, M. Forsyth, B. Hinton, Cerium dibutylphosphate as a corrosion inhibitor for AA2024-T3 aluminum alloys. J. Electrochem. Soc. **153**(9), B392–B401 (2006). doi:10.1149/1.2217260
154. R. Catubig, A.E. Hughes, I.S. Cole, B.R.W. Hinton, M. Forsyth, The use of cerium and praseodymium mercaptoacetate as thiol-containing inhibitors for AA2024-T3. Corros. Sci. **81**, 45–53 (2014). doi:10.1016/j.corsci.2013.12.001
155. R.W. Revie, *Uhlig's Corrosion Handbook*, 2nd edn. Corrosion Control of Steel by Organic Coatings, Chapter 55 (Wiley)
156. M.E. Nanna, G.P. Bierwagen, Mg-rich coatings: a new paradigm for Cr-free corrosion protection of al aerospace alloys. JCT Res. **1**(2), 69–80 (2004)
157. G.P. Bierwagen, M.E. Nanna, D. Battocchi, G. Bierwagen, M. Nanna, Treatment of metal, i.e. aluminum, or preferably aluminum alloy, which is copper-containing aluminum alloy, used in airplanes and other aircraft, by applying, to surface of metal, coating, which comprises magnesium powder and binder. WO2005051551-A1; EP1689534-A1 (2005)
158. S.D. Cramer, B.S. Covino, Jr. ASM handbook, volume 13A—corrosion: fundamentals, testing, and protection. ASM International
159. D. Battocchi, N. Richter, S. Hayes, M. Yan, D.E. Tallmann, G.P. Bierwagen, E. Tallman, G. Bierwagen, Electrochemical study of the galvanic interaction between the Mg rich primer and its Al substrate. 14–18 March 2010
160. D. Battocchi, A.M. Simões, D.E. Tallman, G.P. Bierwagen, Electrochemical behaviour of a Mg-rich primer in the protection of Al alloys. Corros. Sci. **48**(5), 1292–1306 (2006). doi:10.1016/j.corsci.2005.04.008
161. R. Brown, Protecting metal with magnesium. Finish. Today **83**(3), 40–42 (2007)
162. G. Bierwagen, R. Brown, D. Battocchi, S. Hayes, Active metal-based corrosion protective coating systems for aircraft requiring no-chromate pretreatment. Prog. Org. Coat. **67**(2), 195–208 (2010). doi:10.1016/j.porgcoat.2009.10.009
163. B.J.E. Merten, D. Battocchi, G.P. Bierwagen, Aluminum alloy 2024-T3 protection by magnesium-rich primer with chromate-free metal salts. Prog. Org. Coat. **78**, 446–454 (2015). doi:10.1016/j.porgcoat.2014.09.013
164. H. Xu, D. Battocchi, D.E. Tallman, G.P. Bierwagen, Use of magnesium alloys as pigments in magnesium-rich primers for protecting aluminum alloys. Corrosion **65**(5), 318–325 (2009)
165. P. Plagemann, J. Weise, A. Zockoll, Zinc-magnesium-pigment rich coatings for corrosion protection of aluminum alloys. Prog. Org. Coat. **76**(4), 616–625 (2013). doi:10.1016/j.porgcoat.2012.12.001
166. K.N. Allahar, D. Battocchi, M.E. Orazem, G.P. Bierwagen, D.E. Tallman, Modeling of electrochemical impedance data of a magnesium-rich primer. J. Electrochem. Soc. **155**(10), E143–E149 (2008). doi:10.1149/1.2965519
167. S.R. White, N.R. Sottos, P.H. Geubelle, J.S. Moore, M.R. Kessler, S.R. Sriram, E.N. Brown, S. Viswanathan, Autonomic healing of polymer composites. Nature **409**(6822), 794–797 (2001). doi:10.1038/35057232
168. S.J. Garcia, H.R. Fischer, S. van der Zwaag, A critical appraisal of the potential of self healing polymeric coatings. Prog. Org. Coat. **72**(3), 211–221 (2011). doi:10.1016/j.porgcoat.2011.06.016
169. M.G.S. Ferreira, M.L. Zheludkevich, J. Tedim, in *Nanocoatings and Ultra-Thin Films Advanced Protective Coatings for Aeronautical Applications*, ed. by A.S.H.M. Tiginyanu (Woodhead Publishing Series, 2011), pp. 235–279. doi:10.1533/9780857094902.2.235
170. S.J. Garcia, H.R. Fischer, P.A. White, J. Mardel, Y. Gonzalez-Garcia, J.M.C. Mol, A.E. Hughes, Self-healing anticorrosive organic coating based on an encapsulated water reactive silyl ester: synthesis and proof of concept. Prog. Org. Coat. **70**(2–3), 142–149 (2011). doi:10.1016/j.porgcoat.2010.11.021
171. D. Snihirova, S.V. Lamaka, M.F. Montemor, "SMART" protective ability of water based epoxy coatings loaded with CaCO 3 microbeads impregnated with corrosion inhibitors applied on AA2024 substrates. Electrochim. Acta **83**, 439–447 (2012). doi:10.1016/j.electacta.2012.07.102

172. A.C. Balaskas, I.A. Kartsonakis, L.A. Tziveleka, G.C. Kordas, Improvement of anti-corrosive properties of epoxy-coated AA 2024-T3 with TiO_2 nanocontainers loaded with 8-hydroxyquinoline. Prog. Org. Coat. **74**(3), 418–426 (2012). doi:10.1016/j.porgcoat.2012.01.005
173. F. Maia, J. Tedim, A.D. Lisenkov, A.N. Salak, M.L. Zheludkevich, M.G.S. Ferreira, Silica nanocontainers for active corrosion protection. Nanoscale **4**(4), 1287–1298 (2012). doi:10.1039/c2nr11536k
174. T. Stimpfling, F. Leroux, H. Hintze-Bruening, Unraveling EDTA corrosion inhibition when interleaved into Layered Double Hydroxide epoxy filler system coated onto aluminum AA 2024. Appl. Clay Sci. **83–84**, 32–41 (2013). doi:10.1016/j.clay.2013.08.005
175. E.L. Ferrer, A.P. Rollon, H.D. Mendoza, U. Lafont, S.J. Garcia, Double-doped zeolites for corrosion protection of aluminium alloys. Microporous Mesoporous Mater. **188**, 8–15 (2014). doi:10.1016/j.micromeso.2014.01.004
176. F. Gammel, T. Hack, D. Raps, J. Wehr, Lacquer composition, useful for protecting metallic substrates against corrosion, comprises a matrix material, which contains embedded microcapsules comprising an anti-corrosive agent, preferably mercaptobenzothiazole. WO2009087043-A2 (2009)
177. C.J.A. Barbe, N. Caldeira, E. Campazzi, K.S. Finnie, L. Kong, M. Villatte, C. Barbe, N.M. Caldeira, C.H. Barbe, Coating composition, useful for the protection against corrosion in metallic devices, preferably aircraft, comprises ceramic particles comprising a releasable active material homogeneously distributed in each particle. WO2010146001-A1; US2012085261-A1 (2010)
178. D.E. Tallman, Y. Pae, G.P. Bierwagen, Conducting polymers and corrosion: Part 2—polyaniline on aluminum alloys. Corrosion **56**(4), 401–410 (2000)
179. D.E. Tallman, Y. Pae, G.P. Bierwagen, Conducting polymers and corrosion: polyaniline on steel. Corrosion **55**(8), 779–786 (1999)
180. G. Williams, H.N. McMurray, Polyaniline inhibition of filiform corrosion on organic coated AA2024-T3. Electrochim. Acta **54**(17), 4245–4252 (2009). doi:10.1016/j.electacta.2009.02.077
181. M. Kendig, M. Hon, L. Warren, 'Smart' corrosion inhibiting coatings. Prog. Org. Coat. **47**(3–4), 183–189 (2003). doi:10.1016/s0300-9440(03)00137-1
182. M. Kendig, P. Kinlen, Demonstration of galvanically stimulated release of a corrosion inhibitor—basis for "smart" corrosion inhibiting materials. J. Electrochem. Soc. **154**(4), C195–C201 (2007). doi:10.1149/1.2436628
183. M.C. Yan, D.E. Tallman, G.R. Bierwagen, Role of oxygen in the galvanic interaction between polypyrrole and aluminum alloy. Electrochim. Acta **54**(2), 220–227 (2008). doi:10.1016/j.electacta.2008.08.008
184. T. Stimpfling, F. Leroux, H. Hintze-Bruening, Organo-modified layered double hydroxide in coating formulation to protect AA2024 from corrosion. Colloids Surf., A (2014). doi:10.1016/j.colsurfa.2014.01.042
185. P.J. Kinlen, Protection of metal surface against corrosion comprises applying to surface a coating formulation having radiation curable resin and corrosion-responsive agent. WO2003102034-A;US2004035498-A1 (2003)
186. M.W. Kendig, M. Hon, L. Warren, Corrosion inhibition of metallic substrate involves coating metallic substrate with cationic film containing inhibiting anion, reducing cationic film via electron uptake and releasing anion. EP1382721-A; US2004005478-A1; EP1382721-A1; US6942899-B2; EP1382721-B1 (2004)
187. P. Zarras, N. Anderson, C. Webber, J.D. Stenger-Smith, M. Spicer, D. Buhrmaster, Electroactive materials as smart corrosion-inhibiting coatings for the replacement of hexavalent chromium. JCT Coatingstech **8**(1), 40–44 (2011)
188. M. Pawlik, Electrocoat primers for the aerospace industry. Paper presented at the ASETSDefense 2009, Sustainable Surface Engineering for Aerospace and Defense Workshop, Westminster, CO, USA., Aug 31–Sept 3 2009

189. M. Collinet-Fressancourt, N. Nuns, S. Bellayer, M. Traisnel, Characterization by TEM and ToF-SIMS of the oxide layer formed during anaphoretic paint electrodeposition on Al-alloys. Appl. Surf. Sci. **277**, 186–191 (2013). doi:10.1016/j.apsusc.2013.04.023
190. A. Fujimoto, PPG Introduces E-coat primer for aerospace use. PPG Industries (2012). http://www.ppg.com/en/newsroom/news/pages/20120104a.aspx. Accessed 3 Dec 2013
191. E. Charves, Evaluation of powder coatings for application to aircraft component parts. vol Proj. WR02206 (1980)
192. J.L. Duncan, Powder coatings as an environmentally acceptable alternate to the system of MIL-P-53022 and MIL-PRF-22750. STAR **41**(18) (2003)
193. C. Geib, Environmentally advantaged powder coatings for aerospace applications. Paper presented at the SERDP/ESTCP workshop—surface finishing and repair issues for sustaining new military aircraft, Tempe, Arizona, USA, 26–28 Feb (2008)
194. M.J. O'Keefe, W.G. Fahrenholtz, B.S. Curatolo, Multifunctional UV (MUV) curable corrosion coatings for aerospace applications. Met. Finish. **108**(2), 28–31 (2010). doi:10.1016/S0026-0576(10)80038-8
195. J.R. Deantoni, B.S. Curatolo, Coating for surface e.g. military and commercial aircraft and spacecraft, comprises: acrylate oligomer e.g. acrylate, methacrylate or diacrylate; monomer e.g. acrylate, allyl compound or vinyl ether; photoinitiator; and corrosion inhibitor. US2008118659-A1 (2006)
196. N. Moreau, L. Ni, C. Croutxe-Barghorn, A. Chemtob, T. Bouder, G. Hervé, UV-cured hybrid sol-gel coatings for aeronautical and Direct-To-Metal (DTM) applications. Polym. Paint Colour J. **204**(4595), 36–37 (2014)
197. L. Ni, N. Moreau, A. Chemtob, C. Croutxé-Barghorn, Organic-inorganic tandem route to polymer nanocomposites: Kinetic products versus thermodynamic products. J. Sol. Gel. Sci. Technol. **64**(2), 500–509 (2012). doi:10.1007/s10971-012-2882-8
198. T.H. Muster, D. Lau, H. Wrubel, N. Sherman, A.E. Hughes, T.G. Harvey, T. Markley, D.L. J. Alexander, P.A. Corrigan, P.A. White, S.G. Hardin, M.A. Glenn, J. Mardel, S.J. Garcia, J. M.C. Mol, An investigation of rare earth chloride mixtures: combinatorial optimisation for AA2024-t3 corrosion inhibition. Surf. Interface Anal. **42**(4), 170–174. doi:10.1002/sia.3172
199. T.A. Markley, M. Forsyth, A.E. Hughes, Corrosion protection of AA2024-T3 using rare earth diphenyl phosphates. Electrochim. Acta **52**(12), 4024–4031 (2007). doi:10.1016/j.electacta.2006.11.028
200. B.D. Chambers, S.R. Taylor, M.W. Kendig, Rapid discovery of corrosion inhibitors and synergistic combinations using high-throughput screening methods. Corrosion **61**(5), 480–489 (2005)
201. S.J. García, T.H. Muster, O. Özkanat, N. Sherman, A.E. Hughes, H. Terryn, J.H.W. de Wit, J.M.C. Mol, The influence of pH on corrosion inhibitor selection for 2024-T3 aluminium alloy assessed by high-throughput multielectrode and potentiodynamic testing. Electrochim. Acta **55**(7), 2457–2465 (2010). doi:10.1016/j.electacta.2009.12.013
202. J.C. Myland, K.B. Oldham, Anal. Chem. **72**, 3972 (2000); J. Newman, J. Electrochem. Soc. (December), 1235 (1966); W. Zhang, G.S. Frankel, Electrochem. Solid-State Lett. **3**, 268 (2000), 10.1002/sia.3150
203. P.A. White, A.E. Hughes, S.A. Furman, N. Sherman, P.A. Corrigan, M.A. Glenn, D. Lau, S. G. Hardin, T.G. Harvey, J. Mardel, T.H. Muster, S.J. Garcia, C. Kwakernaak, J.M.C. Mol, High-throughput channel arrays for inhibitor testing: proof of concept for AA2024-T3. Corros. Sci. **51**(10), 2279–2290 (2009). doi:10.1016/j.corsci.2009.06.038
204. P.A. White, G.B. Smith, T.G. Harvey, P.A. Corrigan, M.A. Glenn, D. Lau, S.G. Hardin, J. Mardel, T.A. Markley, T.H. Muster, N. Sherman, S.J. Garcia, J.M.C. Mol, A.E. Hughes, A new high-throughput method for corrosion testing. Corros. Sci. **58**, 327–331 (2012). doi:10.1016/j.corsci.2012.01.016
205. T.H. Muster, A. Trinchi, T.A. Markley, D. Lau, P. Martin, A. Bradbury, A. Bendavid, S. Dligatch, A review of high throughput and combinatorial electrochemistry. Electrochim. Acta **56**(27), 9679–9699 (2011). doi:10.1016/j.electacta.2011.09.003

206. P. Visser, Novel totally chrome free corrosion inhibiting coating technology for protection of aluminium alloys. Trans. Inst. Metal Finish. **89**(6), 291–294 (2011). doi:10.1179/174591911X13172025236839
207. P. Visser, Protection for high flyers. Eur. Coat. J. **12**, 88–92 (2010)
208. P. Visser, Chrome free corrosion inhibiting coating technology for the protection of high strength aluminum alloys. In: National Association for Surface Finishing Annual Conference and Trade Show, SUR/FIN 2013, Chicago, IL, (2013), pp. 514–526

Chapter 13
Automotive Coatings

Mark Nichols and Janice Tardiff

Modern automotive coating systems are purposefully designed to provide long-term corrosion protection and aesthetic enhancement to vehicle bodies. The pretreatment (conversion coating) and the electrodeposition coating provide corrosion protection by forming a strong bond with the underlying metal and preventing ion and water transport to the metal surface, while the primer, basecoat and clearcoat provide stone chip resistance, color, and long-term gloss retention. Future enhancements to automotive paint systems could provide additional functionality including active corrosion protection. Further process and quality improvements may be found through the reduction of VOCs, and a reduction in energy usage through the lowering of bake temperatures or a reduction in the number of application steps.

13.1 Introduction

Automotive coatings play a key role in both the aesthetic and functional performance of a vehicle. From a vehicle owner's standpoint, a vehicle's paint serves to enhance its visual appeal, through its color and high gloss appearance. From an automotive OEM's perspective, the paint system serves to entice customers to buy the vehicle, but additionally, the paint system protects the underlying body structure from environmental degradation caused by weather, de-icing salts, water, stones, and environmental fallout [1].

A typical, modern automotive paint system is shown in cross section in Fig. 13.1. The multilayer system is applied in a series of painting steps, with each layer serving specific purposes. The process by which the paint is applied, the chemistry of each layer, and their functional properties are detailed below.

M. Nichols (✉) · J. Tardiff
Ford Motor Company, Dearborn, MI 48121, USA
e-mail: mnichols@ford.com

A.E. Hughes et al. (eds.), *Active Protective Coatings*,
Springer Series in Materials Science 233, DOI 10.1007/978-94-017-7540-3_13

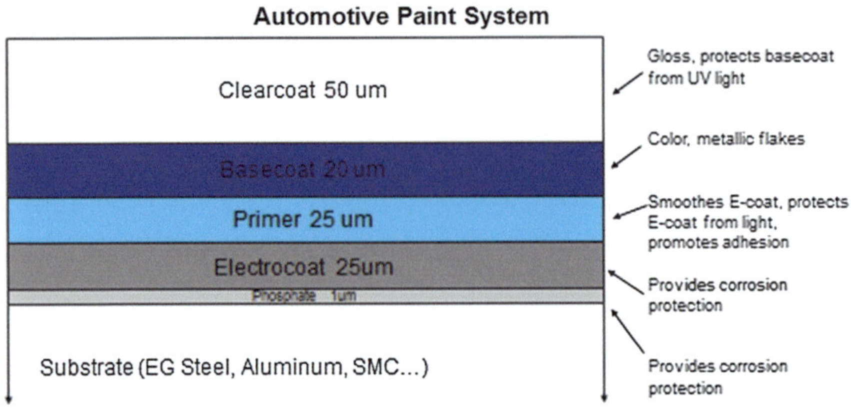

Fig. 13.1 Cross-section of an automotive paint system

13.2 Automotive Paint Application and Materials

13.2.1 Pretreatment

The first coating layer applied to an automotive body is the pretreatment. Traditionally, this layer has been a tri-cationic (zinc, nickel, manganese) phosphate coating on metallic surfaces, which serves to improve the adhesion between the metal substrate and the organic layers that are subsequently applied. This strong adhesive bond also acts to improve the corrosion resistance of the coating system by preventing ionic species from penetrating to the substrate.

Zinc phosphate is applied in a series of steps that begin with the cleaning of the vehicle body. The steps are a mixture of immersions and sprays, depending on the details of each assembly plant. A representative layout of the assembly plant pretreatment system is shown in Fig. 13.2. Cleaning of the metal surface is necessary

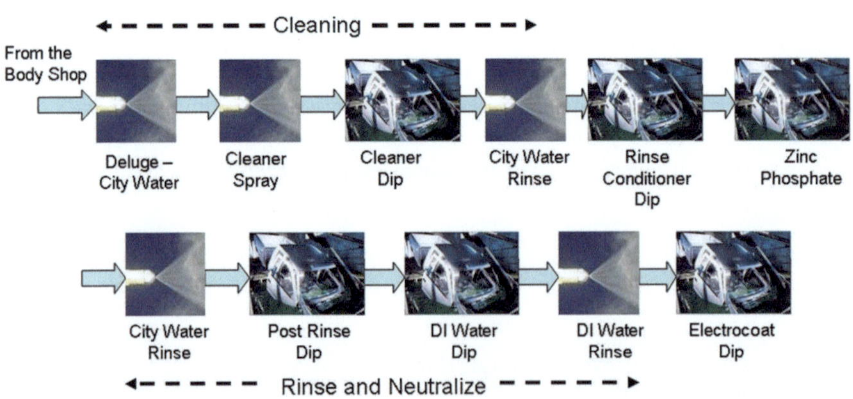

Fig. 13.2 Representative automotive plant pretreatment system layout

to remove lubricants that are used in the stamping process, as well as dirt and manufacturing debris such as weld balls. These contaminants, if left on the surface, can prevent good adhesion of the deposited zinc phosphate to the metallic surface. In automotive applications, an alkaline cleaner is typically used. The alkaline cleaner is generally composed of a builder, such as sodium or potassium hydroxide, and a surfactant.

The cleaning step is followed by water rinsing, which is then followed by a conditioning step. Typically the conditioning step is an aqueous mixture of titanium salts that deposit onto the metallic surface and provide nucleation sites for the zinc phosphate. A more recent alternative to the titanium salts is the use of a sparse layer of zinc phosphate to provide nucleation sites for the full zinc phosphate coating.

The zinc phosphate bath applies the zinc phosphate coating on the surface. When the zinc phosphate solution is exposed to the prepared metallic surface, acid in the bath etches the metallic surface, using up hydronium ions in the acidic solution. As a result, water and hydrogen gas evolve from the solution, causing a pH increase at the interface of the solution and the metal. The pH increase causes the insoluble zinc phosphate crystals to deposit on the nucleation sites. It is necessary to build a multi-layer crystalline coating to ensure adequate protection of the metallic surface.

A scanning electron micrograph of a zinc phosphate coating is shown in Fig. 13.3. The small interlocking crystals are typical of a high quality coating. The final step in the zinc phosphating process is a sealer, which acts to close the pores between the zinc phosphate crystals and enhance corrosion performance. This sealer is typically a zirconium oxide based material which deposits from an aqueous solution as a thin ($\sim$50 nm) coating on top of the zinc phosphate.

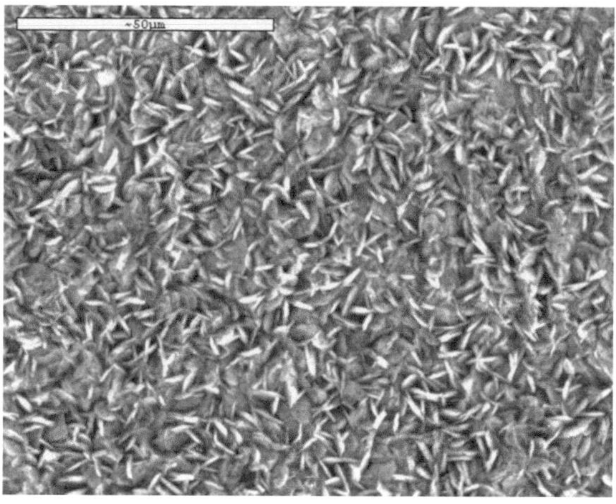

Fig. 13.3 Zinc phosphate on galvanized steel

13.2.2 Electrocoat

The electrocoat (e-coat) layer is applied on top of the pretreatment layer. The e-coat's primary function is to provide outstanding corrosion protection to the underlying metallic substrate. It does this by forming a tight adhesive bond with the pretreatment and by suppressing the transport of ionic species to the substrate [2].

The electrocoat is applied via a cathodic deposition process in a large bath. The vehicle acts as the cathode and anodes are positioned along the sides and bottom of the electrocoat tank. The process is run between 200 and 275 V with a vehicle residence time of approximately 3 min in the bath. The electrocoat solution is a water based material that contains paint resin and pigments. The paint particles are positively charged and migrate to the cathode. At the cathode (the vehicle) water in the solution is reduced, resulting in hydrogen evolution and hydroxyl formation [3].

$$2H_2O + 2e^- \rightarrow H_2 + 2OH^-$$

The hydroxyl ions neutralize the charged paint particles and the particles precipitate onto the surface of the vehicle. As the paint particles build up on the cathode surface, the resistance of the paint film increases and deposition on the surface slows. Nearby unpainted surfaces will then be coated since the surface resistivity on these surfaces is lower than on the painted surface. This process continues while the vehicle remains in the electrocoat bath, with exterior locations coated first, followed by interior locations, and finally recessed cavities of the vehicle body. The efficiency of the process of coating all surfaces of the vehicle body is termed the throw power of the electrocoat material [3].

A challenge in the electrocoating process is to adequately coat all exterior and interior surfaces of the vehicle body. Electrodeposition is dependent on both paint material contact with the surface and the availability of electric current. The exterior surface of a vehicle body sees both the material and current, but there are many locations on a vehicle body that are not in the sight line of the anodes. Deposition of electrocoat in these recessed locations is dependent on access of the electric field. An example of the behavior of the electric field in recessed locations is shown in Fig. 13.4.

The thickness of the deposited film varies as a function of the applied voltage, the bath temperature, and the resin concentration in solution. Evidence indicates that the coagulated material deposits in such a way that creates pores within the deposited material. By varying the electrocoat bath parameters, the size of the pores can be reduced to create a more compact coating that is more resistant to the diffusion of water, salts, and other contaminants to the metal or pretreated surface. If these contaminants reach the metal-coating interface, they can adversely affect the adhesion and corrosion performance of the paint system [3].

There are multiple rinse steps following the electrocoat deposition process, primarily to remove excess coagulated resin material that has not fully deposited onto the surface as a film. The intent is for all electrocoated surfaces to be uniformly

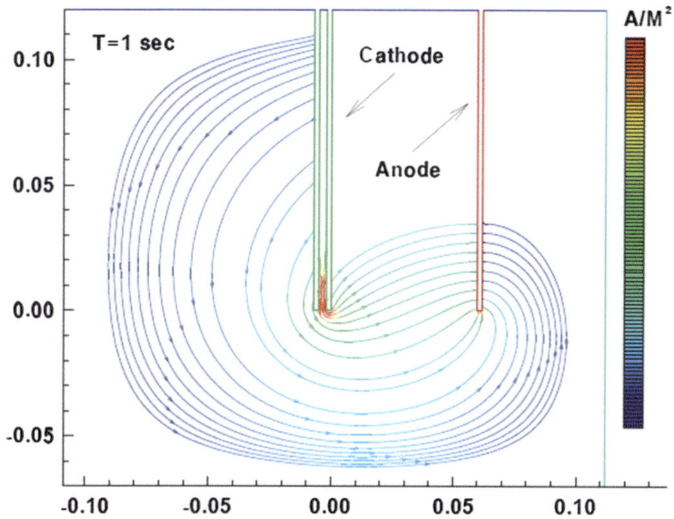

Fig. 13.4 Electric current trajectories on the exterior surface compared to the current within a thin gap immersed in the electrocoat bath

smooth, regardless of differences in the material of construction on exterior parts of the vehicle.

After the rinse steps the vehicle body enters an oven to crosslink the coating. Modern cathodic electrocoat is formed from an epoxy polymer crosslinked by blocked isocyanates. Because of the aromatic nature of the backbone and cross-links, the electrocoat is highly photolabile and must be protected from UV and short wavelength visible light by the paint layers on top of it. Failure to do so can result in catastrophic delamination of the paint system. The electrocoat is generally lightly pigmented, but contains little in the way of corrosion inhibiting pigments. The process is highly efficient with less than 5 % of the electrocoat wasted during the process.

13.2.3 Primer

After the electrocoat is cured the primer is electrostatically applied in a spray booth. During this process the paint is atomized and applied to a thickness of approximately 25 μm onto the vehicle surface using rotary electrostatic atomizers or electrostatic spray guns. While the pretreatment and electrocoat layers are applied to the entire vehicle body, primer, basecoat, and clearcoat are applied only to exterior surfaces and select high visibility interior locations such as under hood and door openings. All substrates that are used on a vehicle body, both metallic and non-metallic, are coated with primer.

The voltage used to apply the primer material at the applicator can vary depending on the material and the application conditions, but generally is 60–80 kV. To keep the atomized paint directed at the target (the vehicle body), shaping air is used. A goal in electrostatic painting is to direct the majority of the paint onto the target with minimal losses to the environment. The transfer efficiency is a measure of the amount of paint that reaches the target compared to the amount of paint that was sprayed. A good transfer efficiency is greater than 80 % but most modern automotive paint applicators provide transfer efficiencies in excess of 85 %.

Although new application processes are being utilized that reduce energy usage in the paint application facilities, the standard practice to date has been to cure the primer layer in an oven after application. As with the electrocoat oven, the primer oven crosslinks the deposited material and provides a smooth and uniform surface for further paint application.

The primer layer serves many functions in an automotive paint system. As mentioned earlier, it must protect the underlying electrocoat from damaging light, thus it is typically highly pigmented and hides very well. The primer also plays a primary role in smoothing the surface of the e-coat to ensure a smooth topcoat appearance. Both the e-coat and substrate metal can display objectionable surface roughness. This roughness must be mitigated to ensure a superior paint system appearance. By bridging and filling the valleys between surface features, the primer reduces the surface texture. In addition, the primer also plays a substantial role in the chip resistance of a paint system. The high speed impact of rocks and stones produced by other vehicles can be extremely damaging to a paint system. A significant amount of energy, concentrated in a small area, must be dissipated by the paint system during a very short impact event. The primer serves to absorb this energy and prevent delamination and cracking of the paint system due to stone impact [4, 5]. Modern automotive primers are typically formulated from acrylic or polyester polymers crosslinked with alkoxylated melamine or from polyurethanes and can be either solventborne, waterborne or powder based. Often primers are "color-keyed" with the subsequently applied basecoats. This means that the color of primer sprayed on a vehicle is matched to the color of basecoat that will subsequently be applied. Thus, the amount of basecoat required to achieve hiding is minimized and the visual impact of stones chips is reduced.

13.2.4 Basecoat and Clearcoat

After primer application and cure, the topcoat, which includes the basecoat and clearcoat layers, is applied. As with the primer, the paint is atomized using rotary electrostatic atomizers or electrostatic spray guns. Basecoat and clearcoat are applied sequentially without a baking step in between the application steps. The time between the basecoat and clearcoat application allows the solvents used in the paint to evaporate. This prevents the clearcoat from striking into the basecoat, which can cause appearance issues by altering the flake orientation in the basecoat.

Changes in flake orientation cause light to be reflected in a non-uniform manner, resulting in an unsatisfactory appearance for the consumer.

The basecoat is the color coat and is applied to a thickness that provides uniform color across the surface while hiding the underlying primer—usually to a thickness of 10–20 μm, depending on the color. Clearcoat is used to protect the color of the vehicle and to provide a glossy surface. The clearcoat is usually applied to a thickness of 40–50 μm. As with the primer, the basecoat and clearcoat are sprayed at voltages of 60–80 kV, with shaping air used to direct the atomized particles to the target. After the basecoat and clearcoat application process, the vehicle body is again baked to crosslink the coating system.

The basecoat is the most expensive layer in the automotive paint system and provides the vibrant array of colors that are available to vehicle buyers. The brilliant colors are the result of a mixture of many high performance pigments and effect flakes encapsulated in the basecoat binder. Basecoats binders are formulated from either acrylic or polyester polymers crosslinked with melamine. Environmental regulations have forced the solids levels in basecoats to continually rise. This had led to both high solids solvent borne basecoats and waterborne basecoats displacing low and medium solids solvent borne basecoats in much of the world. It should be noted that waterborne basecoats still contain appreciable levels of volatile organic compounds (VOCs).

The metallic or sparkle effects seen in many basecoats are the result of light reflecting off of the small flakes of aluminum or mica that are dispersed in the basecoat. The alignment of these flakes parallel to the surface of the coating is critical to achieving excellent appearance and high "travel" (the difference in lightness of the coating when viewed from different angles) of the coating (Fig. 13.5). Alignment of the flakes is achieved by various mechanisms including deformation of the paint droplets on impact with the vehicle, rheology control additives in the paint, and shrinkage of the paint film as VOCs evaporate from the

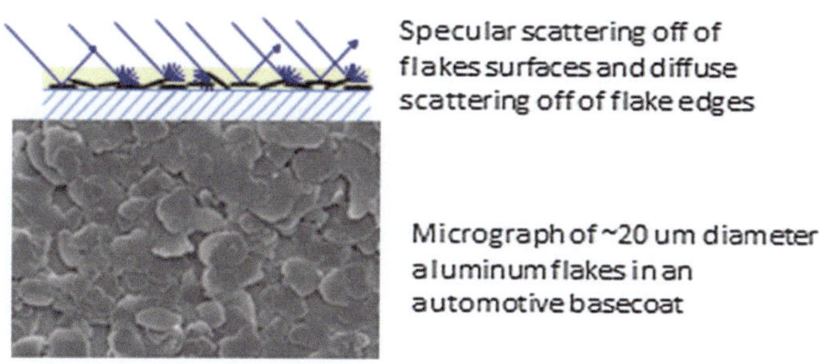

Fig. 13.5 *Upper*—Light rays incident upon a silver basecoat surface reflect off of flakes in both a specular and diffuse manner. Specular reflections give rise to metallic sparkle and increase the head-on brightness of the color. *Lower*—Top down SEM view of a silver metallic automotive basecoat. Note that flakes are nominally aligned with the surface of the coating

wet film [6]. The strive for higher solids has challenged paint formulators to maintain good metallic appearance. In addition to the effect flakes, the pigments in modern basecoats are typically high performance transparent organic pigments (perylenes, quinacridones, ...) and inorganic pigments (TiO_2, carbon black, iron oxides...). These pigments must be highly UV and thermally stable to ensure minimal color fade during the lifetime of a vehicle.

The clearcoat, the uppermost layer in the automotive paint system, functions to maintain the high gloss appearance of the vehicle for many years. It must be able to withstand the severe environmental pressures that a vehicle is exposed to such as UV radiation, temperature, environmental fallout (acid rain and industrial pollutants), mechanical deformation and water. Only the most durable polyurethanes or acrylics functionalized with hydroxyl, carbamate, or epoxy groups are suitable for use. These resins are crosslinked with melamine or aliphatic isocyanates. To enhance their durability and meet 10+ year life requirements, all clearcoats are formulated with ultraviolet light absorbers (UVAs) and hindered amine light stabilizers (HALS). The stabilizer additives inhibit and reduce the rate of photooxidation induced by UV radiation. The UVA also acts to screen the underlying basecoat layer from UV light, thereby enhancing the basecoat's longevity [7].

In addition to UV durability, the clearcoat must protect against mechanical stresses from stone impact and scratch events. Keys, tree branches, fingernails and other asperities are often abraded across the surface of the clearcoat during a vehicle's life. Its ability to resist plastic deformation and fracture is key to minimizing the visual appearance of these scratches [8].

13.2.5 Compact Paint Processes

To reduce the footprint and operational costs of vehicle painting, automotive OEMs have begun to implement "compact painting" processes. In general, compact painting processes eliminate the stand alone primer booth and oven. Instead, the primer is applied as the first step in the basecoat/clearcoat booth. Thus, compact painting necessitates the application of three wet layers on top of each other and the subsequent co-cure of those layers. The elimination of the primer booth and oven reduces the energy consumption by $\sim 15\ \%$, the effective CO_2 emissions by $\sim 15\ \%$, and the VOCs released by $\sim 10\ \%$ [9]. To-date, compact paint processes have been deployed in medium solids solventborne, high solids solventborne, and waterborne coatings. These processes go by various names and are sometimes referred to as "primerless" when the primer layer is replaced by a layer that more closely resembles the basecoat layer. In all compact processes, film build control is critical to maintain opacity at low film builds (to protect the electrocoat) and on the high end to avoid pops and sags.

13.3 Future Technologies Needs

As one looks to the future of automotive painting, three underlying themes will drive most of the innovation: reduced costs, lower environmental footprint, and increased functionality. The first theme, reduced costs, will likely result in the continued roll-out of compact painting processes in all assembly plants where it is feasible. The economic and environmental benefits are too compelling to reverse this trend. There is relatively little that can be done to lower the cost of raw materials and the coating layers themselves. Because of stringent performance criteria, the pigments and binders are relatively costly compared to those used in coating systems for other manufactured goods. Instead, much of the cost opportunities lie in the improved efficiency of the various painting operations.

A reduction in the environmental footprint of automotive painting means that the use of energy and the output of waste must decrease. The main waste streams from automotive painting are wastewater; which comes from the pretreatment operations and from capture of the primer and topcoat overspray; VOCs, which are released during application of both solventborne and waterborne coatings; paint sludge, which is produced during both pretreating and paint spraying; and CO_2, which is produced by the burning of fossil fuels for ovens, the generation of electricity to circulate enormous volumes of air in the spray booths, and the conditioning (cooling, heating, and (de)humidification) of air in the spray booths. Several technologies are now beginning to be implemented that provide significant advantages in these areas.

As zinc phosphate produces significant amounts of metal-containing sludge, particularly when significant aluminum is incorporated into the vehicle body, thin film pretreatments are poised to supplant zinc phosphate in the near future. These thin film pretreatments are based on zirconium oxide chemistry that etch the surface of the vehicle body much less than traditional pretreatments. Reduced etch results in up to a 90 % reduction in solid waste produced. In addition, these pretreatments baths operate at lower temperatures and with fewer process steps. The technical challenges for such pretreatments, however, are not insignificant. In particular, achieving acceptable corrosion performance on mixed metal (steel, galvanized steel, and aluminum) vehicle bodies is difficult due to the need to balance deposition rates on each substrate.

Due to significantly higher fuel economy targets, vehicle weights will decrease, perhaps substantially, in the near future. The incorporation of lighter substrates may drive new coating strategies. For example, magnesium is used sparingly on current vehicles due to its strong tendency to corrode, particularly at joints with dissimilar metals. Most magnesium parts are pretreated and coated, typically powder coated, before careful assembly and isolation onto a vehicle body. The ability to pretreat and coat magnesium parts in parallel with the rest of the vehicle would be an enormous benefit. However, no pretreatment system currently exists that would enable all relevant substrates to be pretreated together.

Similarly, the extensive use of carbon fiber composites could lower the weight of vehicles and improve fuel economy. The current bake temperatures in the electrocoat oven (~180 °C for 20 min) prevent the use of most low cost polymers and composites in vehicle bodies. Thus, only a few thermosetting resins and high temperature thermoplastics can be contemplated for use at either structural or closure panels on vehicles. To enable the use of lower temperature resins in carbon fiber composites components a dramatic shift in the cure chemistry and secondary operations of the automotive paint shop may be necessary. It should be noted that the electrocoat oven, in addition to curing the paint, also cures adhesives and sealers, and serves to bake harden both steel and aluminum components. Thus, a drive to lower the temperature of the e-coat oven has significant secondary and tertiary effects on other parts of the automobile assembly process.

While lower temperature curing is challenging for the electrocoat layer, low temperature curing of the topcoat is attractive. A reduction of the curing temperature would drive significant CO_2 and natural gas reductions. If thermal curing could be eliminated, via ambient cured chemistry or radiation curing, additional significant savings could be obtained due to footprint reduction. Quality would also likely improve, as circulated air in baking ovens is a source of dirt defects. While significant work has been done on UV curing of automotive coatings, line of sight or shadowing issues continue to be problematic.

Higher solids coatings, particularly for basecoats and clearcoats would be beneficial for multiple reasons. Currently, VOCs are abated from both spray booths and ovens to a greater or lesser extent depending on the local regulations and the solids level of the coatings being applied. Higher solids coatings would reduce abatement requirements (and the CO_2 output generated during thermal oxidation of the VOCs) and reduce the costs of painting. Practical limits on basecoat solids are driven by acceptable appearance. Some moderate amount of solvent will always be required to achieve flake orientation in metallic colors, as solvent loss is one of the main mechanisms of flake alignment (Fig. 13.6). However, no such constraint is placed on clearcoats, and 100 % solids clearcoats are technically possible, particularly with 2K formulations. Further advances in oligomer chemistry and viscosity reductions, coupled with lights-out paint shops to reduce worker exposure to monomers, make this goal achievable.

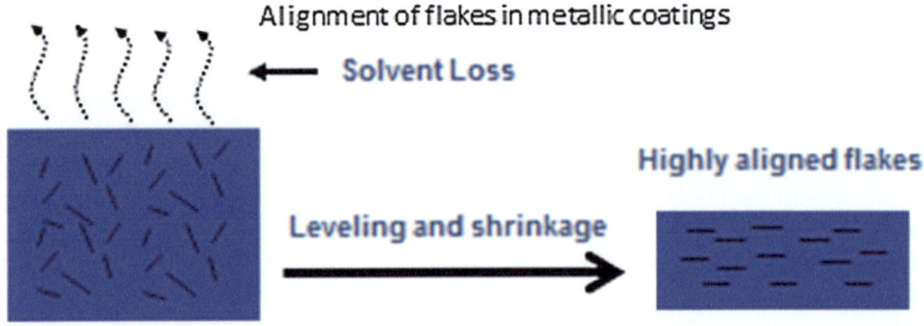

Fig. 13.6 Flakes in metallic basecoats align preferentially with the surface of the coating due to shrinkage in the thickness direction as solvent evaporates from the coating surface. Improved flake alignment leads to improved metallic appearance

Currently, appearance and corrosion protection are the two main functions of an automotive paint system, coating systems in the future may be required to perform additional functions or perform traditional functions in fundamentally different ways. For example, corrosion protection is currently achieved through a combination of sacrificial coatings (zinc galvanized), barrier to ions and water (e-coat) and strong adhesion (zinc phosphate and e-coat). Future corrosion protection may occur through self-healing of coatings. Micro or nano-spheres that release corrosion inhibitors or crack healing agents after pH changes or mechanical stress are active areas of research [10, 11]. Currently, little work has been done to make these coatings electrodepositable or to make the spheres small enough for incorporation into thin paint layers. These deficits represent rich and meaningful research opportunities.

As vehicle electrification grows, some have proposed that the surface of the vehicle would be a valuable solar energy collector, and photovoltaic paints have been developed [12]. However, the practicality of these proposals is doubtful. Unlike traditional solar cells, automotive paint is subjected to significant damage events in terms of scratches and stone impacts. The impact on cell output (short circuits, reduced performance…) has not been addressed. In addition, the available surface area of a vehicle is not sufficient to contribute a significant amount to the vehicle's energy budget. [For example, a typical electric vehicle might have 4 m^2 of surface area that could be used for solar cells with perhaps 50 % ever facing the sun. A 20 % efficient solar cell operating during 8 h of peak sunshine in a typical environment would deliver less than 1.3 kW h of energy to the vehicle ($400\ W/m^2 \times 4\ m^2 \times 0.5 \times 0.2 \times 8$ h). A typical EV or PHEV battery is on the order of 16–20 kW h.] The impact of these coatings on appearance would also likely not be met with customer approval.

Other functional enhancements to automotive paint have also been envisaged. Pigments that respond to external stimuli (pH, temperature, UV light) or that are luminescent have been utilized in less demanding applications such as nail polish or toys. However, their use in automotive coatings is questionable due to durability, and color harmony issues. Also, consumers are less likely to indulge in trendy or whimsical colors on expensive items that last 10+ years. Luminescent pigments are problematic as lighting on the exterior of vehicles is highly regulated by most federal governments for safety reasons.

13.4 Summary

Significant unmet needs will continue to guide innovations in automotive coatings technology. Reductions in cost and environmental footprint, in addition to improved functionality, will drive many of these future changes. The high volume and stringent functional requirements will ensure that only the most robust and practical materials and process solutions will be implemented.

References

1. D.W. Boyd, R.R. Zwack, Prog. Org. Coat. **27** (1996)
2. D.W. Boyd, R.R. Zwack, Predicting electrocoat efficiency: a simple data-based model for predicting electrocoat usage. Prog. Org. Coat. **27**(1–4), 25–32 (1996). doi:10.1016/0300-9440(95)00516-1
3. V. Miskovic-Stankovic, The mechanism of cathodic electrodeposition of epoxy coatings and the corrosion behaviour of the electrodeposited coatings. J. Serb. Chem. Soc. **27**(5), 305–324 (1996)
4. P. Pierce, J. Coat. Technol. **53** (1981)
5. H. Hintze-Bruening, A.-L. Troutier-Thuilliez, F. Leroux, Layered particle-based polymer composites for coatings: Part II—Stone chip resistant automotive coatings. Prog. Org. Coat. **64** (2–3), 193–204 (2009). doi:10.1016/j.porgcoat.2008.09.020
6. D. Bhattacharya, K. Seo, L.T. Germinario, M.D. Clark, K.W. McCreight, C. Williams, Novel techniques to investigate the impact of cellulose esters on the rheological properties and appearance in automotive basecoat systems. J. Coat. Technol. Res. **4**(2), 139–150 (2007)
7. J.L. Gerlock, A.V. Kucherov, M.E. Nichols, On the combined use of UVA, HALS, photooxidation, and fracture energy measurements to anticipate the long-term weathering performance of clearcoat/basecoat automotive paint systems. J. Coat. Technol. **73**(918), 45–54 (2001)
8. C. Seubert, M. Nichols, Scaling behavior in the scratching of automotive clearcoats. J. Coat. Technol. Res. **4**(1), 21–30 (2007). doi:10.1007/s11998-007-9006-3
9. M. Vandelinder, D. Havlin, M.E. Rosenberger, T. Weingartz, in *Proceedings of SURCAR*, Cannes (2007)
10. S.H. Cho, S.R. White, P.V. Braun, Self-healing polymer coatings. Adv. Mater. **21**(6), 645+ (2009). doi:10.1002/adma.200802008
11. B. Ghosh, M.W. Urban, Self-repairing oxetane-substituted chitosan polyurethane networks. Science **323**(5920), 1458–1460 (2009). doi:10.1126/science.1167391
12. M.P. Genovese, I.V. Lightcap, P.V. Kamat, Sun-believable solar paint. A transformative one-step approach for designing nanocrystalline solar cells. ACS Nano **6**(1), 865–872 (2012)

Chapter 14
Polymer Coatings for Oilfield Pipelines

Russell J. Varley and K.H. Leong

14.1 Introduction

Pipes and pipeline systems are an integral part of modern society, from the network of pipes that transport water, gas and sewage of our daily needs to the continuous transmission of hydrocarbons across continents satisfying the energy demands of the global economy. Despite the enormous distances and the harsh environments involved, oil and gas pipelines are the most efficient and environmentally safe form of hydrocarbon transportation currently available today. For this reason, they are critical to continued global prosperity in the 21st century.

The oil and gas (O&G) industry, primarily uses steel pipelines to transport natural gas, crude oil, water, as well as petrochemical feedstocks and petroleum products which are susceptible to corrosion from their surrounding environment, be it soil (buried lines), or seawater (subsea lines) and compromising their service life. The life expectancy of oilfield pipelines ranges typically between 25–35 years so reliable protection during this period is of utmost importance. External coatings are one of the most widely used methods for providing a barrier to the steel surface against air (oxygen) and water (electrolytes), the two critical factors which most impact the electrochemical process of corrosion. Despite this, all coatings deteriorate over time due to inter alia soil stress, moisture and gas ingress, reduced adhesion, chemical attack and mechanical damage, thus imposing the challenge of maintaining the required durability and performance over the intended service life.

R.J. Varley (✉)
CSIRO Manufacturing, Private Bag 10, Clayton South MDC,
VICTORIA 3169, Australia
e-mail: russell.varley@csiro.au

K.H. Leong
PETRONAS Research, Lot 3288/3289 off Jalan Ayer Itam,
Kawasan Institusi Bangi, 43000 Kajang, Selangor DE, Malaysia
e-mail: kh_leong@petronas.com.my

A.E. Hughes et al. (eds.), *Active Protective Coatings*,
Springer Series in Materials Science 233, DOI 10.1007/978-94-017-7540-3_14

Fig. 14.1 3 layered polyethylene (3LPE) coated pipes; note the ends of the pipes are exposed to facilitate jointing by welding

Indeed, the universal strategy accepts that a coating will inevitably deteriorate during its service life, so an additional solution is always required in readiness for when this happens. Cathodic protection (CP) is almost always employed as a back-up in tandem with a physical coating. Together, a coating protects the bulk of the metal pipeline while CP addresses any potential threat of corrosion where the coating has been breached.

This chapter concentrates on the application of multi-layered polyolefin based pipe coatings, as illustrated in Fig. 14.1, the most widely used external coating systems in the O&G industry today. Other coating systems are covered, for completeness, but only relatively briefly. The history of their development and use, the major variants available, the underlying materials chemistry and processing, industry standards, as well as some of the commonest failure mechanisms encountered are reviewed and examined. Finally, some of the newer developments in high performance coatings are also discussed as new oilfields continue to place greater demands upon pipes and consequently the coatings required to protect them.

14.2 A Brief History on the Evolution of Pipe Coatings

Pipes and pipelines have been made from virtually any material to transport fluids and date back to the ancient Egyptians and Chinese where they were used to transport water for drinking and irrigation [1]. With the discovery of oil in the mid-nineteenth century, pipes and pipelines experienced a new importance that grew with global energy consumption and relied upon them for the efficient, low cost and safe transportation of hydrocarbons. At first they were constructed using wrought iron and rivets, but as welding techniques developed this was replaced in the 1930s with more durable steel [2]. Since then, the strength, temperature resistance and durability of steel has made it the logical choice for oilfield pipelines. So began the need for advanced pipe coatings that have become integral to ensuring that steel pipelines function as they were intended over their design life, with no or minimum disruption.

The first pipe coatings to emerge during the 1860s were made from *coal tar/asphalt enamel* and typically were applied manually in the field and thus highly variable in quality and performance [3, 4]. It was soon recognised that their resistance to corrosion left much to be desired, and as well as being highly susceptible to oxidation, cracking and soil stress, were limited to low stress and temperature environments. The need to reduce these deficiencies led to a gradual improvement throughout the first part of the 20th century as better materials and more controlled application methods became available. Asbestos, for example, was used to reinforce coal tar coatings while waxy coatings were also introduced around the same time [5].

In fact, coal tar coatings remain in use today due to its ease of application and minimal surface preparation. As a by-product from refineries, it is often also the cheapest coating option for many pipeline projects further enhancing its usefulness, but its inherent disadvantages, including health-related concerns [6], limited wider usage, particularly as better two layer extruded polyethylene tape coating systems emerged in the 1940s and followed by fusion bonded epoxy systems. Application methods continued to improve and by the 1950s it was recognised that the combination of cathodic protection and pipe coatings was the best long-term defence against corrosion.

Modern oilfield pipeline coatings however, are not just required to provide external protection against corrosion but, where necessary, also protect against structural damage, such as impact during transportation and installation; maintain on-bottom stability; and/or provide insulation. Driven by the need to explore and operate in harsher and more remote locations, pipe coating suppliers face the constant challenge to meet the increasingly demanding requirements of the industry [7]. Indeed, continuous development of pipe coatings over the years has improved their properties, and cost competitiveness, such that coatings encountered in the field today range from being single layered asphaltic or thermosetting polymer systems to multilayered polyolefin-based coatings, depending on budget and service condition requirements [8, 9].

Despite the wide variety of environments, materials and applications available, the majority of oilfield pipe coatings can be generally categorised into a small number of different types of coatings as shown in Table 14.1. Each system is specified by pipeline owners/operators and consultants for particular pipeline projects and based on factors, including short-term cost, long-term investments, captive usage, regional availability of the coating material, control on handling, transportation and installation of pipelines, as well as technical requirements. Matching intended service conditions with the right materials is necessary to ensure coating longevity [10]. Figure 14.2 shows a historical timeline illustrating the development of external and field joint coatings from the 1940s to the present day.

Table 14.1 Typical coatings used for oilfield pipelines, and corresponding maximum service temperature and typical thickness

Coating		Max. svc. temp. (°C)	Typical thickness (mm)
Epoxy	Standard FBE[a]	65 [11]	0.35–0.50 [12]
	Dual layer FBE	130 [11]	0.65–1.30 [12]
	Liquid, 100 % solid	150 [11]	
PE[b]	2LPE	90 [11]	60 [13]
	3LPE	85 [14, 15]	85–90 [13]
PP[c]	3LPP	130 [11]	110–140 [13]
	5LPP	155 [16]	~55 [16]
	7LPP	155 [16]	130 [16]
Concrete		N/A	Max. 25 [17]
Asphaltic mastic		95 [18]	3–5 [18]
PU[d]		85 [19]	40–50 [20]
Internal coatings (Epoxy)		120 [21]	Ave. 0.22 [21]

[a]*FBE* Fusion bonded epoxy
[b]*PE* Polyethylene
[c]*PP* Polypropylene
[d]PU Polyurethane

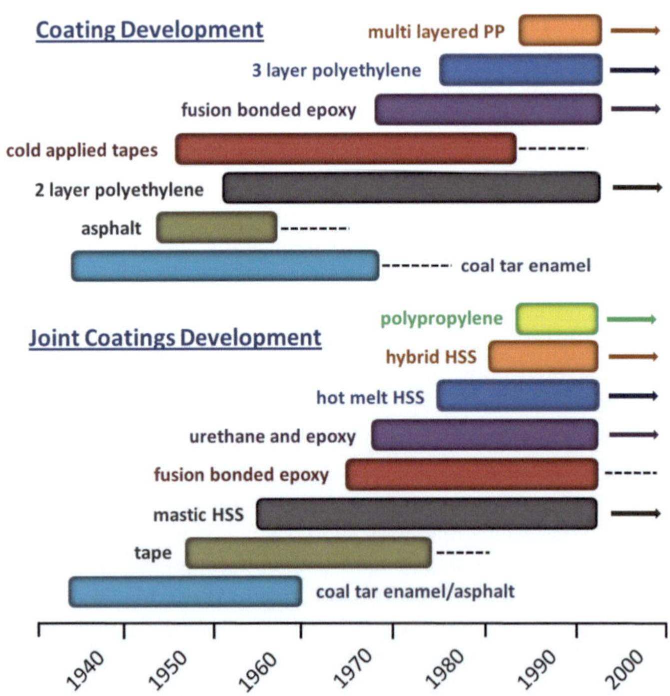

Fig. 14.2 Historical timeline illustrating the development of pipe coatings, and field pipe joint coatings, from the 1940s to the present day [4] (*PP* polypropylene; *HSS* heat shrink sleeve)

Fusion bonded epoxy (FBE) systems are one-part thermosetting formulations generally based upon well known epoxy resin chemistries. As such, they adhere well to a metal substrate and after cure provide a tough, chemically and abrasion resistant single layered coating that is resistant to cathodic disbondment [22, 23] and soil stress [5, 24]. Since the 1960s formulations have been continuously developed further to provide better control of the polymer network and improved properties and performance [23, 25]. These single-layer coating systems use a highly controlled application and installation process and are commonly used when environmental demands and/or operating temperatures are less severe (typically below 65 °C). They are however, known to be susceptible to blistering and external damage such as impact, but when adequate impressed current is used, protection to the pipeline can be very effectively maintained despite the presence of blisters [25].

FBE systems are also applied as the primer layer for many two, three and multi-layered coatings systems [23]. More recently, two-layer FBE systems consisting of an FBE coating applied over an existing FBE layer still in the melt, has been shown to facilitate intimate chemical bonding between the two layers [26]. The benefits of this two-layer FBE approach is an improvement in abrasion resistance and mechanical properties due to plasticization of the FBE topcoat [12], while having inherently favorable compatibility with impressed current and thus improved corrosion resistance.

The first *tape coatings* and *extruded two layer systems* appeared after World War II represented the birth of multilayer coatings as they took advantage of the then newly commercialised inert, non-polar, chemically resistance, flexible and highly impact resistant polyethylene [27]. Both methods wrapped polyethylene around the pipe using a suitable adhesive either via a manual infield method (in the case of tape coating), or a new factory based extrusion process. Tape coatings tended to suffer from poor stress resistance and shielding of cathodic protection, while the adhesives used with the tapes, particularly butyl rubber, tended to biodegrade [28, 29]. Two layer extruded coatings systems continued to develop and remains an important product in the market for applications where temperature requirements are modest.

As temperature requirements increased and greater resistance against impact damage during transportation and installation, in-ground shear forces, chemicals and abrasive soils was required, *three layered polyolefin (3LPO) systems* have become the preferred choices. 3LPO systems consist of an FBE primer, a polyolefin adhesive and a polyolefin topcoat, as shown schematically in Fig. 14.3a. Polyethylene is used in situations where service temperature requirements reach up to 60–80 °C [30], while polypropylene is preferred when service temperatures up to 110 °C are required [30, 31] for more demanding applications. Polypropylene has very low moisture absorption (0.02–0.03 %), excellent processability, low health and environmental risks, excellent mechanical properties (including tensile strength, elongation-to-failure and flexural modulus), flexibility in design for a wide range of requirements and a long track record of use and available service information [32].

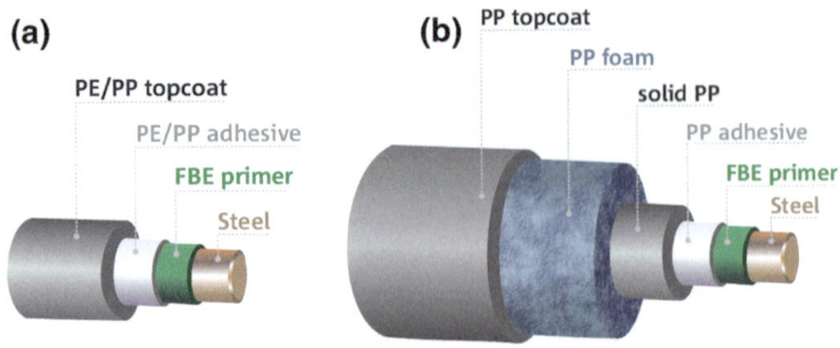

Fig. 14.3 Schematics of multilayer polyolefin pipe coating systems. **a** 3 layer systems: FBE primer (80–100 μm), polyolefin adhesive (350 μm) and polyolefin topcoat (1–3 mm). **b** 5 layer systems: FBE primer, polypropylene adhesive, solid polypropylene layer, polypropylene foam and polypropylene topcoat

Multi-layered coating systems are typically made up of 5–7 layers with the additional layers providing insulation and added protection from damage [33, 34]. Foam insulation layers are often made from polyurethane or polypropylene, though epoxy-based systems are also available albeit less popular [35]. They are primarily designed to control excessive cooling of the crude oil to prevent wax formation and crystallisation of clathrate hydrate crystals which can otherwise adversely affect the flow assurance of the pipeline. These coating systems are normally associated with deeper water fields which also increases the required service temperatures and compressive properties. Figure 14.3b shows an idealised 5 layer pipe coating system, differentiated from the 3 layered coating systems by the addition of a polypropylene foam layer and an additional solid polypropylene layer providing further thermal barrier protection. Table 14.2 lists several subsea insulation coatings systems used in the industry. It is noteworthy that other than wet insulation, pipe-in-pipe, which tends to be a heavier option, and the flexible insulated pipe, often more expensive but cost-effective in niche applications such as flowlines and risers on Floating Production, Storage and Offloading vessels (FPSOs), are also available to afford heat insulation to pipelines [35].

Concrete coatings, typically applied over the aforementioned coating systems, are an important part of the pipe coating family. They are targeted primarily for buoyancy control of subsea pipelines, and/or for rugged outdoor applications subject to potentially heavy impact during and/or after installation. Guan [36] listed four different methods of applying concrete coating to a steel pipe that had been pre-coated with an anti-corrosion coating. The processes are impingement, compressive wrapping, casting and gunniting. A typical concrete coated steel pipe is shown in Fig. 14.4 illustrating how it is used in addition to a 3LPE system.

The types of *internal pipe coating* used are dependent upon the media transported by the pipeline. Generally, they are used to reduce pressure drop due to surface roughness by inhibiting corrosion, preventing precipitation on the pipe wall, and modifying the surface energies of the pipe wall [21]. Internal FBE is sometimes used in combination with a phenolic primer that is applied prior to powder

Table 14.2 Selected subsea insulation coatings systems in use today [20]

Material		Insulation type	K-value (W/m K)	U-value (W/m² K)†	Service temp. capability (°C)	Max. water depth (m)
Type	Insulation					
Thermoplastic	3LPP[b]	Solid	0.22	10.0	140	2000[a]
	4LPP[b]	Foam	0.17	3.0–4.0	115	600
	nLPP[b]	Foam	0.17	3.0–4.0	140	600
	Syntactic PP[b]	Glass syntactic	0.19	3.0	140	2000[a]
Thermosetting elastomer	Solid PU[c]	Solid	0.19	10.0	140	2000[a]
	PU[c] foam	Foam	0.03–0.04	0.7	140	2000[a]
	Syntactic PU[c]	Polymer syntactic	0,12	2.0	115–125	300–600
	Glass syntactic PU[c]	Glass syntactic	0.16	2.0	90–100	2000[a]
Synthetic rubbers/elastomers	PCP[d]	Solid	0.35	30.0	95	2000[a]
	EPDM[e]	Solid	0.35	30.0	150	2000[a]

[a]Also known as the overall heat transfer coefficient (OHTC)
[b]*PP* Polypropylene
[c]*PU* Polyurethane
[d]*PCP* Polychloroprene (Neoprene)
[e]*EPDM* Ethyl propyl diene monomer

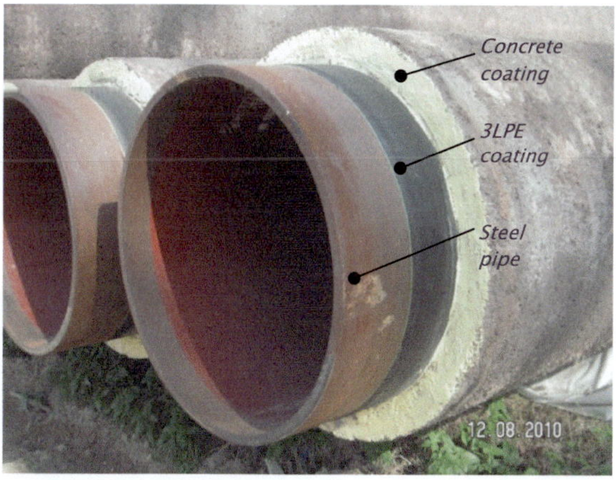

Fig. 14.4 Concrete coating applied over a 3LPE system on a steel pipe

application to enhance the coating performance. These coated pipes are also used for water and corrosive media transportation, while internal cement mortar lined pipes are used predominantly for waste and potable water [37].

14.3 Materials and Processing

This section discusses some of the chemistry and process engineering that underpins the effectiveness of the most commonly used coating systems: FBE; three layered polyolefin; and multilayered thermally insulated coatings.

14.3.1 *Fusion Bonded Epoxy Coatings*

Fusion bonded epoxy coatings are multi-component formulations containing epoxy resins, curing agents, corrosion inhibitors and fillers to control viscosity, color or toughness. Their primary role in either single or multi-layered coatings is to prevent corrosion, but when applied as a single layer are also required to protect the pipe from damage.

Corrosion is an electrochemical reaction, shown schematically in Fig. 14.5, consisting of oxidation at a metal surface in the presence of water and air to form iron oxides. Thus in the context of a coating, a strong and durable polymer and polymer/steel interface is required to separate air and moisture from the metal substrate and inhibit the electrochemical process of corrosion.

Clearly, this can only be achieved through the optimisation of the FBE resin chemistry, cure process and suitable coating deposition methodologies. To achieve this, the pipe is pre-heated prior to FBE application, using an oven or induction methods, so that when it is applied to the pipe, the resin viscosity rapidly decreases

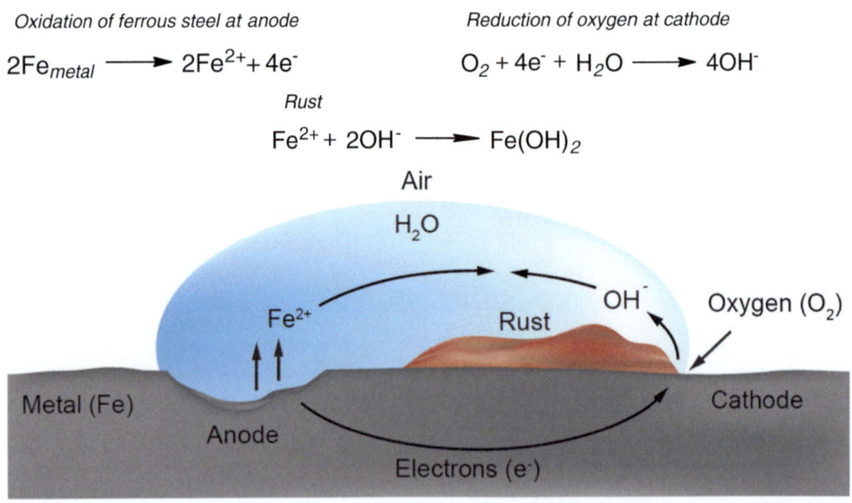

Fig. 14.5 Schematic representation and the chemical reactions of the electrochemical processes occurring during corrosion illustrating the impact of electron movement on the formation of $Fe(OH)_2$

as it melts, promoting resin flow and even wetting over the pipe. It is noteworthy that pipes are heated to a temperature of approximately 180 °C when the FBE is used as primer, or to as high as 250 °C when used as a top coat [36]. At this point, the relatively polar nature of the epoxy monomers in the FBE facilitates good interaction and dissolution with the inorganic metal oxide on the pipe surface also promoting adhesion. The hardener within the FBE, typically dicyandiamide, well known for its latency, initiates resin cure which increases viscosity until the epoxy network forms an intractable gel. Continued cure consumes reactive groups, further increasing crosslink density of the network, to produce a tough, durable and abrasion resistant coating. Final properties therefore are a function of many parameters such as the extent of cure, molecular structure, stoichiometry, cure temperature and time, all of which need to be tuned for optimum performance.

FBE coatings are often made from high molecular weight bisphenol A based epoxy resins partly because of their low cost and availability, but also due to their comparatively high toughness in epoxy networks. This is particularly advantageous for materials that need to accommodate significant flex and to resist impact damage during their lifetime such as pipe coatings. A schematic representation of the physical and chemical transformations occurring during FBE application and cure is shown in Fig. 14.6, along with the chemical structures of the epoxy resin and hardener, dicyandiamide, typically used in advanced FBE formulations.

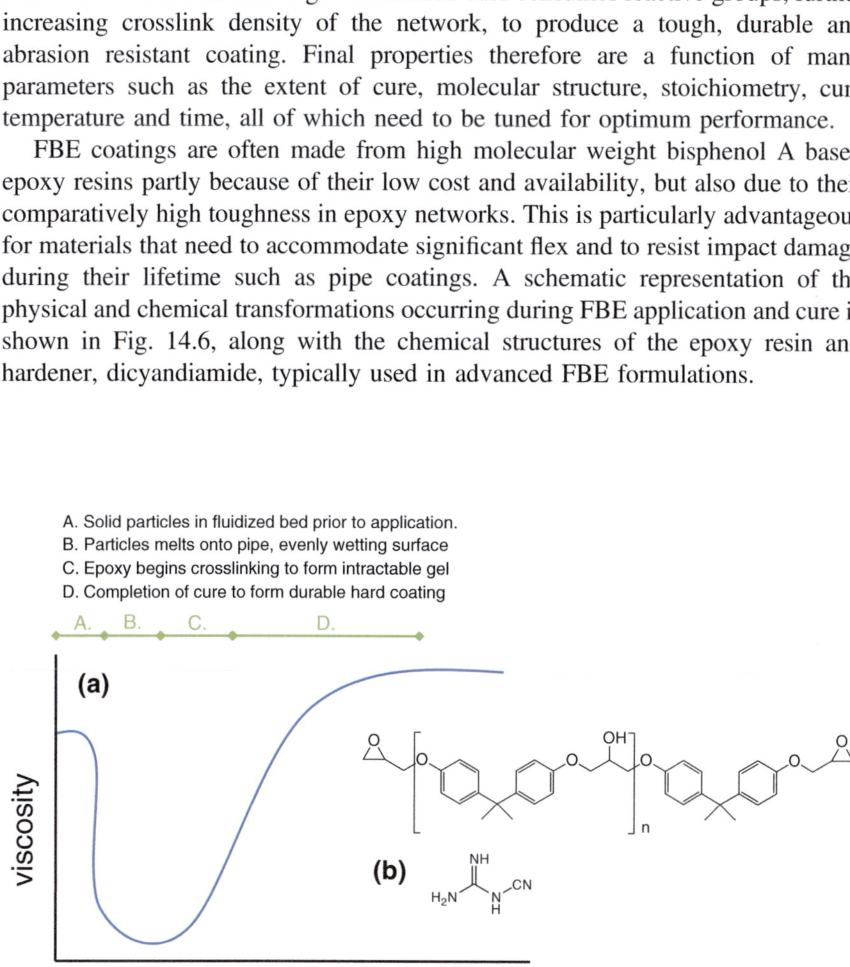

Fig. 14.6 **a** Schematic representation of the rheological transformations of the FBE during pipe coating from the melting and wetting of the FBE, through gelation and finally to full cure. **b** Chemical structures of high molecular weight diglycidyl ether of bisphenol A epoxy resin, and hardener, dicyandiamide

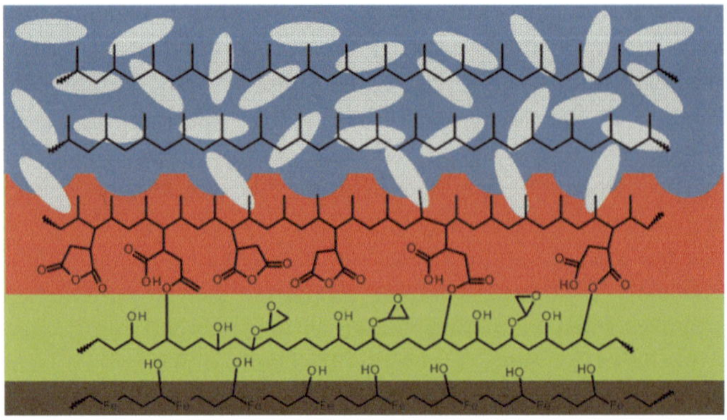

Fig. 14.7 Schematic illustration of the materials science underpinning a 3LPO coating, from the topcoat to the steel surface. Layer 1, the epoxy primer is compatible with the steel pipe surface because of its polarity and solubility with the metal oxides surface. Layer 2, the polyolefin adhesive ties the primer and the topcoat together through chemical reactions with epoxy primer and co-crystallisation and chain entanglement with the topcoat

14.3.2 Three Layer Polyolefin Coatings

Polyolefins are ideal topcoats because of their high impact strength, ease of use and very low moisture absorption characteristics, but their inertness and hydrophobicity provides negligible adhesion to the surface of steel pipes. To overcome this lack of adhesion, a multilayered approach is utilised where each successive layer is designed to adhere to the next, by gradual altering the chemistry to aid compatibility and chemical interaction from the steel pipe surface to the outer layer topcoat. These chemical interactions and physical processes associated with each of these different layers are shown schematically in Fig. 14.7 and discussed in the following sections.

14.3.2.1 Chemistry of Primer

The FBE coating is expected to have good adhesion to the steel pipe, but when, as part of a three layered system, it must also be compatible and adhere well to the polyolefin adhesive subsequently applied over it. This is a challenge because a highly filled epoxy resin will be polar and contain many inorganic components which will not be compatible with the highly hydrophobic non-polar polyolefin adhesive. However, the increasing cure reaction of the epoxy primer illustrated in Fig. 14.8, shows how the uncured epoxy resin is transformed into a less polar and more hydrophobic network through chain branching (Fig. 14.8a) and crosslinking (Fig. 14.8b). Furthermore, residual epoxide groups and other reactive functional groups such as hydroxyls, and secondary and tertiary amines produced during cure create potential reaction sites within the adhesive layer. As cure proceeds, the

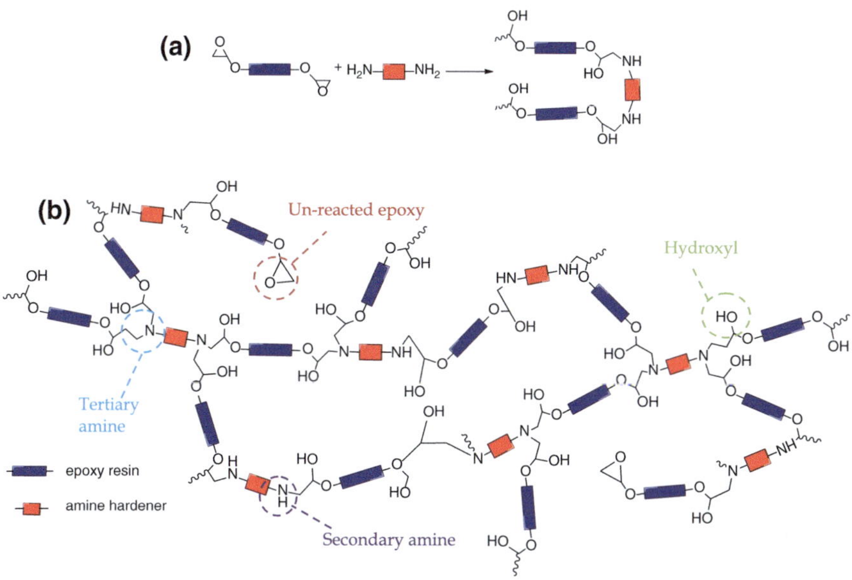

Fig. 14.8 Reaction mechanism of epoxy/amine cure illustrating **a** chain extension reaction occurring during primary amine addition to epoxide groups followed by **b** crosslinking and network formation via secondary amine addition to the epoxide. Formation of hydroxyl groups, secondary and tertiary amine catalysts and unreacted epoxy groups are circled to illustrate potential sites for chemical interactions with subsequent polyolefin adhesive layer. **a** primary amine addition, **b** secondary amine addition

primer forms a gel, representing the initial formation of a cross-linked network that gives some structural integrity and toughness to the coating. Beyond gelation, the reaction continues to increase the strength of the polymer network, but as the concentration of the functional groups decreases, reactivity slows and the surface becomes increasingly inert. This creates a processing window whereby applying the adhesive too soon, or prior to gelation will produce a coating with no structural integrity and poor overall performance, but applying the adhesive too long after curing will produce poor adhesion because of the lack of available reactive functional groups.

14.3.2.2 Chemistry and Synthesis of Adhesives

Polyolefins are inert, non-functional, non-polar hydrocarbons and for this reason adhere very poorly to most other materials. It is somewhat surprising therefore that in certain circumstances they can be successfully used as adhesives in pipe coating applications. Despite their lack of functionality, reactive chemical groups can be chemically grafted onto the polymer using free radical chemistries which promotes compatibility and chemical reaction with the primer. Maleic anhydride (MAH) is a widely used grafting agent for the polyolefins pipe coating industry although many other grafting chemicals can be used to good effect, ranging from methacrylic acid, glycidyl methacrylate and others [38]. MAH is particularly advantageous because of its low cost, relative ease of grafting and excellent adhesion at low levels of grafting for both polyethylene and polypropylene.

Chemical grafting of polyolefins begins with the decomposition of a free radical initiator such as peroxybenzoate (Fig. 14.9a) abstracting a hydrogen from the polymer backbone (Fig. 14.9b) which is then able to attack the alkene group of the MAH. This propagation or grafting of MAH to the polymer backbone continues (Fig. 14.9c) until it is terminated (Fig. 14.9d). Controlling this reaction therefore is critical to the overall performance of the final adhesive as side reactions, chain scission and crosslinking can all interfere and reduce the effectiveness of the grafting process.

The most economically viable method to modify polyolefins for pipe coatings is via reactive extrusion [39, 40] because it uses little or no solvent, there is no need to purify the reaction product, the process is continuous and the infrastructure costs are relatively low. The primary challenge of reactive processing, however, is to develop a method that maximises initiator lifetime and propagation of the grafting reaction

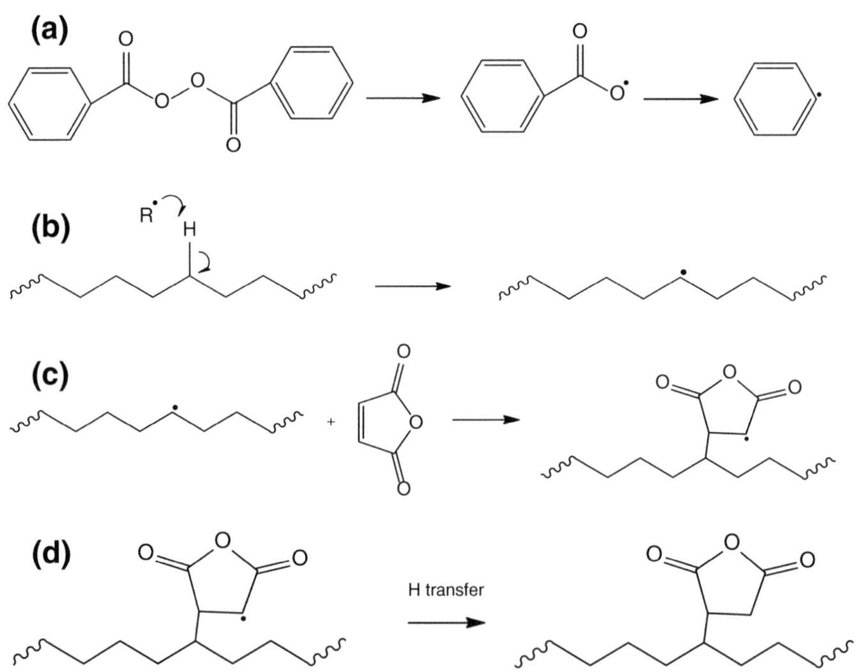

Fig. 14.9 Free radical reaction mechanism showing the formation of maleic anhydride grafted polyethylene. Process begins with **a** decomposition of initiator to form free radicals, followed by **b** initiation (H abstraction) to form free radicals in polyethylene, **c** propagation (chemical modification with maleic anhydride grafting agent), and finally **d** termination

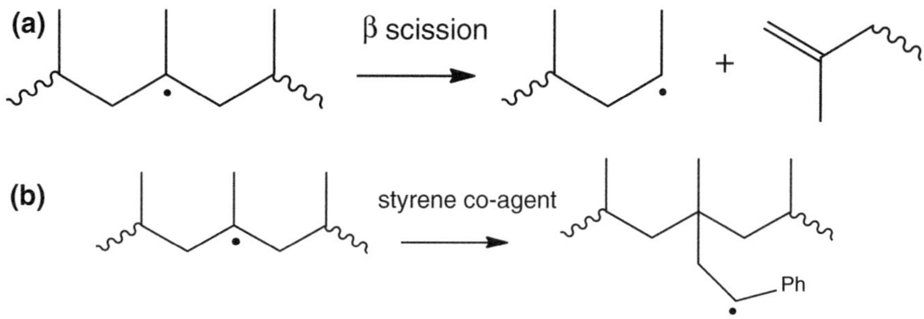

Fig. 14.10 Chemical reaction mechanisms illustrating **a** degradation via β scission arising from the stability of the tertiary carbon free radical and **b** subsequent use of styrene co-agent to minimise degradation

while minimising side reactions such as chain scission or crosslinking. The process parameters most commonly used to optimise grafting include the barrel temperature, screw speed, reactant concentrations, initiator lifetime and screw design. Grafting levels of only 1–2 wt% of MAH onto the polymer backbone are all that is typically required to produce excellent adhesion to the FBE primer [41]. Nonetheless, careful control is required at all times, particularly in the case of polypropylene, as chain scission can easily occur, decreasing molecular weight and reducing overall efficiency and hence performance [42, 43]. In the case of polypropylene the methylene group stabilises the tertiary carbon free radial thus making it more vulnerable to β chain scission, according the reaction mechanism illustrated in Fig. 14.10. This can be minimised or controlled using a co-agent such as styrene, which preferentially adds to the vulnerable tertiary carbon radical but continues propagating the grafting reaction via a radical less susceptible to degradation [44].

14.3.2.3 Composition of Topcoat

Three layered polyolefin (3LPO) coatings are formed through the gradual chemical and physical transformation of each successive layer to maintain adhesion between chemically disparate materials. While the adhesive layer maintains adhesion to the primer primarily through chemical interactions, it is more physical processes such as chain entanglements and/or co-crystallisation that promote adhesion to the topcoat. The primary role of the topcoat is to protect the pipe from the external environment and any damage arising from mechanical impact, during fabrication, transportation, installation or in service. So, while maintaining good adhesion with the adhesive layer is critical, being damage tolerant and chemically resistant will ultimately determine whether the coating system successfully protects the pipeline over its service life.

As previously discussed, polyolefins are used as topcoats because of their ease of processing and excellent properties which derives from their non-functional and non-polar hydrocarbon chemical structure. This however, is somewhat misleading because at a molecular level, polyolefins can vary widely in their level of branching (density) and molecular weight as shown in Fig. 14.11a, and at a microstructural level, are semi-crystalline in nature, containing both amorphous and crystalline regions (Fig. 14.11b) connected by tie layers. Branching, molecular weight and the different kind of crystal structures and how they are interconnected all have a profound impact upon the final polyolefin properties and will ultimately determine suitability as a topcoat. For example, the inherent flexibility of the aliphatic chains in polyethylene ensures it is inherently tough although higher density polyethylene (less branching) tends to be used when greater strength and stiffness is required. In contrast, the semi-crystalline structure of polypropylene provides increased mechanical properties and higher temperature resistance, but the increased brittleness from the crystal structure remains a constant challenge in many pipe coating applications.

Toughening polypropylene without impacting its other desirable properties, is therefore of great interest for pipe coating applications and has been investigated extensively using a variety of strategies [45]. Modifiers ranging from rubbers [46], rigid thermoplastics [47], block copolymers [48] and nano-composite based techniques [49] have all been somewhat successful depending on the specific application. Ideally however, inherently tough polypropylene synthesised in the reactor

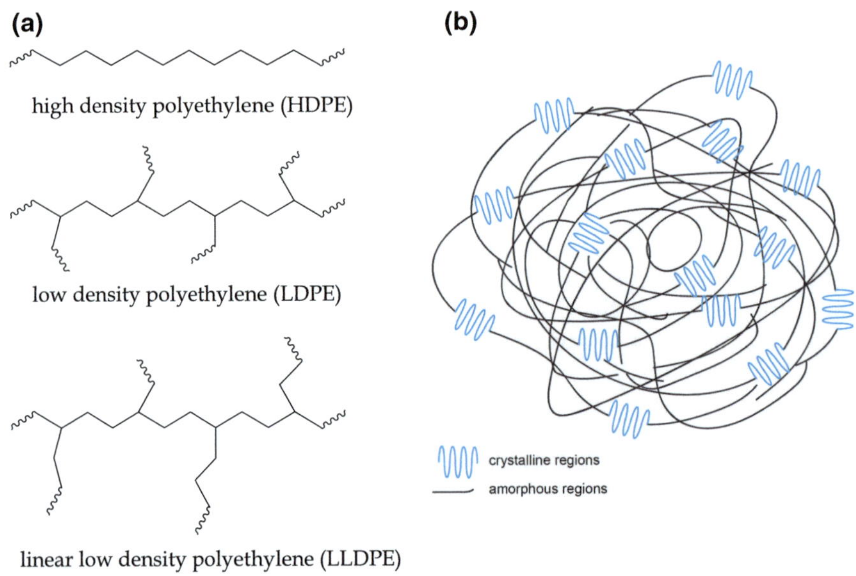

Fig. 14.11 **a** Molecular structure of polyethylene polymers varying only by the extent of branching and the likely impact on density. **b** Micro-structure of semi-crystalline polyolefins showing the crystalline and amorphous regions

without the need for further blending and processing is highly desirable. The development of metallocene catalysts in the 1990s [50] has made this possible through increased selectivity providing greater control of molecular weight and distribution of olefinic species. This has allowed the synthesis of a variety of new polypropylene block copolymers with improved properties that introduce α-olefins or cyclic olefins, for example [51]. Furthermore, if two metallocene catalysts are used in the same reactor, bimodal polymers of varying molecular weight can be synthesised concurrently to produce interpenetrating macromolecular structures with much greater compatibility than would otherwise occur. These developments form the basis of high impact strength polypropylene formulations used in the pipe coatings industry for topcoats and also thermal insulation foams. This technology has been used to independently polymerise polypropylene and ethylene to form highly compatible interpenetrating networks consisting of a biphasic morphology of ethylene propylene elastomer domains of the order of 1 μm dispersed within a continuous polypropylene block copolymer matrix. A schematic representation of the structure and morphology in Fig. 14.12 illustrates the thermodynamic self-assembly of the polyethylene block copolymer into micron sized domains of polypropylene and ethylene propylene elastomeric rich phases. This multiple length scale polymeric structure provides an excellent balance of mechanical properties, stiffness, impact strength and creep resistance over a wide temperature range, whilst providing good chemical and abrasion resistance.

14.3.3 Thermally Insulated Polyolefin Coatings for Deepwater Applications

Offshore oilfields are increasingly being developed in deepwater locations (1000–2000 m and beyond) that inevitably place greater demands upon the functional and

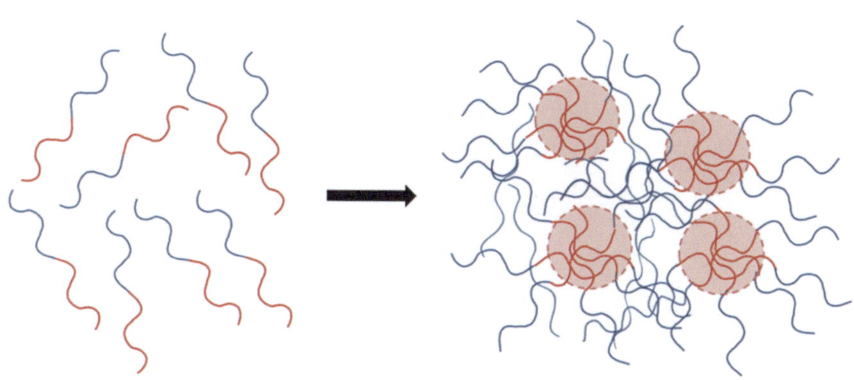

Fig. 14.12 Schematic representation of the self-assembly of high impact strength polypropylene co-polymer into a bi-phasic microstructure of micron sized domains of ethylene-propylene elastomer

mechanical performance of pipe coatings. At such depths, thermal insulation is necessary to prevent crystallisation of paraffin waxes and hydrates during shutdowns and maintain fluid temperature to facilitate separation of the oil at the surface [52]. Hydrates are crystals of methane and water formed during high pressure and low temperature flows and are a major source of pipeline blockages and is exacerbated at greater depths. The oil and gas extracted at these depths is also hotter making pipe coating reliability even more important to the economic recovery of the oil. Thermally insulated coatings, therefore, are necessary for preventing [53]:

- flow difficulties;
- phase separation;
- unplanned shutdown due to formation of hydrate plugs;
- wax formation;
- reduced fluid viscosity; and
- rapid temperature drop.

From a mechanical performance perspective, increased water depth also means higher hydrostatic pressure, so it is important that the pipe coating has the strength to withstand this pressure and not collapse or undergo creep significantly. Pipe coatings at extreme depths, therefore, require high compressive strength and stiffness, as well as sufficient temperature resistance and insulative capacities. Polymer foam, either polypropylene or polyurethane, is widely used for thermal insulation at these depths, although polypropylene is often the material of choice because of its combination of properties and compatibility with the other layers in the coating system.

Foam is typically applied by extruding an additional layer over an existing three layer coating system which produces a five layer system from an additional topcoat over the foam layer, as shown earlier in Fig. 14.3b. The foam is typically created from one of two different routes, either through an in situ gas expansion extrusion process or through the addition of micro-spheres (syntactic polypropylene).

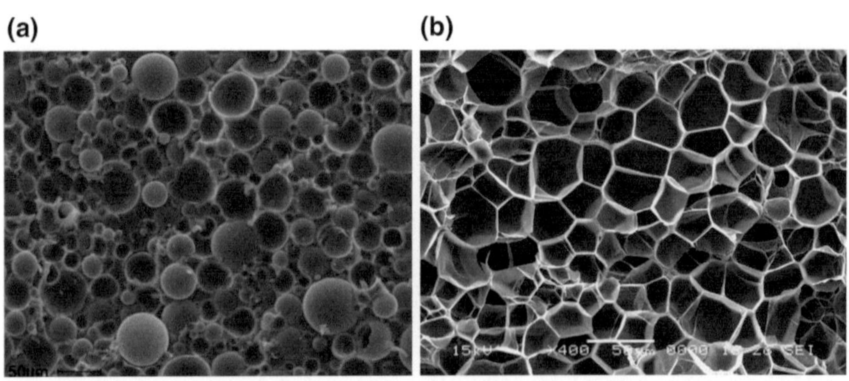

Fig. 14.13 Scanning electron micrographs of polypropylene foam made using **a** glass micro-spheres [54] and **b** in situ foam [55] formation

Scanning electron micrographs of the microstructures of these foams are shown in Fig. 14.13.

The in situ fabrication method uses a chemical foaming agent, such as a low boiling point hydrocarbon, mixed into the blended polypropylene which volatilises into a gas during extrusion. The liberated gas produces bubbles within the polymer melt, which must be strong enough to keep the gas dissolved in the polymer until it exits the extruder. The structural integrity, microstructure of the foam and hence insulation efficiency is therefore highly dependent upon the melt strength of the polypropylene. The density of the foam formed is critical to insulation (i.e. the lower the density, the better the thermal insulation), but reduced density also increases the tendency for it to fail in compression and to creep. If the increased pressure on the pipe collapses the foam, its insulation capacity will inevitably be compromised. Another problem encountered in deepwater applications is that although increasing foam thickness (and hence overall coating thickness) improves thermal insulation, it also reduces seabed stability of the flowline and increases drag on the steel riser [52].

To overcome these operational challenges, more efficient foams that improve insulation without increasing buoyancy or significantly reducing compressive strength are being developed using blends of high impact and melt strength polypropylenes. High melt strength polypropylene can be synthesised from homopolymer polypropylene using the susceptibility of β chain scission degradation as an advantage and effectively crosslink the polypropylene through a re-combination mechanism shown Fig. 14.14a [56]. The highly branched, high melt strength structure promotes strain hardening increasing the drawability of the polymer to provide greater control over bubble growth. As a result, smaller and

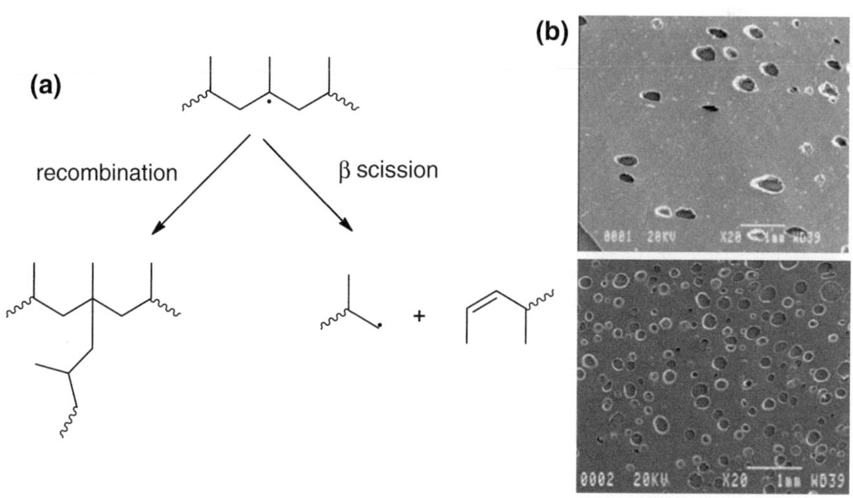

Fig. 14.14 **a** Reaction mechanism showing competition between β scission and chain branching leading to high melt strength polypropylene. **b** Resultant improved morphology [52]

more evenly distributed bubbles within a more closed cellular structure is formed as shown in Fig. 14.14b. High melt strength polypropylene in combination with high impact strength polypropylene block copolymers discussed in Sect. 14.3.2.3 is therefore a very attractive material for thermal insulation for deepwater pipe applications [16, 52].

Despite this, the addition of micro-balloon glass spheres to polypropylene (syntactic polypropylene) has several advantages over the in situ formation approach as it has less impact on the mechanical properties of the polypropylene, in fact it can be further improve mechanical properties, making it highly suited for deepwater applications and hence the most widely used insulation route since the mid-1990s [35]. Manufacturing is more flexible in that the foam can be directly extruded onto the pipe or it can be blended into pellets for subsequent processing. The wide range of hollow glass microspheres available provides much greater control over the density of the foam as well as its thermal properties. The primary disadvantage of this method, however, is that the glass spheres are fragile and there is a tendency for them to fracture, particularly during manufacture, with failures of around 10 and up to 30 %. Polymer and plastic beads are also available but they are restricted to use in shallower waters as they are structurally weaker than glass microspheres [35].

14.3.4 Mill Applied Pipe Coating Manufacture

Without continuous improvements in surface preparation, quality control and manufacturing processes for multilayered coating systems, any improvement to materials would be of little value, such is the importance of the application process. When compared with the advent of 2 and 3 layer coating systems in the 1950s, state of the art pipe coating factories are a sophisticated and high technology manufacturing environment [57–59]. The coating of pipes is now a fully continuous process where rotating pipes traverse through a sequence of tightly controlled treatments and processes shown schematically in Fig. 14.15. The main stages of the coating process comprise the following.

(1) Cleaning of the pipes
(2) Pre-heating of the pipes
(3) Application of coatings
(4) Water cooling
(5) Inspection

Cleaning the pipes starts with a simple pre-treatment step which removes various hydrocarbon and organic based surface contaminants, mill scale and moisture. Salts and soil are removed using high pressure fresh water (sometimes with phosphoric acid [60]) while hydrocarbon solvents, such as xylol or mineral spirits, are used to remove organic contaminants. The pipe surface is then pre-heated to 75 °C to loosen the mill scale (i.e. the flakes on the surface of fabricated steel, comprising

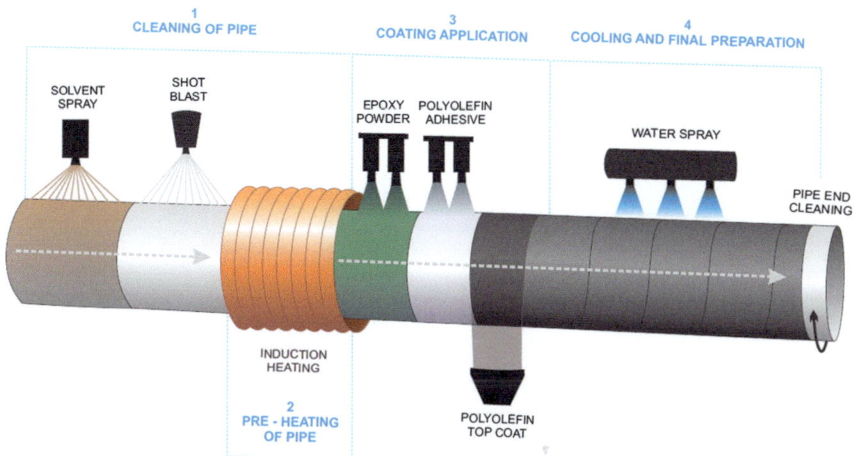

Fig. 14.15 Schematic illustration of a typical factory process to fabricate a 3LPO coating

iron oxide, hematite and magnetite) remove moisture and burn off any remaining organic matter. Final cleaning of the steel surface is carried out by abrasive blasting under high pressure to achieve a SA 2.5 (which states an average profile depth of 85–115 μm) [61].

To apply the primer, the pipe is then typically pre-heated to anywhere between 150 and 220 °C depending on the coating thickness required and the reactivity of FBE coating. Monitoring the exact temperature throughout the process is important to ensure proper wetting of the surface and the curing of the FBE as this will impact final performance of the overall coating system. Upon application of the FBE, the polyolefin adhesive and topcoat are then applied in rapid succession. Earlier methods such as liquid application of the epoxy primer and extrusion wrapping of the adhesive are still in use today, but the state of the art is to use electrostatic spray methods for both the FBE primer and adhesive. Electrostatic application uses powder with particle sizes ranging from 150 to 300 μm to allow suspension in a fluidised bed. The epoxy primer is applied to a thickness of between 80–100 μm while the adhesive film is typically no thicker than 350 μm. The polyolefin topcoat is then applied with a generous overlap of the preceding layer using an extruder to a coating thickness of about 1.5–3 mm.

After the topcoat has been applied, the pipes are then rapidly cooled by quenching using a cold water spray over the length of the pipe. As described, polyolefin topcoats are semi-crystalline polymers, so the rate of cooling is important to achieve a microstructure with the best combination of mechanical and thermal properties. Following quality checks, cutbacks are created at the ends of the pipes by removing a section of the coating using rotary brushes leaving bare endings in readiness for welding in the field using field joint coatings. The pipes are then stored until required. Quality inspection is critically important after coating so a series of on-line tests are performed such as surface inspection, thickness measurements, testing for holidays. Offline destructive quality checks consist of traditional mechanical properties tests including peel adhesion test, impact strength, hardness, elongation but also cathodic disbondment and coating resistivity which are selectively carried out but more typically only during process qualification trials

14.3.5 Field Joint Coatings

Pipelines are typically made from individual pipes of the order of 12 m in length that are welded together in the field during construction. Field joint coatings are applied on-site during pipeline installation and after the welding process. Field joints are potentially the weakest link in the chain because of (i) the inherent difficulty of application in a hostile environment, (ii) process limitations, (iii) time constraints and (iv) the difficulty to achieve a field coating that is completely compatible with the factory applied coating [60, 62]. The challenge of field joint coatings therefore, is to ensure it performs as well as, if not better than the mill applied coating. A variety of field joint coatings and methods of application are available depending upon the type of main line coating involved. Some of the common field joint coating systems include liquid coatings, FBE, heat shrinkable sleeves (HSS), injection-moulded PP/PE/PU systems, and flame-sprayed powders [36, 50, 63].

Generally, during installation of the field joint, the material melts and flows, filling in all surface irregularities as it binds itself to the substrate thus protecting the joint from mechanical damage and corrosion attack. Concurrently the material must also bond itself to the mainline coating, and hence compatibility between the two coating systems is of paramount importance for a durable field joint coating [64]. In the case of stand alone FBE field joint coatings a transportable electrostatic spray machine is used which applies a fluidised epoxy powder to a pre-heated surface after appropriate surface pre-treatment. For a successful field joint it is suggested that a combined coating consisting of a FBE coat and a polyethylene topcoat can provide increased protection due to enhanced mechanical properties [65]. For three layered polyolefin (PO) coated pipes the most common field joint coating systems include heat shrink sleeves, polyurethane coatings, and various two and three layered systems based upon similar factory applied technologies [66].

Heat-shrinkable sleeves (HSS), shown in Fig. 14.16, are one of the most widely used technologies for field girth-weld corrosion protection because they are easy to install while providing excellent protection and are generally very cost-effective. A heat shrink sleeve comprises a crosslinked polyolefin sheet with a hot melt polyolefin adhesive on one side. Prior to use, an FBE primer is typically applied then the sleeve and wrapped around the pipe followed by the application of heat using a flame gun. The coating then shrinks around the pipe until satisfactory adhesion is achieved. This unique behavior is built into the polyolefin by crosslinking it via irradiation during manufacture, and then stretching it prior to cooling. As the polymer recrystallises during cooling the unstretched previous

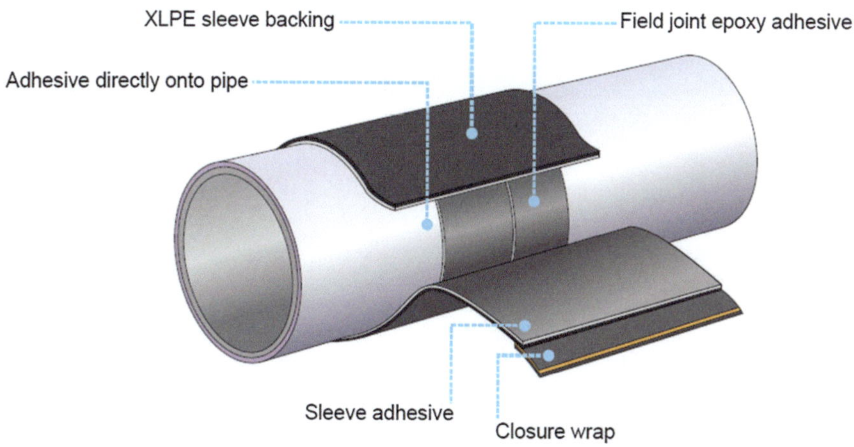

Fig. 14.16 Schematic of the construction of a heat shrink sleeve (HSS) field joint coating

shape will be *remembered* and is known as a shape memory polymer [67]. Shrinking occurs during thermal treatment because as the crystalline structure melts, the covalently crosslinked bonds in the crosslinked polyolefin begin to control the shape of the polymer rather than the crystal structure. Since covalent bonds are shorter than the more complex molecular configurations within a semi crystalline polymer, the polymer shrinks. This technique is used widely, although it requires considerable skill from the operator. Koebsch et al. [50] have reported that significantly lower temperatures are needed (i.e. 170–190 °C) for applying polypropylene HSSs compared to other field joint methods of injection moulded PP (IMPP) or PP tape wrap, or even the mainline 3LPP coating (220–240 °C). Self-sealing systems have reportedly been used to overcome this variability by embedding electrical wires within the heat shrink sleeve which automate the heating of the sleeves [68].

As would be expected, two and three layer field joint coatings require suitable surface preparation and induction heating of the pipe to high temperatures and example of successful field joint coatings applied in the field are shown in Fig. 14.17. In two layer approaches the differences are often more related to how the coatings are applied rather than the materials themselves. Coatings can be applied using a flame spray gun, transports the polypropylene using a mixture of gases, melting before it reaches the pipe surface. This approach requires a high level of skill from the operator, but does not rely on the residual heat of the pipe to achieve the required thickness. Alternatively FBE and sintered polypropylene can be applied using a flocking spray approach, however adhesion between the FBE and sintered polypropylene is determined by the residual heat of the pipe, limiting the thickness that can be achieved. A three layered field joint coating system, again relies on material technologies similar to the sintered polypropylene approach which wraps pre-fabricated polypropylene sheets around FBE and polypropylene

Fig. 14.17 Application of a field coated joint, as exemplified by **a** the flame-sprayed, and **b** the heat shrink sleeve, methods [69]

adhesive layers as it is welded into place longitudinally and circumferentially. This approach has the advantage that the polypropylene sheet is pre-fabricated thus improving uniformity and quality, although it is reported that it cannot be used for bends or tees joints. Field joint coatings today can vary widely in the type of application technology used but they generally follow the same principle of careful surface preparation, a keen awareness not to degrade the adhesion or materials of the surrounding coating and maximise compatibility between the field joint coating and the factory applied coating. Whether the outer wraps are preheated, rolled and pressed, or clamped and welded, they all should ensure excellent adhesion between the FBE, adhesive layer and finally the protective topcoat.

14.4 Performance

14.4.1 Industry Standards

There are now well established industry standards that govern the use of polyolefin external coatings. Table 14.3 lists out some relevant pipe coating standards adopted and used in the oil and gas industry. However, in many instances, end-users have additional or extended requirements based on their experience and/or unique field conditions. Hence, coating suppliers are continuously challenged to improve and enhance their systems and to introduce newer and better products to stay relevant and competitive.

14.4.2 Common Failure Modes

Whilst it is reasonable to expect pipe coatings to protect against corrosion over the design life of the pipeline, a more realistic expectation is that protection should be

Table 14.3 Some relevant standards for external pipe coatings

Origin	Standard	Title/Scope
German	DIN 30670	Polyethylene coating for steel pipes and fittings
	DIN 30678	Polypropylene coating for steel pipes and fittings
ISO	ISO 21809-1:2011	Petroleum and natural gas industries—external coatings for buried or submerged pipelines used in pipeline transportation systems
		Part 1: Polyolefin coatings (3-layer PE and 3-layer PP)
	ISO/DIS 21809-2 (under development)	Petroleum and natural gas industries—external coatings for buried or submerged pipelines used in pipeline transportation systems
		Part 2: Single layer fusion-bonded epoxy coatings
	ISO 21809-3:2008	Petroleum and natural gas industries—external coatings for buried or submerged pipelines used in pipeline transportation systems
		Part 3: Field joint coatings
	ISO 21809-4:2009	Petroleum and natural gas industries—external coatings for buried or submerged pipelines used in pipeline transportation systems
		Part 4: Polyethylene coatings (2-layer PE)
	ISO 21809-5:2010	Petroleum and natural gas industries—external coatings for buried or submerged pipelines used in pipeline transportation systems
		Part 5: External concrete coatings
	ISO/NP 21809-6 (under development)	Petroleum and natural gas industries—external coatings for buried or submerged pipelines used in pipeline transportation systems
		Part 6: Multilayer fusion- bonded epoxy coatings (FBE)
Canadian	CSA Z245.20-06/CSA Z245.21-06	External fusion bond epoxy coating for steel pipe/external polyethylene coating for pipe
	CSA Z245.20-98	External fusion bond epoxy coating for steel pipe
French	AFNOR	
Recommended practice	DNV-RP-F102	Pipeline field joint coating and field repair of linepipe coating

effective for as long as the pipeline remains in service because for most pipelines, they are operated well beyond the original design life. However, coatings inevitably get damaged, sometimes before they even begin service, i.e. whilst in transit or during installation, and they degrade as they age, thus compromising service life. In this section, some of the most common failure modes are described and discussed.

14.4.2.1 Blistering and Disbondment

Blistering and disbondment are the two commonest failure modes encountered in external pipeline coatings. Blistering is known to result from swelling of FBE coatings due to water absorption [70, 71] and from the presence of hygroscopic salts remnant on the pipe surface [72]. In addition, blisters in field applied FBEs are known to form when the force of abrupt vapour expansion exceeds its adhesive strength brought about by uneven heating profiles and excessive heating rates during application [73].

Disbondment, on the other hand, happens primarily when there is separation, due to a loss of adhesion, between the coating and the steel. This loss of adhesion could result from poorly executed coating process, mechanical damage, ageing and deterioration of the coating system over time, the effects of cathodic protection (CP), or a combination of these [30]. The latter effect is called cathodic disbondment (CD), which is an important phenomenon and is covered in more detail elsewhere (i.e. Section "Cathodic Protection Considerations"). Even if originally well-bonded, loss of adhesion over time is extremely likely due to the diffusion of ions, water and oxygen through the inherently permeable organic layers of the coating system [74]. Hence, it is not surprising that partial to more severe cases of complete debonding of mill and field applied coatings from steel surfaces of in-service pipelines have been reported for both PE [75] and PP [63, 64, 76] systems.

When disbondment happens, CP is often considered the next best line of defence against corrosion. Hence, the combined approach of coatings and CP are normally adopted for the task of protecting long stretches of buried and subsea pipelines against external corrosion. However, for CP to provide effective corrosion protection to the pipeline when the coating has disbonded, the CP current must be able to reach the pipe surface [77, 78]. "Non-shielding" coatings readily provides this path, while "shielding" coatings essentially do the opposite. Field data strongly suggests that FBE has superior non-shielding characteristics that contribute to FBE coated pipes being well protected from corrosion [25] despite their tendencies to blister [25, 79, 80]. Melot et al. [64] observed that CP shielding is confined to only buried pipelines onshore. It is postulated that when disbondment of the coating occurs in a seawater environment, the more conductive medium makes it easier for CP current to access the exposed cathode and provide more effective protection against corrosion. Depending on the degree of shielding, increasing CP current is necessary [30], however in so doing it can also create other problems including further disbondment due to, for example, hydrogen evolution, and coating deterioration, on top of introducing potential interference effects and cost impacts [78] (Figs. 14.18 and 14.19).

Fig. 14.18 Crack and disbonded field joint polypropylene [76]

Fig. 14.19 Example of blistering on pipe coating; including a punctured blister [69]

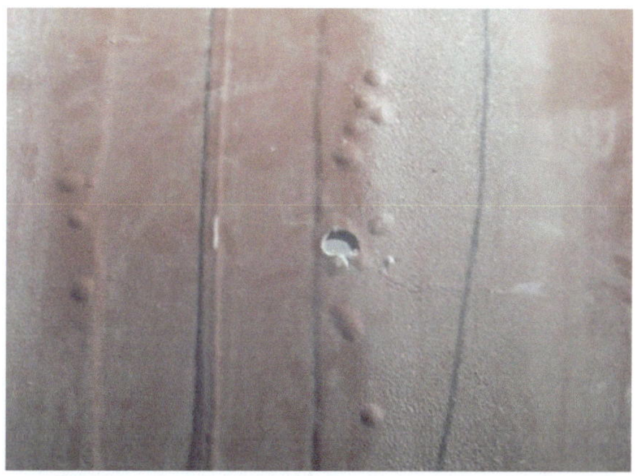

14.4.2.2 Cracking and Wrinkling

Cracking in three layer organic coatings have been attributed to thermo-oxidative degradation and high residual stresses in the topcoat [76]. These cracks, which originate mainly from field joints but propagate to the mill applied coating in the pipe body, have been observed to run circumferentially, spirally along the extruded coating laid-down direction and axially along the pipeline [63, 76]. In addition, cracking has also been observed at stress-raiser locations such as where CP cables are tack welded to the pipe [63]. 3LPP systems are also susceptible to cracking as a result of prolonged exposure to sunlight since they do not have good inherent resistance to ultraviolet (UV) rays. Cracking has been observed to occur during service on both mill applied coatings as well as on field coated joints in 3LPP [63] and 3LPE [6, 64] systems (Fig. 14.20).

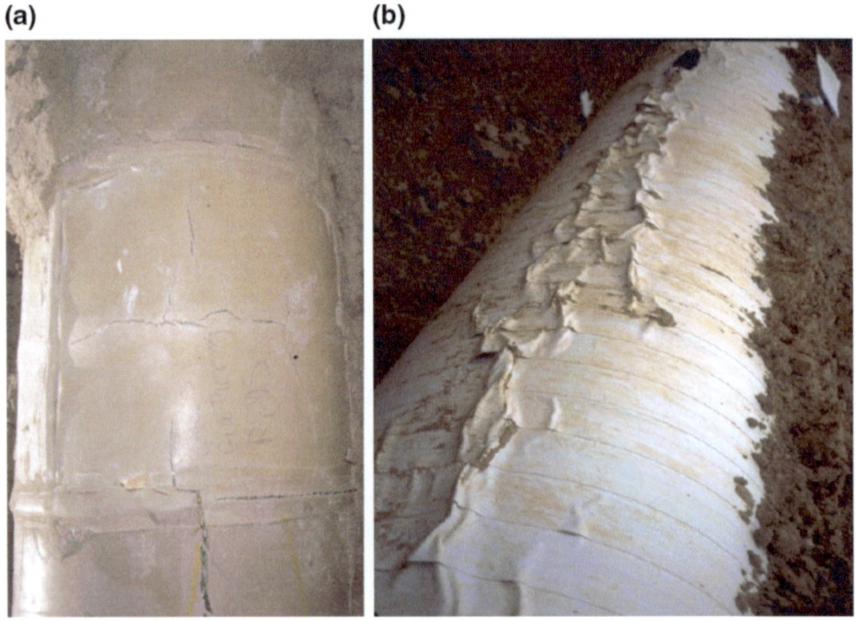

Fig. 14.20 **a** Cracking [76] and **b** wrinkling of three layered polypropylene coatings [78]

14.4.2.3 Field Joint Failures

Additional to meeting the same requirements imposed on the mill applied pipe coating systems, field applied joint coating systems need also to be compatible with the mainline system whilst affording fast and ease of application under field conditions and environments much harsher than those typically encountered in the mill. By and large, pre-mature failures of field joint coatings are attributed to a number of reasons including variable workmanship of applicators, harsh and uncontrolled field conditions, and inability to reproduce in the field coating the monolithic microstructure of the mill applied polymer coating [81, 82]. All these factors lead to poor bonding at the overlap between the field and the mill applied coatings, as typically evident by large scale debonding in in-service pipelines [64]. As a consequence an easy path for moisture and other fluids to ingress to the steel surface is produced thereby causing corrosion of the pipe. Risk of premature corrosion is also high due to free-edge debonding when this overlap is too small [64].

14.4.2.4 Testing and Performance Evaluation

Table 14.4 summarises some of the major measurements and requirements based on five widely accepted industry standards for polyolefin pipe coatings. There are over 100 tests to evaluate some 30 different coating properties [83], and a full coverage

Table 14.4 Main coating measurements and requirements based on five widely accepted industry standards

Test	DIN 30670 (PE system)	DIN 30678 (PP system)	CSA Z245.21 (PE system)	NF A 49-711 (PP system)	ISO 21809-1: 2011E (PP and PE systems)
Coating thickness (ASTM xx; ASTM xxx	Min 2.8 mm	Min 2.2 mm	Within 20 % of purchaser's specification	Depending on pipe diameter 1.2–2.5 mm)	Based on service condition and pipe dimension
Holiday detection	At 25 kV no electrical breakdown	At 25 kV no electrical breakdown	At 15 kV no electrical breakdown	At 10 kV no electrical breakdown	At 25 kV no electrical breakdown
Peel strength	20 °C: ≥35 N/cm 50 °C: ≥25 N/cm	50 °C: ≥100 N/cm 90 °C: ≥80 N/cm	20 °C: ≥150 N	23 °C: ≥15 N/mm 110 °C: ≥4 N/mm	Hanging mass: 23 °C: ≥15 N/mm (HDPE) ≥ 25 N/mm (PP) 60 °C: ≥3 N/mm (HDPE) 90 °C: ≥4 N/mm (PP)
Cathodic disbondment	No requirement	No requirement	20 °C at 28 days: 12 mm max radius	≤7 mm after: 2 days at 65 °C 7 days at 40 °C 28 days at 23 °C	23 °C @ 28 days ≤ 7 mm 65 °C @ 24 h ≤ 7 mm
Specific electrical insulation resistance	Insulation resistance (100 days) > 10^8 Ω m^2	Insulation resistance (100 days) > 10^8 Ω m^2	No requirement	Insulation resistance (100 days) > 10^8 Ω m^2	No requirement
Dielectric strength					
Impact resistance	After 30 impacts no electrical breakdown E = 5.1 Nm/mm	After 30 impacts no electrical breakdown E = 5.1 Nm/mm	Min 3.0 j/mm of actual total coating thickness	Based on pipe diameter	23 °C: ≥7 J/mm (HDPE) ≥10 J/mm (PP)

(continued)

Table 14.4 (continued)

Test	DIN 30670 (PE system)	DIN 30678 (PP system)	CSA Z245.21 (PE system)	NF A 49-711 (PP system)	ISO 21809-1: 2011E (PP and PE systems)
Indentation resistance	23 °C: ≤0.2 mm	90 °C: ≤0.3 mm	No requirement	20 °C: ≤0.1 mm	23 °C: ≤0.2 mm (HDPE)
					≤ 0.1 mm (PP)
				110 °C: ≤0.4 mm	HT: ≤0.4 mm (HDPE)
	70 °C: ≤0.3 mm				≤0.4 mm (PP)
Elongation at break	At 23 ° C: ≥ 200 %	At 23° C: ≥ 300 %	Min 600 %	Min 400 %	PP ≥ 400 %
					HDPE ≥ 400 %
Thermal ageing	MFR ≤ ±35 % of initial value (200 days at 100°)	No embrittlement (100 days at 140 °C)	At least 65 % of original tensile stress at yield.	ΔMFR ≤ 50 % of initial value (300 days at 150 °C)	ΔMFR ≤ 35 %
			Elong. > 150 % (100 °C for 100 days)		
(UV) Light ageing	MFR ≤ ±35 % of initial value	No requirement	No requirement	Value of elongation should be minimum 75 % of the value initial sample	ΔMFR ≤ 35 %
Environmental stress cracking resistance	No requirement	No requirement	>300 h	No test piece crack after 1000 h of immersion at 50 °C (ASTM D 1693)	No test piece crack after 1000 h of immersion at 50 °C (ASTM D 1693)—only for PE system
			Method B : 100 % Igepal		
Resistance to micro-organism	No requirement	No requirement	No requirement	Loss of adhesion shall not be less than 80 % of the initial value (after 6 month of burying test) and no defect ≥20 mm	No requirement

(continued)

Table 14.4 (continued)

Test	DIN 30670 (PE system)	DIN 30678 (PP system)	CSA Z245.21 (PE system)	NF A 49-711 (PP system)	ISO 21809-1: 2011E (PP and PE systems)
Bendability/flexibility	No requirement	No requirement	No requirement	Able to withstand bending with radius of curvature equal to 20 times the diameter of the tube without visible damage	No cracking at an angle of 2.0° per pipe diameter length
Hot water immersion	No requirement	No requirement	No requirement	No requirement	Average ≤ 2 mm
					Maximum ≤ 3 mm

of them are beyond the scope of this chapter. Suffice to say that these tests cover aspects of surface preparation, raw material properties and performance of the complete coating system. Only the main properties are presented and discussed here, and for a more comprehensive coverage of the tests, the reader is referred to the review articles published by Papavinasam and colleagues [84].

Adhesion

The quality of adhesion between a virgin 3LPO coating system and steel substrate is evaluated using a peel test. The test involves measuring the force necessary to detach the coating from its substrate when subjected to a constant rate of pull (viz. 10–300 mm min^{-1}, depending on specification [11]). Hence the test is carried out by gripping and pulling at 90° to the tangent of the pipe surface a peel strip (or, "tongue"). The strip is of a pre-determined width (typically 25 mm) which is formed by carefully cutting and chiselling into the coated pipe [85–87], see Fig. 14.21. Alternatively, the rate at which the coating is peeled from the metal substrate due to a constant load, effected by means of a hanging mass, can also be used as the measure of adhesion strength [88]. Standards generally recommend that the test be carried out at different temperatures, covering the field conditions of room/ambient, low and high temperatures, although the exact recommended temperatures may vary somewhat between different standards [85, 86, 88]. Similarly, different hang-off masses are also used per specification in the relevant standard [88]. The test requires cohesive failure of the adhesive layer for values to be deemed valid.

Note that the peel test only measures the level of adhesion of a virgin coated pipe. Long-term behaviour and properties of the bond of the coating to the steel substrate cannot be predicted reliably, and, hence, should not be, inferred from the results of this test. The test is normally used to qualify a coating system with respect to the process, including the associated conditions and parameters.

Cathodic Protection Considerations

In order to provide effective corrosion protection, the electrochemical process of CP is required to alkalinise (to a pH range of 9–13) the surrounding environment around the cathode (i.e. pipe, tank, etc.) [77]. During the process hydrogen is evolved, which is coupled with other potentially damaging electrochemical changes in the environment that could lead to disbondment of the 3LPO coating from its substrate. This degradation in adhesion of the coating due to CP is more commonly referred to as cathodic disbondment (CD) (see also Sect. 14.4.2.1). Papavinasam and colleagues [70] attributed CD to factors such as pH, cathodic potential, stability of the interfacial oxide, substrate surface roughness, defect geometry, coating formulations, wet-dry fluctuations, and water ingress. The complex array of factors

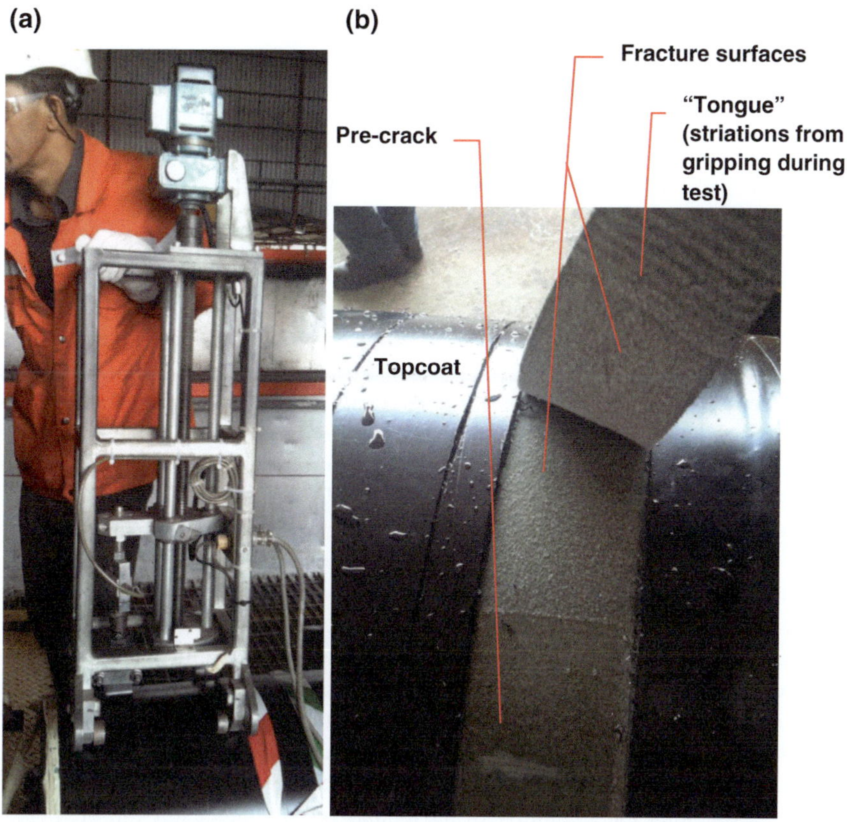

Fig. 14.21 The peel test is one of the standard quality assessment carried out after procedure qualification trial (PQT), carried out prior to full production. **a** During testing: an instrumented test rig is used to carry out the peel test. **b** Post-testing: observe the cohesive fracture surface morphology

necessitates that coatings pass a CD test before they are qualified for field application.

Cathodic disbondment test

The effectiveness of coatings to withstand CD is evaluated using the CD test. In simple terms, a CD test is used to give some indication of how well the adhesion of the coating will withstand the electrochemical process of CP. In accordance to ISO 21809-1:2011 (E), for example, debonding must not be greater than 7 mm after a 28 day exposure at 23 °C and this applies to 3LPE and 3LPP. In other standards, for example the CSA Z245.21-06, debonding due to CD testing for a 3LPE system must not exceed 12 mm after a 28 day exposure at 20 °C. It has been argued [89] that in order to predict the long-term behaviour of coatings, a better approach is perhaps to generate CD rates for an array of system and service conditions, rather than specifying CD size for particular (short-term) periods of time.

It is noteworthy that there is no justification for the limits to disbondment set out in specifications; limits are just ideas of what it is thought maximum disbondment tolerable [90]. In fact, over a period of 30 years, from 1965, CD resistance is reported to have improved from approximately 65 mm to less than 10 mm, for FBE systems [23], thus showing tremendous enhancement in the performance of coating systems in line with ever demanding field requirements.

Hot CD test

With the advent of high temperature coatings, normally required for deepwater applications, the need for CD tests to be conducted at elevated temperatures has become increasingly necessary. One of the limitations of standard CD tests is that the electrolyte used to instigate disbondment is water-based and hence not suitable for high temperature (beyond 100 °C) assessment. Buchanan [11] highlighted that published data have indicated that a CD test carried out above 95 °C is indeed not practical, nor is it representative of in-service conditions while CD testing at a lower temperature of 65 °C can actually be more severe. Others [91, 92] have pointed out that CD tests should be carried out at temperatures beyond the limitations of the boiling point of water, to better simulate and represent the conditions to which high temperature structural coatings are exposed. This is because other degradation mechanisms could come into play when the service conditions reach these higher temperatures [92]. Attempts to carry out CD testing at temperatures above 100 °C have been investigated using autoclaves [93] and by heating the test specimen above 100 °C while maintaining the temperature of the test (aqueous) solution at below boiling point of water [94]. In addition, using concentrated salt brines that are thermodynamically stable above 100 °C has also been proposed [95]. Much work remains to ascertain the suitability of these modified test procedures in predicting long-term behaviours of high temperature polymer coatings. Research into the effects of coating thickness with respect to CD at “low” [89] and “high” [92] temperatures is also being pursued.

Dielectric strength and Insulation resistance

All coatings have varying dielectric properties that reduce the tendency of the electrolyte to complete the electrical circuit between adjacent anodic and cathodic sites on a substrate, that is necessary to mitigate corrosion in the event of localised coating failure [78].

Ideally, a coating should have a high enough dielectric strength to resist corrosion while the coating is bonded, but at the same time sufficiently low to allow protective CP current to reach the pipe when disbonded. Such coatings are classified as non-shielding. FBE is known to provide excellent non-shielding characteristics which explains how well in-service FBE coated pipes protected against corrosion [25] despite their tendencies to blister [79, 80].

Presence of Holidays

The presence of defects such as voids, pores and pinholes, or holidays, that result in discontinuities are unavoidable in polymeric coatings. Such discontinuities in the coating are frequently minute and not readily visible. ASTM D5162–08 [96] recommends a method for picking up unacceptable levels of these discontinuities by applying low voltage electric circuit (which is also connected in series with an LED or noise-producing component such as a piezoelectric speaker) so that if electrical flow is detected then the affected area is termed as conductive thus indicating the presence of holidays. The flaws are picked up by an electric arc that forms between the sensor electrode and the metallic support. The allowable number of discontinuities should be pre-determined prior to conducting the test since the acceptable quantity of discontinuities will vary depending on coating film thickness, design, and service conditions.

14.4.2.5 Long-Term Exposure to Elevated Temperatures

Pipe coatings are increasingly required to service elevated temperature conditions. These services include transmission of oil sands and heavy oils which, at 25 °C, have viscosities, that are in the order of a million times higher than conventional crude oil. For transmission at parity with crude oil, these highly viscous hydrocarbon need to be transported at temperatures around 200 °C [95]. Hence, thermal ageing of coatings need to be better assessed using a wet medium, rather than restricted to only dry environments, and at higher temperatures that are either akin to actual field conditions, or even to higher to accelerate the effects of long term, elevated temperature exposures. This is an area of on-going research and development.

Thermal ageing

Moosavi et al. [76] have revealed discoloration and severe spiral cracking in 3LPP when subjecting it to 1000-h thermal ageing in an oven at 150 °C. They suggested the use of 150 °C for the thermo-oxidative test in order to accelerate thermal degradation, and as a minimum qualification requirement, the specimen should survive a 4000 h continuous exposure.

Hot wet soak test

The electrical resistance of polymeric materials are high, in the region of 10^{14} Ω, and 3LPO coatings can hence be expected to provide effective insulation for the steel pipeline to which it provides corrosion protection. However, the resistance is expected to degrade over time with the ingress of fluids into the coating thus deteriorating its corrosion protection capability [83]. An example of the hot wet soak test can be referenced to ISO 21809-1: 2011E (Annex M) [88]. However, the maximum temperature is set at only 95 °C and it has been proposed that this test conditions could misrepresent the actual ones experienced by high temperature coatings [95].

14.5 New Developments

14.5.1 Toughness and Durability

From fabrication, transportation, installation to in-service conditions, pipe coatings are subject to various potential damage events and as pipes are deployed in more demanding environments their susceptibility to damage increases. Thus there is a on-going need for tougher and more damage resistant as well as more durable coatings whether single layered FBE, dual or multi-layered coatings. Some innovative approaches and new materials which seek to address some of this challenge are discussed below.

Commonly referred to simply as a high performance composite coating (HPCC) [16, 97], this polyethylene based coating is a multi-component system applied in a single step using spray coating application methods that claims to be less susceptible to internal stress development over wide temperature variations and less likely to delaminate under cyclic conditions. Applied as a single coating, the multi-component system essentially phase separates into a three layered FBE, adhesive and topcoat structure after application and cure. The lack of a sharp or well defined boundary between the layers is understood to produce strongly interlocking and compatible layers which provide this improved resistance to delamination, penetration and abrasion, all of which are existing problems for three layer polyethylene coatings [98]. Its primary target application is situations where FBE coatings have been determined to be inadequate [99] and it has been demonstrated to be extremely beneficial in onshore underground situations where the resistance to impact during installation is particularly important [100].

This technology has been further developed [101] to overcome common problems such as "weld tenting" and low coating thickness over welds. A polyethylene coating, minimum 2.2 mm in thickness, is applied using a side extruder wrapping process over the multi-component HPCC system. In this case HPCC is typically used to replace the FBE and polyolefin adhesive layers in three layered systems with the polyolefin topcoat being subsequently wrapped over the HPCC

In addition to the use of block copolymers and related systems, polypropylene can also be further toughened via β nucleation [102, 103], a technique which catalyses the formation of an alternative and more inherently tougher β crystal structure at the expense of the more thermodynamically favorable yet brittle α crystalline phase. The transformation is illustrated in Fig. 14.22 as large α spherulites with sharp, distinct boundaries become much smaller, more diffuse β crystallites and hard to observe boundaries at only 0.1 wt% of nucleant. This approach is very attractive for pipe coatings applications because only very small concentrations of β nucleant are required to significantly improve impact strength ensuring little impact upon processability [104]. Recent work has combined the β nucleation approach with elastomer modification, in the search for synergistic improvement in properties [105, 106] with some success, thus improving performance even further [107, 108].

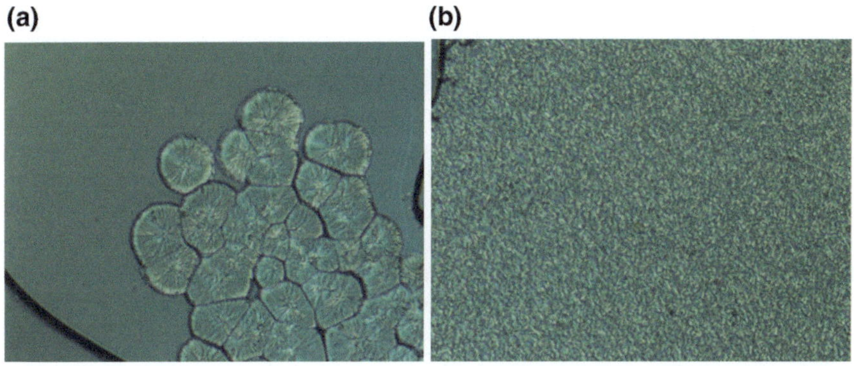

Fig. 14.22 Illustration of the impact of β nucleants, transforming polypropylene from the a) α phase consisting of large spherulitic structures with sharp boundaries and little interaction, into the β crystal phase with its very small crystals which are much more interconnected [109]

FBE coatings, whether applied as single, dual or three layered coatings, will always remain susceptible to damage to some degree because of their crosslinked network structure. As such a range of approaches have been tried to improve their toughness and in-service durability. Block copolymers are one such approach because they are known to self assemble into nanostructured morphologies with advantageous properties [110]. This technology produces a gradient of interactions with the curing epoxy resin to produce nano and microscale morphologies with enhanced performance. The particular advantages of this is that both the flexibility and impact strength of the coating are increased dramatically with little impact upon other properties such as glass transition temperature, flowability and corrosion resistance. Another nanotechnology based strategy [9, 111] that also claims to simultaneously improve the impact strength and flexibility of FBE coatings does so by applying a second abrasion resistance outer (ARO) layer. Although the nano-modifier is not identified this dual layer coating method is reported to improve resistance to crack propagation and maintain cold temperature flexibility similar to that of a single layer FBE, without compromising impact or gouge resistance.

14.5.2 Self-healing Capabilities

Pipelines and pipe coatings are expected to last many years in hostile environments without the need for constant repair or major interventions once they are installed. However, despite the best use of available materials and engineered design, materials deteriorate and age over time, eventually failing. Failure of coatings could mean under-protected pipelines which would subsequently lead to premature failure of the pipelines themselves. Repair of pipelines are typically intrusive in nature, and is a costly and time consuming process, made even more difficult by the need to achieve factory quality finishes in the field. If the need for field repairs could be eliminated or transformed into a simple process with minimal intrusion, the effective service life of a pipe coating and pipeline could be enhanced thereby bringing enormous economic benefits. Self-healing and/or in situ activated healing coatings are part of an emerging technology that despite being in their infancy have already been shown to improve pipe coating lifetimes through corrosion prevention and the use of viscoelastic materials [112].

A type of self-healing coating based upon a poly-isobutene viscoelastic adhesive substrate and a mechanical polyolefin protective topcoat is already on the market today. The self-healing adhesive is fully amorphous and remains soft and tacky yet resistant to weathering, chemicals, moisture ingress and bacteria [113]. It can restore adhesion to a metallic surface after failure through its ability to flow even at low temperatures such that all pores, anomalies and cavities would eventually be filled. Conversely, despite its ability to flow over time, its modulus of elasticity provides high strength solid like behaviour which further resists failure during deformation. This self-healing mechanism has been adapted for use in rehabilitation of buried steel pipelines using a three ply tape consisting of a central polyethylene carrier film and a butyl rubber "self-amalgamating" adhesive used on both sides [114]. When the tape is wrapped around the pipe the butyl adhesive intermingles in the overlapping areas and because of the viscous nature of the butyl rubber, the interfacial regions ultimately disappears completely.

Viscoelastic modifiers have been used in polyethylene adhesive formulations to transform three layered coatings into a system that can heal the adhesive via in situ thermal activation. This method is an adaptation of the pressure delivery healing mechanism created by poly(ethylene-*co*-methacrylic acid) (EMAA) in epoxy networks [115]. Meure et al showed that when EMAA is added to the polyolefin adhesive, a two phase morphology consisting of EMAA particles dispersed within a polyolefin adhesive is produced. If good adhesion between the matrix and particles is achieved and the particles are sufficiently large, complete restoration of adhesion can be achieved after thermal activation. Scanning electron micrographs of the fracture surfaces are shown in Fig. 14.23 which suggests that the healing mechanism is derived from the ability of the phase separated EMAA functioning effectively as reservoirs which provide polymer that can flow into any cavity during thermal activation. Upon cooling, the inherent adhesive performance of the EMAA then binds both the primer and polyolefin surfaces together via hydrogen bonding and van der Waals interactions.

14.5.3 High Temperature Adhesives

Within a three layered polyolefin coating system, the polyolefin adhesive is the most susceptible to failure at high temperature because reactive processing inevitably causes some degradation or chain scission during grafting. An alternative approach to traditional reactive grafted adhesives is to use an intimately blended

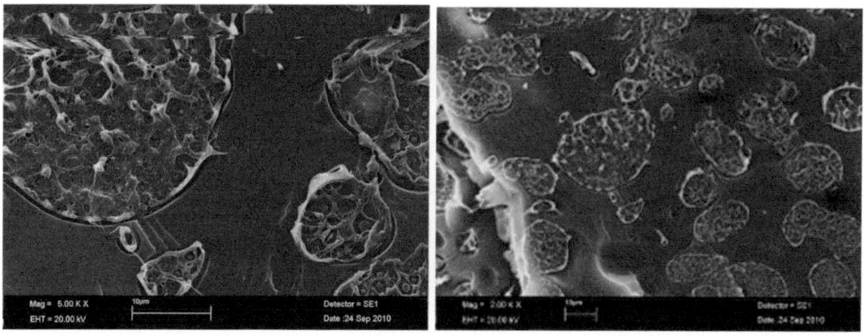

Fig. 14.23 Scanning electron micrographs showing the bi-phasic microstructure of the formulation and "reservoirs" of healing agent produced

semi-interpenetrating network (sIPN) consisting of a linear polyolefin and a crosslinkable monomeric epoxy, commercially referred to as a protective network coating (PNC) [62, 116, 117]. This avoids the use of reactive extrusion to synthesise the adhesive thus minimising any potential polymer degradation as it simply is a blend of an epoxy primer and polyolefin adhesive. This is achieved by blending the epoxy monomer and α olefin monomers together in solution and polymerising the olefin while the epoxy monomer is largely unaffected. The resulting sIPN comprises a highly compatible blend of polyolefin and a monomeric or low molecular weight poly-epoxide resin. The excellent adhesion derives from the strong interfacial interactions between the two components, but also their respective interactions with the epoxy primer and polyolefin topcoat. The epoxide functional groups promote covalent bonding between the primer, while the non-polar polyolefin would be fully compatible with the polyolefin topcoat. A schematic representation of how a sIPN can promote interfacial adhesion in a three layered coating system is shown in Fig. 14.24. This technology has been applied as a field joint coating where the lack of polymer degradation of the adhesive facilitates its usage as a high temperature adhesive [118].

Similarly, the glass transition temperature (Tg) of an FBE adhesive in a single, dual or three layered coating system is also critically important in determining the usage of a coating system. Their Tg's are typically less than 110 °C, yet in modern deepwater oil wells, the operating temperature requirements may well be about 140 °C or more, thus creating performance problems for standard FBE coatings. To overcome this limitation, epoxy adhesives with higher Tg values have been produced using brominated epoxy resins [119] which at the same time also has improved flexibility and impact resistance, weatherability, and cathodic disbondment properties.

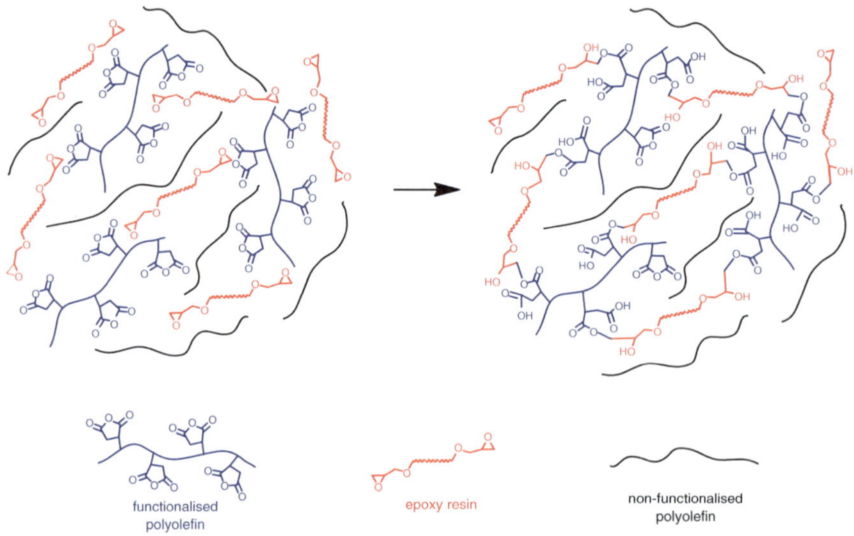

Fig. 14.24 Schematic representation of the semi interpenetrating networks of polyolefins and crosslinkable monomeric epoxy resin. The blending of functionalised polyolefin helps to create a more compatible network through interconnections allowing the epoxy resin to cure while remaining miscible with the polyolefin

14.6 Concluding Remarks

Oil and gas pipelines are critical assets for the industry. Large quantities of hydrocarbon are transported around the clock in order to fulfil the ever increasing needs for energy to power the growing economies of the world. Globally, huge networks of pipelines run through vast lands and seas and they are expected to provide the hydrocarbon resources to end-users in a timely and reliable fashion. The pipelines, in many cases designed to last no less than 25 years, are constantly threatened by corrosion attack due to the environments in which they operated. Hence, almost universally, these pipelines are, by an arsenal of techniques, encapsulated within the current generation of advanced pipe coatings and cathodic protection. However, increasingly, these coatings are required to withstand harsh service environments that continually challenge their integrity and effectiveness to provide corrosion protection to the pipelines. Hence, from the days when the first polymer pipe coatings were used, they have evolved from highly inefficient manually applied single coating systems based on coal tar, to sophisticated multilayered and multifunctional systems that can be equally applied in remote and extreme environments or in pristine state of the art factories. A wide variety of pipe coating systems have been developed to suit specific service conditions, starting with epoxy coatings, for environments where service temperature requirements are modest (<65 °C) while the likelihood of impact events and moisture ingress are low. Where service temperatures reach up to about 110 °C, three layer coating systems consisting of an epoxy primer, applied in conjunction with a polyolefin adhesive and topcoat, are more commonly used. First and foremost against corrosion, but also against exposure to physical damage incurred during fabrication, transportation, installation and actual service. In service conditions where temperatures of up to 140 °C and thermal insulation becomes important for flow assurance, multi-layered (5 or more) systems comprising layers of foam that are interspersed between the adhesive and topcoat layers are more commonly used.

The need to understand the underpinning materials science and engineering of each and every aspect of the fabrication and application of these materials therefore is clear whether for factory or in the field applied coatings. Without this knowledge, pipe coatings that protect pipelines against corrosion and impact damage throughout their lifetime would not be achieved. Standardisation and specifications are equally critical to this process, as it will ensured quality control, reproducibility and a level of confidence to the industry enabling it to continue making better pipe coatings.

As the O&G industry continues to explore in more challenging oilfields, e.g. deepwater offshore, desert or arctic, the need for improved performance and durability at extreme temperatures and pressures will continue unabated. Developments in materials and processes are continuing to evolve, ranging from self healing systems, high temperature adhesives to high compression strength foams and tougher adhesives and top coats, to meet these demands. Similarly, developments in testing methods are actively pursued so that the performance of coatings can be more accurately assessed and determined consistent with more extreme service conditions. Indeed the amount of new innovations seen in the pipe coating industry have been tremendous in recent times, particularly in this new millennium, in order to keep up with the demands of the end-users. It is anticipated that due to its versatility and effectiveness, both economically as well as technically, polymer based coatings will continue to dominate as one of the most important corrosion protection systems for oilfield pipelines.

Acknowledgements The authors are grateful for the many useful discussions with colleagues from both within and outside their respective organisations, with particular thanks to M. Dell'Olio of CSIRO, and M.S.F. Samsudin and Z. Ahamid of PETRONAS. The help of MSFS in compiling Table 14.4 is also gratefully acknowledged.

References

1. P. Hopkins, *Oil and Gas Pipelines: Yesterday and Today*, International Petroleum Technology Institute, special article written for American Society of Mechanical Engineers (2007)
2. D. Howell, *A Brief History of Steel Pipe, Pipeline Equities, Pipeline Appraisals and Recovery* (2010)
3. M. Romano, The ins and outs of pipeline coatings. J. Protect. Coat. Linings 41–47 (2005)

4. R. Buchanan, Pipeline coatings & joint protection: a brief history, conventional thinking & new technologies, in *Rio Pipeline Conference and Expo, 22–24 October 2003, Rio de Janeiro, Brazil* (2003)
5. R.N. Sloan, Fifty years of pipe coatings—we've come a long way. Mater. Perform. **36**(6), 42–47 (1993) (Energy Coatings Company, Houston Texas)
6. M. Roche, D. Melot, G. Paugam, Recent experience with pipeline coating failures. J. Protect. Coat. Linings 18–28 (2006)
7. N. Tsalic, Trends and development in the market for pipe coating, plastic pressure pipe 2012, 22–23 February 2012, Cologne Germany, Applied Market Information
8. J.W. Cox, Multi-coat systems of HDPE for exterior anti-corrosion. Protection of steel pipelines based on 50 year service life, in *Technical Presentation for Pemex and Mexican Petroleum Industry, 24–26 Nov 1999* (1999)
9. J.A. Kehr, M. Mallozzi, Fast, worry free pipeline installation with dual-layer fbe coatings, *Corrosion and Prevention 2007 Paper 010* (2007)
10. A. Jackson, E. Simonsen, T. Stark, R. Nixon, B. Melve, D. Ali, Long term ageing of polypropylene foam. The results of testing performed after 18 years of exposure in a subsea environment, in *MERL Oilfield Engineering with Polymers, London, UK, October 2012* (2012)
11. R. Buchanan, Feeling the heat, World Pipelines, March 2006, 5 pp
12. K. Varughese, Dual powder FBE systems for directional drill and high temperature pipelines —field experience, in *Proceedings of the Conference on 15th International Conference on Pipeline Protection, Aachen Germany, October 2003, pp. 199–207* (2003)
13. W. Hodgins, R. Buchanan, High Temperature Pipeline Coatings—Field Joint Challenges in Remote Construction, Canusa-Canada
14. J.P.L. Fontaine, Trends in Pipe Coating—An Operator's Perspective. Exxon Mobile Development
15. U. Anees, S.A. Malik, I. Andijani, *Corrosion Protection Evaluation of Some Organic Coatings in Water Transmission Pipes*. Research & Development Center, Kingdom of Saudi Arabia (1999)
16. S.W. Guan, N. Gritis, G. Bick, A. Jackson, P. Singh, Advanced onshore and offshore pipeline coating technologies, in *China International Oil and Gas Pipe Technology Conference & Expo, 14–17 September 2005, Shanghai, China* (2005)
17. Hevicote Product Datasheet CWC, BrederoShaw
18. R.J. Chi Ming Lee, P. Morgan, R. Weatherhead, International experiences with cathodic protection of Offshore pipeline and flowlines. TWI Report, November 2007 (2007)
19. M. Batallas, H. Yih, P. Singh, Determining the performance of polyurethane foam pipe insulation for high temperature service, in *NACE International Northern Area Western Conference, Calgary Alberta* (2006)
20. S.W. Guan, Insulation for flowlines and risers, in *PETRONAS—ETRAD-INTSOK-CCOP workshop on Deepwater Subsea Tieback Kuching, Malaysia Jan 2011* (2011)
21. O.Ø. Knudsen, A. Bjørgum, A.K. Kvernbråten, Internal coating of multiphase pipelines—requirements for the Coating, *Paper 10004, NACE Corrosion 2010, March 14–18 San Antonio* (2010)
22. J.A. Kehr, M. Dabiri, R. Hislop, Dual-layer fusion-bonded epoxy (FBE) coatings protect pipelines, BHR Group, in *15th International Conference on Pipeline Protection, 29–31 October 2003, Aachen, Germany* (2003)
23. J.A. Kehr, D.G. Enos, FBE, a foundation for pipeline corrosion coatings, *Paper 00757, NACE Corrosion 2000, March 26–31 Orlando Fl* (2000)
24. J.A. Kehr, FBE pipeline and rebar corrosion coatings, in *PICon Internet Conference on Pipeline Reliability* (2000)
25. T. Fore, K. Varughese, FBE found effective after 30 years of service. Pipeline Gas J. 64–65 (2006)

26. J.A. Kehr, Dual-layer fusion bonded epoxy (FBE) coatings protect pipelines, in *15th international conference on Pipeline Protection, 29–31 October 2003 Aachen, Germany* (2003)
27. M. Demirors, in *Chap 9: The History of Polyethylene in 100+ Years of Plastics. Leo Baekeland and Beyond*, ACS Symp. Series, vol. 1080 (2011), pp. 115–156
28. History of traditional pipeline coatings, assessing the future of coating work, in *Proceedings of the PCE 2000 Conference and Exhibition, 8–10 March 2000, Genoa, Italy* (2000)
29. M. Wilmott, J. Highams, R. Ross, A. Kopystinski, Coating and thermal insulation of subsea or buried pipelines. J. Protect. Coat. Linings **17**(11), 47–54 (2000)
30. C. Argent, D. Norman, Fitness for purpose issues relating to FBE and three layer PE coatings, in *Paper 05034, NACE Corrosion 2005, 3–7 April 2005, Houston, Texas* (2005)
31. G.P. Guidetti, G.L. Rigosi, R. Marzola, The use of polypropylene in pipeline coatings. Prog. Org. Coat. **27**, 79–85 (1996)
32. B.J. Turner, Polypropylene coatings for high temperature and thermal insulation of pipelines, im *Paper 11030, Corrosion 2011, 13–17 March 2011, Houston, Texas* (2011)
33. A.B. Hansen, A. Delesalle, Cost effective thermal insulation systems for deep water West Africa in combination with direct heating, in *Offshore West Africa 2000 Conference and Exhibition, 21–23 March 2000, Abidjan, Cote d-Ivoire* (2000)
34. C. Rydin, L. Sjoberg, Polypropylene thermal insulation systems for offhsore pipeline application. J. Protect. Coat. Linings **19**(10), 42–49 (2002)
35. R. Hunter, Wet insulation evolves to meet subsea flowline performance demands. Offshore **68**(10), 122–129 (2008)
36. S.W. Guan, Challenges and New Developments in Pipeline Coating Technology, Pipeliner 2011, pp. 36
37. www.bauhuis.com/Pipe-coating
38. G. Moad, The synthesis of polyolefin graft copolymer be reactive extrusion. Prog. Polym. Sci. **24**, 81–92 (1999)
39. S.B. Brown, in *Reactive Extrusion*, ed. by M. Xanthos (Hanser, Munich, 1992), p. 75
40. R.C. Kowalski, in *Reactive Extrusion*, ed. by M. Xanthos (Hanser, Munich, 1992), pp. 7–32
41. S. Sanchez-Valdes, F. Orona-Villarreal, M. Lopez-Quintanilla, I. Yanez-Flores, L.F.R.D. Valle, C. Guerrero-Salazar, Performance of multilayer films using maleated linear low-density polyethylene blends. Polym. Eng. Sci. **38**, 150 (1998)
42. M. Xanthos (ed.), *Reactive Extrusion* (Hanser, Munich, 1992), p. 33
43. M. Dorn, Modification of molecular weight and flow properties of thermoplastics. Adv. Polym. Tech. **5**, 87–97 (1985)
44. X. Li, X.M. Xie, B.H. Guo, Study on styrene-assisted melt free radical grafting of maleic anhydride onto polypropylene. Polymer **42**, 3419–3425 (2001)
45. J.Z. Liang, R.K.Y. Li, Rubber toughening in polypropylene: a review. J. Appl. Polym. Sci. **77**, 409 (2000)
46. P.R. Hornsby, K. Premphet, Fracture toughness of multiphase polypropylene composites containing rubbery and particulate inclusions. J. Mater. Sci. **32**, 4767 (1997)
47. J. Lu, G.X. Wei, H.J. Sue, J. Chu, Toughening mechanisms in commercial thermoplastic polyolefin blends. J. Appl. Polym. Sci. **76**, 311 (2000)
48. H. Mae, M. Omiya, K. Kishimoto, Material ductility and toughening mechanism of polypropylene blended with bimodal distributed particle size of styrene-ethylene-butadiene-styrene triblock copolymer at high strain rate. J. Appl. Polym. Sci. **110**, 3941 (2008)
49. D.N. Bikiaris, G.Z. Papageorgiou, E. Pavlidou, N. Vouroutzis, P. Palatzoglou, G. P. Karayannidis, Preparation by melt mixing and characterization of isotactic polypropylene/SiO2 nanocomposites containing untreated and surface-treated nanoparticles. J. Appl. Polym. Sci. **100**, 2684 (2006)
50. A. Koebsch, A.L. Lemuchi, R.R. Oliveira, R. Buchanan, Advantages of polypropylene based coatings for buried or submerged pipelines operating in aggressive environments, in *Rio*

Pipeline Conference and Expo, 20–22 September 2011, Rio de Janeiro, Brazil. Paper IBP1094_11
51. M. Ratzsch, Special PP's for a developing and future market. J. Macromol. Sci. Part A: Pure Appl. Chem. **31**(11), 1587–1611
52. A.B. Hansen, C. Rydin, Development and qualification of novel thermal insulation systems for deepwater flowlines and risers based on polypropylene, in *Offshore Technology Conference, Houston Texas, USA*, 6–9 May 2002
53. B. Cavalie, APIA
54. N. Gupta, V.C. Shunmugasamy, High strain rate compressive response of syntactic foams: trends in mechanical properties and failure mechanisms. Mater. Sci. Eng. A **528**, 7596–7605 (2011)
55. M. Aureli, M. Porfiri, N. Gupta, Effect of polydispersity and porosity on the elastic properties of hollow particle filled composites. Mech. Mater. **42**, 726–739 (2010)
56. G. Shearer, C. Tzonganakis, Free radical hydrosilylation of polypropylene. J. Appl. Polym. Sci. **65**, 439–447 (1997)
57. J.D. Kellner, A.J. Doheny, B.B. Patil, A 3 layer polyethylene coating for plant application. Mater. Perform. **37**, 6 (1998)
58. C. Bates, Chemical Treatment Enhances Pipe Coating Quality, Pipe Line Industry, September 1989
59. 3-Layer Polyethylene Coating Systems, PRD, Pipe Manufacturing Systems, www.prdcompany.com
60. A.N. Moosavi, Advances in field joint coating for underground pipelines, in *NACE Corrosion 2000 Conference and Expo. Paper No. 00754* (2000)
61. SA 2.5 refers to Swedish Standard SIS 05 59 00 (1967) Pictorial Surface Preparation Standards for Painting Steel Surface, that prescribes a specific level of cleanliness referred to as near white metal 85 % clean
62. M. Mallozzi, M. Perez, A new 3 LPP offhsore field joint coating, in *NACE Corrosion 2010 paper 10010, March 14–18 2010, San Antonio, TX, USA* (2010)
63. A.N. Moosavi, S. Al Mutawa, S. Balboul, M. Saady, Hidden problems with three layer polypropylene pipeline coatings, in *Corrosion NACE 2006, Paper 06057* (2006)
64. D. Melot, G. Paugam, M. Roche, Disbondments of pipeline coatings and their effects on corrosion risk. J. Protect. Coat. Linings 18–31 (2009)
65. V. Rodriguez, E. Perozo, E. Alvarez, Coating application and evaluation for heavy wall thickness, temperature and pressure pipe. Mater. Perform. **37**, 2 (1998)
66. A.N. Moosavi, Advances in field joint coatings for underground pipelines. Mater. Perform. **39**, 8 (2000)
67. www.canusaps.com
68. Y. Okano, N. Shoji, T. Namioka, M. Komura, The development of auto-sealing systems for field joints of polyethylene coated pipelines, in *Corrosion 97, Paper 389*
69. M. Surkein, J. Didas, Pipe coating and failure modes of mainline and field joint coatings—pictorial learnings, in *Presentation to the OGP Coatings Committee, London, England. 16 November 2011*, http://info.ogp.org.uk/standards/11LondonCoatings/Presentations/Presentation06.pdf
70. S. Papavinasam et al. External polymeric pipeline coatings failure modes. Mater. Perform. 28–30 (2006)
71. K. Varughese, Dual powder FBE systems for directional drill and high temperature pipelines—field experience, in *15th International Conference on Pipeline Coating, Oct 2003, p 199–207. BHR Group 2003 Pipeline Protection* (2003)
72. D. Norman, Excellent pipeline coatings require excellent pipeline substrates, in *NACE Corrosion 2004, Paper 04035, 28 March–1 April 2004, New Orleans, La* (2004)
73. S.W. Guan, petromin Pipeliner, p 36
74. K. Varughese, Improving adhesion properties of three layer polyethylene systems. Corrosion 6 (2003)
75. al FFLe. Paper 10003, NACE 2010

76. A.N. Moosavi, B.T.A. Chang, K. Morsi, Failure analysis of three layer polypropylene pipe coatings, in *NACE Corrosion 2010 Conference and Expo, March 14–18, 2010, San Antonio, TX, Paper 10002* (2010)
77. R. Norsworthy, Study examines coating compatibility with CP. J. Oil Gas **107**(20), 52 (2009)
78. R. Norsworthy, Coatings used in conjunction with cathodic protection—shielding vs non-shielding pipeline coatings, in *NACE Corrosion 2009, Paper 09043. March 22–261, 2009, Atlanta, GA* (2009)
79. Non-shielding coating systems now part of new US regulatory requirement, Next Generation Oil and Gas MENA, Issue 5, No 67 (2010)
80. R. Norsworthy, Is your pipeline coating "fail safe"? Pipeline Gas J. 62–63 (2006)
81. M. Surkein, J. Didas, Pipe Coating and Failure Modes of Mainline & Field Joint Coatings. OGP Coatings Committee–London
82. C. Dasgupta, New field joint coating technology: eliminates "weak link" in the pipeline, in *Presented at the India O&G Pipeline Infrastructure Conference, New Delhi* (2011)
83. S. Papavinasam, R.W. Revie, Review of standards for evaluating coatings to control external corrosion of pipelines. Corros. Rev. **26**(5&6), 295–371 (2008)
84. S.S. Papavinasam, N. Zaver, J. Pollock, Comparison of international standards to evaluate field-applied pipeline girth weld coatings. Corros. Rev. **30**(5&6), 171–197 (2012)
85. DIN 30670
86. DIN 30678
87. CSA Z245.21
88. ISO 21809–1:2011 (E)
89. S.W. Guan, Long-term performance, World Pipelines, November 2008
90. J.M. Leeds, S.S. Leeds, Coating properties and test procedures. Pipeline Gas J. **237**(3) (2010)
91. A.A. Borno, M. Brown, S. Rao, High temperature cathodic disbondment tests, in *NACE Corrosion Conference and Expo 2010, Paper 10008, 14–18 March 2010, San Antonio, TX, USA* (2010)
92. O.Ø. Knudsen, T.G. Eggen, K.K. Brende, Test method for studying cathodic disbonding at high temperature, in *NACE Corrosion Conference and Expo 2010, Paper 10007, 14–18 March 2010, San Antonio, TX, USA* (2010)
93. B. Melve, D. Ali, Corrosion coating for high temperature water immersion service, in *Corrosion 2006, Paper 06021, Houston, TX, USA* (2006)
94. M. Batallas, P. Singh, Evaluation of anticorrosion coatings for high temperature service, in *Corrosion 2008, Paper 08039, Houston TX, USA* (2008)
95. P.K. Shukla, R. Pabalan, L. Yang, M.A. Smith, On development of accelerated testing methods for evaluating organic coating performance above 100 °C, *in Corrosion 2010, Paper 10006, 14–18 March 2010, San Antonio*, *TX* (2010)
96. ASTM D5162-08 Standard Practice for Discontinuity (Holiday) Testing of Nonconductive Protective Coating on Metallic Substrates
97. G.R. Howell, Y.F. Cheng, Characterisation of high performance composite coating for the northern pipeline application. Prog. Org. Coat. **60**, 148–152 (2007)
98. I. Williamson, P. Singh, J. Hancock, The resistance of advanced pipeline coatings to penetration and abrasion by hard rock, in *Proceedings of IPC 2000: International Pipeline Conference, The Palliser, Calgary, Alberta, Canada, 1–5 Oct 2000* (2000)
99. P. Singh, S. Haberer, N. Gritis, R. Worthington, M. Cetiner, New developments in high performance coatings, BHR Group 2005. Pipeline Protect. **16**, 55–66 (Paphos, Cyprus) (2005)
100. P. Singh, J. Cox, Development of coat effective powder coated multi-component coating for underground pipelines
101. C.N.C. Lam, D.T. Wong, R.E. Steele, S.J. Edmonson, A new approach to high performance polyolefin coatings, in *Corrosion 2007, Conference and Expo, NACE International paper 07023* (2007)
102. J. Karger-Koscis, How does phase transformation toughening work in semi-crystalline polymer. Polym. Eng. Sci. **36**, 203 (1996)

103. E. Nezbedova, V. Pospisil, P. Bohaty, B. Vlach, Fracture Behaviour of beta polypropylene as a function of processing conditions. Macromol. Symp. **170**(1), 349–357 (2001)
104. P. Jacoby, Beta nucleating masterbatch offers enhanced properties in polypropylene products. Plast. Add. Compd. **9**, 32 (2007)
105. M. Slouf, J. Kolarik, J. Kotek, Rubber-toughened polypropylene/acrylonitrile-co-butadiene-styrene blends: morphology and mechanical properties. Polym. Eng. Sci. **47**, 482 (2007)
106. J. Kotek, J. Scudla, M. Slouf, M. Raab, Combined effect of specific nucleation and rubber dispersion on morphology and mechanical behaviour of iso-tactic polypropylene. J. Appl. Polym. Sci. **103**, 3539 (2007)
107. C. Grein, M. Gahleitner, On the influence of nucleation on the toughness of iPP/EPR blends with different rubber molecular architectures. Exp. Polym. Lett. **392**, 2 (2008)
108. W. Bai, Y. Wang, B. Song, X. Fan, Z. Zhou, Y.L. Li, Nucleating agent induced impact fracture behaviour change in PP/PEO blend. Polym. Bull. **62**, 405 (2009)
109. R.J. Varley, M. Dell'Olio, Y. Qiang, S. Khor, K. Leong, S. Bateman, different beta nucleants and the resultant microstructural, fracture, and tensile properties for filled and unfilled iso polypropylene. J. Appl. Polym. Sci. **128**(1), 619–627 (2013)
110. H. Pham, F. Aguiie, M. Dettloff, N. Verghese, Development of Novel Toughening Technology for Fusion Bonded (FBE) Powder Coatings, Paint and Coatings Industry, October 2007
111. J.K. Pratt, M. Mallozzi, A. D'Souza, Advances in damage resistant coating technology, in *NACE Corrosion 2011 Conference and Expo* (2011)
112. S.J. García, H.R. Fischer, P.A. White, J. Mardel, Y. González-García, J.M.C. Mol et al., Self-healing anticorrosive organic coating based on an encapsulated water reactive silyl ester: synthesis and proof of concept. Prog. Org. Coat. **70**(2–3), 142–149 (2011)
113. J.F. Doddema, The use of viscoelastic self healing pipeline coating, in *NACE Corrosion 2010 Conference and Expo paper 10042* (2010)
114. M. Quast, Rehabilitation of corrosion proactive coatings on buried steel pipelines with self-amalgamating three ply tapes
115. S. Meure, R.J. Varley, D.Y. Wu, S. Mayo, K. Nairn, S. Furman, Confirmation of the healing mechanism in a mendable EMAA-epoxy resin. Eur. Polym. J. **48**(3), 524–531 (2012)
116. Proprietary product of 3M, USA
117. J.A. Kehr, S. Attaguile, M. Smith, M. Perez, New technology helps develop coatings for high temperature operating pipelines, in *NACE International Corrosion Conference 2008* (2008)
118. M. Mallozzi, M. Perez, A new three layer polypropylene offshore field-joint coating. Mater. Perform. **49**(10), 36 (2010)
119. W. Zhou, T.E. Jeffers, O.H. Decker, Properties of a new high Tg FBE coating for high temperature service. Mater. Perform. **46**, 4 (2007)

MIX
Papier aus verantwortungsvollen Quellen
Paper from responsible sources
FSC® C105338
FSC
www.fsc.org

If you have any concerns about our products,
you can contact us on
ProductSafety@springernature.com

In case Publisher is established outside the EU,
the EU authorized representative is:
Springer Nature Customer Service Center GmbH
Europaplatz 3, 69115 Heidelberg, Germany

Printed by Libri Plureos GmbH
in Hamburg, Germany